Achieving sustainable production of milk

Volume 1: Milk composition, genetics and breeding

It is widely recognised that agriculture is a significant contributor to global warming and climate change. Agriculture needs to reduce its environmental impact and adapt to current climate change whilst still feeding a growing population, i.e. become more 'climate-smart'. Burleigh Dodds Science Publishing is playing its part in achieving this by bringing together key research on making the production of the world's most important crops and livestock products more sustainable.

Based on extensive research, our publications specifically target the challenge of climate-smart agriculture. In this way we are using 'smart publishing' to help achieve climate-smart agriculture.

Burleigh Dodds Science Publishing is an independent and innovative publisher delivering high quality customer-focused agricultural science content in both print and online formats for the academic and research communities. Our aim is to build a foundation of knowledge on which researchers can build to meet the challenge of climate-smart agriculture.

For more information about Burleigh Dodds Science Publishing simply call us on +44 (0) 1223 839365, email info@bdspublishing.com or alternatively please visit our website at www.bdspublishing.com.

Related titles:

Achieving sustainable production of milk Volume 2: Safety, quality and sustainability
Print (ISBN 978-1-78676-048-7); Online (ISBN 978-1-78676-050-0, 978-1-78676-051-7)

Achieving sustainable production of milk Volume 3: Dairy herd management and welfare
Print (ISBN 978-1-78676-052-4); Online (ISBN 978-1-78676-054-8, 978-1-78676-055-5)

Improving organic animal farming
Print (ISBN 978-1-78676-180-4); Online (ISBN 978-1-78676-182-8, 978-1-78676-183-5)

Chapters are available individually from our online bookshop: https://shop.bdspublishing.com

climate•SMART•publishing
IN AGRICULTURAL SCIENCE

BURLEIGH DODDS SERIES IN AGRICULTURAL SCIENCE

NUMBER 08

Achieving sustainable production of milk

Volume 1: Milk composition, genetics and breeding

Edited by Dr Nico van Belzen,
Director General of the International Dairy Federation (IDF), Belgium

burleigh dodds
SCIENCE PUBLISHING

Published by Burleigh Dodds Science Publishing Limited
82 High Street, Sawston, Cambridge CB22 3HJ, UK
www.bdspublishing.com

Burleigh Dodds Science Publishing, 1518 Walnut Street, Suite 900, Philadelphia, PA 19102-3406, USA

First published 2017 by Burleigh Dodds Science Publishing Limited

Library of Congress Control Number: 2016962692

British Library Cataloguing in Publication Data
A catalogue record for this book is available from the British Library

ISBN 978-1-78676-044-9 (print)
ISBN 978-1-78676-046-3 (online)
ISBN 978-1-78676-047-0 (online)
ISSN 2059-6936 (print)
ISSN 2059-6944 (online)

Typeset by Deanta Global Publishing Services, Chennai, India
Printed by Lightning Source

Contents

Series list

Title	Series Number
Achieving sustainable cultivation of maize - Vol 1 From improved varieties to local applications Edited by: Dr Dave Watson, CGIAR Maize Research Program Manager, CIMMYT, Mexico	001
Achieving sustainable cultivation of maize - Vol 2 Cultivation techniques, pest and disease control Edited by: Dr Dave Watson, CGIAR Maize Research Program Manager, CIMMYT, Mexico	002
Achieving sustainable cultivation of rice - Vol 1 Breeding for higher yield and quality Edited by: Prof. Takuji Sasaki, Tokyo University of Agriculture, Japan	003
Achieving sustainable cultivation of rice - Vol 2 Cultivation, pest and disease management Edited by: Prof. Takuji Sasaki, Tokyo University of Agriculture, Japan	004
Achieving sustainable cultivation of wheat - Vol 1 Breeding, quality traits, pests and diseases Edited by: Prof. Peter Langridge, The University of Adelaide, Australia	005
Achieving sustainable cultivation of wheat - Vol 2 Cultivation techniques Edited by: Prof. Peter Langridge, The University of Adelaide, Australia	006
Achieving sustainable cultivation of tomatoes Edited by: Dr. Autar Mattoo, USDA-ARS, USA & Prof. Avtar Handa, Purdue University, USA	007
Achieving sustainable production of milk - Vol 1 Milk composition, genetics and breeding Edited by: Dr Nico van Belzen, International Dairy Federation (IDF), Belgium	008
Achieving sustainable production of milk - Vol 2 Safety, quality and sustainability Edited by: Dr Nico van Belzen, International Dairy Federation (IDF), Belgium	009
Achieving sustainable production of milk - Vol 3 Dairy herd management and welfare Edited by: Prof. John Webster, University of Bristol, UK	010
Ensuring safety and quality in the production of beef - Vol 1 Safety Edited by: Prof. Gary Acuff, Texas A&M University, USA & Prof.James Dickson, Iowa State University, USA	011
Ensuring safety and quality in the production of beef - Vol 2 Quality Edited by: Prof. Michael Dikeman, Kansas State University, USA	012
Achieving sustainable production of poultry meat - Vol 1 Safety, quality and sustainability Edited by: Prof. Steven C. Ricke, University of Arkansas, USA	013
Achieving sustainable production of poultry meat - Vol 2 Breeding and nutrition Edited by: Prof. Todd Applegate, University of Georgia, USA	014
Achieving sustainable production of poultry meat - Vol 3 Health and welfare Edited by: Prof. Todd Applegate, University of Georgia, USA	015

Introduction

Milk and associated dairy products constitute the world's most important agricultural commodity by value, particularly if dairy ingredients in other food products are taken into account. The dairy sector provides livelihoods for 1 billion people and is key to enriching diets the world over, although global consumption of dairy still falls short of national dietary guidelines. At the same time, dairy production is also a significant user of land and other resources, and is responsible for 2.7% of total anthropogenic greenhouse gas (GHG) emissions. There is therefore an urgent need to improve the efficiency of dairy production so that it can meet the nutritional needs of a growing population in a more environmentally sustainable way.

The two volumes of *Achieving sustainable production of milk* address this challenge. Volume 1 starts by summarizing current research on the composition of milk, both as a source of nutrition and as a vital nutritional, nutraceutical or structural ingredient in many other food products. It also considers factors affecting the sensory quality of milk.

Part 1 The composition and quality of milk

Chapters 1–3 provide a comprehensive review of the most important components of milk, their nutraceutical and technological properties and uses as ingredients in further processing. As Chapter 1 indicates, the proteins of milk are its most important constituents from a nutritional and technological point of view. Milk and milk processing have been researched for many years and, today, milk proteins are probably the best characterized of all food proteins. In recent years, numerous new milk protein ingredients have been developed, making this a particularly active area of research. Written by one of the leading research groups in the field, Chapter 1 provides a comprehensive and authoritative overview of the composition and properties of the major and minor milk proteins, the methods used to prepare milk protein fractions in the laboratory and the production of milk proteins on an industrial scale. This topic is then explored in more detail in Chapter 3. Some of the factors determining milk protein synthesis are also further explored in Chapter 9.

Chapter 1 starts by briefly summarizing the various methods used in protein analysis, including polyacrylamide gel electrophoresis (PAGE), sodium dodecyl sulphate (SDS)-PAGE, lab-on-a-chip techniques, capillary electrophoresis and the more recent development of advanced proteomic approaches such as high-resolution two-dimensional electrophoresis, as well as multidimensional high-performance liquid chromatography.

Around 78% of milk proteins belong to a unique group of milk-specific proteins, the caseins. The review of key trends in casein research starts with factors affecting the microheterogeneity of caseins (such as genetic polymorphism) which has an important effect on processing properties. The chapter also summarizes what we know about the other distinct characteristics which determine the functionality and stability of caseins such as degree of insolubility, susceptibility to proteolysis, heat stability and amino acid composition. Caseins in milk have long been known to exist as large colloidal particles known as casein micelles which affect properties such as colour. The stability of milk and many of its technologically important properties are related to the properties of the casein micelles, which have, therefore, been the focus of considerable research. Chapter 3 looks at properties such as micelle formation, structure and function as well as degree of stability in response to milk processing operations.

Apart from the caseins, the remaining 22% of milk proteins are referred to as whey (serum) proteins. The major components of whey are β-lactoglobulin and α-lactalbumin. Other components include blood serum albumin, immunoglobulins and lactoferrin. Whey also contains many minor proteins and enzymes. The chapter looks at composition and functional properties of each of these components of whey including the proteose-peptone fraction of milk protein and caseinomacropeptide (CMP).

Finally, the chapter also discusses what we know about the minor proteins, enzymes and other components in milk, some of which have attracted considerable attention as nutraceuticals, a topic picked by in Chapter 2. These include:

- metal-binding proteins
- β_2-microglobulin
- osteopontin (OPN)
- vitamin-binding proteins
- angiogenins
- kininogens
- glycoproteins
- growth factors
- indigenous milk enzymes
- biologically active cryptic peptides (such as phosphopeptides, angiotensin-converting enzyme (ACE)-inhibitory peptides)
- non-protein nitrogen (NPN)

Chapter 1 concludes by summarizing key methods in the laboratory-scale preparation of whey and casein proteins as well as the preparation of industrial milk protein products using membrane and other technologies. Whey protein products, for example, are key ingredients in several growth areas of the food industry, such as infant formulae, clinical nutrition and sports nutrition. They include sweet whey, whey powder, demineralized whey, whey protein concentrates, serum protein concentrates, whey protein isolates, and enriched and isolated whey protein fractions. The casein market is dominated by products produced by renneting or acid precipitation of milk. However, casein-derived ingredients manufactured using membrane filtration, such as milk protein concentrates (MPC) and micellar casein concentrates, now have a significant presence in the global market for casein. The chapter also reviews liquid/gelled casein concentrates, β-casein and hydrolysates. The topic of dairy-derived ingredients is discussed in more detail in Chapter 3.

Apart from the nutritional value of milk, milk-borne biologically active compounds such as proteins, peptides, lactoferrin, enzymes, lipids and carbohydrates have been shown to be increasingly important for physiological and biochemical functions that affect human metabolism and health beyond nutrition. In recent decades, major progress has been made in the science, technology and commercial applications of the many bioactive components in bovine milk and colostrum. Chromatographic and membrane separation techniques have been developed to fractionate and purify these components on an industrial scale. Production of bioactive milk ingredients by fractionation has thus emerged as a lucrative new sector for the dairy industry.

Building on Chapter 1, Chapter 2 provides a comprehensive review of bioactives in milk and research on their nutraceutical properties. Ingredients include bioactive proteins such as caseins, whey proteins such as α-lactalbumin and β-lactoglobulin, and enzymes such as lactoperoxidase and lysozyme, as well as bioactive peptides and bioactive lipids such as conjugated linoleic acid (CLA), phospholipids and cholesterol. The chapter also discusses bioactive carbohydrates such as lactose and oligosaccharides, other bioactive compounds such as growth factors, cytokines, nucleosides and nucleotides as well as bioactive minerals and vitamins.

The chapter starts by looking at bioactive proteins, beginning with caseins. Digested or catabolized caseins produce a variety of bioactive peptides, including antihypertensive and immuno-stimulating peptides. Many bioactive compounds are generated from different casein (CN) fractions, including casomorphins, casokinins, phosphopeptides, immunopeptides, isracidin, casocidin, casoxins and casoplatelins. As research shows, these exert different bioactive functionalities such as reduction of hypertension (ACE inhibitory), mineral-binding, immunomodulatory, antimicrobial and antithrombotic activities as well as pain management (with opioid agonist/antagonist properties).

The chapter then reviews individual whey protein components and their hydrolysed peptide fragments which exhibit various bioactive properties including opioid agonist, antimicrobial and antiviral actions, immune system stimulation, anticarcinogenic activity and other metabolic functions. The chapter reviews recent research on nutraceutical properties and potential applications of α-lactalbumin, β-lactoglobulin, lactoferrin, immunoglobulins, glycomacropeptide (GMP) (including caseinomacropeptide (CMP), the non-glycosylated form of GMP). As an example, α-lactalbumin hydrolysates and their specific peptides have been shown to have antihypertensive, antimicrobial, anticarcinogenic, immunomodulatory, opioid and prebiotic properties. Similarly, β-lactoglobulin from milk has proven to be an excellent source of peptides with a wide range of bioactivities, such as antihypertensive, antimicrobial, antioxidative, anticarcinogenic, immunomodulatory, opioid, hypocholesterolemic and other metabolic effects.

After reviewing enzymes such as lactoperoxidase and lysozyme, the chapter goes on to discuss bioactive peptides. There are more than 200 biologically and functionally active peptides that exist in milk and dairy products. Bioactive peptides affect functions in the body such as gastrointestinal, cardiovascular, endocrine, immune and nervous systems. The chapter summarizes the wealth of research on antihypertensive (ACE inhibitory) peptides with applications in the control of blood pressure, antioxidative peptides, antithrombotic peptides, hypocholesterolemic peptides with the potential to reduce blood cholesterol levels and opioid peptides (with opiate-like effects which influence pain, mood and appetite). The chapter also reviews mineral-binding peptides including phosphopeptides, caseinophosphopeptides (CPPs) and calcium-binding phosphopeptides (CCPs) with the ability to improve the absorption of minerals such as calcium, before going on to cover anti-appetizing peptides, which can help to reduce energy intake and promote a healthy body composition with less body fat due to their positive effects on satiation/satiety. Other peptides include antimicrobial peptides (such as lactoferricins) which are able to modulate inflammatory responses in addition to killing microorganisms, immunomodulatory peptides with the potential to boost immune cell function and cytomodulatory peptides able to suppress cancer cell activity.

Chapter 2 then assesses bioactive lipids such as CLA, phospholipids, cholesterol and minor lipids (which include gangliosides, glycolipids, glycosphingolipids and cerebrosides as well as alkylglycerol). It goes on to survey research on bioactive carbohydrates such as lactose (which affects the adsorption of minerals and vitamins), lactose-derived compounds (including lactulose, lactitol, lactobionic acid and galacto-oligosaccharides) and oligosaccharides.

The chapter concludes with a review of other bioactive compounds in milk such as growth factors (including epidermal growth factor, IGF-I and IGF-II (insulin-like growth factor), FGF1 and FGF2 (fibroblast growth factor), TGF-β1 and TGF-β2 (transforming growth factor), BTC (β-cellulin) and platelet-derived growth factor). It also covers cytokines (which include chemokines, interferons, interleukins and lymphokines), as well as main milk hormones: gonadal hormones (oestrogens, progesterone, androgens), adrenal (glucocorticoids), pituitary (prolactin, growth hormone) and hypothalamic hormones (gonadotropin-releasing hormone, luteinizing hormone-releasing hormone, thyrotropin-releasing hormone and somatostatin). After discussing nucleosides, nucleotides and polyamines, it reviews organic acids (such as lactic acid, citric acid, pyruvic acid, uric acid, orotic acid, nucleic acid and neuraminic acid). The final group of bioactive ingredients surveyed are bioactive minerals and vitamins, from calcium, phosphorus and potassium to trace minerals such as iron, zinc, iodine, selenium and manganese, as well as vitamins. Particularly, riboflavin (B_2) and vitamin B_{12}. In each case the chapter provides a valuable summary of key clinical research.

Building on both Chapters 1 and 2, Chapter 3 concludes the first group of chapters by looking at the range of dairy-derived ingredients for use in dairy and non-dairy foods, as well as non-food products. These ingredients range from commodity ingredients, such as milk and whey powder, to milk protein ingredients, such as caseins, caseinates, whey protein ingredients, MPC and milk protein hydrolysates. Lactose and lactose derivatives, including lactulose, lactobionic acid and the prebiotic galactooligosaccharides, and milk fat globule membrane (MFGM) material fractions are also produced as ingredients. This chapter reviews the main dairy-derived ingredients and their physical and nutritional functionalities and range of applications. The focus is primarily on ingredients produced on an industrial scale. The chapter shows how scientific and technical innovations have created a new range of products driven by the demand for dairy ingredients for nutritional products for infants and the elderly, performance nutritional snacks and ingredients with nutraceutical properties in preventing or managing a range of chronic diseases.

As highlighted in Chapter 2, milk proteins in particular not only are a source of amino acids, but can also confer immunity and are a carrier of calcium phosphate, which is essential for bone growth. In addition, some milk proteins contain bioactive sequences which may be released upon hydrolysis during digestion. The proteins in the MFGM are also known to have antimicrobial and antiviral properties. In addition to the main carbohydrate lactose, milk also contains smaller amounts of oligosaccharides. These are known to aid the development of the intestinal flora of the neonate, which provides important anti-infection properties and is an important factor stimulating postnatal development. Since they have a high value, milk proteins have received extensive attention with respect to preparing functional ingredients. Desired functionality may either be physical (in improving process functionality or product quality), nutritional or nutraceutical. Techniques used include selective precipitation, membrane filtration and chromatography. In addition, enzymatic hydrolysis of proteins may be used to improve physical, nutritional or nutraceutical functionality.

The chapter starts by reviewing techniques for refining casein and caseinates as well as their wide range of applications. As an example, sodium and potassium caseinate are excellent emulsifiers and foamers, and also have high heat stability, strong water-binding functionality and excellent nutritional properties. They are therefore widely applied in coffee creamers and other high fat products, cream liqueurs, bakery products, whipped toppings, soups, sauces, ice cream, meat products, and infant and clinical nutrition.

The chapter also looks at the preparation of whey protein concentrates and isolates using techniques such as ultrafiltration, as well as fractionated whey protein ingredients (such as α-lactalbumin, β-lactoglobulin and lactoferrin) developed for the infant formula industry. The chapter then discusses processing of MPCs used for standardization of cheese milk, protein fortification of yogurt, ice cream mixes, and clinical and infant nutrition products. Because micellar calcium phosphate is largely retained in the micelles during ultrafiltration, MPCs contain high levels of encapsulated bioavailable calcium, thus making them interesting ingredients for nutritional products.

Milk protein hydrolysates from caseins and whey proteins are a class of milk protein ingredients that have attracted more and more interest in the last few decades. Milk protein hydrolysates can be divided into three main categories based on their designated application: hydrolysates or specific peptides with biological activity, hydrolysates for consumers with specific nutritional needs and hydrolysates for improved protein functionality. As an example, and as discussed in Chapter 2, antihypertensive peptides are one of the most well-known categories of milk peptides with biological activity, manufactured by enzymatic hydrolysis of milk protein ingredients or fermentation by proteolytic bacteria. The chapter also discusses the three main categories of milk protein hydrolysates designated to address specific nutritional needs: milk protein hydrolysates used as ingredients in hypoallergenic infant milk formulae developed for infants that suffer from cow's milk protein allergies; low-phenylalanine hydrolysates for consumers that suffer from phenylketonuria; and mildly hydrolysed milk proteins for easier digestion developed either for infants or for the elderly. Finally, hydrolysis of whey proteins can increase their processing functionality such as solubility, viscosity, surface activity, emulsifying and foaming ability, as well as increase thermal stability.

The carbohydrate fraction of milk has also been a rich source of ingredients such as lactose and lactose derivatives. Lactose is a widely used carbohydrate in food products in baking and confectionery, but also as an excipient in pharmaceutical products. In addition, lactose can also be converted into functional ingredients such as lactulose, lactitol and lactobionic acid used in sweetener and other applications. Furthermore, prebiotic galactooligosaccharides (GOS) can be produced from lactose and have found wide application, particularly for infant nutrition products. In each case, Chapter 3 reviews the key processing steps in developing ingredients on a commercial scale. Finally, building on Chapter 1, the chapter reviews the preparation and use of MFGM material fractions. MFGM material has useful properties for emulsion stabilization and controlling protein interactions and has therefore been used in ice cream, evaporated milk, cheese and processed cheese, and nutritional products.

Whilst the first group of chapters concentrates on ingredients derived from milk, the following two chapters discuss milk itself, starting with the key issue of spoilage. As Chapter 4 shows, milk spoilage is essentially a result of inadequate control of the growth of microorganisms, combined with the activity of enzymes which have found their way into milk from production and processing environments. As the chapter shows, psychrotrophic, thermophilic or thermoduric and spore-forming microorganisms (PTS) can contaminate milk, grow in chilled bulk tanks and survive heat treatments to reduce shelf life. They also produce thermotolerant lipolytic and proteolytic enzymes that can survive the pasteurization process to cause spoilage. Spoilage microorganisms can be classified by their heat resistance (thermoduric) as well as their preferred growth temperature as psychrotrophic, mesophilic or thermophilic, with thermoduric psychrotrophs a particular challenge.

The chapter summarizes what we know about conditions favouring growth and survival. It also reviews sources of contamination such as the mammary glands, the external surfaces of the udder and teats, the farm environment (e.g. bedding), milking equipment and personnel, as well as tankers and the transport chain through to the processor. Based on this foundation, the chapter then describes best practice in monitoring and prevention of contamination by spoilage bacteria. Preventing spoilage is a question of preventing contamination through hygiene and sanitation, proper cooling and understanding the conditions specific to the processes leading to the wide variety of dairy products and ingredients. Control measures on the farm include mastitis control, udder hygiene, milking routine, environmental sanitation (including feed and bedding), tank and truck sanitation, processing conditions and equipment sanitation. Udder hygiene and teat preparation (cleaning and drying) are considered critical points. However, the concentration of spores in silage and feed during housing periods is now regarded as having significant impact on the spore load of milk. The chapter concludes by identifying future trends in this area, including the prospect of better detection and typing methods for identifying problem areas as well as improved technologies for ensuring milk quality all along the value chain. It also identifies the need for more research in areas such as determining critical points in the origin of spore formers on the farm as well as in the processing plant and the synergistic effects of combining control technologies.

Chapter 5 builds on Chapter 4 by first reviewing the causes of off-flavours in milk and the importance of good management practices (including herd size, milking routine and bedding) in determining levels of mesophilic and thermophilic spores in milk. It also looks at good practice in monitoring for and identifying different categories of off-flavour and their likely causes, as well as ways they can be prevented. As the chapter shows, preventing absorbed off-flavours generally involves good cow nutrition (appropriate feeds, balanced rations) and management (ventilation, health monitoring, manure management) practices. Preventing bacterial off-flavours hinges on good training of staff that prepare teats for milking and proper maintenance of equipment, temperature control, proper selection of application of cleaning and sanitizing chemicals, and prompt milk processing. Preventing chemical off-flavours involves keeping milk away from light, reactive metals, and excessive agitation and using appropriate processing controls. Preventing delinquency off-flavours relies on attentive care by all who handle milk, from cow to consumer. Finally, the chapter reviews techniques for instrumental and sensory shelf life testing.

Part 2 Genetics, breeding and other factors affecting quality and sustainability

The next group of chapters in Part 2 looks at ways of balancing milk yield and quality with other factors affecting the sustainability of milk production. Chapter 6 gives an authoritative overview of some of the key developments in breeding dairy cattle in recent decades. As the chapter shows, there has been a large increase in the productivity of dairy cows over the last half century, with the yield per cow more than doubling. This is substantially due to the use of genetic selection in dairy cattle breeding programmes. Early gains were achieved through progeny test schemes supported by artificial insemination (AI) and embryo transfer (ET) technologies. AI and ET technologies made possible a strong international trade in genetic material, resulting in the large-scale introduction of Holstein Friesian genes into many dairy populations throughout the world. The chapter also covers the use of multiple ovulation and ET and juvenile *in vitro* fertilization and ET to further improve breeding efficiency.

The chapter then reviews the development of genomic selection which has resulted in halving of the generation interval and doubling of the rate of genetic gain. An example is the DGAT1 mutation with a significant effect on milk fat and other milk constituents. Initial results found a relatively low number of individual quantitative trait loci (QLT) useful in in marker-assisted selection. Subsequent studies based on increasingly dense marker panels have revealed that most of the observed genetic variation on quantitative traits is due to a large number of genes, each with a small effect. This resulted in a shift to the use of all marker information across the whole genome in a single analysis to predict breeding values, using single-nucleotide polymorphism chips to genotype individuals for numerous genetic markers.

Finally, the chapter shows that genomic selection is opening powerful new opportunities to select for more complex traits such as phenotypes associated with feed efficiency, methane production, fertility and health traits that have previously been difficult or expensive to measure, but which are important to the sustainability of dairy production. Accuracy of trait prediction is determined by the size of the reference population, or the number of bulls and cows for which both genotype and phenotype are known. This has pushed countries to increase the size of reference populations and to share their data. Many breeding programmes still lack good phenotypic information about non-production traits and, as a result, genetic change remains dominated by an increase in yield, in spite of an increased selection emphasis on other traits. More intensive measurement of a wider range of traits is needed in dedicated resource herds, and these can serve as a training population to allow for genomic selection of bulls.

Chapter 6 provides a context for Chapter 7. As it notes, up until recently most of the emphasis of breeding in dairy cattle was for traits such as milk production, fat and protein content. Inclusion of secondary or 'functional' traits in breeding objectives only began to develop in the 1990s. Although there is a nearly complete consensus about the economic importance of functional traits such as fertility, health and longevity, genetic evaluation and inclusion of these traits in selection indices has been hindered by factors such as difficulties of definition and measurement, low heritability and negative correlations with milk production.

The chapter reviews progress in breeding for functional traits which started with efforts to compute genetic evaluations for a range of traits, using different models, for inclusion in a selection index. The chapter also reviews the range of studies of the heritability of key functional traits: fertility and calving traits, health traits such as susceptibility to mastitis, hoof disorders, udder oedema, milk fever, retained placenta, metritis, ketosis, lameness, cystic ovaries and displaced abomasum, as well as growth rate and longevity. The chapter also looks at alternative approaches such as evaluation and selection of traits that are genetically correlated with 'functional' traits, but are more amenable to genetic evaluation, or crossbreeding breeds that are superior for production to breeds with economically higher genetic levels for secondary traits. The chapter also reviews what we know about genetic parameters and genetic and phenotypic trends for these traits. Finally, the chapter discusses methodologies for detection and analysis of the actual segregating genes that affect functional traits, including mapping and identification of quantitative trait loci related to functional traits.

Chapter 8 builds on Chapter 6 by looking in more detail at ways of improving breeding efficiency in dairy cattle whilst maintaining high milk production. As noted earlier, for decades genetic selection of dairy cattle was largely performed with a focus on traits relating to milk production with a corresponding decline in fertility. This is despite the importance of high reproductive efficiency to sustainable dairy farming, and the fact that reproductive failure is the primary reason for culling dairy cows in many countries. As the chapter shows, many factors, either independently or through their interactions, can influence reproductive efficiency. The main factors affecting reproduction can be broadly grouped into four categories, namely human (managerial), animal (intrinsic and extrinsic), nutritional and environmental.

Looking first at managerial factors, the chapter shows that poor oestrus detection efficiency is a primary cause of reduced reproductive efficiency in dairy herds. In herds using AI, accurate detection of oestrus is extremely important for reproductive success. Other factors include identification of non-pregnant cows as soon as possible after breeding, feeds and feeding, disease management (e.g. vaccination) and environmental management (e.g. heat abatement during hot summers) which can have major impacts on reproductive efficiency. The chapter also looks at the relative importance of animal factors such as breed, genotype and age as well as nutritional factors. Inadequate energy intake during the early postpartum period is common in high-producing dairy cows, resulting in negative energy balance, with mobilization of fat and high concentrations of non-esterified fatty acids (NEFA). High NEFA concentrations have negative effects on oocyte function and embryo quality which likely contribute to subfertility in dairy cows.

With this foundation, the chapter reviews the various strategies that can be used to improve reproductive efficiency in dairy herds. Since poor oestrus detection efficiency is a major factor decreasing reproductive inefficiency, the first strategy should be to improve oestrus detection efficiency. Other strategies include oestrus synchronization and synchronization of ovulation (OvSynch), the use of voluntary waiting periods, shortening the dry period to minimize negative energy balance and techniques to minimize embryo loss such as use of supplemental exogenous progesterone and boosting endogenous progesterone. Although high-protein diets have a detrimental effect on fertility, supplemental fats and specific polyunsaturated long-chain fatty acids have positive effects.

The final chapter in Volume 1 looks at ways of improving nutritional efficiency in cows to both optimize milk quality and improve sustainability. As Chapter 9 shows, the conversion of feed nitrogen into milk nitrogen often has only 20% efficiency which results in significant losses of nitrogen to the environment, contributing to the degradation of air and water systems. The chapter considers the two nitrogen-utilizing systems in the cow: the rumen microbiota and ruminant tissues. The more efficient both the rumen and tissue systems are, the lower the urinary nitrogen excretion. The rumen microbiota considerably alters feed inputs producing microbial protein for digestion and metabolism. Rumen fermentation of feeds provides energy (primarily from carbohydrates) and nitrogen (primarily from protein) for microbial protein synthesis. A balance of rumen-available carbohydrate and rumen-degradable feed protein results in efficient microbial protein synthesis, minimizing nitrogen wastage from the rumen. The proportion of essential amino acids and total metabolizable protein absorbed from the small intestine then determines the efficiency of tissue utilization of protein for milk protein synthesis and maintenance. Absorption of an ideal proportion of essential amino acids improves efficiency of milk protein synthesis, enabling a reduction of total protein in the diet. Both efficiency of nitrogen utilization in the rumen and nitrogen utilization by the mammary gland influence nitrogen losses.

The chapter reviews approaches to estimating rumen microbial protein synthesis, the protein and energy requirements of dairy cows and the use of milk urea nitrogen to assess the nitrogen efficiency of dairy cows. It then considers the development of nutritional systems which account for rumen microbial synthesis, rumen-degradable and -undegradable feed protein, and endogenous protein supplies of amino acids based on utilization of feed inputs. The goal of precision protein feeding programmes is to capture as much dietary nitrogen into milk nitrogen as possible and reduce urinary nitrogen losses. It is now possible to significantly reduce dietary crude protein and maintain reasonable milk production levels in dairy herds. Forage quality, appropriate protein and energy supplements are necessary to ensure adequate rumen-available energy and nitrogen for microbial protein synthesis. Nutritional systems are thus evolving which account for rumen microbial synthesis, rumen-degradable and -undegradable feed protein, and endogenous protein supplies of amino acids based on utilization of feed inputs. Accurate prediction of essential amino acid supply to the mammary gland by ration models will facilitate improved conversion of feed nitrogen to milk protein nitrogen, reducing urinary nitrogen. Future work will more fully describe influences of feed nutrients on rumen fermentation, better characterize endogenous protein supplies and incorporate hindgut models of nutrient utilization to improve the precision of ration formulation models. Efficient grouping of cattle combined with more precise ration formulation will further influence the performance of nutritional models to reduce environmental pollution from dairy farms.

Summary

As the wealth of material in Volume 1 shows, research is continuing to improve our understanding of what a rich resource bovine milk is in meeting the nutritional and wider health needs of a growing population, as well as an important ingredient in a wide range of other food products. It also shows what is being done to preserve milk quality and yield whilst, at the same time, making milk production more efficient, whether in terms of developments in breeding for functional traits such as reproductive efficiency or management strategies for optimizing nutrition to reduce environmentally damaging waste. Other aspects of sustainability are discussed in Volume 2.

Part 1

The composition and quality of milk

Chapter 1

The proteins of milk

Shane V. Crowley, James A. O'Mahony and Patrick F. Fox, University College Cork, Ireland

1 Introduction

Bovine milk contains ~3.0–3.5% protein. The proteins, as well as the lipids, lactose and ash, are subject to inter-species differences (Table 1). Around 78% of the protein is a unique group of milk-specific proteins called the caseins, which are insoluble at their isoelectric point, pH ~4.6. The remaining 22% of the protein (those soluble at pH 4.6) is referred to as the whey or serum proteins. The two principal whey proteins, β-lactoglobulin (β-lg) and α-lactalbumin (α-la), which represent ~10 and ~4% of total milk proteins, respectively, are also milk-specific.

The remaining whey proteins comprise a very heterogeneous mixture of proteins and peptides, mostly present at low levels, and are obtained from the blood or secretory tissue. Many of these minor proteins are biologically active, including immunoglobulins, binding/carrier proteins (for metals or vitamins) and about 60 enzymes. Colostrum (the mammary secretion during the first few days post-partum) contains a high level of Igs (mainly IgG1 in bovine milk) which are absorbed *via* the intestine of the neonate and impart passive immunity. The composition and properties of colostrum were reviewed by McGrath et al. (2016), and will not be considered further in this chapter. The casein and whey proteins can

http://dx.doi.org/10.19103/AS.2016.0005.03

Table 1 Composition (%) of the milk of different species

Species	Solids	Fat	Protein	Lactose	Ash
Human	12.2	3.8	1.0	7.0	0.2
Cow	12.7	4.5	3.0	4.1	0.8
Sheep	19.3	7.4	4.5	4.8	1.0
Pig	18.8	6.8	4.8	5.5	–
Horse	11.2	1.9	2.5	6.2	0.5
Donkey	11.7	1.4	2.0	7.4	0.5
Reindeer	33.1	16.9	11.5	2.8	–
Domestic rabbit	32.8	18.3	11.9	2.1	1.8
Bison	14.6	3.5	4.5	5.1	0.8
Indian elephant	31.9	11.6	4.9	4.7	0.7
Polar bear	47.6	33.1	10.9	0.3	1.4
Grey seal	67.7	53.1	11.2	0.7	–

be separated from each other in the laboratory by various approaches (see Section 6) which facilitate their characterization.

The fat globules in milk are surrounded and stabilized by a membrane (the milk fat globule membrane; MFGM) consisting of polar lipids and proteins, the latter of which represents ~1% of the total protein in bovine milk. About 60 proteins have been identified by polyacrylamide gel electrophoresis (PAGE) in the MFGM, the principal ones being butyrophilin, which is unique to milk, and xanthine oxidoreductase (see Mather, 2000). The composition and structure of the MFGM has been reviewed by Keenan and Mather (2006). The stability of the MFGM, and, hence, MFGM proteins, is critically important in several aspects of dairy technology. As buttermilk contains MFGM, it is considered a good source of polar lipids and proteins, which have good emulsification properties (see Ward et al., 2006; Corredig et al., 2003; O'Connell and Fox, 2000).

The natural function of mature milk is a nutritional one. Milk supplies energy (mainly through lipids and lactose), amino acids (proteins), essential fatty acids (lipids), vitamins and inorganic elements to the neonate of the species. The milk of cattle, buffalo, sheep and goats, and to a lesser extent of camel, yak, horse, donkey and reindeer, has been a component of the human diet for several thousand years, consumed fresh or fermented and as cheese or butter/ghee. As a result, the dairy industry is of major importance in a number of countries. In Europe, North America, Australia and New Zealand, about 30% of dietary protein is supplied by milk and dairy products. Milk and milk processing have been researched for many years and, today, milk proteins are probably the best characterized of all food proteins. This chapter provides an overview of major and minor milk proteins, the methods used to prepare milk protein fractions in the laboratory and the milk protein products that are produced on an industrial scale.

2 Analytical methods for the study of milk proteins

Starch gel electrophoresis (SGE) was first used to study the caseins by Wake and Baldwin (1961), followed by PAGE, which was used by Peterson (1963). These methods give similar results but since PAGE is easier to use, it has become the standard electrophoretic method for the analysis of caseins. Gel electrophoretic methods for the analysis of milk proteins have been reviewed by Swaisgood (1975), Shalabi and Fox (1987), IDF (1991), Strange et al. (1992), Van Hekken and Thompson (1992), O'Donnell et al. (2004) and Chevalier (2011a, b). Urea-PAGE is used for analysis of caseins. The procedure of Andrews (1983), incorporating a stacking gel, is commonly used with good results. Fig. 1 illustrates the use of such methodology for evaluating inter-species differences in milk protein profiles. Sodium dodecyl sulphate (SDS)-PAGE is effective for the resolution of whey proteins (Laemmli, 1970; Creamer and Richardson, 1984). The availability of pre-cast, mini-gels for the analysis of proteins by SDS-PAGE has made the technique easier and faster to perform.

Capillary electrophoresis is becoming more widely used, owing to its low sample and solvent volumes, in addition to its short analysis time. Good resolution of both casein and whey proteins is achievable, and this method can be coupled with online detectors to facilitate quantitative analysis. The ability to automate capillary electrophoresis procedures allows high-throughput routine analysis (Dupont et al., 2013). The microheterogeneity of casein can be investigated effectively by capillary electrophoresis. Other developments in electrophoretic methods for milk protein analysis include microfluidic techniques. Anema (2009), who compared 'lab-on-a-chip' microfluidic SDS electrophoresis with standard

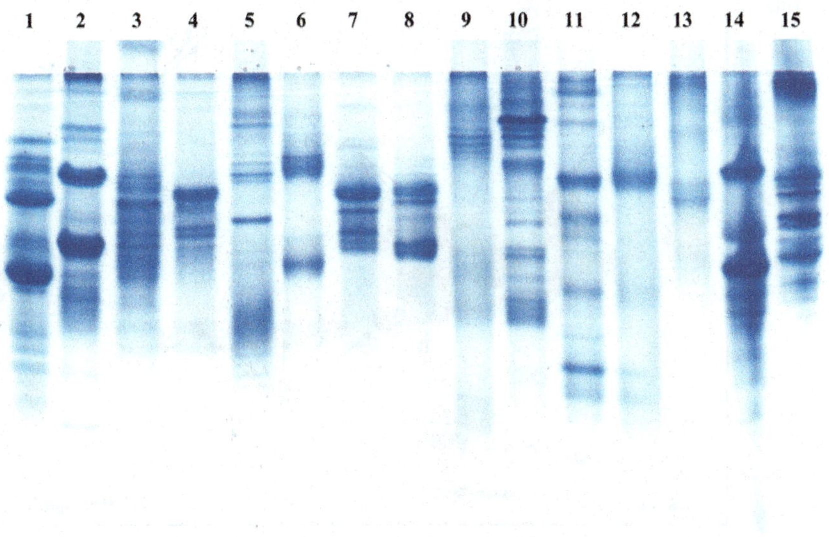

Figure 1 Urea polyacrylamide gel electrophoretogram of milk from 15 species. Lanes: 1, Bovine; 2, Camel; 3, Equine; 4, Asinine; 5, Human; 6, Rhinoceros; 7, Caprine; 8, Ovine; 9, Asian elephant; 10, African elephant; 11, Vervet monkey; 12, Macaque monkey; 13, Rat; 14, Canine; 15, Porcine.

Source: From Thérèse Uniacke-Lowe, unpublished data.

SDS-PAGE, found that the microfluidic technique resolved most milk proteins to the same degree as standard SDS-PAGE. The resolution of minor whey proteins (e.g. immunoglobulins, serum albumin, lactoferrin) by the lab-on-a-chip technique was found by Anema (2009) to be inferior to standard SDS-PAGE, but microfluidic SDS electrophoresis is nonetheless considered a rapid and reproducible technique (Anema and de Kruif, 2016).

The recent development of advanced proteomic approaches based on traditional gel electrophoresis, such as high-resolution two-dimensional electrophoresis (possibly with mass spectrometry identification/quantification), have proven very effective for protein profiling and for assessing the pattern and extent of protein hydrolysis in more complex milk protein systems (Mann et al., 2001; Yamada et al., 2002; Manso et al., 2005; Armaforte et al., 2010; Chevalier, 2011b). Figure 2 shows a two-dimensional electrophoretogram for detailed analysis of the protein system of bovine milk.

For protein analysis, high-performance liquid chromatography (HPLC) is generally a more powerful quantitative tool than gel electrophoresis. Other methods for quantification of individual proteins include fast-protein liquid chromatography (FPLC) (Davies and Law, 1987). Specific methodologies have long been in place for the quantification of casein or whey proteins. More recently, HPLC methods which allow the simultaneous quantification of individual proteins in the casein and whey fractions have been developed (Bordin et al., 2001). For the resolution of casein and whey proteins by HPLC, samples are mixed with dissociating agents such as guanidine hydrochloride (GuHCl), urea and trisodium

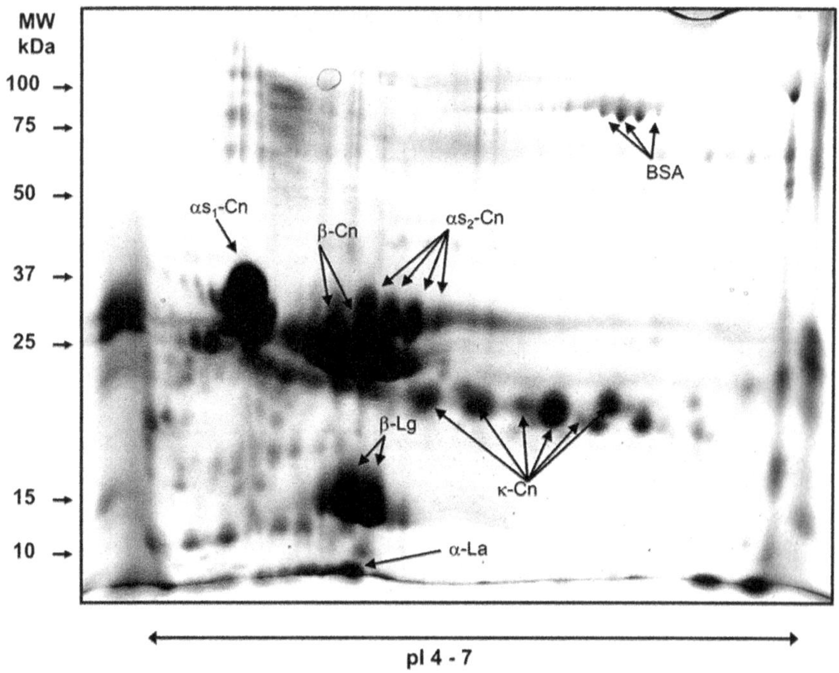

Figure 2 Two-dimensional electrophoretogram of bovine milk under reducing conditions using isoelectric focusing in the range pH 4 to 7 for the first dimension and a 12% acrylamide gel for the second dimension.

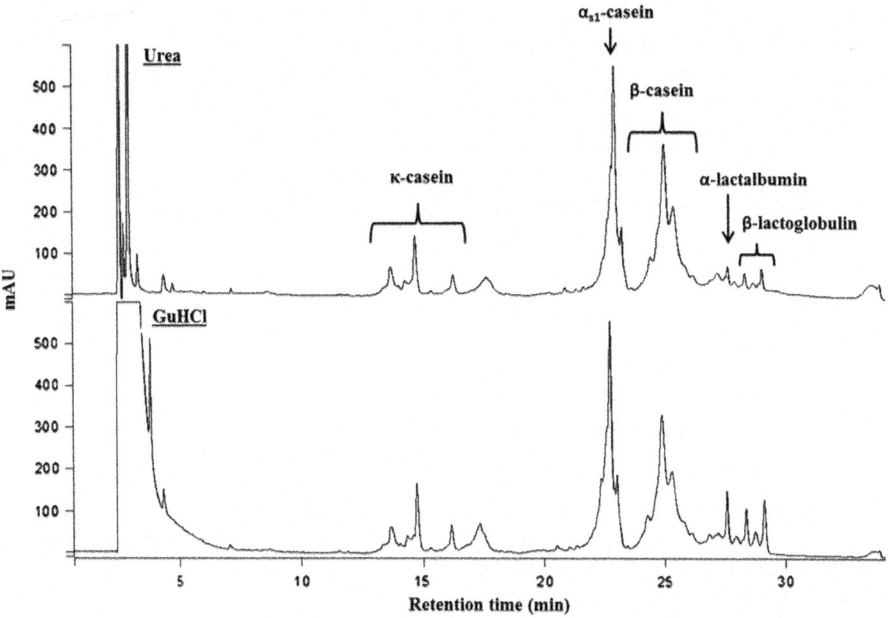

Figure 3 Influence of dissociating buffer, urea or guanidine hydrochloride (GuHCl), on the resolution of the major proteins in skim milk using reversed-phase high-performance liquid chromatography with a C18 column.

Source: From Veronica Caldeo, unpublished data.

citrate, on their own or in combination, and a reducing agent (usually dithiothreitol). Prepared samples are then injected onto a reversed-phase (RP-) HPLC column, with photodiode array detection. In our laboratory, a modification of the method of Bobe et al. (1998a, b) is used, where samples are diluted in a solution containing 0.1 M BisTris buffer, 6 M GuHCl, 5.37 mM sodium citrate and 19.5 mM dithiothreitol (Crowley et al., 2015b). By using this method, excellent resolution of most of the principal casein (CN) proteins (α_{s1}, β, κ) and whey proteins (β-lg and α-la) is obtained with a C_{18} column, though other workers have also successfully used C_4 or C_8 columns. Replacement of urea with GuHCl was an important development in the simultaneous quantification of major milk proteins by RP-HPLC, and it resolved issues associated with the use of urea as a dissociating agent, including poor resolution of β-lg and α-la, co-elution of α-la with caseins and inaccuracies associated with the quantification of β-CN (Dupont et al., 2013). Examples of chromatograms obtained from skim milk dissociated in either urea or GuHCl are shown in Fig. 3. Quantification of α_{s2}-CN by RP-HPLC remains challenging (Hinz et al., 2012).

3 Caseins

Bovine casein consists of four principal proteins, representing the following percentages of whole casein: α_{s1} (38%), α_{s2} (10%), β (35%) and κ (12%). However, SGE, PAGE or capillary electrophoresis indicate much greater heterogeneity, which is due to relatively small variations in one or more of the four principal caseins. These minor variations are referred to as *microheterogeneity*. Capillary zone electrophoresis, which is more easily quantifiable than SGE or PAGE, is particularly useful for quantifying microheterogeneity (see Heck et al., 2008).

3.1 Microheterogeneity of caseins

Microheterogeneity arises from five factors:

- Variability in the degree of phosphorylation
- Genetic polymorphism
- Disulphide bonding
- Variations in the degree of glycosylation
- Hydrolysis of the primary caseins by plasmin

These are discussed below.

The significance of genetic polymorphism of milk proteins for the technological quality of milk has been recognized for a long time (see Martin et al., 2013) and the significance of variations in the levels of phosphorylation and glycosylation are now attracting attention (Bijl et al., 2014a, b, c).

3.1.1 Variability in the degree of phosphorylation

All the caseins are phosphorylated but to a different extent, with each showing variability in the degree of phosphorylation:

Casein	Number of phosphate residues per mole
α_{s1}	8, occasionally 9
α_{s2}	10, 11, 12 or 13
β	5, occasionally 4
κ	1, occasionally 2 or perhaps 3

The number of phosphate residues is indicated thus:

α_{s1}-CN 8P, β-CN 5P, etc.

Before the true relationship of the caseins was established, α_{s1}-CN 8P and α_{s1}-CN 9P were referred to as α_{s1} and α_{s0}, respectively, and α_{s2}-CN13P, α_{s2}-CN 12P, α_{s2}-CN 11P and α_{s2}-CN 10P as α_{s2}-, α_{s3}-, α_{s4}- and α_{s6}-, respectively.

3.1.2 Genetic polymorphism

Aschaffenburg and Drewry (1955) showed that the whey protein, β-lg, exists in two forms (A and B), differing by two amino acids. The protein in milk is genetically controlled and may be AA, AB or BB, depending on the genetic profile of the parents. This phenomenon is known as *genetic polymorphism*, and has since been observed for all milk proteins (Martin et al., 2013). Since PAGE differentiates on the basis of charge, only a small proportion of the genetic polymorphs of milk proteins have been detected. Genetic polymorphs can also be identified by RP-HPLC or mass spectrometry, but few studies using these techniques have been conducted so far. The genetic polymorph(s) present is indicated by a Latin letter as follows: β-CN A 5P, etc.

3.1.3 Disulphide bonding

The α_{s1}- and β-CNs do not contain cysteine, while the α_{s2}- and κ-CNs contain two cysteine residues linked by intermolecular disulphide bonds. α_{s2}-CN exists as a disulphide-linked dimer, while up to ten κ-CN molecules may be linked by disulphide bonds (Farrell et al., 1998). The κ-CN of all species for which data are available contains at least one cysteine residue near the N-terminus (at position 9, 10 or 11). Ruminant κ-CNs have a second cysteine residue near the chymosin-sensitive bond (Bouguyon et al., 2006). Disulphide-linked caseins and their role in casein micelles were discussed by Rasmussen et al. (1999).

3.1.4 Variations in the degree of glycosylation

κ-CN is the only member of the casein family which is glycosylated. It contains galactose, *N*-acetylgalactosamine and *N*-acetylneuraminic (sialic) acid, which occur as tri- or tetra-saccharides, the number of which varies from 0 to 4 per molecule of protein (i.e. a total of nine variants).

3.1.5 Hydrolysis of the primary caseins by plasmin

Milk contains several indigenous proteinases, the most important of which is plasmin, a trypsin-like, serine-type proteinase from blood with a high specificity for peptide bonds containing lysine or arginine (Kelly and McSweeney, 2003; Fox and Kelly, 2006a, b; O'Mahony et al., 2013). The preferred casein substrates are β and α_{s2}. All the caseins contain several lysine and arginine residues, but only a few bonds are hydrolysed rapidly. β-CN is hydrolysed at the bonds Lys28-Lys29, Lys105-His106 and Lys107-Glu108. The resulting C-terminal peptides are the γ-CNs (γ_1: β-CN f29–209; γ_2: β-CNf106–209; γ_3: β-CN f108–209; Visser et al., 1989a; Kelly and McSweeney, 2003). The γ-CNs represent ~3% of total casein and are identifiable by PAGE (Aimutis and Eigel, 1982). The other plasmin-produced peptides are either too small to be easily detected by standard techniques (such as urea-PAGE) or their concentrations are very low relative to the principal caseins. In solution, α_{s2}-CN is also susceptible to hydrolysis by plasmin (Le Bars and Gripon, 1989; Visser et al., 1989b), but peptides derived from this protein seem to appear at concentrations that are too low to be detected in milk, although a few have been isolated (O'Flaherty, 1997). α_{s1}-CN in solution is also hydrolysed readily by plasmin (Le Bars and Gripon, 1993; McSweeney et al., 1993). El-Negoumy (1973) identified a minor casein fraction, λ-CN, produced from α_{s1}-CN by plasmin.

3.2 Major characteristics of the caseins

Caseins have a number of distinct characteristics which determine their functionality, stability and isolation:

- insolubility at pH 4.6
- susceptibility to proteolysis
- heat stability

- amino acid composition
- site of biosynthesis

3.2.1 Insolubility at pH 4.6

The caseins are, by definition, insoluble at pH 4.6. The whey proteins are soluble under the ionic conditions of milk; however, they are least soluble around pH 4.6, with isoelectric points ranging from approximately pH 4.2 to 5.4 (Gordon, 1971). The isoelectric precipitation of casein is of industrial importance since it facilitates production of caseins and caseinates, fermented milk products and acid-coagulated cheese (see Section 7.2.2).

3.2.2 Susceptibility to proteolysis

Unlike the whey proteins, the caseins are rendered coagulable by proteolysis using enzymes such as chymosin. The open, flexible 'rheomorphic' structure of the caseins makes them susceptible to proteolysis. The susceptibility of the caseins (in micellar form) to coagulation after hydrolysis of κ-CN is used in the production of rennet-coagulated cheese and rennet casein. Owing to their high hydrophobicity, the peptides generated from caseins by enzymatic hydrolysis (especially β-CN) can produce a bitter taste in cheese and hydrolysates (requiring the need for masking).

3.2.3 Heat stability

The caseins are very heat-stable compared to the whey proteins. Milk at pH 6.7 may be heated at 100°C for 24 h without coagulation and may withstand heating at 140°C for up to 20–25 min. Removal of whey proteins from milk by membrane filtration renders the resultant casein-rich material more stable to cooking/boiling (Heino and Huumonen, 2015). The high heat stability of the caseins also allows production of heat-sterilized dairy products with relatively little physical changes. Casein-dominant milk ingredients (e.g. milk protein concentrates, MPCs) are commonly used for infant and clinical nutrition products. As the protein:lactose ratio of MPCs increases, they become less stable to heating (Crowley et al., 2015a). Caseinates, on the other hand, exhibit very high heat stability. The manufacture of MPCs, caseinates and other ingredients is discussed in more detail in Section 7.2.

3.2.4 Amino acid composition

The caseins are very flexible molecules, due mainly to the high content of the structure-disrupting amino acid, proline (Holt and Sawyer, 1993). β-CN is particularly rich in proline, which contributes 35 residues of its total 209 (Table 2). Whole isoelectric casein contains ~0.8% phosphorus with varying degrees of phosphorylation among individual caseins. The presence of phosphate groups affects hydration, solubility and heat stability.

Casein is a very effective binder of metals. Most of the calcium, zinc and inorganic phosphorus in milk are associated with the caseins, which influences their physico-chemical, functional and nutritional properties. The metal-binding properties of casein enable a high concentration of calcium phosphate to be carried in milk in a stable suspension (to supply the requirements of the neonate); in addition, binding of calcium phosphate by casein micelles prevents precipitation-induced blockage of the ducts of the mammary gland. As a consequence of metal binding, most of the caseins can be precipitated by polyvalent

Table 2 Amino acid (aa) profile, total residues and molecular weight (MW, in kDa) of the major proteins in the milk of Western cattle

aa	α_{s1}-Casein B	α_{s2}-Casein B	β-Casein A^2	κ-Casein B	γ_1-Casein A^2	γ_2-Casein A^2	γ_3-Casein A	β-Lacto-globulin A^2	α-Lact-albumin B
Asp	7	4	4	4	4	2	2	11	9
Asn	8	14	5	7	3	1	1	5	12
Thr	5	15	9	14	8	4	4	8	7
Ser	8	6	11	12	10	7	7	7	7
SerP	8	11	5	1	1	0	0	0	0
Glu	24	25	18	12	11	4	4	16	8
Gln	15	15	21	14	21	11	11	9	5
Pro	17	10	35	20	34	21	21	8	2
Gly	9	2	5	2	4	2	2	3	6
Ala	9	8	5	15	5	2	2	14	3
½ Cys	0	2	0	2	0	0	0	5	8
Val	11	14	19	11	17	10	10	10	6
Met	5	4	6	2	6	4	4	4	1
His	11	11	10	13	7	3	3	10	8
Leu	17	13	22	8	19	14	14	22	13
Tyr	10	12	4	9	4	3	3	4	4
Phe	8	6	9	4	9	5	5	4	4
Trp	2	2	1	1	1	1	1	2	4
Lys	14	24	11	9	10	3	3	15	12
His	5	3	5	3	5	3	3	2	3
Arg	6	6	4	5	2	2	2	3	1
PyroGlu	0	0	0	1	0	0	0	0	0
Residues	199	207	209	169	181	104	102	162	123
MW	24	25	24	19	21	12	12	18	14

Source: Adapted from Swaisgood (1982).

cations at sufficiently high concentrations, which is essential for the rennet coagulation of milk, as in cheese manufacture.

The caseins are generally regarded as very hydrophobic proteins, but, as shown in Table 2, with the exception of β-CN, they do not contain a high proportion of hydrophobic amino acids. Instead, they have a high surface hydrophobicity because, due to their lack of stable secondary and tertiary structures, most of their hydrophobic residues are exposed to the solvent.

The caseins contain a low level of sulphur (0.8%), which occurs mainly in methionine, with little cystine or cysteine present. The principal caseins (α_{s1}- and β-CN) contain no cystine or cysteine. In milk heated at a temperature >70°C, β-lg (and to a lesser extent α-la) interacts with caseins *via* disulphide bonding between free sulphydryl groups exposed on β-lg (due to heat-induced unfolding of the globular whey protein) and κ-CN. These protein–protein interactions have important implications for the heat stability, physical stability and sensory properties of heat-treated dairy products. The caseins also contain low levels of branched-chain amino acids, such as isoleucine, leucine and valine, which are important in muscle growth and repair.

One of the more notable features of the amino acid sequence of the caseins is that the hydrophobic and hydrophilic residues are not distributed uniformly, thereby giving the caseins a distinctly amphipathic structure (Huppertz, 2013). This feature, coupled with their flexible structure and hydrophobicity, gives the caseins good surface activity, foaming and emulsifying properties, making casein the functional protein of choice for many applications.

3.2.5 Site of biosynthesis

The caseins are synthesized in the mammary gland and are unique to this organ. Presumably, they are synthesized to meet the amino acid requirements of the neonate and as carriers of important minerals required by the neonate.

4 Casein micelles

Caseins in milk have long been known to exist as large colloidal particles (Kastle and Roberts, 1909). These particles are now known as *casein micelles*, a term introduced by M. Beau in 1921 (see Fox and Brodkorb, 2008). As described by Fox and Brodkorb (2008), the term 'micelle' has been used for different types of multi-molecular structures and the casein micelle is quite different from the classical 'soap micelle'. To emphasize this difference, the casein particles should be referred to as 'casein micelles', not just as 'micelles'. McMahon and Oommen (2008) described casein micelles as 'supramolecules', a term which the International Union of Pure and Applied Chemists (IUPAC) defines '*as a system consisting of multiple entities held together and organized by means of intermolecular non-covalent binding interactions*'. However, Horne (2010) explained that the casein micelle does not strictly conform to the IUPAC definition of a 'supramolecule' and that the term should not be used.

About 95% of the casein of milk exists as micelles, the dry matter of which is ~94% protein and ~6% low molecular mass species called micellar or colloidal calcium phosphate (CCP), which consists mainly of calcium and phosphate, with small amounts of magnesium and citrate and trace amounts of other elements. The micelles are highly hydrated, binding ~2.0–4.0 g H_2O g^{-1} protein (the value depends on how hydration is measured). CCP is an integral component of the casein micelle and the milk salt system; although CCP will not be covered in detail in the present chapter, reviews on milk salts/CCP are available (Pyne, 1934, 1962; Holt, 1985a, 1997; Lucey and Horne, 2009).

The micelles scatter light and are mainly responsible for the white colour of milk. The milk of all species is white, to different degrees, suggesting that all contain casein micelles. The white colour is lost if the micelle structure is disrupted, for example, by dissolving CCP by the addition of citrate, ethylenediaminetetraacetate (EDTA) or oxalate, by acidification at <4°C to pH 4.6, followed by exhaustive dialysis against cold bulk milk, by increasing pH to >9.5 or by adding urea (>5 M) or ethanol (~35% at 70°C). Some of the principal properties of the casein micelle are summarized in Table 3.

Casein micelles are typically spherical in shape. The diameter of bovine casein micelles ranges from 50 to 500 nm (average ~150–180 nm), and they have a mass ranging from ~10^6 to 3×10^9 Da (average ~10^8 Da). On a number basis, there are many small micelles, but these represent only a small proportion of the mass. The micelles in human milk are small (~60 nm in diameter), while those in equine (or asinine) milk, which contains only ~1% casein, of which only ~1.8–2.6% is κ-CN, are very large (mean ~255 nm). The casein micelles in camel milk are even larger (~380 nm). Bovine milk contains 10^{14}–10^{16} micelles mL^{-1} milk, and they are roughly two micelle diameters apart, that is, they are quite tightly packed. Since the milk of lagomorphs contains up to 20% protein, the micelles must be very closely packed; the micelles in lagomorph milk appear generally similar to those in the milk of other species (Buchheim et al., 1989). Generally, the level of κ-CN in casein micelles is inversely related to the size of the micelles, but this does not apply across all species (see Table 4). Methods for determining the size distribution of casein micelles were reviewed by Holt (1985b), who also considered the biological significance of the variation

Table 3 Average characteristics of the casein micelles in cow milk

Characteristic	Value
Diameter	150–182 nm
Surface	8×10^{-10} cm^2
Volume	2.1×10^{-15} cm^3
Density (hydrated)	1.0632 g cm^3
Mass	2.2×10^{-15} g
Water content	63%
Hydration	3.7 g H_2O g protein
Voluminosity	4.4 cm^3 g
Molecular weight (hydrated)	1.3×10^9 Da
Molecular weight (dehydrated)	5×10^4 Da
Number of peptide chains (MW; 30 kDa)	10^4
Number of particles per mL milk	10^{14}–10^{16}
Whole surface of particle	5×10^4 cm^2 mL milk
Mean free distance	240 nm

Table 4 Relationship between casein micelle size and the proportion of κ-casein in the casein fraction of the milk of different species

Species	Total casein (gL^{-1})	κ-casein in total casein (%)	Casein micelle size (nm)
Cow	24.6–28	9.3–11.8	150–182
Human	2.4–4.2	3.3–5.8	64–80
Goat	23.3–46.3	12.0–28.9	260
Sheep	41.8–46	7.7–9.3	180–210
Buffalo	32–40	12.8–13.5	180
Horse	9.4–13.6	1.8–2.6	255
Camel	22.1–26	5.0⁺	380
Yak	34.3–45.8	14.3–18.6	212*

Source: Data is adapted from Claeys et al. (2014), except where denoted with *(Wang et al., 2013) and ⁺(Ramet, 2001).

in micelle size. Dynamic light-scattering is very commonly used for determining the mean size and distribution of casein micelles.

Casein micelles can be prepared in their native form by ultracentrifugation (e.g. Dalgleish et al., 1981), microfiltration with flat-sheet membranes of ~0.1−0.2 μm (Hernández and Harte, 2009; Crowley et al., 2015b) or chromatography on controlled pore glass or silica gel (McGann et al., 1979; Robson et al., 1985). These techniques are discussed in more detail in Section 6.

The stability of milk and many of its technologically important properties are related to the properties of the casein micelles, which have, therefore, been the focus of considerable research, especially during the past 60 years.

4.1 Association of the caseins and formation of casein micelles

All the major caseins associate with themselves and with each other. In unreduced form, κ-CN is present largely as disulphide-linked polymers; it also forms hydrogen and hydrophobic bonds with itself and with other caseins, but these associations have not been studied in detail.

At 4°C, β-CN exists in solution as monomers of molecular mass ~24 kDa. As the temperature is increased, the monomers polymerize to form long thread-like chains of about 20 units at 8.5°C and to still larger aggregates at higher temperatures. The degree of association depends on protein concentration, ionic strength, [Ca^{2+}], temperature and pH. The ability to form thread-like polymers may be important in the structure of the casein micelle. Apparently, β-CN forms 'disc-like' micelles below its isoelectric point (Moitzi et al., 2008). β-CN also undergoes a temperature-dependent conformational change in which the content of poly-L-proline helix decreases with increasing temperature; the transition temperature is about 20°C, that is, very close to the temperature at which β-CN becomes insoluble in Ca.

α_{s1}-CN polymerizes to form tetramers of molecular mass ~113 kDa; the degree of polymerization increases with increasing protein concentration and increasing temperature. The literature on the self-association of the caseins was reviewed by Rollema (1992). Euston and Horne (2005) concluded that α_{s1}-CN forms linear micelles and flower-like circular micelles and at high concentrations forms worm-like micelles. α_{s2}-CN forms small linear micelles and β-CN forms circular micelles with a dense core. Euston and Nicolosi (2007) simulated the self-association of the caseins individually and in mixtures leading to the formation of casein micelles. They concluded that the formation of micelles in their system was a random association process but considered the formation of casein micelles in the mammary gland to be a non-random process and that more information is required on the order of addition of the caseins during the formation of micelles in the secretory tissue.

As a result of their high number of phosphate residues, the α_{s1}-, α_{s2}- and β-CNs, which represent ~85% of total casein, are very sensitive to Ca^{2+} and are precipitated by $[Ca^{2+}] > 6$ mM. Since bovine milk contains ~30 mM calcium, it would be expected that these caseins would precipitate in milk, but they react with and are stabilized by κ-CN, which is not sensitive to Ca^{2+}, and facilitates the formation of stable casein micelles. The Ca-sensitive caseins are located mainly in the core of the micelles with κ-CN predominantly on the surface. The N-terminal 2/3 of κ-CN is strongly hydrophobic, while the C-terminal is hydrophilic and is oriented into the surrounding aqueous serum phase. The structure of κ-CN has been described in detail by Creamer et al. (1998), Huppertz (2013) and Farrell et al. (2013).

4.2 The structure of casein micelles

Key aspects of the structure of micelle are

- κ-CN (representing ~12–15% of total casein) must be located so that it can stabilize the calcium-sensitive α_{s1}-, α_{s2}- and β-CNs (representing ~85% of total casein).
- Chymosin and similar proteinases, which hydrolyse most of the κ-CN and cause destabilization of micelles, are relatively large molecules (~35 kDa).
- κ-CN and β-lg (MW of 36 kDa in milk) which, when heated in the presence of each other, interact to form a disulphide-linked complex which modifies the properties (e.g. hydrodynamic diameter) of the micelles.

These features suggest that a surface layer of κ-CN surrounds the Ca-sensitive caseins, similar to a lipid emulsion in which the triglycerides are surrounded by a layer of emulsifier.

If CCP is removed, the micelles disintegrate into particles of MW ~10^6 Da, suggesting that CCP is a major integrating factor in the micelles. The properties of a CCP-free system are very different from normal milk, but these properties can be partially restored by an increased concentration of calcium. Other factors that can lead to dissociation of micelles include temperature, urea, SDS, ethanol or alkaline pH. As the temperature is lowered, casein (particularly β-CN) dissociates from the micelles (Rose, 1968; Downey and Murphy, 1970; Downey, 1973; Bansal, 2011), although the integrity of the micelle structure is maintained. On the other hand, addition of 5 M urea causes complete dissociation of casein.

Various models of casein micelle structure have been proposed over the last 60 years and have been refined progressively. Recent studies include Holt (1992, 1994, 1998), Walstra (1990, 1999), Holt and Horne (1996), Horne (1998, 2002, 2006, 2011), McMahon and McManus (1998), de Kruif (1999, 2014), de Kruif and Holt (2003), Farrell et al. (2006),

Qi (2007, 2009), Phadungath (2005), McMahon and Oommen (2008), Dalgleish (2011), Dalgleish and Corredig (2012) and Holt et al. (2013).

Early models (e.g. Morr, 1967) proposed that micelles are composed of sub-micelles of MW ~10^6 Da and 10–15 nm in diameter (Fig. 4). The sub-micelles were believed to be linked together by CCP, thereby giving the micelle an open, porous structure. Later models suggest differing structures for the sub-micelle; for example, Payens (1966), Rose (1969), Waugh et al. (1970), Slattery and Evard (1973), Slattery (1976), Schmidt (1980, 1982), Walstra and Jenness (1984), Rollema (1992) and Walstra (1990, 1999).

Farrell et al. (2006) suggest that casein associations, triggered by conserved protein sequences, begin in the endoplasmic reticulum (ER), preventing protein accumulation and destruction by the endoplasmic-reticulum-associated protein degradation (ERAD) process. α_{s1}-CN appears to have a crucial role in this process. For species with a heavy secretory load, this protein acts as a 'molecular detergent' to prevent large particle accumulation and ERAD removal. On transport to the Golgi apparatus, these pre-formed proteinacious particles (sub-micelles) are phosphorylated and calcium and phosphate are incorporated into these particles. Micelle self-assembly occurs in the Golgi vesicles through the condensation of the sub-micelles complexed with calcium and phosphate. Further rearrangement of the casein micelles occurs here and the outer loose layer of sub-micellar particles of the system may overlap (interdigitate) with adjacent sub-micelles. Rearrangement of the casein monomers, particularly κ-CN, may also occur here. Finally, the introduction of molecular oxygen at the time of milking may activate sulphydryl oxidase and finalize the formation of the micelles through surface-oriented disulphide bonds of κ-CN.

Chanat et al. (1999) showed that α_{s1}-CN interacts with the other caseins in the ER and proposed that the formation of this complex is essential for the export of the caseins out of the ER. However, neither α_{s1}- nor β-CN is absolutely necessary for the formation of casein micelles, which is essentially based on hydrophobic and electrostatic interactions which are achieved on aggregation of the caseins following neutralization of their negative charges by calcium and the formation of electrostatic bridges between phosphoseryl residues by CCP. The role of α_{s1}-CN in the transport of β- and κ-CNs from the ER to the Golgi was confirmed by Le Parc et al. (2010), who concluded that α_{s1}-CN plays a key role in the early stages of casein micelle formation.

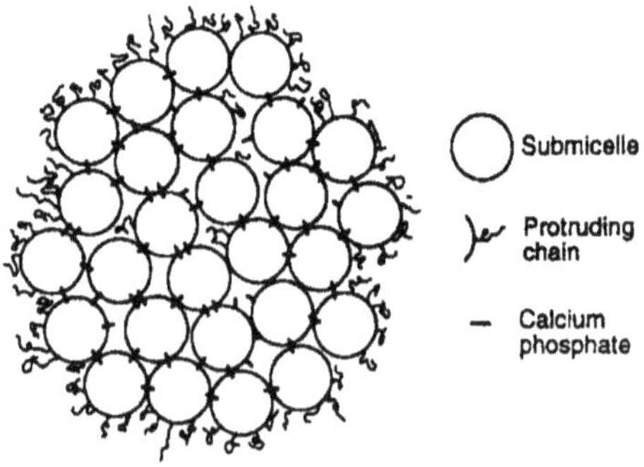

Figure 4 Sub-micelle model of the casein micelle.
Source: Adapted from Walstra and Jenness (1984).

The principal features of the sub-micelle models were described by Phadungath (2005), Farrell et al. (2006) and O'Mahony and Fox (2013). The concept that the hydrophilic C-terminal segment of κ-CN protrudes from the surface of the micelles, forming a layer 5–10 nm thick and giving the micelles a hairy appearance, has received support from many researchers. This hairy layer is considered to be responsible for micelle stability through major contributions to zeta potential (−20 mV) and steric stabilization (Hill and Wake 1969). Removal of the hairy layer (e.g. through specific hydrolysis of κ-CN) or its collapse (e.g. by ethanol) destroys the colloidal stability of the micelles, and they coagulate or precipitate (see Holt and Horne, 1996). Variations of the sub-unit model are those of Kimura et al. (1979) and Ono and Obata (1989).

A contrasting model of the casein micelle is that of Parry and Carroll (1969), who suggested that κ-CN forms the core (nucleus) of the micelle and is surrounded by α- (α_{s1}- and α_{s2}-CNs) and β-CN. The model of Garnier and Ribadeau-Dumas (1970) might be regarded as a variant of this: κ-CN was considered to form nodes in a three-dimensional network in which the branches were proposed to consist of copolymers of α_s- (α_{s1}- and α_{s2}-) and β-CNs.

Although the sub-micelle model of the casein micelle adequately explains many of the principal features of, and physico-chemical reactions undergone by, the micelles, and has been supported widely, it has never enjoyed unanimous support, and alternative models have been proposed. Visser (1992) proposed that the micelles are spherical structures of casein molecules randomly aggregated and held together partly by salt bridges in the form of amorphous calcium phosphate and partly by other forces (e.g. hydrophobic bonds) with a surface layer of κ-CN.

In 1992, Carl Holt commenced a series of publications on the structure of the casein micelle. Holt (1992, 1994) depicted the casein micelle as a tangled web of flexible casein molecules forming a gel-like structure in which nanoclusters of CCP are an integral feature, from the surface of which the C-terminal region of κ-CN extends, forming a 'hairy layer', 12 ± 2 nm thick (Holt and Dalgleish, 1986) (Fig. 5). The concept of the hairy casein micelle and its implications for dairy technology have been reviewed by Holt and Horne (1996), Holt (1998) and Holt et al. (2003).

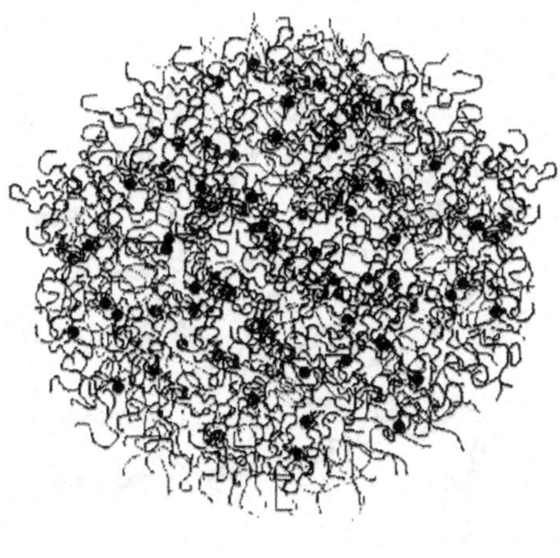

Figure 5 Holt model of the casein micelle.

Source: Modified from de Kruif and Holt (2003).

de Kruif (1998) supported the structure of the casein micelle as depicted by Holt (1992, 1994) and described the behaviour and properties of the micelles in terms of adhesive hard spheres. The structure, functions and interactions of the casein micelle were the subject of a thorough review by de Kruif and Holt (2003). Holt's model retains two of the key features of the sub-micellar model (i.e. the cementing role of CCP and the predominantly surface location and micelle-stabilizing role of κ-CN) and differ from it mainly with respect to the internal structure of the micelle.

A 'dual-binding' or 'block copolymer' model of casein micelle structure has been proposed and developed by Horne (1998, 2002, 2003, 2006, 2011, 2014). This model suggests that micelle structure is governed by a balance between hydrophobic and electrostatic interactions and CCP-mediated cross-linking of hydrophilic regions, which contain seryl phosphate residues (Fig. 6). All four caseins have phosphoseryl residues, with the phosphoseryl residues in α_{s1}-, α_{s2}- and β-CNs occurring as clusters; these caseins have 2, 3 and 1 cluster(s) of phosphoryl residues, respectively; the caseins form a network by intermolecular hydrophobic interactions and CCP crosslinks. κ-CN, which has only 1 phosphoseryl residue, has no such cluster and functions as a polymerization blocker. Horne has argued strongly in support of his proposed structure and explained how it meets the various characteristics of the casein micelle. In the opinion of the authors, the model of Horne is a variant of that of Holt, but describes in greater detail how α_s- and β-CNs interact in forming the matrix of the casein micelle.

Dalgleish (1998) concluded that the micellar surface is only partially covered with κ-CN, which is distributed non-uniformly on the surface; the spaces between the κ-CN molecules may permit the passage of other molecules, for example, chymosin. This surface coverage provides steric stabilization against the approach of large particles, such as other micelles, but the small-scale heterogeneities and the gaps between κ-CN molecules provide relatively easy access for molecules with dimensions of individual proteins or smaller. He

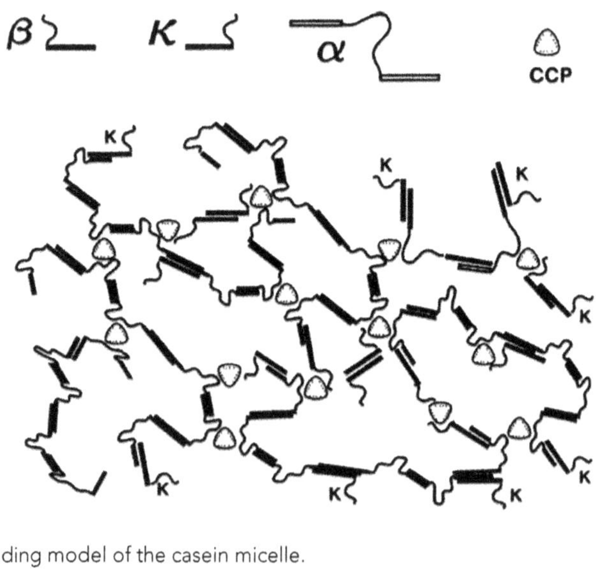

Figure 6 Dual-binding model of the casein micelle.
Source: Adapted from Horne (1998).

concluded that the hairy layer is about 10 nm thick. Using a cold-field emission ultra-high-resolution electron microscope, without metal coating, Dalgleish et al. (2004) showed that the surface of casein micelles is formed from cylindrical or tubular structures, 10–20 nm in diameter and 40 nm long, protruding from the particle, rather than a hairy layer. These structures may give the impression of 'classical' sub-micelles. The tubules might continue through the micelle which might be considered as a 'bicontinuous' water/protein system. The authors describe how various micelle characteristics, for example, accessibility of κ-CN to rennet, decrease in the size of the micelles on renneting and liberation of β-CN on cooling can be explained by such a system. A new improved method for immobilizing casein micelles developed by Martin et al. (2006) showed the same type of micelle structure as shown by Dalgleish et al. (2004).

The reactivity of casein micelles was discussed by Dalgleish (2007). The various proposed structural models of the casein micelle were reviewed by Dalgleish (2011) and a new model was proposed (Fig. 7). The effects of various processing operations on the casein micelle were discussed by Dalgleish and Corredig (2012).

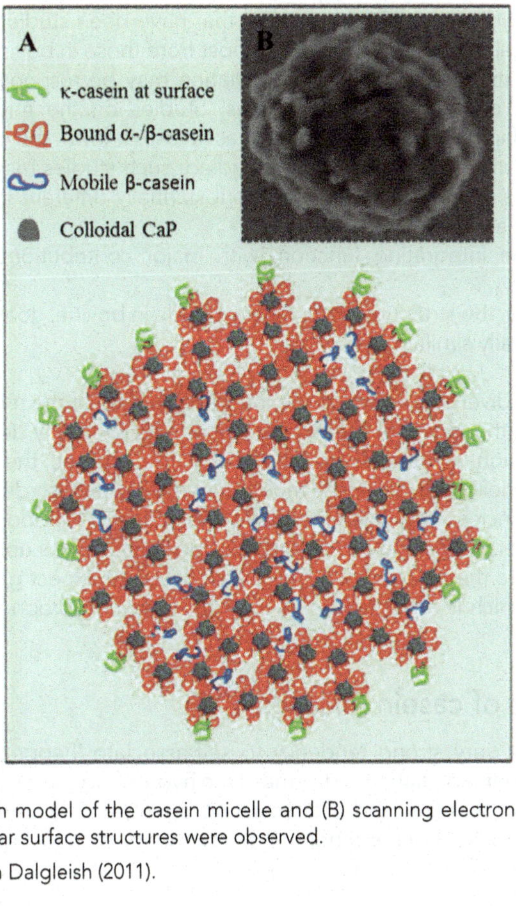

Figure 7 (A) Dalgleish model of the casein micelle and (B) scanning electron microscopy image of micelle in which tubular surface structures were observed.

Source: Adapted from Dalgleish (2011).

In summary, there is general agreement on certain aspects of the structure of the casein micelle:

- They are spherical particles, ranging in diameter from ~50 to 600 nm, averaging about 150–182 nm, for bovine milk, ~60 nm for human milk and ~255 nm for equine milk.
- The micelles are highly hydrated, up to ~4 g water per g protein for bovine micelles. The degree of hydration exhibits inter-species differences, with the casein micelles in human milk containing ~6.5 g water per g protein, for example.
- The micelles are stabilized by κ-CN, the hydrophilic C-terminal segment of which protrudes from the surface as a 'hairy' layer, about 10 nm thick (see McMahon and Oommen, 2008); however some investigators (e.g. Marchin et al., 2007) failed to find evidence for a hairy layer.
- The size of the casein micelles is inversely related to their κ-CN content. Indeed, milk homologous for κ-CN B, which contains a greater proportion of κ-CN than that containing the A variant, has smaller micelles. The proportion of β-CN varies inversely to the proportion of κ-CN, while the proportions of α_{s1}- and α_{s2-} are independent of casein micelle size.
- The casein micelles in milk of all species are generally similar, but there are substantial inter-species differences. Of the species that have been studied in sufficient detail, the casein micelles in human milk differ most from those in bovine milk; the low level of α_s-CN and its low level of phosphorylation may be responsible for the unusual characteristics of human casein micelles. Studies on the milk of primates were reviewed by Uniacke and Fox (2011), but as far as is known, the casein micelles in the milk of other primates have not been studied; such studies may be rewarding. The casein micelles in wallaby milk are also substantially different from those in bovine milk (Horne et al., 2007).
- CCP serves an integrating function, with major contributions from hydrophobic interactions.
- Apart from size, the structure of the casein micelle in bovine, goat, human and equine milk are generally similar.

However, there are divergent views on the internal structure of the micelles, but all recent proposals preclude the idea that the micelle is made up of clearly defined proteinaceous sub-micelles, although a type of sub-micelle may be present; the calcium phosphate nanoclusters with their shell of casein molecules, might be regarded as a type of sub-micelle. It remains unclear why, given that κ-CN is attached to the body of the micelles only by hydrophobic interactions and has an exposed position, it does not dissociate from the micelle on cooling to the extent that β-CN does; one might expect greater dissociation of κ-CN than β-CN, which is attached at three points, two *via* hydrophobic interactions and one *via* CCP.

4.3 Properties of casein micelles

The caseins have a very strong tendency to self-associate (Pepper, 1972; Pepper and Farrell, 1982). This self-association is due mainly to hydrophobic interaction. This can make it hard to isolate the caseins, and a dissociating agent, such as urea or SDS is usually required for this purpose. On the other hand, a tendency to associate is important in the formation and stabilization of casein micelles.

Owing to their open structure, caseins have a high specific volume, and, consequently, form highly viscous solutions, particularly when concentrated, which can cause difficulties during the production of casein-dominant ingredients. For example, the viscosity of sodium caseinate solutions is so high that it is not possible to spray-dry solutions containing >20% protein. However, this high viscosity is desirable in certain applications, such as the stabilization of emulsions.

The stability/instability of casein micelles in milk has been reviewed by Walstra (1990), Tuinier and de Kruif (2002) and Dalgleish (2007). The effects of various processing operations on the casein micelle were discussed by Dalgleish and Corredig (2012). The casein micelles are quite stable to standard milk processing (Ono et al. 1999). They are very stable at high temperatures, coagulating only after heating for 15–20 min at 140°C at the normal pH of milk. The micelles are also stable to compaction. They can be sedimented by ultracentrifugation and redispersed readily in solvent by mild agitation. They are stable to conventional homogenization but are partially disrupted by high-pressure homogenization (Sandra and Dalgleish, 2005; Roach and Harte, 2008). Casein micelles are also unstable to high-pressure processing, particularly at pressures of >200 MPa (Desobry-Banon et al., 1994; Gaucheron et al., 1997; Needs et al., 2000; Scollard et al., 2000; Huppertz et al., 2002).

On cooling skim milk to a temperature in the range 0–5°C, up to ~50% of total β-CN (and some of the other caseins also) dissociates from the micelles (Rose, 1968; Downey and Murphy, 1970; Downey, 1973; Creamer et al., 1977; Swaisgood, 2003). The dissociation of β-CN from the β-CN micelles is prevented by treatment with transglutaminase (O'Connell and de Kruif, 2003). Slow freezing and storage of milk at a temperature in the range −10°C to −20°C can cause some destabilization (cryo-destabilization) due to increased [Ca^{2+}] in the serum phase and decreased pH. Cryo-destabilized casein can be dispersed in water to give particles with micelle-like properties (Moon et al., 1988).

Combinations of unit operations for milk powder production which include concentration of milk proteins/casein by ultrafiltration/microfiltration, concentration of total solids by evaporation and spray-drying into a powder can cause interlinking of casein micelles (Oldfield et al., 2005; Havea, 2006; Karlsson et al., 2007; Martin et al., 2007; Fox and Brodkorb, 2008). If it occurs, this destabilization can have serious negative implications for the rehydration characteristics of casein-dominant powders (Crowley et al., 2016).

Casein micelles are stable to high Ca^{2+} concentrations, up to at least 200 mM at a temperature up to 50°C. However, high Ca-ion activity can have a strong negative effect on the stability of casein micelles at temperatures >120°C (Crowley et al., 2015a). In addition, the reactivity of casein micelles to certain environmental conditions is strongly affected by Ca, for example, stability to ethanol and renneting. These areas have been the subject of a series of studies over many years with some recent reports including those of Tsioulpas et al. (2007) and Philippe et al. (2005).

When the pH is reduced to the isoelectric point of casein (pH ~4.6), the caseins aggregate and precipitate from suspension. Precipitation at this pH is dependent on temperature. It does not occur at temperatures of <5–8°C. At higher temperatures, it occurs over a wide pH range, perhaps 3.0–5.5. CCP dissolves as the pH of milk is reduced and is fully dissolved <pH 4.9. Acidification of cold (4°C) milk to pH 4.6, followed by dialysis against cold bulk milk, is a convenient and widely used technique for changing the CCP content of milk (Pyne and McGann, 1960). If undialysed, acidified cold milk is readjusted to pH 6.7, the micelles reform, provided that the pH had not been reduced below 5.2, suggesting

that most of the CCP can be dissolved (removed) without destroying the structure of the micelles.

Chymosin and other acid proteinases specifically hydrolyse κ-CN at, or close to, the Phe_{105}-Met_{106} bond, as a result of which the micelles coagulate or gel in the presence of Ca^{2+} (see Horne and Banks, 2004; Lucey, 2011; Horne, 2016). This is the key step in the manufacture of rennet-coagulated cheese varieties. Coagulation is very temperature-dependent and does not occur below ~18°C, although light-scattering properties indicate that aggregation does occur at a lower temperature (Bansal et al., 2008) and this low-temperature aggregation is very dependent on $[Ca^{2+}]$, pH and ionic strength.

The *N*-acetylneuraminic residues of κ-CN are significant but not the most important factor for preventing the destabilization of casein micelles (Minkiewicz et al., 1993). Milk with smaller casein micelles forms a stronger rennet-induced gel than milk containing larger micelles (Glantz et al., 2010), presumably due to the high surface area of the micelle population which increases the number of interaction sites.

The size of casein micelles is related to the genetic variant of κ-CN; milk containing κ-CN B has smaller micelles than milk containing κ-CN A. Small micelle milk is also associated with the level of glycosylation of κ-CN, which appears to be the most important factor affecting the size of casein micelles (Bijl et al., 2014a). The level of glycosylated κ-CN may influence the formation of casein micelles. Park et al. (1999) reported that the zeta potential of casein micelles is inversely related to micelle size, suggesting that small micelles could aggregate more readily than larger micelles and form a firmer gel.

The micelles are destabilized by ~40% ethanol at pH 6.7 at room temperature, but can be destabilized by a lower concentration of ethanol if the pH is reduced (Horne, 2003). The precipitate dissolves and the system becomes translucent when the system is heated to ~70°C. The white appearance of milk is restored if the mixture is cooled, and a gel is formed if the ethanol-milk mixture is held at 4°C. Removal of the ethanol by evaporation results in the formation of large aggregates (average diameter ~3,000 nm) with significantly different properties from those of natural micelles (O'Connell et al., 2001a, b). The aggregates can be dispersed by agitation. Increasing the pH promotes the dissociating effect of ethanol (35% ethanol causes dissociation at 20°C at pH 7.3) or adding NaCl. Methanol and acetone have a dissociating effect similar to that of ethanol. Propanol causes dissociation at ~25°C.

The micelles can be dissociated by

- Exhaustive dialysis against phosphate-free buffer, which dissolves the CCP (Holt, 1998)
- Urea (>5 M), which disrupts hydrophobic attractions (McGann and Fox, 1974; Holt, 1998); smaller micelles reform on removing the urea
- SDS (≥10 g L^{-1}), an anionic detergent which, like urea, dissociates micelles by disrupting hydrophobic interactions (Lefebvre-Cases et al., 1998a, b; Smiddy et al., 2006)
- Raising the pH to >9, which increases electrostatic repulsion between proteins and increases solvent quality for caseins; reformation of micelles occurs on reacidification (de Kruif and Holt, 2003; Vaia et al., 2006)
- Calcium-binding agents: addition of phosphate, citrate, oxalate or phytate salts of sodium or potassium results in the binding of ionic calcium and colloidal calcium, which, in turn, can result in the swelling and dissociation of casein micelles (Morr, 1967; Mizuno and Lucey, 2005; de Kort et al., 2011).

Casein micelles can be stabilized to dissociation by urea, acidification or high pressure by the formation of internal inter-protein covalent bonds by transglutaminase (O'Sullivan et al., 2002a, b; Smiddy et al., 2006).

Casein micelles reformed from alkaline-dissociated (pH 10.0) micelles by reacidification of the dispersion to pH 6.6 were characterized by Huppertz et al. (2008). The reformed micelles were smaller than the native micelles, had a slightly lower zeta potential, a lower ethanol stability and similar heat stability. However, unlike native micelles, heat stability did not recover at pH $\geq$ 7.2, which is similar to the behaviour observed for CCP-reduced bovine casein micelles (Fox and Hoynes, 1975).

Casein micelles can react with charged polysaccharides, with consequent changes in their stability. κ-Carrageenan reacts with casein micelles *via* κ-CN, forming a weak casein-carrageenan network which resists the tendency to phase separation in long shelf-life products (Spagnuolo et al. 2005; Martin et al., 2006). Acidified milk can be stabilized by pectin, which is exploited in the manufacture of drinking yoghurt and other types of acidified milk beverages; pectin forms a layer around the casein micelles to provide steric and electrostatic stabilization to the particles.

5 Whey proteins

It was recognized early that acid whey (i.e. the pH 4.6-soluble fraction from isoelectric precipitation of casein) contains two well-defined groups of proteins:

- Lactoglobulins, which are salted-out in 50% saturated $(NH_4)_2SO_4$ or saturated $MgSO_4$, and comprise mainly of immunoglobulins,
- Lactalbumins, which are soluble under these conditions.

The lactalbumin fraction was considered homogeneous until Palmer (1934) isolated and crystallized a protein that behaved as an albumin in that it was soluble in half-saturated $(NH_4)_2SO_4$ or saturated $MgSO_4$ but had some characteristics of globulins (i.e. was insoluble in pure water at its isoelectric point (pH 5.2) but was soluble in dilute salt solutions). This protein was identified as the β-peak in free-boundary electrophoretograms of milk proteins; initially it was called *β-lactalbumin*, but was later renamed *β-lactoglobulin*. The major components of whey are thus:

- β-Lactoglobulin (~50%)
- α-Lactalbumin (~20%)

Other components include:

- Blood serum albumin (<10%)
- Immunoglobulins (~10%)
- Lactoferrin (~1%)

Whey also contains many minor proteins and enzymes.

5.1 β-Lactoglobulin

β-Lactoglobulin (β-Lg) represents ~50% of the whey proteins and ~12% of the total protein in bovine milk. The extensive literature on this protein has been reviewed by, among others, Sawyer (2003, 2013) and Creamer et al. (2011). β-Lg is the principal whey protein in the milk of the cow, buffalo, sheep and goat, although there are slight inter-species differences. Bovine β-lg consists of 162 amino acid residues per monomer, with a MW of ~18 kDa. Its amino acid sequence and that of several other species has been established (Martin et al., 2013). Its isoelectric point is pH ~5.2–5.4. It contains two intramolecular disulphide bonds and 1 mol of cysteine per monomer. The cysteine residue is important due to its reaction, following thermal denaturation, with the intermolecular disulphide of κ-CN, which affects the rennet coagulation and heat stability of milk. It is also responsible for the cooked flavour of heated milk. Some β-lg (e.g. porcine, equine) lack a free sulphydryl group, while the milk of other species is devoid of β-lg (e.g. human, camel).

β-Lg is a highly structured protein: in the pH range 2–6, 10–15% of the molecule exists as α-helices, 43% as β-sheets and 47% as unordered structures, including β-turns; the β-sheets occur in a β-barrel-type calyx. The molecule has a very compact globular structure; each monomer approximates a sphere, which is about 3.6 nm in diameter and exists as a dimer of MW ~36 kDa in the pH range 5.5–7.5, as a monomer at pH < 3.5 and at pH > 7.5 and as a tetramer (MW ~144 kDa) in the pH range 3.5–5.5. Porcine and other β-lgs which lack a free thiol do not form dimers, a property that is probably not due to the absence of a thiol group. β-Lg is very resistant to proteolysis in its native state; this feature suggests that its primary function is not solely as a source of amino acids.

β-Lg binds retinol (vitamin A) in a hydrophobic pocket, protects it from oxidation and transports it through the stomach to the small intestine where the retinol is transferred to a retinol-binding protein, which has a similar structure to that of β-lg. β-Lg can bind many hydrophobic molecules and hence its ability to bind retinol may be incidental. β-Lg is a member of the lipocalin family, all of which have hydrophobic ligand-binding properties (Akerstrom et al., 2000). Through its ability to bind fatty acids, β-lg stimulates lipase activity, which may be its most important physiological function. β-Lg is one of the most allergenic protein in bovine milk for human infants and there is interest in producing whey protein products free of, or depleted in, β-lg for use in infant formulae. β-Lg has very good thermo-gelling properties and dominates the gelation of whey protein concentrates (WPCs).

5.2 α-Lactalbumin

About 20% of the protein of bovine whey (4% of total milk protein) is α-lactalbumin (α-la), which is the principal protein in human milk. It is a small protein, containing 123 amino acid residues, with a mass of ~14 kDa, and has been well characterized; the literature has been reviewed by, among others, McKenzie and White (1991) and Brew (2003, 2011, 2013).

α-La contains 4 tryptophan residues per mole, giving it a specific absorbance at 280 nm of 20. Tryptophan, because of its role in synthesis of the neurotransmitter serotonin, is a key driver for the use of α-la-enriched whey protein ingredients in infant and clinical nutritional products, and is also exploited in the development of α-la-enriched whey protein ingredients developed to promote sleep. α-La contains four intramolecular disulphide bonds per mole but no cysteine, phosphate or carbohydrate. Its isoelectric point is ~pH 4.8. α-La has been isolated from the milk of the cow, sheep, goat, sow, human, buffalo, rat, guinea pig, horse and many other species; there are minor inter-species differences in its composition and properties.

The primary structure of α-la is homologous with lysozyme; out of a total of 123 amino acid residues in α-la, 54 are identical to corresponding residues in chicken egg white lysozyme and 23 others are structurally similar. α-La is a compact, highly structured globular protein. The tertiary structure of α-La is similar to that of lysozyme (McKenzie and White, 1991), and X-ray crystallography-based analysis of *apo*- and *holo*- bovine α-La have been published (Chrysina et al., 2000).

α-La is a component of lactose synthetase, the enzyme that catalyses the final step in the biosynthesis of lactose, and there is a direct correlation between the concentrations of α-la and lactose in milk.

A metalloprotein, α-la contains two Ca ions per molecule in a pocket containing four Asp residues. Most lysozymes do not bind Ca, but equine lysozyme is an exception (Nitta et al., 1987) and seems to be an intermediate in the evolution of lysozyme to α-la (see Nitta and Sugai, 1989). Ca-containing α-la is the most heat-stable of the principal whey proteins, or, more correctly, the protein has the ability to renature following heat denaturation, which occurs at a relatively low temperature, as indicated by differential scanning calorimetry. When the pH is reduced to <5, the Asp residues become protonated and lose their ability to bind Ca. The Ca-free protein is denatured at quite a low temperature and does not renature on cooling; this characteristic has been exploited to isolate α-la from whey (Pearce, 1983). The high heat stability of α-la results in it having poor thermo-gelling properties.

α-La is synthesized in the mammary gland, but a very low level is transferred, probably *via* leaky mammocyte junctions, into blood serum, in which the concentration of α-la increases during pregnancy or following administration of steroid hormones to male or female animals (Akers, 2000). The concentration of α-la in blood serum is a reliable, non-invasive indicator of mammary gland development and of the potential of an animal for milk production.

A high MW form of α-La, isolated from human acid-precipitated casein, has anti-carcinogenic activity; it was named HAMLET (human α-la made lethal to tumour cells). Bovine α-la can be converted to a form with similar activity, called BAMLET; it is the molten globular state formed from *apo* α-la and *cis*Δ9-octadecenoic acid (see Svensson et al., 2000; Chatterton et al., 2006 and Pettersson et al., 2006; Lisková et al., 2010). Both forms are reported to have comparable cytotoxic activity against three different cancer cell lines (Brinkmann et al., 2011).

5.3 Blood serum albumin

Normal bovine milk contains 0.1–0.4 g L^{-1} of blood serum albumin (BSA; 0.3–1.0% of total nitrogen), presumably as a result of leakage from blood; it has no known biological function in milk. BSA has been studied extensively; for reviews, see Fox (2003) and O'Mahony and Fox (2013). Owing to its low concentration in milk, BSA has little effect on the physico-chemical properties of whey-protein-based ingredients such as WPC and whey protein isolate (WPI).

5.4 Immunoglobulins

Mature bovine milk contains 0.6–1 g immunoglobulins (Ig) L^{-1} (~3% of total nitrogen). Colostrum contains ≈10% (w/v) Ig, the level of which decreases rapidly *post-partum*. IgG1 is the principal Ig in bovine, buffalo, caprine or ovine milk, with lesser amounts of IgG2,

IgA and IgM; IgA is the principal Ig in human milk. The cow, buffalo, sheep and goat do not transfer Ig to the foetus *in utero* and the neonate is born without Ig in its blood; consequently, it is very susceptible to bacterial infection with a very high risk of mortality. The young of these species can absorb Ig from the intestine for several days after birth and thereby acquire passive immunity until it synthesizes its own Ig, within a few weeks of birth. The human mother transfers Ig *in utero* and the offspring is born with a broad spectrum of antibodies. Although the human baby cannot absorb Ig from the intestine, the ingestion of colostrum is still very important because its Igs prevent intestinal infections. Some species, for example, the horse, transfer Ig both *in utero* and *via* colostrum (see Hurley, 2003, Marnila and Korhonen, 2011 and Hurley and Thiel, 2013).

The modern dairy cow produces colostrum far in excess of the requirements of its calf; surplus colostrum is available for the recovery of Ig and other nutraceuticals (Pakkanen and Aalto, 1997). Some work has been done on hyperimmunizing cows against certain human pathogens, for example, rota virus, for the production of antibody-rich milk for human consumption, especially by infants (Campbell and Peterson, 1959). Milk powder manufactured from 'immune milk' is commercially available in several markets (e.g. Stolle Milk Biologics Inc., Cincinnati, Ohio, USA).

5.5 Whey acidic protein

Whey acidic protein (WAP) was identified first in the milk of the mouse and has since been found in the milk of the rat, rabbit, camel, wallaby, possum, echidna and platypus. Since the milk of all of these species lacks β-lg, it was thought that these proteins were mutually exclusive. However, porcine milk, which contains β-lg, was recently found to contain WAP also (see Saidi et al., 2007). The presence of WAP in human milk (which contains no β-lg) has yet to be conclusively demonstrated, although the human genome indicates that it may be present (Martin et al., 2013). The MW of WAP is 14–30 kDa (the variation may be due to differences in degrees of glycosylation) and it contains two or three 4-disulphide domains. WAP functions as a proteinase inhibitor, is involved in terminal differentiation in the mammary gland and has antibacterial activity (for more detailed discussion, see Simpson and Nicholas, 2002, Hajjoubi et al., 2006 and Martin et al., 2011; O'Mahony and Fox, 2013).

5.6 Proteose peptone 3

The proteose-peptone (PP) fraction of milk protein is a very complex mixture of peptides, most of which are produced by the action of plasmin, but some are indigenous to milk. The fraction has been only partially characterized (Fox, 2003; O'Mahony and Fox, 2013). PP fractions 5, 8_{slow} and 8_{fast} have little or no technological significance, but proteose peptone 3 (PP3) has several interesting technological functionalities.

Bovine PP3 is a heat-stable phosphoglycoprotein that was first identified in the PP (heat-stable, acid-soluble) fraction of milk. Unlike the other peptides in this fraction, PP3 is an indigenous milk protein, synthesized in the mammary gland. Bovine PP3 consists of 135 amino acid residues, with five phosphorylation and three glycosylation sites. When isolated from milk, the PP3 fraction contains at least three components of MW (28, 18 and 11 kDa; the largest of these is PP3, while the smaller components are fragments thereof generated by plasmin (see Girardet and Linden, 1996). PP3 is present mainly in acid whey, but some is present in the MFGM.

Owing to its strong surfactant properties (Campagna et al., 1998), PP3 can prevent contact between milk lipase and its substrates, thus preventing spontaneous lipolysis; its emulsifying properties have also been evaluated in dairy products such as ice cream and recombined dairy cream (Vanderghem et al., 2007; Innocente et al., 2011). Although its amino acid composition suggests that PP3 is not a hydrophobic protein, it behaves hydrophobically, possibly owing to the formation of an amphiphilic α-helix, one side of which contains hydrophilic residues while the other side is hydrophobic. The biological role of PP3 is unknown.

5.7 Lactoferrin

Lactoferrin, a non-haem iron-binding glycoprotein, is a member of a family of iron-binding proteins (see Section 6.1), which includes transferrin and ovotransferrin (conalbumin) (see Lonnerdal, 2003, Lonnerdal and Suzuki, 2013). Lactoferrin is a major protein in human milk, where it is present at a concentration approximately tenfold higher (≈20% of total N) than in bovine milk. Infant formulae supplemented with lactoferrin are commercially available in Asia (Tomita et al., 2009) and lactoferrin has also found application in cosmetic and health-care products (El-Loly and Mahfouz, 2011). Lactoferrin can be isolated from milk or whey by using ion-exchange chromatography. Lactoferrin improves the bioavailability of Fe, is bacteriostatic, has antioxidant, antiviral, anti-inflammatory, immunomodulatory and anti-carcinogenic activity. Hydrolysis of lactoferrin by pepsin yields peptides called lactoferricins, which are more bacteriostatic than lactoferrin, and their activity is independent of iron status.

5.8 Caseinomacropeptide (CMP)

CMP is the hydrophilic fragment of κ-CN liberated through hydrolysis by chymosin, and can be purified by ion-exchange of whey (Etzel, 1999; McMahon et al., 2006). CMP is a very valuable ingredient, particularly as a nutrient source for sufferers of phenylketonuria (PKU). PKU is characterized by an inability to metabolize the essential amino acid, phenylalanine, which precludes proteins as a source of amino acids in the PKU diet. Protein substitutes in PKU diets are based mostly on phenylalanine-free blends of amino acids. Life-long compliance with a restricted diet is required in PKU cases, but this is made difficult by the low palatability of amino acid blends (van Calcar and Ney, 2012). Failure to comply with the restricted diet can result in severe neurological impairment. CMP is free of aromatic amino acids, including phenylalanine, making it a viable alternative to amino acid blends (van Calcar et al., 2009). Compared to amino acid blends, CMP-based formulae (supplemented with essential amino acids) are more palatable, reduce the level of phenylalanine in blood plasma, are associated with an increased satiating effect ('feeling of fullness') after consumption and result in generally improved dietary outcomes for PKU sufferers (LaClair et al., 2009; Ney et al., 2009; van Calcar and Ney, 2012).

6 Minor proteins, enzymes and other components

Milk contains several proteins at very low or trace levels, many of which are biologically active (see Schrezenmeir et al., 2000; Wynn and Sheehy, 2013); some are regarded as highly significant and have attracted considerable attention as neutraceuticals. When potential methods for increasing the value of milk proteins are discussed, the focus is usually on these minor proteins but they are, in fact, of little economic value to the overall dairy industry. They are found mainly in the whey, but some are also located in the MFGM. Reviews on the minor proteins include Fox and Flynn (1992), Haggarty (2003), Wynn et al. (2011), O'Mahony and Fox (2013) and Wynn and Sheehy (2013).

6.1 Metal-binding proteins

Milk contains several metal-binding proteins, listed below with the bound metals in parentheses:

- The caseins (Ca, P, Mg, Zn),
- α-La (Ca),
- Xanthine oxidase (Mo, Fe),
- Alkaline phosphatase (Zn, Mg),
- Lactoperoxidase (Fe),
- Catalase (Fe),
- Ceruloplasmin (Cu),
- Glutathione peroxidase (Se),
- Lactoferrin (Fe),
- Transferrin (Fe).

Milk contains a copper-binding glycoprotein, ceruloplasmin, also known as ferroxidase (EC 1.16.3.1). Ceruloplasmin is an α_2-globulin with a MW of ~126 kDa; it binds six atoms of copper per molecule and may play a role in delivering essential copper to the neonate.

Glutathione peroxidase (GTPase) is a Se-containing protein. It has been reported that milk contains GTPase and that it binds 30% of the total Se in milk (see Fox and Kelly, 2006b; O'Mahony et al., 2013). GTPase has no known enzymatic function in milk and the activity attributed to GTPase in milk may be due to sulphydryl oxidase (Stagsted, 2006).

6.2 β_2-Microglobulin

β_2-Microglobulin, initially called lactollin, was first isolated from acid-precipitated bovine casein by M. L. Groves in 1963. Lactollin, reported to have a MW of 43 kDa, is a tetramer of β_2-microglobulin, which consists of 98 amino acids, with a calculated MW of 11 636 Da. β_2-Microglobulin, a component of the immune system, is probably produced by proteolysis of a larger protein, mainly within the mammary gland; it has no known significance in milk.

6.3 Osteopontin (OPN)

OPN is a highly phosphorylated acidic glycoprotein, consisting of 261 amino acid residues with a calculated MW of 29283 Da (total MW of the glycoprotein is 60000 Da). OPN has 50 potential Ca-binding sites, about half of which are saturated under the normal physiological concentrations of Ca and Mg. OPN occurs in bone (it is one of the major non-collagenous proteins in bone), in many other normal and malignant tissues, in milk and urine and can bind to many cell types. It is believed to have a diverse range of functions (Bayless et al., 1997; Christensen and Sørensen, 2016), but its role in milk is not clear.

A rapid method for the isolation of OPN from milk was reported by Azuma et al. (2006) who showed that it binds lactoferrin, lactoperoxidase and Igs and may serve as a carrier for these proteins. An acidic whey protein fraction, which contains OPN, reduces bone loss in ovariectomized rats (Kruger et al., 2006; Wynn et al., 2011; O'Mahony and Fox, 2013; Wynn and Sheehy, 2013).

6.4 Vitamin-binding proteins

Milk contains binding proteins for at least the following vitamins: retinol (vitamin A, i.e. β-lg), biotin, folic acid and cobalamine (vitamin B_{12}). The precise role of these proteins is not clear, but they probably improve the absorption of vitamins from the intestine or act as antibacterial agents by rendering vitamins unavailable to bacteria. The concentration of these proteins varies during lactation, but the influence of other factors such as individuality, breed and nutritional status is not known. The activity of these proteins is reduced or destroyed on heating at a temperature somewhat higher than pasteurization (see Wynn et al., 2011; Wynn and Sheehy 2013).

6.5 Angiogenins

Angiogenins induce the growth of new blood vessels, that is, angiogenesis. They have high-sequence homology with members of the RNase A superfamily of proteins and have RNase activity. Two angiogenins (ANG-1 and ANG-2) have been identified in bovine milk and blood serum; both strongly promote the growth of new blood vessels in an *in vitro* assay. The function(s) of the angiogenins in milk is/are unknown. They may be part of a repair system to protect either the mammary gland or the intestine of the neonate and/or part of the host-defence system (see Wynn et al., 2011; Wynn and Sheehy, 2013).

6.6 Kininogen

Two kininogens have been identified in bovine milk, a high (88–129 kDa, depending on the level of glycosylation) and a low (16–17 kDa) MW form. Bradykinin, a biologically active peptide containing nine amino acids which is released from the high MW kininogen by the action of the enzyme, kallikrein, has been detected in the mammary gland, and is secreted into milk, from which it has been isolated. Plasma kininogen is an inhibitor of thiol proteinases and has an important role in blood coagulation. Bradykinin affects smooth muscle contraction and reduces hypertension. The biological significance of bradykinin and kininogen in milk is unknown (see Wynn et al., 2011; Wynn and Sheehy, 2013).

6.7 Glycoproteins

Many of the minor proteins discussed above are glycoproteins; in addition, several other minor glycoproteins have been found in milk and especially in colostrum, the functions of which have not been elucidated. One of the high MW glycoproteins in bovine milk is prosaposin, a neurotrophic factor which plays an important role in the development, repair and maintenance of nervous tissue. It is a precursor of saposins A, B, C and D, which have not been detected in milk. The physiological role of prosaposin in milk is not known, although saposin C, released by digestion, could be important for the growth and development of the young (Patton et al., 1997; Patton, 1999; Campana et al., 1999).

6.8 Proteins in the milk fat globule membrane

About 1% of the total protein in milk is in the MFGM. Most of the proteins are present at trace levels, including many of the indigenous enzymes in milk. The principal proteins in the MFGM include mucins, adipophilin, butyrophilin and xanthine oxidoreductase. Comprehensive reviews on the MFGM include Spitsberg (2005), Keenan and Mather (2006), Singh (2006) and Lopez (2011).

6.9 Growth factors

Milk contains many peptide hormones, including epidermal growth factor, insulin, insulin-like growth factors 1 and 2, transforming growth factors (α and β), two mammary-derived growth factors (I and II), colony-stimulating factor, nerve growth factor, platelet-derived growth factor and bombasin. It is not clear whether these factors play a role in the development of the neonate or in the development and functioning of the mammary gland, or both (see Fox and Flynn, 1992; Gauthier et al., 2006, Wynn and Sheehy, 2013).

6.10 Indigenous milk enzymes

Milk contains about 60 indigenous enzymes, which are quantitatively minor but nonetheless very important members of the milk protein system (see Fox and Kelly, 2006a,b and O'Mahony et al., 2013). The enzymes originate from the secretory cells or the blood; many are concentrated in the MFGM and originate in the Golgi membranes of the cell or the cell cytoplasm, some of which becomes entrapped as crescents inside the encircling membrane during exocytosis. Plasmin and lipoprotein lipase are associated with the casein micelles and several enzymes are present in the milk serum; many of the latter are derived from the MFGM which is shed as the milk ages. Certain indigenous enzymes are known to be significant for different reasons:

- Deterioration of product quality: plasmin, lipoprotein lipase, acid phosphatase, xanthine oxidoreductase,
- Bactericidal agents: lactoperoxidase and lysozyme,
- Indices of the thermal history of milk: alkaline phosphatase, γ-glutamyltransferase or lactoperoxidase,
- Indices of mastitic infection: catalase, acid phosphatase and especially N-acetylglucosaminidase.

6.11 Biologically active cryptic peptides

One of the most exciting recent developments in milk proteins is the discovery that many milk proteins contain sequences that have biological/physiological activities when released by proteolysis. The best studied are phosphopeptides, angiotensin-converting enzyme (ACE)-inhibitory peptides, platelet-modifying peptides, opiate peptides, immunomodulating peptides and caseino-macropeptides which have many biological properties (see FitzGerald and Meisel, 2003; Korhonen, 2006; Korhonen and Pihlanto, 2006; Mills et al., 2011; Raikos and Dassios, 2014).

6.12 Non-protein nitrogen (NPN)

The NPN fraction of milk contains those nitrogenous compounds soluble in 12% TCA. It represents ~5% of total nitrogen (~300 mg L^{-1}; 230–420 mg L^{-1}; Harland et al., 1955; Journet et al., 1975). The principal components of NPN are summarized in Table 5. Human milk contains a high level of taurine ($H_2NCH_2CH_2SO_3H$) which can be converted to cysteine and may be nutritionally important for infants (interestingly, taurine is an essential amino acid for cats); most modern infant formulae are fortified with taurine for this reason.

The amino acids in milk support the growth of micro-organisms, including lactic acid bacteria (LAB) used as cultures in the production of cheese and fermented milks. However, the concentration of free amino acids in milk is sufficient to support the growth of LAB to only ~20% of the number required for fermented dairy products. Consequently, LAB depend on a cell-envelope-associated proteinase, a complex transport system for peptides and amino acids and a battery of intracellular peptidases to obtain their essential amino acids from casein (Thomas and Pritchard, 1987). Heating to a high temperature (>100°C) leads to the formation of some small peptides from caseins which can support LAB (White and Davies, 1966; Hindle and Wheelock, 1970; Gaucheron et al., 1999; McGrath, 2014). Removal of NPN from milk in the production of milk protein or casein concentrates promotes their storage stability by removing it as a nutrient source for bacteria (Amelia and Barbano, 2013). The concentration of free amino acids in milk is shown in Table 6, and more on this topic can be found in Pellegrino et al. (2015).

Urea, the principal constituent of NPN (~50% of NPN; ~1.5 mM), has a very significant effect on the heat stability of milk (Muir and Sweetsur, 1976). The concentration of urea in milk varies considerably, being highest when cows are on fresh pasture, which is reflected

Table 5 Non-protein nitrogen of cow milk

Component	N (mg L^{-1})
Ammonia	6.7
Urea	83.8
Creatinine	4.9
Creatine	39.3
Uric acid	22.8
α-Amino nitrogen	37.4
Unaccounted	88.1

Table 6 The free amino acids (μm L^{-1}) of bovine and human milk

Amino acid	Bovine	Human
Alanine	30	227.5
Arginine	10	35.4
Aspartic acid	15	183.2
Cystine	21	56
Glutamic acid	117	1184.1
Glutamine	12	284.8
Glycine	88	124.6
Histidine	9	7.7
Isoleucine	3	33.4
Leucine	3	55.6
Lysine	15	39
Methionine	~0	8.8
Phenylalanine	3	23.6
Proline	–	64.3
Serine	23	273.7
Taurine	13	301.1
Threonine	16	97.6
Tyrosine	0.3	2.5
Valine	5	72.7
Total	578	3019.7

Source: Adapted from Uniacke-Lowe et al. (2010).

in seasonal variations in the heat stability of milk. For the impact of dietary protein on the urea level in, and the heat stability of, milk, see Reid et al. (2015).

7 Laboratory-scale preparation of casein and whey proteins

The two major milk protein groups are most commonly prepared from skimmed milk; whole milk may also be used for this purpose, but the fat which is present becomes occluded in the protein precipitate and can interfere with further characterization of the proteins. Before preparation of casein and whey proteins, the fat can be removed by centrifugation, for example, 3000 × g for 30 min, and residual fat may be removed by washing the precipitated proteins with ether. Where used in the following sections, the term 'milk' refers to skimmed bovine milk, unless stated otherwise.

7.1 Routine methods

7.1.1 Isoelectric precipitation of casein

When the pH of milk is reduced to its isoelectric point (pH ~ 4.6), the caseins aggregate and ultimately form a coagulum under quiescent conditions. In laboratories, HCl is typically used for acidification, but lactic or acetic acid are also sometimes used. The extent of precipitation is temperature-dependent and the procedure is preferably performed at >20°C.

An appropriate protocol allows the preparation of casein which is essentially free of CCP. A protocol for the preparation of casein is as follows: acidification of milk to pH 4.6 at ~4°C with a moderately dilute acid (~1 M), followed by holding for ≥30 minutes and subsequent warming to ~35°C with holding for ~30 minutes (fine aggregates are formed at 4°C which allow effective dissolution of CCP during equilibration, while the aggregates formed at 35°C are more coarse and precipitate readily); removal of whey by filtration through cheesecloth or other suitable material, and thorough washing of the casein by multiple re-suspensions in distilled water, followed by filtration, to remove residual lactose and salts. The resultant isoelectric casein may be frozen or dried for storage.

7.1.2 Ultracentrifugation

As the casein micelles are quite large (MW, of ~$10^8 - 10^9$ Da), the majority (90–95%) of the casein in milk can be sedimented by centrifugation at 100 000 × g for 1 h. The whey proteins, which are molecularly dispersed, are not sedimentable and remain in the supernatant; heat-denatured whey proteins can, however, be co-sedimented with the micellar material. More casein is sedimented at 35°C than at 0–4°C, as a greater proportion of casein (primarily monomeric β-CN) dissociates from micelles at low temperatures (Rose, 1968), rendering it non-sedimentable. Casein prepared by ultracentrifugation at the native pH of milk contains the original level of CCP and if redispersed in a suitable buffer (e.g. milk permeate or simulated milk ultrafiltrate) exhibits physico-chemical characteristics similar to those of native micelles in milk.

7.1.3 Salting-out methods

Casein can be precipitated from solution by a number of salts. Addition of 50% saturated $(NH_4)_2SO_4$ to milk causes precipitation of the caseins, together with some of the whey proteins (mainly Igs). $MgSO_4$ or NaCl may also be used. Saturation with NaCl allows a cleaner fractionation of the caseins from most of the whey proteins. This method is used to separate caseins, Igs and denatured lactalbumins from undenatured whey proteins in the characterization of milk powders (McKenzie, 1971).

7.1.4 Gel filtration

Caseins can be separated from whey proteins by gel permeation chromatography. It is also possible to resolve individual whey proteins by gel permeation, though the technique is currently used only in the laboratory (Wang and Lucey, 2003; Roufik et al., 2005; Kehoe et al., 2007; Lisková et al., 2010).

7.1.5 Membrane filtration

Separation with cartridges containing flat-sheet membranes can be used for the preparation of milk proteins at a laboratory scale. Both casein and whey proteins can be retained by small-pore, semi-permeable membranes. This process, referred to as ultrafiltration (UF), may be used to concentrate proteins in the retentate while lactose, soluble salts and other small molecules pass into the permeate stream. UF membranes with a molecular-weight cut-off of ~10 kDa are commonly used for the concentration of whey proteins in whey or milk proteins in skim milk. For the fractionation of milk into casein and whey streams, membranes of larger pore sizes (i.e. 0.1–0.2 μm) are used in a process referred to as microfiltration (MF). The casein fraction produced by using such technology is referred to as 'micellar casein concentrate' (MCC), 'phosphocaseinate' or 'native micellar/micellular casein', while the whey protein fraction is referred to as 'serum protein concentrate' (SPC), 'native', 'ideal' or 'virgin' whey (Pierre et al., 1992; Kelly et al., 2000; Rizvi and Brandsma, 2002). Proteins fractionated in this manner are considered to have much the same physico-chemical and functional properties as the proteins in the original whey or milk and, as such, make useful model systems to study properties such as the physicochemistry of casein micelles (Famelart et al., 1999), casein micelle structure (Salami et al., 2013; Gonzalez-Jordan et al., 2015) and droplet-particle transformation during drying of milk proteins (Sadek et al., 2015). Industrial-scale manufacture of MPCs, MCCs, WPCs, SPCs, etc. will be discussed in Section 7.

7.2 Other methods

7.2.1 Preparation of casein after enrichment with calcium

Addition of $CaCl_2$ to milk (~0.2 M) causes sufficient aggregation of casein to allow its sedimentation by low-speed centrifugation. When Ca-fortified milk is heated to ~90°C, the casein aggregates and precipitates without centrifugation. If whey proteins are present, they become denatured at 90°C, which results in their association and co-precipitation with casein. Casein co-precipitates are produced on a commercial scale, but have enjoyed only limited commercial success, with poor solubility being a significant limitation to their use in many applications.

7.2.2 Precipitation by ethanol

The caseins may be precipitated from milk by ~40% ethanol (a lower concentration of ethanol may be used at a lower pH) (Hewedi et al., 1985; Horne, 2003). Precipitation of caseins by ethanol is not used routinely for the preparation of casein at either laboratory or industrial scale.

7.2.3 Cryoprecipitation

Caseins, primarily in the micellar form, may be destabilized and precipitated by freezing milk or, more effectively, concentrated milk, at about −10°C. Cryoprecipitated casein is reported to have good solubility and curd-forming properties but inferior emulsifying properties to caseinates (Moon et al., 1988, 1989). This method is not routinely used to prepare casein.

7.2.4 Rennet coagulation

The casein micelles are destabilized by specific, limited proteolysis and precipitate/coagulate in the presence of Ca^{2+}. This is an altered form of micellar casein, with properties very different from native or isoelectric casein (Mulvihill and Ennis, 2003). The resultant material is essentially insoluble. Renneting of milk in the laboratory is typically used to investigate cheesemaking properties in simple systems rather than to purify casein.

7.3 Laboratory-scale preparation of enriched or purified whey protein fractions

Whey streams are obtained by using the methods for casein described in the previous section. Whey may be enriched/purified by salting-out or by removing the non-protein constituents by dialysis, crystallization and/or UF. Some whey proteins may be co-precipitated with the caseins. Cheese or 'sweet' whey contains casein-derived peptides, particularly the CMP liberated by the action of rennet.

The composition and properties of products prepared by these various methods differ slightly: Acid whey contains the PP fraction, but no CMP produced from κ-CN by rennet action; Igs are precipitated along with the caseins by saturated NaCl; rennet whey contains the CMP, plus small amounts of casein; MF permeates may contain casein monomers (particularly β-CN), in addition to whey proteins, if MF is conducted at <10°C; and small casein micelles remain in the ultracentrifugal serum, especially if Ca is not added. The salt composition of the serum differs very significantly in whey produced by various methods. The whey obtained by any of the methods in Section 7.1, except gel filtration, contains lactose and soluble salts. Purified whey proteins may be prepared from such whey by dialysis or UF and freeze-drying the retentates.

Several procedures for the isolation of α-la and β-lg have been developed. Early methods were based on salting-out from whole milk, skim milk, rennet or acid whey (Aschaffenburg and Drewry, 1957; Armstrong et al., 1967). Fox et al. (1967) exploited the solubility of β-lg in ~3% TCA to produce highly purified β-lg. Both β-lg and α-la may be purified by cation-exchange chromatography (McKenzie, 1971). Methods for the isolation of whey proteins were reviewed by Imafidon et al. (1997).

Most researchers have used pre-treatment (i.e. high heat and/or acidification) of whey to selectively precipitate either α-la or β-lg and enhance the fractionation of these whey proteins (Pearce, 1995; de Wit and Bronts, 1995; Wu, 2001; Lozano et al., 2008). However, because of its slightly lower molecular weight (MW), native α-la tends to permeate wide-pore UF membranes (50–300 kDa) to a greater degree than β-lg, which was used to generate an α-la-rich fraction from acid whey (Muller et al., 1999) or sweet whey (Marella et al., 2011).

There are large differences between the isoelectric points of the whey proteins (Fig. 8), making charge-based separation attractive. Etzel (1999) used ion-exchange technology to fractionate whey (pH ~ 4.5) by selective elution of β-lg and α-la from a cation-exchanger using alkalization. Positively charged wide-pore (>100 kDa) UF membranes, which separate whey proteins based on charge, have been applied successfully to fractionate β-lg and α-la at much higher purities (80–87%) than has been achieved with uncharged wide-pore membranes (Arunkumar and Etzel, 2013, 2014).

There is significant interest in the production of major and minor whey proteins for nutritional or functional applications, and many of these processes have been developed

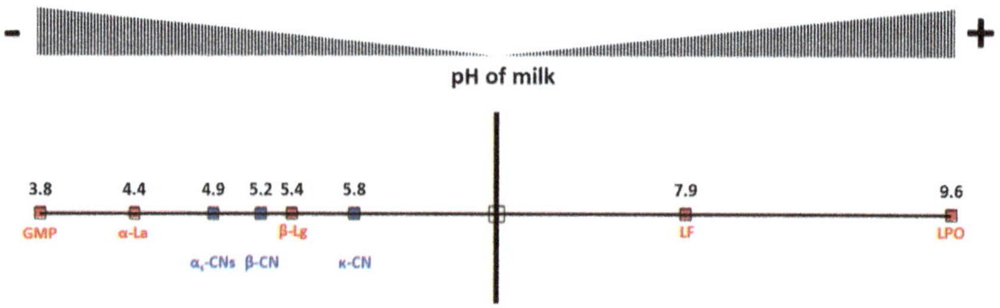

Figure 8 Isoelectric points of select proteins in milk and the relative charge of the proteins at pH 6.7. Source: Adapted from data presented in Etzel (2004).

at laboratory scale and sometimes pilot scale. Approaches that are based on ion-exchange chromatography, membrane filtration technology and thermal, physical or chemical treatments have been used (e.g. Amundson et al., 1982; Pearce, 1983; Stack et al., 1998; Kristiansen et al., 1998; Cheang and Zydney, 2004; Andersson and Mattiasson, 2006 and Marella et al., 2011). Many of these techniques, as applied in industrial-scale production of whey-protein-enriched ingredients, were discussed by Mulvihill and Ennis (2003).

8 Industrial milk protein products

The main types of milk protein ingredients are described in the following sections. The focus is placed on products containing concentrated, enriched or purified or otherwise modified protein fractions. Ingredients such as skim milk powder and whole milk powder are not included as their protein profile is unchanged relative to regular milk; information on these ingredients can be found in Kelly and Fox (2016). The products discussed are divided into whey-protein- or casein-dominant classes based on the relative contribution of whey proteins and casein to the protein component. Some of the principal processes used to manufacture whey-protein- and casein-dominant ingredients are shown in Fig. 9 and 10, respectively. Note that the pure β-CN fractions produced by MF as shown in Fig. 10 generate SPC, a major whey protein-dominant ingredient, as a co-product. Membranes of 10 kDa are typically used industrially for concentration of whey proteins in whey or milk protein in skim milk. Wide-pore, negatively charged membranes have been developed in the laboratory and successfully adapted to pilot scale for concentration of whey proteins (Arunkumar et al., 2016); these charged membranes have superior permeate flux and fouling mitigation properties compared to traditional UF membranes (Arunkumar and Etzel, 2015). On the other hand, positively charged wide-pore membranes have been used in the laboratory for fractionation of the two major whey proteins on a laboratory scale (Arunkumar and Etzel, 2013, 2014). Due to its promise, charged membrane technology is included in Fig. 10 as a next-generation membrane technology for the dairy industry.

8.1 Whey-protein-dominant products

Whey protein products are key ingredients in several growth areas of the food industry, such as infant formulae, clinical nutrition and sports nutrition. These ingredients contribute

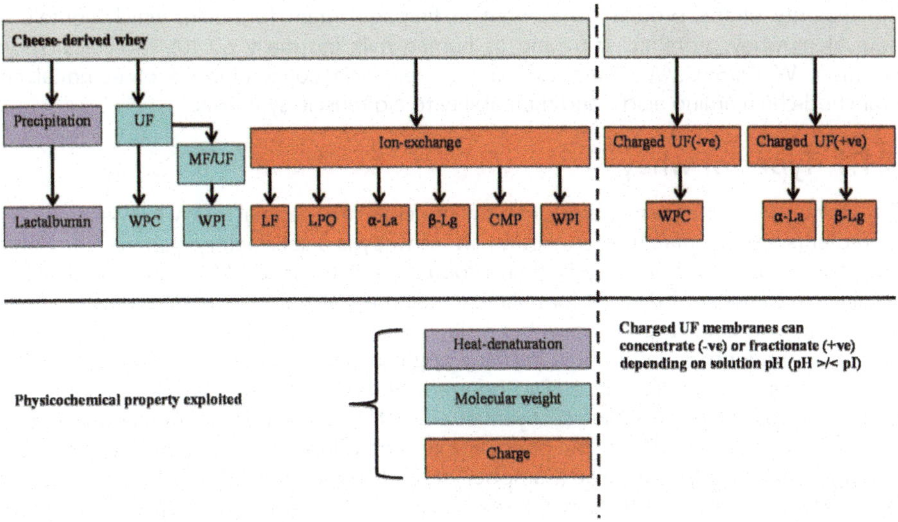

Figure 9 Schematic showing some of the principal processes by which whey proteins or whey protein–dominant fractions are concentrated or fractionated. Note: only the principal processing steps are shown (i.e. diafiltration, evaporation, spray drying and other steps are excluded), by-products are not included and cheese-derived ('sweet') whey is shown only as an example of whey type.

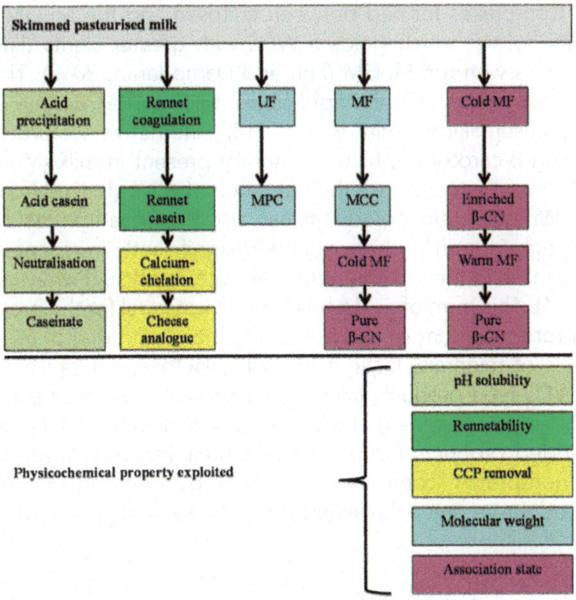

Figure 10 Schematic showing some of the principal processes by which casein or casein-dominant fractions are concentrated or fractionated. Note: only the principal processing steps are shown (i.e. diafiltration, evaporation, spray drying, milling and other steps are excluded) and by-products are not included.

the majority of the protein component in first-age infant formulae, which contain an equivalent whey protein:casein ratio as human milk (normally 60:40). In clinical/sports nutrition, WPCs/SPCs/WPIs are valued for their high concentrations of essential and branched-chain amino acids, and their ability to aid muscle synthesis.

8.1.1 Types of whey

Sweet whey from cheese manufacture is the principal raw material from which whey protein ingredients are manufactured. Although sweet whey is a satisfactory starting material for the production of many whey protein products, the presence of colours or bleaching residues is a concern for infant formula manufacturers. Annatto is often added to cheese milk to yield a cheese product with a yellow/orange colour. Residual annatto in whey protein ingredients is undesirable in many applications (Kang et al., 2010), with a clear, colourless whey ingredient typically preferred. Bleaching of cheese whey is sometimes performed to address this issue (Listiyani et al., 2011). There are strict regulatory limits on the concentration of norbixin, the principal carotenoid in annatto, in ingredients destined for application in infant formulae (Campbell et al., 2014), while the use of bleaching agents (e.g. benzoyl peroxide or hydrogen peroxide) in infant formula ingredients is also heavily regulated, and often not permitted. Benzoyl peroxide reacts with annatto to form benzoic acid, which is the subject of restrictions in the United States and Europe, and is banned in China (Kang et al., 2010). These regulatory hurdles are accelerating efforts to eliminate or reduce carry-over of colourant and/or bleach residues. Some recent research has focused on the production of cheese-whey-derived ingredients that are low in norbixin. These products could satisfy regulatory requirements for norbixin while limiting the need to bleach. Complexes formed between chitosan and the MFGM can be removed from cheese whey by MF, which yields a WPC with greater clarity (Lucey et al., 2009), as norbixin associates with the MFGM (Zhu and Damodaran, 2012). The non-polar form of norbixin, bixin, associates with the lipid phase to a greater degree and is retained in the curd during cheesemaking (Smith et al., 2014). Alternative colourants have also been proposed, including β-carotene, which is naturally present in milk (Moeller et al., 2014). Another approach may be to increase the utilization of whey derived from MF of milk, as it is colourant free. Whey protein ingredients can also be made from acid whey. The Greek and Greek-style yoghurt industry has expanded significantly in recent years, resulting in a considerable increase in the volume of the deproteinized acid whey by-product (Bansal and Bhandari, 2016). This form of acid whey is often not ideal for the production of protein products, as its protein content is <0.1% (i.e. less than 1/6th that of regular sweet whey), and its high levels of minerals, lactic acid and galactose can cause processing issues. Manufacture of WPC-type products from higher protein acid whey streams, such as that from Cottage cheese, generates a similarly deproteinized acid whey stream. Efforts to produce value-added products from these deproteinized acid whey streams are often focused on the non-protein components, such as calcium and phosphorus, which are present at two- and three- times the levels, respectively, in regular whey.

8.1.2 Whey powder

Whey powder is whey which has been clarified, concentrated in total solids and dried without removal of non-protein components or fractionation of proteins; the lactose is typically crystallized prior to drying to prevent stickiness. The solids composition of whey powder is broadly similar to liquid whey, but differs depending on the source (sweet or acid), with sweet whey powder typically having higher lactose, less minerals and lower acidity; these whey powders are also available in reduced-lactose and demineralized formats (Bansal and Bhandari, 2016).

8.1.3 Demineralized whey

Demineralized whey is an important ingredient in the infant formula industry, where there are strict regulatory limits on the levels of individual minerals in the final product. Acid whey, in particular, is rich in minerals and needs to undergo demineralization prior to use in infant formulae. Up to 90% demineralization can be achieved by processing of whey using electrodialysis, while ion-exchange reduces total minerals by up to 97%; nanofiltration is also sometimes used to simultaneously pre-concentrate and partially demineralize liquid whey prior to more effective demineralization processes (O'Mahony and Tuohy, 2013; Bansal and Bhandari, 2016). Only monovalent ions are removed by nanofiltration, which limits its demineralization capacity; however, it is a cost-effective means of supplementing the more expensive ion-exchange/electrodialysis procedures. Electrodialysis of pre-concentrated whey to achieve demineralization of 50–70% is typical for the manufacture of demineralized whey (O'Mahony and Tuohy, 2013).

8.1.4 Whey protein concentrates

WPCs are a range of whey protein ingredients in which non-protein components have been reduced to different degrees in order to produce a product concentrated in protein. Manufacture of WPCs is typically carried out with UF or combined UF/diafiltration (DF) of whey, followed by concentration (by evaporation, sometimes combined with upfront nanofiltration or reverse osmosis) of the total solids and spray drying. DF is performed with deionized water and is necessary if a protein content >50% is desired (Bansal and Bhandari, 2016). The lactose:protein ratio decreases as the protein content (dry basis) increases from ~35% (WPC35) to 80% (WPC80). Higher protein concentration factors and degree of DF are required to make the highest protein WPCs. A filtration temperature of ~50°C is common for WPC production, due to a high permeate flux and a low risk of protein denaturation (Bansal and Bhandari, 2016); however, at this temperature, there are problems associated with the growth of thermophiles and membrane fouling (O'Mahony and Tuohy, 2013) which has led to the increasing adoption of low-temperature (i.e. <15°C) processes. High-protein WPCs have increased in popularity in recent years due to the proliferation of nutritional products (beverages, bars) which are often rich in protein but low in carbohydrate.

8.1.5 Serum protein concentrates

SPCs are similar products to WPCs, with the important distinction that they are manufactured from the whey-protein-enriched permeate derived from microfiltered milk. The difference in source material results in SPCs having no CMP, colour, starter culture and other cheesemaking residues, lower levels of fat/phospholipids and less-denatured whey proteins (O'Mahony and Tuohy, 2013). SPCs have attracted interest due to their clarity, 'clean' sensory profile, low level of denatured whey proteins and negligible fat/phospholipid content compared to WPCs prepared from cheese whey; in addition, they do not contain CMP or annatto. The protein profile and functionality of SPCs can be tailored by adjusting the temperature of the MF process, with a low filtration temperature yielding a SPC higher in β-CN. Coppola et al. (2014), who compared an SPC rich in β-CN produced by cold MF of milk to a regular SPC and several WPCs, reported that the β-CN-rich SPC was less turbid, more stable during storage and had better foaming characteristics. The absence of both CMP and annatto and the ability to increase the β-CN level make SPCs an attractive option for infant formula manufacturers. CMP is rich in threonine, which has been linked with hyperthreoninemia in infants who consume infant formula (Rigo et al., 2001), while increasingly stringent regulations are limiting the use of ingredients that contain norbixin/annatto (Campbell et al., 2014) and/or bleaching residues (Kang et al., 2010) in infant formula. SPCs may also be attractive for beverage formulators due to reduced storage-induced changes compared to those of WPCs. Self-association of CMP has been implicated in the turbidity development which occurs during the storage of acidic whey protein beverages (Villumsen et al., 2015). The primary factor limiting the production and utilization of SPCs is uncertainty regarding the use of its co-product (MCC).

8.1.6 Whey protein isolates

WPIs can be manufactured by using ion exchange, in which the proteins are adsorbed on an ion-exchange resin, washed free of lactose and salts, and then selectively eluted with acid or alkali (Bansal and Bhandari, 2016). Alternatively, a combined MF/UF/DF process can be used, much like the production of a high-protein WPC except that MF is performed to remove lipids, which otherwise limits the protein content to the WPC range (Ramchandran and Vasiljevic, 2013). In such processes, UF is used to first concentrate the proteins in liquid whey. During UF, any residual fat and phospholipid material from the liquid whey are concentrated along with the protein, but are subsequently removed by MF. A second UF step is then used to concentrate the whey further. WPIs with a protein content of 90–95% are available commercially, and are typically used in premium nutritional products. The composition of WPIs made by ion-exchange or membrane filtration is very similar; however, the latter contains a significant quantity of CMP (Bansal and Bhandari, 2016).

8.1.7 Enriched and isolated whey protein fractions

Enriched fractions of individual whey proteins are attracting increasing industrial interest. WPCs enriched in α-la can be produced in two ways, (1) where α-la enrichment is achieved by concentrating α-la resulting in an increased α-la:β-lg ratio or (2) where CMP has been depleted resulting in an increased α-la level on a weight basis but an unchanged α-la:β-lg ratio. Such ingredients are of interest for humanized infant formula products. Hilmar™ 8800 is an example of an α-la-enriched WPC; it is manufactured by Hilmar Ingredients (Hilmar, CA, USA). Large-scale purification of CMP from sweet whey has been achieved with the most successful methods by using ion exchange of whey (Etzel, 1999; McMahon et al., 2006). The whey fraction generated during renneting of pure micellar casein is also highly enriched in CMP. CMP is marketed mainly as an amino acid source for PKU sufferers and a sialic acid source for infants being formula-fed. LACPRODAN® CGMP-10 is a CMP ingredient manufacture by Arla Foods Ingredients (Viby, Denmark). Osteopontin (OPN) can be enriched in whey by using different combinations of membrane filtration and ion-exchange (see O'Mahony and Tuohy, 2013). This ingredient is considered to have potential health benefits when added to infant formula. LACPRODAN® OPN-10 is an OPN ingredient from Arla Foods Ingredients (Viby, Denmark).

Lactoferrin and lactoperoxidase are produced from whey by using ion exchange; these proteins have very high isoelectric points and are positively charged at the natural pH of milk/whey (Fig. 8), which allows their adsorption/elution with the use of cation-exchange resins. To produce 1 kg of lactoferrin, a large volume of whey (~10000 L) is needed (Etzel, 2004); this means that the final ingredient often demands a high price (~$300 per kg) to be profitable for the manufacturer, which has limited the potential consumer base to those formulating high-value products. Increasing the concentration of lactoferrin in milk may reduce the volume of whey required for its purification in sufficient quantities. One approach to achieve this may be milking strategies which involve defined periods of cessation of milking followed by re-milking. Davis and South (2015) investigated whether cessation of milking in mid- and late-lactation cows (which were usually milked twice a day) for two, four or seven days (mid-lactation) or two or four days (late-lactation) prior to subsequent milking of cows and further milking twice a day; this strategy was effective for both lactation stages when two days of milking cessation were applied (no advantage was observed for longer cessation times), increasing lactoferrin concentration over six-fold to a level approaching that found in human milk. Lactoferrin has attracted interest for a broad range of product applications, including cosmetics, oral rehydration solutions and infant formula. Bioferrin® is a lactoferrin ingredient from Glanbia Nutritionals (Evanston, IL, USA).

8.2 Casein-dominant products

The casein market has been dominated traditionally by those ingredients manufactured by renneting or acid-precipitation of milk. These processes are based on treatments applied to skimmed milk that reduce the zeta potential of casein with concomitant formation of a curd, which for many years was the only practical method to separate casein from whey; as a result of this destabilization, the resultant products are insoluble. The wide-scale adoption of cross-flow membrane filtration technologies by dairy processors has facilitated the concentration of casein micelles without modifying their structure or rendering them insoluble. The two casein-dominant ingredients manufactured by using membrane filtration, MPC and MCC now have a significant presence in the global market for casein (Carr and Golding, 2016). These processes also afford a greater degree of flexibility with regard to product composition, with MPCs available in multiple different formats based on the protein concentration achieved during membrane filtration.

8.2.1 Rennet casein

Rennet casein is manufactured by drying of the insoluble casein fraction generated by renneting. Rennet casein has a much higher content of calcium and phosphorus than acid casein, as the CCP level is maintained when the renneting step is performed at approximately the natural pH of milk. Although very insoluble, some properties of rennet casein make it suitable for certain food applications, such as the manufacture of analogue cheese, where rennet casein is dispersed, and ultimately hydrated, in hot solutions of calcium-binding salts before cooling to induce matrix formation (Ennis and Mulvihill, 1999; O'Sullivan and Mulvihill, 2001).

8.2.2 Acid casein

Acid casein is manufactured by isoelectric precipitation of the casein from milk. HCl is the most widely used direct acidulant commercially; however, lactic acid produced *in situ* by LAB cultures is commonly used in New Zealand. Whey from lactic casein production contains lower levels of counterions (i.e. chloride, sulphate) and lactose than that generated using mineral acid, but a considerable incubation period is required to achieve a sufficient degree of acidification. Glucono-δ-lactone (GDL) may also be used to achieve isoelectric precipitation of casein; when added to milk systems, GDL hydrolyses to form gluconic acid, with the rate and degree of acidification depending on the GDL to protein ratio (Braga et al., 2006). GDL is considered to have some benefits (depending on the application) compared to LAB, including the absence of lactose hydrolysis and reduced variability associated with culture type, and is, therefore, used to manufacture certain types of yoghurt, Cottage cheese and Feta cheese (Chen et al., 2016).

8.2.3 Caseinates

Acid casein is insoluble in water and is generally considered a precursor ingredient for the manufacture of caseinates; it may be converted to water-soluble caseinates by dispersion in water and adjusting the pH to ~6.7 with alkali, usually NaOH to yield sodium caseinate. KOH, NH_4OH or $Ca(OH)_2$ may also be used, giving the corresponding caseinate (Mulvihill and Ennis, 2003). Caseinates are usually spray-dried in industrial-scale production. The casein present in rehydrated caseinates is not micellar, as in milk, MPCs or MCCs, which results in caseinates having markedly different functional attributes such as high viscosity, excellent heat stability, interfacial activity and high ethanol stability.

8.2.4 Milk protein concentrates

MPCs are manufactured from milk by UF or combined UF/DF, followed by concentration of the total solids (typically by evaporation) and spray drying. The lactose:protein ratio decreases as the protein content is increased from ~35% (MPC35) to 80% (MPC80). High concentration factors and degree of DF can be used to produce milk protein isolate (MPI) products with a protein content >90%. High-protein MPCs/MPIs have been used in a wide variety of food products, ranging from traditional dairy products (e.g. cheese, yoghurt) to targeted nutritional formulations (e.g. high-protein beverages for therapeutic use or infant formula for lactose-intolerant infants), where their functional attributes (contributing opacity, imparting viscosity/mouthfeel, binding calcium phosphate) and clean label ('milk protein', 'milk protein concentrate') are desirable. It has been demonstrated in a number of studies that MPCs have poor rehydration/solubility characteristics (see review by Crowley et al., 2016). This defect is not due to insolubility (as in the case of rennet casein) but, rather, extremely slow rates of rehydration. For this reason, methods that aim to modify MPCs to yield a faster-dissolving product (Carr and Golding, 2016) have been developed; however, many of these techniques involve considerable dissociation of the micellar component, meaning the casein in the final MPC is often altered to some degree.

8.2.5 Micellar casein concentrates

When micellar casein (with associated minerals) is separated from the serum proteins in milk, without significantly altering micellar structure, the resultant material is referred to as MCC. Less than 10% of the protein (usually 5–8%) in MCCs are whey proteins; however, there is no standard of identity for MCCs, and the distinction between MCCs and MPCs is not definitive. The production of MCCs involves the use of MF membranes, with wider pores than the UF membranes used to produce MPCs. The MF process allows whey proteins to be removed in the permeate, along with lactose and other soluble components (Pouliot and Pouliot, 1996). DF with water facilitates further removal of these components, although a certain proportion of whey proteins (~5% of protein) remains in the retentate with the micellar fraction. The factor that limits complete removal of whey proteins may be the progressive formation of a fouling layer comprised of casein micelles, which increases rejection of whey proteins by a combination of electrostatic and steric repulsion (Gésan-Guiziou et al., 2013). When MF is performed at a temperature <15°C, β-CN is depleted from the MCC, with the degree of depletion increasing as temperature is decreased. MCCs share many of the physico-chemical and techno-functional properties of MPCs, as well as some of their challenges (i.e. poor solubility), but have been less widely researched in general. In the food industry, MCCs have found application mainly in 'slow release' sports nutrition products and have also been investigated for the standardization of cheese milk. MCCs have been the subject of research on their application in traditional dairy products such as yoghurt (Karam et al., 2012), as well as in newer applications such as the encapsulation of bioactive substances which are hydrophobic or hydrophilic (Yazdi et al., 2014). Regular MCCs and β-CN-depleted MCCs are major co-products of SPC and β-CN, respectively, produced by membrane filtration. An active area of non-food application-based research for MCCs is their use as semen extenders, where they act to preserve the fertility of animal sperm during storage. Artificial insemination plays a very prominent role in modern breeding programmes for horses, cows and other species. In these applications, semen is often stored in liquid form in the presence of an extender prior to insemination or transit to another location. Common extenders include combinations of sugars, electrolytes, egg yolk, skim milk and, recently, MCC. A number of studies have demonstrated that an MCC-based extender developed at INRA (INRA96®) performs better than other, more chemically heterogeneous milk-based extenders, such as skim milk (Batellier et al., 1998, 2001), and is highly effective when combined with cold storage and centrifugation to remove seminal plasma (Pagl et al., 2006; Webb et al., 2009).

8.2.6 Liquid/gelled casein concentrates

High-protein powders in which micellar casein is the dominant component, such as MCCs and MPCs, exhibit exceptionally poor dispersibility and require very long rehydration times (Anema et al., 2006; Gaiani et al., 2005, 2009; Gazi and Huppertz, 2014; Havea, 2006; Ji et al., 2016; Le et al., 2011a, b, 2013). Use of incompletely dispersed MPCs in cheese milk is associated with the presence of 'nuggets' in cheese (Carr and Golding, 2016). Strategies to accelerate the dispersion of these powders, which do not involve partial or total destruction of the micellar state, were reviewed by Crowley et al. (2016) and include the application of high shear or an elevated temperature during reconstitution, and the addition of powders into solvents of higher ionic strength than water (e.g. milk, milk permeate, NaCl or KCl solutions). Due to defects associated with the functionality and flavour of milk powders, alternative methods for preserving milk protein fractions without dehydration are being sought. The formation of a gel of milk proteins can also greatly enhance shelf-life compared to milk, with a traditional example of such a product being cheese. More recently, issues encountered with high-protein casein-dominant powders (e.g. poor solubility) and a demand for 'fresh'/'minimally-processed' products have contributed to the growing interest in liquid protein concentrates, some of which undergo temperature-dependent gelation. A few studies have been carried out on the manufacture of high-solids MCC suspensions which are not dried, but rather refrigerated or frozen as gels. Amelia and Barbano (2013) produced an 18% protein MCC from skim milk by performing UF to remove ~50% of lactose and NPN, followed by MF/DF to remove the majority of serum proteins, soluble minerals and remaining lactose and NPN, and finally UF to concentrate the protein. The MCC formed a thermo-reversible gel at 4°C which reverted to a liquid at >22°C. The gelled MCC of Amelia and Barbano (2013) could be stored at 4°C for 16 weeks without significant microbial growth. Lu et al. (2015) produced similar MCCs (19–23% protein) and subjected them to frozen storage. After thawing, MCCs were diluted to 3% protein and analysed for various solubility parameters. Much like MCC powders, these systems needed either high shear/temperature or extended rehydration times to achieve acceptable solubilization.

8.2.7 β-Casein

The manufacture of enriched or purified casein fractions is less advanced compared to whey proteins, and is limited mainly to β-CN. β-CN is a highly functional protein and is being investigated in the food industry for a number of applications, the most notable of which is as a potential ingredient in humanized infant formulae. β-CN can be purified from milk, MCC or casein curds by using low-temperature processing. Many dairy manufacturers are using lower temperatures (from ~50°C to <15°C) in membrane filtration plants; this change has resulted in a higher level of β-CN in some SPC products. To the authors' knowledge, the only commercially available β-CN-enriched product suitable for food applications is that (Ultranor Beta™) manufactured by Kerry Ingredients, Tralee, Ireland).

8.3 Other milk-protein-containing dairy products

Hydrolysates are produced by enzymatic hydrolysis of casein or whey protein ingredients, and sometimes combinations of both, in order to reduce allergenicity and/or increase digestibility. Depending on the degree of hydrolysis, hydrolysates can be characterized as partially- or extensively hydrolysed. In addition, selectively hydrolysed ingredients have been developed recently, in which one major whey protein (e.g. β-lg) is hydrolysed while the other (e.g. α-la) remains intact.

Ingredients derived from milk of non-bovine sources are also manufactured, with goat milk and goat whey powders being prominent examples. These products are typically marketed as being more tolerable to individuals with sensitivity to bovine milk. Goat-milk-derived ingredients are manufactured mainly in New Zealand to supply an expanding global goat milk infant formula industry.

Milk protein ingredients are important components of staged and clinical nutrition products, as well as infant formula. Infant formula is a nutrient blend, usually manufactured from milk protein ingredients combined with a multitude of other ingredients (e.g. lactose, vegetable oil, minerals, vitamins, probiotics). Protein is present in these products at a relatively low level compared to the other macronutrients, but the use of milk protein ingredients (demineralized whey, WPC, WPI, MPI) in combination with skim milk is an important aspect of reproducing the nutrient profile of human milk. In addition, milk proteins are important contributors of innate minerals (i.e. not introduced through mineral fortification) in the final formulation (O'Callaghan et al., 2011).

Ingredients derived from colostrum, which is rich in bioactive proteins such as Igs (see McGrath et al., 2016), are also available.

9 Summary and future trends

The proteins of milk are its most important constituents from both the nutritional and technological viewpoints. Many dairy products exist because of the unique functional properties of its proteins. Due to their technological and nutritional importance, the proteins of milk, which have been the subject of research for 200 years, are probably the best characterized food proteins. However, gaps in knowledge concerning this protein system remain, and this system continues to be an active area of research.

In this chapter, the current state of knowledge of the bovine milk protein system is described and the principal gaps discussed. The protein system in the milk of the other commercial dairy species, buffalo, goat and sheep, and of human milk, are also fairly well characterized, but the milk protein systems of the vast majority of mammalian species have yet to be investigated.

Developments in the enrichment and fractionation of milk proteins continue, as new methods to exploit differences in their physico-chemical properties are discovered. Progress in the separation of milk proteins has also benefited from the application of new separation technologies. However, some practical challenges exist, and these need to be addressed if current innovation in milk protein ingredients is to be commercially successful.

Identifying opportunities for the application of the by-products generated, which are typically protein-enriched themselves, is essential if these processes are to be sustainable and profitable. In this way, manufacturers aim to develop multiple milk protein 'co-products' rather than one primary product and a waste stream.

In recent years, numerous milk protein ingredients have been developed. These materials have attracted interest for their unique nutritional profile and functional properties, which are distinct from more traditional dairy-derived ingredients. These milk protein ingredients pose unique challenges, in relation to their processing, storage, handling and application. This is a very active area of current research.

There is a need for further study of milk proteins. Opportunities exist for investigations of both the fundamental and applied aspects of these systems, against the strong background of many decades of underpinning research in the area.

10 Where to look for further information

Several textbooks are available on the general subject of Dairy Science, which include chapters on both the fundamental and applied aspects of milk proteins. Examples of books which are useful as introductory material are *Principles of Dairy Chemistry* (Jenness and Patton, 1959 – John Wiley), *Dairy Chemistry and Biochemistry* (Fox, Uniacke-Lowe, McSweeney and O'Mahony, 2016 – Springer) and the multivolume set *Encylcopedia of Dairy Sciences* (Fuquay, Fox and McSweeney, 2011 – Academic Press). A valuable online resource on the fundamentals of dairy science is also made available by Prof. Doug Goff of the University of Guelph, who maintains and updates the website at: https://www.uoguelph.ca/foodscience/book-page/dairy-science-and-technology-ebook

For readers who are familiar with the fundamentals of dairy chemistry, more detailed overviews of the protein component can be found in H. McKenzie's two-volume *Milk Proteins: Chemistry and Molecular Biology* (1970, 1971 – Academic Press), *Dairy Chemistry and Physics* (Walstra and Jenness, 1984 – Wiley), *Dairy Technology: Principles of Milk Properties and Processes* (Walstra, Geurts, Noomen, Jellema and van Boekel, 1999 – CRC Press), the 'Proteins' volume of *Advanced Dairy Chemistry* (McSweeney and Fox, 2013; McSweeney and O'Mahony, 2016 – Springer) and *Milk Proteins: From Expression to Food* (Singh, Boland and Thompson, 2014 – Academic Press). The *Handbook of Milk of Non-Bovine Animals* (Park and Haenlein, 2006 – Wiley-Blackwell; new edition expected in 2016) is a unique resource for information on caprine, ovine and buffalo milk, among many others, which are of importance industrially in specific regions of the world, and also human milk, the study of which informs strategies to manufacture humanized infant formula from bovine milk. Several sections in the aforementioned *Encylcopedia of Dairy Sciences* also deal with the milk of various mammals. A comprehensive summary of findings from many years of research on the composition of bovine and human milk can be found in R. G. Jensen's *Handbook of Milk Composition* (1995 – Academic Press).

Important foundational research papers in the field include that of Waugh and von Hippel (1956), who isolated κ-CN and proposed that this protein stabilized the micellar phase of milk, a principle which remains central to modern models of the structure, stability and gelation of casein micelles, and reports on the application of starch gel electrophoresis (SGE) by Wake and Baldwin (1961) and polyacrylamide gel electrophoresis (PAGE) by Peterson (1963), which have had a lasting influence on the analytical procedures used to study the heterogeneity of milk proteins. With continued research and increasingly sensitive analytical techniques, the protein system of milk grows in complexity; many of these developments can be tracked through the reports of the American Dairy Science Association Committee on the *Nomenclature, Classification, and Methodology of Milk Proteins*, the most recent revision of which was published in 2004 with H. M. Farrell Jr. as lead author.

Relevant review articles and book chapters, which provide more in-depth treatments of important areas than was possible in this chapter, include those on the casein micelle (Horne, 1998; Holt et al., 2013), milk minerals (Gaucheron, 2004; Lucey and Horne, 2009; Holt, 1985a), rennet gelation (Lucey, 2002) and heat stability (Singh, 2004) of milk, the fractionation of proteins using ion-exchange and cross-flow filtration (Etzel, 2004; Pouliot, 2008; O'Mahony and Tuohy, 2013) and the quantification of milk proteins (Dupont et al., 2013). Volume 18 of the *International Dairy Journal* contains two review articles on each of the two major protein fractions in milk, one describes developments in the understanding of casein micelles (Fox and Brodkorb, 2008) and the other the evolution of whey from a low-value by-product of cheese making to its current status as the primary raw material for the billion-dollar global whey protein ingredients market (Smithers, 2008).

In those countries with a significant dairy industry there are associations and societies which support and promote the work of academic and industrial scientists active in the field. In addition, these organizations are responsible for convening national and international conferences on the key research themes of the day. Examples of organizations dedicated to dairy science and technology are the *Society of Dairy Technology* in the UK, the *American Dairy Science Association* in the US and the *International Dairy Federation*, which is based in Belgium but international in nature and standing. Organizations which also support research in this area, without being solely associated with the field include, but are not limited to, the *Institute of Food Technology*, the *European Food Safety Authority* (EFSA), the *World Health Organization* (WHO) and the *Food and Agriculture Organization of the United Nations* (FAO).

There are many dedicated journals on the biology, chemistry and physics of milk, which frequently publish findings related to milk proteins, some of the prominent examples are the *International Dairy Journal, Journal of Dairy Science, Dairy Science and Technology, Journal of Dairy Research* and the *International Journal of Dairy Technology*. Research related to milk proteins can also be found frequently in general food journals (*Food Biophysics, Journal of Agricultural and Food Chemistry, Food Research International, Food Structure, LWT – Food Science and Technology, Food Chemistry, Journal of Food Engineering, Food Hydrocolloids, Innovative Food Science and Emerging Technologies*), colloid and soft matter journals (*Journal of Colloid and Interface Science, Langmuir, Soft Matter, Colloids and Surfaces B: Biointerfaces, Biomacromolecules*), journals that specialise in separation sciences (*Separation and Purification Technology, Journal of Membrane Science, Desalination*), journals covering topics in bulk solids handling (*Powder Technology*) and journals on basic science (*Biochimica et Biophysica Acta, European Biophysics Journal*).

Universities around the world with groups who actively undertake research in milk and milk proteins are too numerous to list; however there are several prominent research centres and institutes in this area. In Europe, examples include the *Federal Dairy Research Centre, Kiel* (Germany), *NIZO food research* (the Netherlands), *ICTAN-CSIC* (Spain), *Teagasc Food Research Centre, Moorepark* (Ireland), *Leatherhead Food Research* (UK) and *INRA* (France). The *Riddet Institute* is a key research organization in New Zealand, while *CSIRO* is an important organization in Australia. There are several research centres focused on milk processing associated with strong food research departments in US universities, including the *Centre for Dairy Research* partnered with the University of Wisconsin-Madison, the *Institute for Dairy Ingredient Processing* at South Dakota State University, the *Cornell University Dairy Research Center*, and the *Dairy Products Technology Center* at California Polytechnic State University. Elsewhere in North America, the *STELA Dairy Research Centre* is associated with Laval University in Canada, while the *Elora/Ponsonby Dairy Research Centre* supports dairy research at the University of Guelph. The *Saint-Hyacinthe Research and Development Centre* in Quebec is also active in Canadian dairy research.

11 References

Aimutis, W. R. and Eigel, W. N. (1982). Identification of λ-casein as plasmin-derived fragments of bovine α_{s1}-casein. *J. Dairy Sci.*, 65, 175–81.

Akers, R. M. (2000). Selection for milk production from a lactation biology viewpoint. *J. Dairy Sci.*, 83, 1151–8.

Akerstrom, B., Flower, D. R. and Salier, J. P. (2000). Lipocalins 2000. *Biochim. Biophys. Acta.*, 1482, 1–356.

Amelia, I. and Barbano, D. (2013). Production of an 18% protein liquid micellar casein concentrate with a long refrigerated shelf life. *J. Dairy Sci.*, 96, 3340–9.

Amundson, C. H., Watanawamichakorn, S. and Hill, C. G. (1982). Production of enriched protein fractions of β-lactoglobulin and α-lactalbumin from cheese whey. *J. Food Proc. Preserv.*, 6, 55–71.

Andersson, J. and Mattiasson, B. (2006). Simulated moving bed technology with a simplified approach for protein purification: separation of lactoperoxidase and lactoferrin from whey protein concentrate. *J. Chrom. A.*, 1107, 88–95.
Andrews, A. T. (1983). Proteinases in normal bovine milk and their action on casein. *J. Dairy Res.*, 50, 45–55.
Anema, S. G., Pinder, D. N., Hunter, R. J. and Hemar, Y. (2006). Effects of storage temperature on the solubility of milk protein concentrate (MPC85). *Food Hydrocoll.*, 20, 386–93.
Anema, S. G. (2009). The use of 'lab-on-a-chip' microfluidic SDS electrophoresis technology for the separation and quantification of milk protein. *Int. Dairy J.*, 19, 198–204.
Anema, S. G. and de Kruif, C. G. (2016). Phase separations and composition of coacervates of lactoferrin and caseins. *Food Hydrocoll.*, 52, 670–7.
Armaforte, E., Curran, E., Huppertz, T., Ryan, C. A., Caboni, M. F., O'Connor, P., Ross, P., Hirtz, C., Sommerer, N., Chevalier, F. and Kelly, A. L. (2010). Proteins and proteolysis in pre-term and term human milk and possible implications for infant formulae. *Int. Dairy J.*, 20, 715–23.
Armstrong, J. McD., McKenzie, H. A. and Sawyer, W. H. (1967). On the fractionation of β-lactoglobulin and α-lactalbumin. *Biochim. Biophys. Acta*, 147, 60–72.
Arunkumar, A. and Etzel, M. R. (2013). Fractionation of α-lactalbumin and β-lactoglobulin using positively charged tangential flow ultrafiltration membranes. *Sep. Purif. Technol.*, 105, 121–8.
Arunkumar, A. and Etzel, M. R. (2014). Fractionation of α-lactalbumin and β-lactoglobulin from bovine milk serum using staged, positively charged, tangential flow ultrafiltration membranes. *J. Membrane Sci.*, 454, 486–95.
Arunkumar, A. and Etzel, M. R. (2015). Negatively charged tangential flow ultrafiltration membranes for whey protein concentration. *J. Membrane Sci.*, 475, 340–8.
Arunkumar, A., Molitor, M. and Etzel, M. R. (2016). Comparison of flat-sheet and spiral-wound negatively-charged wide-pore ultrafiltration membranes for whey protein concentration. *Int. Dairy J.*, 56, 129–33.
Aschaffenburg, R. and Drewry, J. (1955). Occurrence of different beta-lactoglobulins in cow's milk. *Nature*, 176, 218–19.
Aschaffenburg, R. and Drewry, J. (1957). Improved method for the preparation of crystalline β-lactoglobulin and α-lactalbumin from cow's milk. *Biochem. J.*, 65, 273–7.
Azuma, N., Maeta, A., Fukuchi, K. and Kanno, C. (2006). A rapid method for purifying osteopontin from bovine milk and interaction between osteopontin and other milk proteins. *Int. Dairy J.*, 16, 370–8.
Bansal, N. (2011). *Aggregation of rennet-altered casein micelles at low temperatures and the retention and inactivation of coagulants in cheese curd*. PhD Thesis, National University of Ireland, Cork.
Bansal, N. and Bhandari, B. (2016). Functional milk proteins: production and utilization: whey-based ingredients. In P. L. H. McSweeney and J. A. O'Mahony (eds.), *Advanced Dairy Chemistry, Vol. 1 B: Proteins: Applied Aspects*. New York: Springer Science and Business Media, pp. 67–98.
Bansal, N., Fox, P. F. and McSweeney, P. L. H. (2008). Factors that affect the aggregation of rennet-altered casein micelles at low temperatures. *Int. J. Dairy Technol.*, 61, 56–61.
Batellier, F., Duchamp, G., Vidament, M., Arnaud, G., Palmer, E. and Magistrini, M. (1998). Delayed insemination is successful with a new extender for storing fresh equine semen at 15°C under aerobic conditions. *Theriogenology*, 50, 229–36.
Batellier, F., Vidament, M., Fauguant, J., Duchamp, G., Arnaud, G., Yvon, J. M. and Magistrini, M. (2001). Advances in cooled semen technology. *Anim. Reprod. Sci.*, 68, 181–90.
Bayless, K. J., Davis, G. E. and Meininger, G. A. (1997). Isolation and biological properties of osteopontin from bovine milk. *Prot. Expr Purif.*, 9, 309–14.
Bijl, E., de Vries, R., van Valenberg, H. and Huppertz, T. (2014a). Factors influencing casein micelle size in milk of individual cows. *Int. Dairy J.*, 34, 135–41.
Bijl, E., van Valenberg, H. J. F., Huppertz, T., van Hooijdonk, T. and Bovenhuis, H. (2014b). Phosphorylation of α_{s1} casein is regulated by different genes. *J. Dairy Sci.*, 97, 7240–6.

Bijl, E., van Valenberg, H., Sikkes, S., Jumelet, S., Sala, G., Olieman, K., van Hooijdonk, T. and Huppertz, T. (2014c). Chymosin-induces hydrolysis of caseins: influence of degree of phosphorylation of α_{s1} casein and genetic variants of beta-casein. *Int. Dairy J.*, 39, 215–21.

Bobe, G., Beitz, D. C., Freeman, A. E. and Lindberg, G. L. (1998a). Separation and quantification of bovine milk proteins by reversed phase high performance liquid chromatography. *J. Agric. Food Chem.*, 46, 458–63.

Bobe, G., Beitz, D. C., Freeman, A. E. and Lindberg, G. L. (1998b). Sample preparation affects separation of whey proteins by reversed phase high performance liquid chromatography. *J. Agric. Food Chem.*, 46, 1321–5.

Bordin, G., Cordeiro Raposo, F., de la Calle, B. and Rodriguez, A. R. (2001). Identification and quantification of major bovine milk proteins by liquid chromatography. *J. Chrom. A*, 928, 63–76.

Bouguyon, E., Beauvallet, C., Huet, J.-C. and Chanat, E. (2006). Disulphide bonds in casein micelles from milk. *Biochem. Biophys. Res. Commun.*, 343, 450–8.

Braga, A. L. M., Menossi, M. and Cunha, R. L. (2006). The effect of the glucono-δ-lactone/caseinate ratio on sodium caseinate gelation. *Int. Dairy J.*, 16, 389–98.

Brew, K. (2003). α-Lactalbumin. In P. F. Fox and P. L. H. McSweeney (eds.), *Advanced Dairy Chemistry – Volume 1: Proteins*, 3rd edn. New York: Kluwer Academic/Plenum Publishers, pp. 387–419.

Brew, K. (2011). α-Lactalbumin. In J. W. Fuquay, P. F. Fox and P. L. H. McSweeney (eds.), *Encyclopedia of Dairy Sciences, Volume 3*, 2nd edn. Oxford: Academic Press, pp. 780–6.

Brew, K. (2013). α-Lactalbumin. In P. L. H. McSweeney and P. F. Fox (eds.), *Advanced Dairy Chemistry, Volume 1A: Proteins: Basic Aspects*, 4th edn. New York: Springer, pp. 261–73.

Brinkmann, C. R., Thiel, S., Larsen, M. K., Petersen, T. E., Jensenius, J. C. and Heegaard, C. W. (2011). Preparation and comparison of cytotoxic complexes formed between oleic acid and either bovine or human α-lactalbumin. *J. Dairy Sci.*, 94, 2159–70.

Buchheim W., Lund, S. and Scholtissek, J. (1989). Vergleichende Untersuchungen zur Struktur und Grosse von Caseinmicellen in der Milch verschiedener Species. *Kieler Milchw. Forsch.*, 41, 253–65.

Campagna, S., Vitoux, B., Humbert, G., Girardet, J.-M., Linden, G., Haertle, T. and Gaillard, J.-L. (1998). Conformational studies of a synthetic peptide from the putitive lipid-binding domain of bovine milk component PP3. *J. Dairy Sci.*, 81, 3139–48.

Campana, W. M., O'Brien, J. S., Hiraiwa, M. and Patton, S. (1999). Secretion of prosaposin, a multifunctional protein, by breast cancer cells. *Biochim. Biophys. Acta*, 1427, 392–400.

Campbell, R. E., Boogers, I. A. and Drake, M. A. (2014). Development of a novel method for the extraction of norbixin from whey and its subsequent quantification via high performance liquid chromatography. *J. Dairy Sci.*, 97, 1313–18.

Campbell, B. and Petersen, W. E. (1959). Antibodies in milk for protection against human disease. *Milchwissenschaft*, 14, 469–73.

Carr, A. and Golding, M. (2016). Functional milk proteins production and utilisation: casein-based ingredients. In P. L. H. McSweeney and J. A. O'Mahony (eds.), *Advanced Dairy Chemistry: Volume 1: Proteins, Part B, Applied Aspects*. New York: Springer, pp. 35–66.

Chanat, E., Martin, P. and Ollivier-Bousquet, M. (1999). α_{s1}-casein is required for the efficient transport of β- and κ-casein from the endoplasmic reticulum to the Golgi apparatus of mammary epithelial cells. *J. Cell Sci.*, 112, 3399–412.

Chatterton, D. E. W., Smithers, G., Roupas, P. and Brodkorb, A. (2006). Bioactivity of β-lactoglobulin and α-lactalbumin – technological implications for processing. *Int. Dairy J.*, 16, 1229–40.

Cheang, B. and Zydney, A. L. (2004). A two-stage ultrafiltration process for fractionation of whey protein isolate. *J. Membrane Sci.*, 231, 159–67.

Chen, Y.-C., Chen, C.-C., Chen, S.-T. and Hsleh, J. F. (2016). Proteomic profiling of the coagulation of milk proteins induced by glucono-delta-lactone. *Food Hydrocoll.*, 52, 137–43.

Chevalier, F. (2011a). Analytical methods: electrophoresis. In J. W. Fuquay, P. F. Fox and P. L. H. McSweeney (eds.), *Encyclopedia of Dairy Sciences, Volume 1*, 2nd edn. San Diego, CA: Academic Press, pp. 185–92.

Chevalier, F. (2011b). Milk proteins: proteomics. In J. W. Fuquay, P. F. Fox and P. L. H. McSweeney (eds.), *Encyclopedia of Dairy Sciences, Volume 3*, 2nd edn. San Diego, CA: Academic Press, pp. 843–7.

Christensen, B. and Sørensen, E. S. (2016). Structure, function and nutritional potential of milk osteopontin. *Int. Dairy J.*, 57, 1–6.

Chrysina, E. D., Brew, K. and Acharya, K. R. (2000). Crystal structures of *apo*- and *holo*-bovine α-lactalbumin at 2.2 Å resolution reveal an effect of Ca^{2+} on inter-lobe interactions. *J. Biol. Chem.*, 275, 37021–9.

Claeys, W. L., Verraes, C., Cardoen, S., De Block, J., Huyghebaert, A., Raes, K., Dewettinck, K. and Herman, L. (2014). Consumption of raw or heated milk from different species: an evaluation of the nutritional and potential health benefits. *Food Control*, 42, 188–201.

Coppola, L. E., Molitor, M. S., Rankin, S. A. and Lucey, J. A. (2014). Comparison of milk-derived whey protein concentrates containing various levels of casein. *Int. J. Dairy Tech.*, 67, 467–73.

Corredig, M., Roesch, R. R. and Dalgleish, D. G. (2003). Production of a novel ingredient from buttermilk, *J. Dairy Sci.*, 86, 2744–50.

Creamer, L. K., Berry, G. P. and Mills, O. E. (1977). A study of the dissociation of β-casein from the bovine casein micelle at low temperature. *N. Z. J. Dairy Sci. Technol.*, 12, 58–66.

Creamer, L. K., Loveday, S. M. and Sawyer, L. (2011). β-Lactoglobulin. In J. W. Fuquay, P. F. Fox and P. L. H. McSweeney (eds.), *Encyclopedia of Dairy Sciences, Volume 3*, 2nd edn. Oxford: Academic Press, pp. 787–94.

Creamer, L. K. and Richardson, T. (1984). Anomalous behavior of bovine α_{s1}- and β-CNs on gel electrophoresis in sodium dodecyl sulfate buffer. *Arch. Biochem. Biophys.*, 234, 476–86.

Creamer, L. K., Plowman, J. E., Liddell, M. J., Smith, M. H. and Hill, J. P. (1998). Micelle stability: κ-casein structure and function. *J. Dairy Sci.*, 81, 3004–12.

Crowley, S. V., Boudin, M., Chen, B., Gazi, I., Huppertz, T., Kelly, A. L. and O'Mahony, J. A. (2015a). Stability of milk protein concentrate suspensions to in-container sterilisation heating conditions. *Int. Dairy J.*, 50, 45–9.

Crowley, S. V., Caldeo, V., McCarthy, N. A., Fenelon, M. A., Kelly, A. L. and O'Mahony, J. A. (2015b). Processing and protein-fractionation characteristics of different polymeric membranes during filtration of skim milk at refrigeration temperatures. *Int. Dairy J.*, 48, 23–30.

Crowley, S. V., Jeantet, R., Schuck, P., Kelly, A. L. and O'Mahony, J. A. (2016). Rehydration and solubility characteristics of high-protein dairy powders. In P. L. H. McSweeney and J. A. O'Mahony (eds.), *Advanced Dairy Chemistry: Volume 1: Proteins, Part B, Applied Aspects*. New York: Springer, pp. 99–131.

Dalgleish, D. G. (1998). Casein micelles as colloids: surface structures and stabilities. *J. Dairy Sci.*, 81, 3013–18.

Dalgleish, D. G. (2007). The casein micelle and its reactivity. *Lait*, 87, 385–7.

Dalgleish, D. G. (2011). On the structural models of bovine casein micelles – review and possible improvements. *Soft Matter*, 7, 2265–72.

Dalgleish, D. G., Brinkhuis, J. and Payens, T. A. J. (1981). The coagulation of differently-sized casein micelles by rennet. *Eur J. Biochem.*, 119, 257–61.

Dalgleish, D. G. and Corredig, M. (2012). The structure of the casein micelle of milk and its changes during processing. *Annu. Rev. Food Sci. Technol.*, 3, 449–67.

Dalgleish, D. G., Spagnuolo, P. A. and Goff. H. D. (2004). A possible structure of the casein micelle based high-resolution field-emission scanning electron microscopy. *Int. Dairy J.*, 14, 1025–31.

Davies, D. T. and Law, A. J. R. (1987). Quantitative fractionation of casein mixtures by fast protein liquid chromatography. *J. Dairy Res.*, 54, 369–76.

Davis, S. R. and South, C. R. (2015). Suspension of milking in dairy cows produces a transient increase in milk lactoferrin concentration and yield after resumption of milking. *J. Dairy Sci.*, 98, 7823–30.

de Kort, E., Minor, M., Snoeren, T., van Hooijdonk, T. and van der Linden. (2011). Effect of calcium chelators on physical changes in casein micelles in concentrated micellar casein solutions. *Int. Dair. J.*, 21, 907–13.

de Kruif, C. G. (1998). Supra-aggregates of casein micelles as a prelude to coagulation. *J. Dairy Sci.*, 81, 3019–38.

de Kruif, C. G. (1999). Casein micelle interactions. *Int. Dairy J.*, 9, 183–8.

de Kruif, C.G. (2014). The structure of casein micelles: a review of small-angle scattering data. *J. Appl. Crystallogr.*, 47, 1479–89.

de Kruif, C. G. and Holt, C. (2003). Casein micelle structure, functions and interactions. In P. F. Fox and P. L. H. McSweeney (eds.), *Advanced Dairy Chemistry*, Volume 1 A – *Proteins*; *Basic Aspects*, 3rd edn. New York: Kluwer Academic/Plenum Publishers, pp. 233–76.

de Wit, J. and Bronts, H. (1995). Process for the recovery of α-lactalbumin and β-lactoglobulin from a whey protein product. Patent, US005420249A.

Desobry-Banon, S., Richard, F. and Hardy, J. (1994). Study of acid and rennet coagulation of high pressurized milk. *J. Dairy Sci.*, 77, 3267–74.

Downey, W. K. (1973). Structure of bovine casein micelles. *Neth. Milk Dairy J.*, 27, 218–19.

Downey, W. K. and Murphy, R. F. (1970). The temperature-dependant dissociation of β-casein from bovine casein micelles and complexes. *J. Dairy Res.*, 37, 361–75.

Dupont, D., Croguennec, T., Brodkorb, A. and Kouaouci, R. (2013). Quantification of proteins in milk and milk products. In P. L. H. McSweeney and P. F. Fox (eds.), *Advanced Dairy Chemistry 1 A, Proteins: Basic Aspects.* New York: Springer Science+Business media, pp. 87–134.

El-Loly, M. M. and Mahfouz, M. B. (2011). Lactoferrin in relation to biological functions and applications: a review. *Int. J. Dairy Sci.*, 6, 79–111.

El-Negoumy, A. M. (1973). Separation of lambda casein and some of its properties. *J. Dairy Sci.*, 56, 1486–91.

Ennis, M. P. and Mulvihill, D. M. (1999). Compositional characteristics of rennet caseins and hydration characteristics of the caseins in a model system as indicators of performance in Mozzarella cheese analogue manufacture. *Food Hydrocoll.*, 13, 325–37.

Etzel, M. R. (1999). Isolating β-lactoglobulin and α-lactalbumin by eluting from a cation exchanger without sodium chloride. Patent, US005986063A.

Etzel, M. R. (2004). Manufacture and use of dairy protein fractions. *J. Nutr.*, 134, 9965–10025.

Euston, S. R. and Horne, D. S. (2005). Simulating the self-association of caseins. *Food Hydrocoll.*, 19, 379–86.

Euston, S.R. and Nicolosi, M. (2007). Simulating the self-association of caseins: towards a model of the casein micelle. *Lait*, 87, 389–412.

Famelart, M. H., Le Graet, Y. and Raulot, K. (1999). Casein micelle dispersions into water, NaCl and $CaCl_2$: physicochemical characteristics of micelles and rennet coagulation. *Int. Dairy J.*, 9, 293–7.

Farrell, H. M. Jr., Brown, E. M. and Malin, E. L. (2013). Higher order structures of the caseins: a paradox? In P. L. H. McSweeney and P. F. Fox (eds.), *Advanced Dairy Chemistry*, volume 1 A – *Proteins*; *Basic Aspects*, 4th edn. New York: Springer, pp. 161–84.

Farrell, H. M., Jr., Jimenez-Flores, R., Bleck, G. T., Brown, E. M., Butler, J. E., Creamer, L. K., Hicks, C. L., Hollar, C. M., Ng-Kwai-Hang, K. F., and Swaisgood, H. E. (2004). Nomenclature of the proteins of cows' milk – Sixth revision. *J. Dairy Sci.*, 87, 1641–74.

Farrell, H. M. Jr., Malin, E. L., Brown, E. M. and Qi, P. X. (2006). Casein micelle structure: what can be learned from milk synthesis and structural biology? *Curr. Opin. Colloid Interface Sci.*, 11, 135–47.

Farrell, H. M., Jr., Wickham, E. D. and Groves, M. L. (1998). Environmental influences on purified κ-casein: disulphide interactions. *J. Dairy Sci.* 81, 2974–84.

FitzGerald, R. J. and Meisel, H. (2003). Milk protein hydrolysates and bioactive peptides. In P. F. Fox and P. L. H. McSweeney (eds.), *Advanced Dairy Chemistry* – Volume 1: *Proteins*, 3rd edn. New York: Kluwer Academic/Plenum Publishers, pp. 675–98.

Fox, K. K., Holsinger, V. H., Posati, L. P. and Pallansch, M. J. (1967). Separation of β-lactoglobulin from other milk serum proteins by trichloroacetic acid. *J. Dairy Sci.*, 50, 1363–7.

Fox, P. F. (2003). Milk proteins: general and historical aspects. In P. F. Fox and P. L. H. McSweeney (eds.), *Advanced Dairy Chemistry, Volume 1 – Proteins*, 3rd edn. New York: Kluwer Academic-Plenum Publishers, pp. 1–48.

Fox, P. F. and Brodkorb, A. (2008). The casein micelle: historical aspects, current concepts and significance. *Int. Dairy J.*, 18, 677–84.

Fox, P. F. and Flynn, A. (1992). Biological properties of milk proteins. In P. F. Fox (ed.), *Advanced Dairy Chemistry, Volume 1, Proteins*, 2nd edn. London: Elsevier Applied Science, pp. 255–84.

Fox, P. F. and Hoynes, M. C. T. (1975). Heat stability of milk: influence of colloidal calcium phosphate and β-lactoglobulin. *J. Dairy Res.*, 42, 427–35.

Fox, P. F. and Kelly, A. L. (2006a). Indigenous enzymes in milk: overview and historical aspects – Part 1. *Int. Dairy J.*, 16, 500–16.

Fox, P. F. and Kelly, A. (2006b). Indigenous enzymes in milk: overview and historical aspects – Part 2. *Int. Dairy J.*, 16, 517–32.

Gaiani, C., Banon, S., Scher, J., Schuck, P. and Hardy, J. (2005). Use of a turbidity sensor to characterise micellar casein powder rehydration: influence of some technological effects. *J. Dairy Sci.*, 88, 2700–6.

Gaiani, C., Schuck, P., Scher, J., Ehrhardt, J. J., Arab-Tehrany, E., Jacquot, M. and Banon, S. (2009). Native phosphocaseinate powder during storage: lipids released onto the surface. *J. Food Eng.*, 94, 130–4.

Garnier, J. and Ribadeau-Dumas, B. (1970). Structure of the casein micelle. *J. Dairy Res.*, 37, 493–504.

Gaucheron, F. (2004). The minerals of milk. *Reprod. Nutr. Dev.*, 45, 473–83.

Gaucheron, F., Famelart, M. H., Mariette, F., Raulot, K., Michel, F. and Le Graet, Y. (1997). Combined effects of temperature and high-pressure treatments on physicochemical characteristics of skim milk. *Food Chem.*, 59, 439–47.

Gaucheron, F., Molle, D. and Leonil, B. J. (1999). Identification of low molar mass peptides released during sterilisation of milk. *Int. Dairy J.*, 9, 515–21.

Gauthier, S. F., Pouliot, Y. and Maubois, J.-L. (2006). Growth factors in bovine milk and colostrum: composition, extraction and biological activities. *Lait*, 86, 99–125.

Gazi, I. and Huppertz, T. (2014). Influence of protein content and storage conditions on the solubility of caseins and whey proteins in milk protein concentrates. *Int. Dairy J.*, 46, 22–30.

Gésan-Guiziou, G., Jimenez-Lopez, A., Lambrouin, F. and Leconte, N. (2013). Transmission of soluble proteins during microfiltration (0.1 μm) of skimmed milk: impact of casein micelles deposit and critical conditions. *Journal of Membrane Science*, Presented at International Membrane Science and Technology, IMSTEC, Melbourne, Australia (25 November 2013 to 29 November 2013). http://prodinra.inra.fr/record/252545.

Girardet, J.-M. and Linden, G. (1996). PP 3 component of bovine milk: a phosphorylated whey glycoprotein. *J. Dairy Res.*, 63, 333–50.

Glantz, M., Devold, T. G., Vegarud, G. E., Lindmark Mansson, H., Stalhammer, H. and Paulsson, M. (2010). Importance of casein micelle size and milk composition for milk gelation. *J. Dairy Sci.*, 93, 1444–51.

Gonzalez-Jordan, A., Thomar, P., Nicolai, T. and Dittmer, J. (2015). The effect of pH on the structure and phosphate mobility of casein micelles in aqueous solution. *Food Hydrocoll.*, 51, 88–94.

Gordon, W. G. (1971). α-Lactalbumin. In H. A. McKenzie (ed.), *Milk Proteins: Chemistry and Molecular Biology*, New York: Academic Press, pp. 331–65.

Haggarty, N. W. (2003). Milk proteins: minor proteins, bovine serum albumin and vitamin-binding proteins. In H. Roginski, J. W. Fuquay and P. F. Fox (eds.), *Encyclopedia of Dairy Sciences*. London: Academic Press, pp. 1939–46.

Hajjoubi, S., Rival-Gervier, S., Hayes, H., Floriot, S., Eggen, A., Pivini, F., Chardon, P., Houdebine, L.-M. and Thepot, D. (2006). Ruminants genome no longer contains whey acidic protein gene but only a pseudogene. *Gene*, 370, 104–12.

Harland, H. A., Coulter, S. T. and Jenness, R. (1955). Natural variation of the milk serum proteins as a limitation of their use in evaluating the heat treatment of milk. *J. Dairy Sci.*, 38, 858–69.

Havea, P. (2006). Protein interactions in milk protein concentrate powders. *Int. Dairy J.*, 16, 415–22.

Heck, J. M. L., Olieman, C., Schennink, A., van Valenberg, H. J. F., Visker, M. H. P. W., Meuldijk, R. C. R. and van Hooijdonk, A. C. M. (2008). Estimation of variation in concentration, phosphorylation and genetic polymorphism of milk proteins using capillary zone electrophoresis. *Int. Dairy J.*, 18, 548–55.

Hernández, A. and Harte, F. M. (2009). Isolation of caseins from whey proteins by microfiltration modifying the mineral balance in skim milk. *J. Dairy Sci.*, 92, 5357–62.

Heino, A. and Huumonen, J. (2015). Method for preventing scorching of milk-based product. Patent: US 8968814B2.

Hewedi, M. M., Mulvihill, D. M. and Fox, P. F. (1985). Recovery of milk protein by ethanol precipitation. *Irish J. Food Sci. Technol.*, 9, 11–23.

Hill, R. J. and Wake, R. J. (1969). Amphiphilic nature of κ-casein as the basis for its micelle stabilizing property. *Nature*, 221, 635–9.

Hindle, E. J. and Wheelock, J. V. (1970). The release of peptides and glycopeptides by the action of heat on cow's milk. *J. Dairy Res.*, 37, 397–405.

Hinz, K., Huppertz, T. and Kelly, A. L. (2012). Susceptibility of the individual caseins in reconstituted skim milk to cross-linking by transglutaminase: influence of temperature, pH and mineral equilibria. *J. Dairy Res.*, 79, 414–21.

Holt, C. (1985a). The milk salts: their secretion, concentration and physical chemistry. In P. F. Fox (ed.), *Developments in Dairy Chemistry* – Volume 3, London: Elsevier Applied Science Publishers, pp. 143–81.

Holt, C. (1985b). The size distribution of bovine casein micelles: a review. *Food Struct.*, 4, 1–10.

Holt, C. (1992). Structure and stability of bovine casein micelles. *Adv. Prot. Chem.*, 43, 63–151.

Holt, C. (1994). The biological function of casein. In E. Taylor (ed.), *Yearbook 1994*, Ayr, Scotland: The Hannah Institute, pp. 60–8.

Holt, C. (1997). The milk salts and their interaction with casein. In P. F. Fox (ed.), *Advanced Dairy Chemistry, Volume 3, Lactose, Water, Salts and Vitamins*, 2nd ed. London: Chapman and Hall, pp. 233–56.

Holt, C. (1998). Casein micelle substructure and calcium phosphate interactions studied by Sephacryl column chromatography. *J. Dairy Sci.*, 81, 2994–3003.

Holt, C., Carver, J. A., Ecroyd, H. and Thorn, D. C. (2013). Casein and the casein micelle: their biological functions, structures and behaviour in foods, *J. Dairy Sci.*, 96, 1–20.

Holt, C. and Dalgleish, D. G. (1986). Electrophoretic and hydrodynamic properties of bovine casein micelles interpreted in terms of particles with an outer hairy layer. *J. Colloid Interface Sci.*, 114, 513–24.

Holt, C., de Kruif, C. G., Tuinier, R. and Timmins, P. A. (2003). Substructure of bovine casein micelles by small-angle X-ray and neutron scattering, *Colloids Interfaces A: Physicochem. Eng. Aspects.*, 213, 275–84.

Holt, C. and Horne, D. (1996). The hairy casein micelle: evolution of the concept and its implications for dairy technology. *Neth. Milk Dairy J.*, 50, 85–111.

Holt, C. and Sawyer, L. (1993). Caseins as rheomorphic proteins: interpretation of primary and secondary structures of α_{s1}-, β- and κ-CNs. *J. Chem. Soc. Faraday Trans.*, 89, 2683–92.

Horne, D. S. (1998). Casein interactions: casting light on the *black boxes*, the structure in dairy products. *Int. Dairy J.*, 8, 171–7.

Horne, D. S. (2002). Casein structure, self-assembly and gelation. *Curr. Opin. Colloid Interface Sci.*, 7, 456–61.

Horne, D. S. (2003). Ethanol stability. In P. F. Fox and P. L. H. McSweeney (eds.), *Advanced Dairy Chemistry – Volume 1: Proteins*, 3rd edn. New York: Kluwer Academic/Plenum Publishers, pp. 975–99.

Horne, D. S. (2006). Casein micelle structure: models and muddles. *Curr. Opin. Colloid Interface Sci.*, 11, 148–53.

Horne, D. S. (2010). Casein micelles or casein supramolecules? *J. Dairy Sci.*, 93, 3403.

Horne, D. S. (2011). Casein, micellar structure. In J. W. Fuquay, P. F. Fox and P. L. H. McSweeney (eds.), *Encyclopedia of Dairy Sciences. Volume 3*, 2nd edn. San Diego, CA: Academic Press, pp. 772–9.

Horne, D. S. (2014). Casein micelle structure and stability. In H. Singh, M. Boland and A. Thompson (eds.), *Milk Proteins: From Expression to Food*, 2nd edn. Amsterdam: Elsevier, pp. 169–200.

Horne, D. S. (2016). Rennet-induced coagulation of milk. In P. L. H McSweeney, P. F. Fox, P. Cotter and D. W. Everatt (eds.), *Cheese: Chemistry, Physics and Microbiology*, 4th edn. Oxford: Elsevier, [in press].

Horne, D. S., Anema, S., Zhu, X., Nicholas, K. R. and Singh, H. (2007). A lactational study of the composition and integrity of casein micelles from the milk of the tamer wallaby (*Macropus eugenii*). *Arch Biochem. Biophys.*, 467, 107–18.

Horne, D. S. and Banks, J. M. (2004). Rennet-induced coagulation of milk. In P. F. Fox, P. L. H McSweeney, T. M. Cogan and T. P. Guinee (eds.), *Cheese: Chemistry, Physics and Microbiology*, 3rd edn. Oxford: Elsevier, pp. 47–70.

Huppertz, T. (2013). Chemistry of the caseins. In P. L. H. McSweeney and P. F. Fox (eds.), *Advanced Dairy Chemistry, Volume 1 A – Proteins; Basic Aspects*, 4th edn. New York: Springer, pp. 135–60.

Huppertz, T., Kelly, A. L. and Fox, P. F. (2002). Effects of high pressure on constituents and properties of milk. *Int. Dairy J.*, 12, 561–72.

Huppertz, T., Vaia, B. and Smiddy, M. A (2008). Reformation of casein particles from alkaline-disrupted casein micelles. *J. Dairy Res.*, 75, 44–7.

Hurley, W. L. (2003). Immunoglobulins in mammary secretions. In P. F. Fox and P. L. H. McSweeney (eds.), *Advanced Dairy Chemistry, Volume 1 – Proteins*, 3rd edn. New York: Kluwer Academic-Plenum Publishers, pp. 421–47.

Hurley, W. L. and Thiel, P. K. (2013). Immunoglobulins in mammary secretions. In P. L. H. McSweeney and P. F. Fox (eds.), *Advanced Dairy Chemistry, Volume 1 A: Proteins: Basic Aspects*, 4th edn. New York: Springer, pp. 275–94.

IDF (1991). *Chemical Methods for Evaluation of Proteolysis in Cheese Maturation*, Bulletin 261, Brussels: International Dairy Federation.

Imafidon, G. F., Farkye, N. Y. and Spanier, A. M. (1997). Isolation, purification and alteration of some functional groups of major milk proteins: a review. *CRC Crit. Rev. Food Sci. Nutr.*, 37, 663–89.

Innocente, N., Biasutti, M. and Blecker, C. (2011). HPLC profile and dynamic surface properties of the proteose-peptone fraction from bovine milk and from whey protein concentrate. *Int. Dairy J.*, 21, 222–8.

Ji, J., Fitzpatrick, J., Cronin, K., Maguire, P., Zhang, H. and Miao, S. (2016). Rehydration behaviours of high protein dairy powders: the influence of agglomeration on wettability, dispersibility and solubility. *Food Hydrocoll.*, 58, 194–203.

Journet, M., Verite, R. and Vignon, B. (1975). L'azote non proteique du lait: factures de variation. *Lait.*, 55, 212–23.

Kang, E. J., Campbell, R. E., Bastian, E. and Drake, M. A. (2010). Invited review: annatto usage and bleaching in dairy foods. *J. Dairy Sci.*, 93, 3891–901.

Karam, M. C., Gaiani, C., Barbar, R., Hosri, C. and Scher, J. (2012). Effect of dairy powder rehydration state on gel formation during yogurt process. *J. Dairy Res.*, 79, 280–6.

Karlsson, A. O., Ipsen, R. and Ardo, Y. (2007). Observations of casein micelles in skim milk concentrate by transmission electron microscopy. *LWT – Food Sci. Technol.*, 40, 1102–7.

Kastle, J. H. and Roberts, N. (1909). The chemistry of milk. In *Milk and Relation to Public Health*, Washington, DC, USA: Bulletin No. 56, Hygienic Laboratory, Treasury Department, Government Printing Office, pp. 315–423.

Keenan, T. W. and Mather, I. H. (2006). Intracellular origin of milk fat globules and the nature of the milk fat globule membrane. In P. F. Fox and P. L. H. McSweeney (eds.), *Advanced Dairy Chemistry, Volume 2*, 3rd edn. New York: Springer, pp. 137–71.

Kehoe, J. J., Morris, E. R. and Brodkorb, A. (2007). The influence of bovine serum albumin on β-lactoglobulin denaturation, aggregation and gelation. *Food Hydrocoll.*, 21, 747–55.

Kelly, A. L. and Fox, P. F. (2016). Manufacture and properties of dairy powders. In P. L. H. McSweeney and J. A. O'Mahony (eds.), *Advanced Dairy Chemistry, Volume 1: Proteins, Part B, Applied Aspects.*, New York: Springer, pp. 1–22.

Kelly, A. L. and McSweeney, P. L. H. (2003). Indigenous proteinases in milk. In P. F. Fox and P. L. H. McSweeney (eds.), *Advanced Dairy Chemistry, Volume 1 – Proteins*, 3rd edn. New York: Kluwer Academic-Plenum Publishers, pp. 495–521.

Kelly, P. M., Kelly, J., Mehra, R., Oldfield, D. J., Raggett, E. and O'Kennedy, T. (2000). Implementation of integrated membrane processes for pilot scale development of fractionated milk components. *Lait*, 80, 139–53.

Kimura, T., Taneya, S. and Kanaya, K. (1979). Observation of internal structure of casein submicelles by means of ion beam sputtering. *Milchwissenschaft*, 34, 521–4.

Korhonen, H. (2006). Technological and health aspects of bioactive components of milk. *Int. Dairy J.*, 16, 1227–428.

Korhonen, H. and Pihlanto, A. (2006). Bioactive peptides: production and functionality. *Int. Dairy J.*, 16, 945–69.

Kristiansen, K. R., Otte, J., Ipsin, R. and Qvist, K. B. (1998). Large-scale preparation of β-lactoglobulin A and B by ultrafiltration and ion-exchange chromatography. *Int. Dairy J.*, 8, 113–18.

Kruger, M., Poulsen, R. C., Schollum, L., Haggarty, N., Ram, S. and Palmano, K. (2006). A comparison between acidic and basic protein fractions from whey or milk for reduction of bone loss in the ovariectomised rat. *Int. Dairy J.*, 16, 1149–56.

LaClair, C. E., Ney, D. M., MacLeod, E. L. and Etzel, M. R. (2009). Purification and use of glycomacropeptide for nutritional management of phenylketonuria. *J. Food Sci.*, 74, 199–206.

Laemmli, U. K. (1970). Cleavage of structural proteins during the assembly of the head of bacteriophage T4. *Nature*, 227, 196–204.

Le, T. L., Bhandari, B. R. and Deeth, H. C. (2011a). Chemical and physical changes in milk protein concentrate (MPC80) powder during storage. *J. Agric. Food Chem.*, 59, 5465–73.

Le, T. L., Bhandari, B., Holland, J. W. and Deeth, H. C. (2011b). Maillard reaction and protein cross-linking in relation to the solubility of milk powders. *J. Agric. Food Chem.*, 59, 12473–9.

Le, T. L., Holland, J. W., Bhandari, B. R., Alewood, P. F. and Deeth, H. C. (2013). Direct evidence for the role of Maillard reaction products in protein cross-linking in milk powder during storage. *Int. Dairy J.*, 31, 83–91.

Le Bars, D. and Gripon, J.-C. (1989). Specificity of plasmin towards bovine α_{s2}-casein. *J. Dairy Res.*, 56, 817–21.

Le Bars, D. and Gripon, J.-C. (1993). Hydrolysis of α_{s1}-casein by bovine plasmin. *Lait.*, 74, 337–44.

Lefebvre-Cases, E., Gastaldi, E. and de la Fuente, B.T. (1998a). Influence of chemical agents on interactions in dairy products: effect of SDS on casein micelles. *Colloids Surf. B.*, 11, 281–5.

Lefebvre-Cases, E., Gastaldi, E., Vidal, V., Marchesseau, S., Lagaude, A., Cuq, J.-L. and Tarodo de la Fuente, B. (1998b). Identification of interactions among casein gels using dissociating chemical agents. *J. Dairy Sci.*, 81, 932–8.

Le Parc, A., Leonil, J. and Chanat, E. (2010). α_{s1}-casein, which is essential for efficient ER-to-Golgi casein transport, is also present in a tightly bound membrane-associated form. *BMC Cell Biol.*, 11, 65.

Lisková, K., Kelly, A. L., O'Brien, N. and Brodkorb, A. (2010). Effect of denaturation of α-lactalbumin on the formation of BAMLET (Bovine α-Lactalbumin Made Lethal to Tumor Cells). *J. Agric. Food Chem.*, 58, 4421–7.

Listiyani, M. A. D., Campbell, R. E., Miracle, R. E., Dean, L. O. and Drake, M. A. (2011). Influence of bleaching on flavor of 34% whey protein concentrate and residual benzoic acid concentration in dried whey proteins. *J. Dairy Sci.*, 94, 4347–59.

Lonnerdal, B. (2003). Lactoferrin. In P. F. Fox and P. L. H. McSweeney (eds.), *Advanced Dairy Chemistry, Volume 1, Proteins*, 3rd edn. New York: Kluwer Academic/Plenum Publishers, pp. 449–66.

Lonnerdal, B. and Suzuki, Y. A. (2013). Lactoferrin. In P. L. H. McSweeney and P. F. Fox (eds.), *Advanced Dairy Chemistry, Volume 1 A: Proteins: Basic Aspects*, 4th edn. New York: Springer, pp. 295–315.

Lopez, C. (2011). Milk fat globules enveloped by their biological membrane: unique colloidal assemblies with a specific composition and structure. *Curr. Opin. Colloid Interface Sci.*, 16, 391–404.

Lozano, J. M., Giraldo, G. I. and Romero, C. M. (2008). An improved method for the isolation of β-lactoglobulin. *Int. Dairy J.*, 18, 55–63.

Lu, Y., McMahon, D. J., Metzger, L. E., Kommineni, A. and Vollmer, A. H. (2015). Solubilization of rehydrated frozen highly concentrated micellar casein for use in liquid food applications. *J. Dairy Sci.*, 98, 5917–30.

Lucey, J. A., (2002). Formation and physical properties of milk protein gels. *J. Dairy Sci.*, 85, 281–94.

Lucey, J. A. (2011). Rennet-induced coagulation of milk. In J. W. Fuquay, P. F. Fox and P. L. H. McSweeney (eds.), *Encyclopedia of Dairy Sciences. Volume 1*, 2nd edn. San Diego, CA: Academic Press, pp. 579–84.

Lucey, J. A. and Horne, D. S. (2009). Milk salts: technological significance. In P. L. H. McSweeney and P. F. Fox (eds.), *Advanced Dairy Chemistry: Volume. 3. Lactose, Water, Salts and Minor Constituents*, 3rd ed. New York: Springer, pp. 351–89.

Lucey, J. A., Zhu, D. and Damodaran, S. (2009). Production of protein-polysaccharide conjugates. Patent Application US 20090311407 A1.

Mann, M., Hendrickson, R. C. and Pandey, A. (2001). Analysis of proteins and proteomes by mass spectrometry. *Ann. Rev. Biochem.*, 70, 437–73.

Manso, M. A., Leonil, J., Jan, G. and Gagnaire, V. (2005). Application of proteomics to the characterisation of milk and dairy products. *Int. Dairy J.*, 15, 845–55.

Marchin, S., Putaux, J.-L., Pignon, F. and Leonil, J. (2007). Effects of the environmental factors on the casein micelle structure studied by cryo transmission electron microscopy and small-angle X-ray scattering/ultrasmall-angle X-ray scattering. *J. Chem. Phys.*, 126, 045101-1–045101-10.

Marella, C., Muthukumarappan, K. and Metzger, L. E. (2011). Evaluation of commercially available, wide-pore ultrafiltration membranes for production of α-lactalbumin-enriched whey protein concentrate. *J. Dairy Sci.*, 94, 1165–75.

Marnila, P. and Korhonen, H. (2011). Immunoglobulins. In J. W. Fuquay, P. F. Fox and P. L. H. McSweeney (eds.), *Encyclopedia of Dairy Sciences, Volume 3*, 2nd edn. Oxford: Academic Press, pp. 807–15.

Martin, A. H., Goff, H. D., Smith, A. Dalgleish, D. G. (2006). Immobilization of casein micelles for probing their structure and interactions with polysaccharides using scanning electron microscopy (SEM). *Food Hydrocoll.*, 20, 817–24.

Martin, P., Cebo, C. and Miranda, G. (2011). Interspecies comparison of milk proteins: quantitative variability and molecular diversity. In J. W. Fuquay, P. F. Fox and P. L. H. McSweeney (eds.), *Encyclopedia of Dairy Sciences, Volume 3*, 2nd edn. Oxford: Academic Press, pp. 821–42.

Martin, P., Cebo, C. and Miranda, G. (2013). Interspecies comparison of milk proteins: quantitative variability and molecular diversity. In P. L. H. McSweeney and P. F. Fox (eds.), *Advanced Dairy Chemistry, Volume 1 A: Proteins: Basic Aspects*, 4th ed. New York: Springer Science+Business Media, pp. 387–430.

Martin, G. J. O., Williams, R. P. W. and Dunstan, D. E. (2007). Comparison of casein micelles in raw and reconstituted skim milk. *J. Dairy Sci.*, 90, 4543–51.

Mather, I. H. (2000). A review and proposed nomenclature for major proteins of the milk-fat globule membrane. *J. Dairy Sci.*, 83, 203–47.

McGann, T. C. A. and Fox, P. F. (1974). Physicochemical properties of casein micelles reformed from urea-treated milk. *J. Dairy Res.*, 41, 45–53.

McGann, T. C. A., Kearney, R. D. and Donnelly, W. J. (1979). Developments in column chromatography for the separation and characterization of casein micelles. *J. Dairy Res.*, 46, 307–11.

McGrath, B. A. (2014). *Generation and characterisation of biologically active milk-derived protein and peptide fractions*. PhD Thesis, University College Cork.

McGrath, B. A., Fox, P. F., McSweeney, P. L. H. and Kelly, A. L. (2016). Composition and properties of bovine colostrum: a review. *Dairy Sci. Tech.*, 96, 133–58.

McKenzie, H. A. (1971). Whole casein: isolation, properties, and zone electrophoresis. In H. A. McKenzie (ed.), *Milk Proteins: Chemistry and Molecular Biology*, vol. 2. New York: Academic Press, pp. 87–116.

McKenzie, H. A. and White, F. H. (1991). Lysozyme and α-lactalbumin: structure, function, and interrelationships. *Adv. Prot. Chem.*, 41, 173–315.

McMahon, D. J. and McManus, W. R. (1998). Rethinking casein micelle structure using electron microscopy. *J. Dairy Sci.*, 81, 2985–93.

McMahon, D. J. and Oommen, B. S. (2008). Supramolecular structure of the casein micelle. *J. Dairy Sci.*, 91, 1709–21.

McMahon, R. J., Wang, B., Brand-Miller, J. and Rumsay, S. C. (2006). Method of improving learning and memory in mammals. Patent application, US20060247153A1.

McSweeney, P. L. H., Olson, N. F., Fox, P. F., Healy, A. and Højrup, P. (1993). Proteolytic specificity of plasmin on bovine α_{s1}-casein. *Food Biotechnol.*, 7, 143–58.

Mills, S., Ross, R. P., Hill, C., Fitzgerald, G. F. and Stanton, C. (2011). Milk intelligence: mining milk for bioactive substances associated with human health. *Int. Dairy J.*, 21, 377–401.

Minkiewicz, P., Dziuba, J. and Muzińska, B. (1993). The contribution of N-acetylneuraminic acid in the stabilization of micellar casein. *Polish J. Food Nutr. Sci.*, 2/43, 39–48.

Mizuno, R. and Lucey, J. A. (2005). Effects of emulsifying salts on the turbidity and calcium-phosphate–protein interactions in casein micelles. *J. Dairy Sci.*, 88, 3070–8.

Moeller, J., Madsen, B. and Lauridsen, D. (2014). Carotenoid coloring composition. Patent application US 20140113027 A1.

Moitzi, C., Portnaya, I., Glatter, O., Ramon, O., Danino, D. (2008). Effect of temperature on self-assembly of bovine β-casein above and below isoelectric pH. Structural analysis by cryogenic-transmission electron microscopy and small-angle X-ray scattering. *Langmuir*, 24, 3020–9.

Moon, T. W., Peng, I. C. and Lonergan, D. A. (1988). Chemical properties of cryocasein. *J. Food Sci.*, 53, 1687–93.

Moon, T. W., Peng, I. C. and Lonergan, D. A. (1989). Functional properties of cryocasein. *J. Dairy Sci.*, 72, 815–28.

Morr, C. V. (1967). Effect of oxalate and urea upon ultracentrifugation properties of raw and heated skim milk casein micelles. *J. Dairy Sci.*, 50, 1744–51.

Muir, D. D. and Sweetsur, A. W. M. (1976). The influence of naturally occurring levels of urea on the heat stability of bulk milk. *J. Dairy Res.*, 43, 495–9.

Muller, A., Daufin, G. and Chaufer, B. (1999). Ultrafiltration modes of operation for the separation of α-lactalbumin from acid casein whey. *J. Membrane Sci.*, 153, 9–21.

Mulvihill, D. M. and Ennis, M. P. (2003). Functional milk proteins: production and utilisation. In P. F. Fox and P. L. H. McSweeney (eds.), *Advanced Dairy Chemistry – Volume 1: Proteins*, 3rd edn. New York: Kluwer Academic/Plenum Publishers, pp. 1175–228.

Needs, E. C., Stenning, R. A., Gill, A. L., Ferragut, V. and Rich, G. T. (2000). High-pressure treatment of milk: effects on casein micelle structure and on enzymic coagulation. *J. Dairy Res.*, 67, 31–42.

Ney, D. M., Gleason, S. T., van Calcar, S. C., MacLeod, E. L., Nelson, K. L., Etzel, M. R., Rice, G. M. and Wolff, J. A. (2009). Nutritional management of PKU with glycomacropeptide from cheese whey. *J. Inherit. Metab. Dis.*, 32, 32–9.

Nitta, K. and Sugai, S. (1989). The evolution of lysozyme and α-lactalbumin. *Eur. J. Biochem.*, 182, 111–18.

Nitta, K., Tsuge, H., Sugai, S. and Shimazaki, K. (1987). The calcium-binding properties of equine lysozyme. *FEBS Lett.*, 223, 405–8.

O'Callaghan, D. M., O'Mahony, J. A., Ramanujam, K. S. and Burgher, A. M. (2011). Dehydrated dairy products: infant formulae. In J. W. Fuquay, P. F. Fox and P. L. H. McSweeney (eds.), *Encyclopedia of Dairy Sciences*, 2nd edn., Vol. 2. Oxford: Academic Press, pp. 135–45.

O'Connell, J. E. and de Kruif, C. G. (2003). β-casein micelles; cross-linking with transglutaminase. *Colloids Surf. A Physicochem. Eng. Asp.*, 216, 75–81.

O'Connell, J. E. and Fox, P. F. (2000). Heat stability of buttermilk. *J. Dairy Sci.*, 83, 1728–32.

O'Connell, J. E., Kelly, A. L., Fox, P. F. and de Kruif, C. G. (2001a). Ethanol-dependent heat-induced dissociation of casein micelles. *J. Agric. Food Chem.*, 49, 4420–3.
O'Connell, J. E., Kelly, A. L., Fox, P. F. and de Kruif, C. G. (2001b). Mechanism for the ethanol-dependent heat-induced dissociation of casein micelles. *J. Agric. Food Chem.*, 49, 4424–8.
O'Donnell, R., Holland, J. W., Deeth, H. C. and Alewood, P. (2004). Milk proteomics. *Int. Dairy J.*, 14, 1013–23.
O'Flaherty, F. (1997). *Characterization of Some Minor Caseins and Proteose Peptones of Bovine Milk.* MSc Thesis, National University of Ireland, Cork.
Oldfield, D. J., Taylor, M. W. and Singh, H. (2005). Effect of preheating and other process parameters on whey protein reactions during skim milk powder manufacture. *Int. Dairy J.*, 15, 501–11.
O'Mahony, J. A. and Fox, P. F. (2013). Milk proteins: introduction and historical aspects. In P. L. H. McSweeney and P. F. Fox (eds.), *Advanced Dairy Chemistry, Volume 1 A: Proteins: Basic Aspects*, 4th edn. New York: Springer, pp. 43–85.
O'Mahony, J. A. and Fox, P. F. (2014). Milk: an overview. In M. Boland, H. Singh and A. Thompson (eds.), *Milk Proteins: From Expression to Food*, 2nd edn. London: Academic Press (Elsevier), pp. 19–73.
O'Mahony, J. A., Fox, P. F. and Kelly, A. L. (2013). Indigenous enzymes of milk. In P. L. H. McSweeney and P. F. Fox (eds.), *Advanced Dairy Chemistry, Volume 1 A: Proteins: Basic Aspects*, 4th edn. New York: Springer, pp. 337–85.
O'Mahony, J. A., and Tuohy, J. J. (2013). Further applications of membrane filtration in dairy processing. In A. Y. Tamine (ed.), *Membrane Processing: Dairy and Beverage Applications.* Oxford: Wiley-Blackwell, pp. 225–61.
Ono, T. and Obata, T. (1989). A model for the assembly of bovine casein micelles from F2 and F3 subunits. *J. Dairy Res.*, 56, 453–61.
Ono, T., Yoshida, M., Tanaami, H. and Ohkosi, H. (1999). Changes in casein micelle size induced by heating. *Int. Dairy J.*, 9, 405–6.
O'Sullivan, M. M., Kelly, A. L. and Fox, P. F. (2002a). Influence of transglutaminase treatment on some physico-chemical properties of milk. *J. Dairy Res.*, 69, 433–42.
O'Sullivan, M. M., Kelly, A. L. and Fox, P. F. (2002b). Influence of transglutaminase on the heat stability of milk. *J. Dairy Sci.*, 85, 1–7.
O'Sullivan, M. M. and Mulvihill, D. M. (2001). Influence of some physico-chemical characteristics of commercial rennet caseins on the performance of the casein in Mozzarella cheese analogue manufacture. *Int. Dairy J.*, 11, 153–63.
Pagl, R., Aurich, J. E., Müller-Schlösser, Kankofer, M. and Aurich, C. (2006). Comparison of an extender containing defined milk protein fractions with a skim milk-based extender for storage of equine semen at 5°C. *Theriogenology*, 66, 1115–22.
Pakkanen, R. and Aalto, J. (1997). Growth factors and antimicrobial factors in bovine colostrums. *Int. Dairy J.*, 7, 285–97.
Palmer, A. H. (1934). The preparation of a crystalline globulin from the albumin fraction of cow's milk. *J. Biol. Chem.*, 104, 359–72.
Park, S.-Y., Niki, R. and Sano, Y. (1999). Size effects of casein micelles on rennet gels in the presence of β-lactoglobulin. *Int. Dairy J.*, 9, 379–80.
Parry, R. M., Jr. and Carroll, R. J. (1969). Location of κ-casein on milk micelles. *Biochim. Biophys. Acta.*, 194, 138–50.
Patton, S. (1999). Some practical implications of the milk mucins. *J. Dairy Sci.*, 82, 1115–17.
Patton, S., Carson, G. S., Hiraiwa, M., O'Brien, J. S. and Sano, A. (1997). Prosaposin, a neurotrophic factor: presence and properties in milk. *J. Dairy Sci.*, 80, 264–72.
Payens, T. A. J. (1966). Association of caseins and their possible relation to structure of the casein micelle. *J. Dairy Sci.*, 49, 1317–24.
Pearce, R. J. (1983). Thermal separation of β-lactoglobulin and α-lactalbumin in bovine Cheddar cheese whey. *Austr. J. Dairy Technol.*, 38, 144–9.
Pearce, R. J. (1995). Enriched whey protein fractions and method for the production thereof. Patent, US005455331A.

Pellegrino, L., Rosi, V., D'Incecco, P., Stroppa, A. and Hogenboom, J. A. (2015). Changes in the soluble nitrogen fraction of milk throughout PDO Grana-Padono cheese-making. *Int. Dairy J.*, 47, 128–35.

Pepper, L. (1972). Casein interactions as studied by gel chromatography and ultracentrifugation. *Biochim. Biophys. Acta*, 278, 147–54.

Pepper, L. and Farrell, H. M. Jr. (1982). Interactions leading to formation of casein submicelles. *J. Dairy Sci.*, 65, 2259–66.

Peterson, R. F. (1963). High resolution of milk proteins obtained by gel electrophoresis. *J. Dairy Sci.*, 46, 1136–9.

Pettersson, J., Mossberg, A. K. and Svanborg, C. (2006). α-Lactalbumin species variation, HAMLET formation and tumor cell death. *Biochem. Biophys. Res. Commun.*, 345, 260–70.

Phadungath, C. (2005). Casein micelle structure: a concise review. *Songklanakarin J. Sci. Technol.*, 27, 201–12.

Philippe, M., Le Graët, Y. and Gaucheron, F. (2005). The effect of different divalent cations on the physicochemical characteristics of casein micelles. *Food Chem.*, 90, 673–83.

Pierre, A., Fauquant, J., Le Graet, Y., Piot, M. and Maubois, J.-L. (1992). Préparation de phosphocaséinate natif par microfiltration sur membrane. *Lait*, 72, 461–74.

Pouliot, Y. (2008). Membrane processes in dairy technology – From a simple idea to a worldwide panacea. *Int. Dairy J.*, 18, 735–40.

Pouliot, M. and Pouliot, Y. (1996). On the conventional cross-flow microfiltration of skim milk for the production of native phosphocaseinate. *Int. Dairy J.*, 6, 105–11.

Pyne, G. T. (1934). The colloidal phosphate of milk. *Biochem. J.*, 28, 940–8.

Pyne, G. T. (1962). Some aspects of the physical chemistry of the salts in milk. *J. Dairy Res.*, 29, 101–30.

Pyne, G. T. and McGann, T. C. A. (1960). The colloidal phosphate of milk. *J. Dairy Res.*, 27, 9–17.

Qi, P. X. (2007). Studies on casein micelle structure: the past and the present. *Lait*, 87, 363–83.

Qi, P. X. (2009). Studies on the biological function and structure of casein micelles, and future implications. In *Dairy-derived Ingredients, Food Science Technology Nutrition*, Woodward Publishers, 182, 147–69.

Raikos, V. and Dassios, T. (2014). Health-promoting properties of bioactive peptides derived from milk proteins in infant food: a review. *Dairy Sci. Tech*, 94, 91–101.

Ramchandran, L. and Vasiljevic, T. (2013). Whey processing. In A. Y. Tamine (ed.), *Membrane Processing: Dairy and Beverage Applications*, Oxford: Wiley-Blackwell, pp. 193–207.

Ramet, J.-P. (2001). FAO Animal Production and Health Paper 113: The technology of making cheese from camel milk (*Camelus dromedaries*). Rome: FAO.

Rasmussen, L. K., Johnsen, L. B., Tsiora, A., Sorensen, E. S., Thomsen, J. K., Nielsen, N. C., Jakobsen, H. J. and Petersen, T. E. (1999). Disulphide-linked caseins and casein micelles. *Int. Dairy J.*, 9, 215–18.

Reid, M., O'Donovan, M., Elliot, C. T., Bailey, J. S., Watson, C. J., Lalor, S. T. J., Corrigan, B., Fenelon, M. A. and Lewis, E. (2015). The effect of dietary crude protein and phosphorus on grass-fed dairy cow production, nutrient status, and milk heat stability. *J. Dairy Sci.*, 98, 517–31.

Rigo, J., Boehm, G., Georgi, G., Jelinek, J., Nyambugabo, K., Sawatzki, G. and Studzinski, F. (2001). An infant formula free of glycomacropeptide prevents hyperthreoninemia in formula-fed preterm infants. *J. Pediatr. Gastroenterol. Nutr.*, 32, 127–30.

Rizvi, S. S. H. and Brandsma, R. L. (2002). Microfiltration of skim milk for cheesemaking and whey proteins. US Patent 6,485,762 B1.

Roach, A. and Harte, F. (2008). Disruption and sedimentation of casein micelles and casein micelle isolates under high-pressure homogenisation. *Innovative Food Sci. Emerg. Technol.*, 9, 1–8.

Robson, E., Horne, D. S. and Dalgleish, D. G. (1985). Chromatography of casein micelles and particles from homogenized milk on columns of controlled pore glass and large-pore silica. *J. Dairy Res.*, 52, 391–400.

Rollema, H. S. (1992). Casein association and micelle formation. In P. F. Fox (ed.), *Advanced Dairy Chemistry, vol. 1, Proteins*, London: Elsevier Applied Science, pp. 111–40.

Rose, D. (1968). Relation between micellar and serum casein in bovine milk. *J. Dairy Sci.*, 51, 1897–902.
Rose, D. (1969). A proposed model of micelle structure in bovine milk. *Dairy Sci. Abstr.*, 31, 171–5.
Roufik, S., Paquin, P. and Britten, M. (2005). Use of high-performance size exclusion chromatography to characterise protein aggregation in commercial whey protein concentrates. *Int. Dairy J.*, 15, 231–41.
Sadek, C., Pauchard, L., Schuck, P., Fallourd, Y., Pradeau, N., Le Floch-Fouéré, C. and Jeantet, R. (2015). Mechanical properties of milk protein skin layers after drying: understanding the mechanisms of particle formation from whey protein isolate and native phosphocaseinate. *Food Hydrocoll.*, 48, 8–16.
Saidi, S., Rival-Gervier, S., Daniel-Carlier, N., Thepot, D., Morgenthaler, C., Viglietta, C., Prince, S., Passet, B., Houdebine, L.-M. and Jolivet, G. (2007). Distal control of the pig whey acidic protein (WAP) locus in transgenic mice. *Gene*, 401, 97–107.
Salami, S., Rondeau-Mouro, C., van Duynhoven, J. and Mariette, F. (2013). PFG-NMR self-diffusion in casein dispersions: effects of probe size and protein aggregate size. *Food Hydrocoll.*, 31, 248–55.
Sandra, S. and Dalgleish, D. G. (2005). Effects of ultra-high-pressure homogenisation and heating on structural properties of casein micelles in reconstituted skim milk powder. *Int. Dairy J.*, 15, 1095–104.
Sawyer, L. (2003). β-Lactoglobulin. In P. F. Fox and P. L. H. McSweeney (eds.), *Advanced Dairy Chemistry – Volume 1: Proteins*, 3rd edn. New York: Kluwer Academic/Plenum Publishers, pp. 319–86.
Sawyer, L. (2013). β-Lactoglobulin. In P. L. H. McSweeney and P. F. Fox (eds.), *Advanced Dairy Chemistry, Volume 1 A: Proteins: Basic Aspects*, 4th edn. New York: Springer, pp. 211–59.
Schmidt, D. G. (1980). Colloidal aspects of casein. *Neth. Milk Dairy J.*, 34, 42–64.
Schmidt, D. G. (1982). Association of caseins and casein micelle structure. In P. F. Fox (ed.), *Developments in Dairy Chemistry, vol. 1, Proteins*. London: Applied Science Publishers, pp. 63–110.
Schrezenmeir, J., Korhonen, H., Williams, C. M., Gill, H. S. and Shah, N. P. (2000). Beneficial natural bioactive substances in milk and colostrums. *Br. J. Nutr*, 84 (Suppl. 1), S1–S166.
Scollard, P. G., Beresford, T. P., Needs, E. C., Murphy, P. M. and Kelly, A. L. (2000). Plasmin activity, β-lactoglobulin denaturation and proteolysis in high pressure treated milk. *Int. Dairy J.*, 10, 835–41.
Shalabi, S. I. and Fox, P. F. (1987). Electrophoretic analysis of cheese: comparison of methods. *Irish J. Food Sci. Technol.*, 11, 135–51.
Simpson, K. J. and Nicholas, K. R. (2002) Comparative biology of whey proteins. *J. Mammary Gland Biol. Neoplasia.*, 7, 313–26.
Singh, H. (2004). Heat stability of milk. *Int. J. Dairy Technol.*, 57, 111–19.
Singh, H. (2006). The milk fat globule membrane – a biophysical system for food applications. *Curr. Opin. Colloid Interface Sci.*, 11, 154–63.
Slattery, C. W. (1976). Review: casein micelle structure; an examination of models. *J. Dairy Sci.*, 59, 1547–56.
Slattery, C. W. and Evard, R. (1973). A model for the formation and structure of casein micelles from subunits of variable composition. *Biochim. Biophys. Acta.*, 317, 529–38.
Smiddy, M. A., Martin, J.-E. G. H., Kelly, A. L., de Kruif, C. G. and Huppertz, T. (2006). Stability of casein micelles cross-linked by transglutaminase. *J. Dairy Sci.*, 89, 1906–14.
Smith, T. J., Li, X. E. and Drake, M. A. (2014). Norbixin and bixin partitioning in Cheddar cheese and whey. *J. Dairy Sci.*, 97, 3321–7.
Smithers, G. W. (2008). Whey and whey proteins – from 'gutter-to-gold'. *Int. Dairy J.*, 18, 695–704.
Spagnuolo, P. A., Dalgleish, D. G., Goff, H. D. and Morris, E. R. (2005). Kappa-carrageenan interactions in systems containing casein micelles and polysaccharide stabilizers. *Food Hydrocoll.*, 19, 371–7.
Spitsberg, V. L. (2005). Invited review: bovine milk fat globule membrane as a potential nutraceutical. *J. Dairy Sci.*, 88, 2289–94.

Stack, F. M., Hennessy, M., Mulvihill, D. and O'Kennedy, B. T. (1998). Process for the fractionation of whey constituents. United States Patent US5747647.

Stagsted, J. (2006). Absence of both glutathione peroxidase activity and glutathione in bovine milk. *Int. Dairy J.*, 16, 662–8.

Strange, E. D., Malin, E. L., van Hekken, D. L. and Basch, J. J. (1992). Chromatographic and electrophoretic methods used for analysis of milk proteins. *J. Chromatogr.*, 624, 81–102.

Svensson, M., Hakansson, A., Mossberg, A. K., Linse, S. and Svanborg, C. (2000). Conversion of alpha-lactalbumin to a protein inducing apoptosis. *Proc. National Acad. Sci.* USA, 97, 4221–6.

Swaisgood, H. E. (1975). *Methods of Gel Electrophoresis of Milk Proteins*. Champaign, IL: American Dairy Science Association.

Swaisgood, H. E. (1982). Chemistry of milk proteins. In P. F. Fox (ed.), *Developments in Dairy Chemistry, vol. 1, Proteins*, London: Applied Science Publishers, pp. 1–59.

Swaisgood, H. E. (2003). Chemistry of the caseins. In P. F. Fox and P. L. H. McSweeney (eds.), *Advanced Dairy Chemistry*, Vol. 1, 3rd edn. New York: Kluwer Academic/Plenum Publishers, pp. 139–201.

Thomas, T. D. and Pritchard, G. G. (1987). Proteolytic enzymes of dairy starter cultures. *FEMS Microbiol. Lett.*, 46, 245–68.

Tomita, M., Wakabayashi, H., Shin, K., Yamauchi, K., Yaeshima, T. and Iwatsuki, K. (2009). Twenty-five years of research on bovine lactoferrin applications. *Biochimie*, 91, 52–7.

Tsioulpas, A., Lewis, M. J. and Grandison, A. S. (2007). Effect of minerals on casein micelle stability in cows' milk. *J. Dairy Res.*, 74, 167–73.

Tuinier, R. and de Kruif, C. G. (2002). Stability of casein micelles in milk. *J. Chem Phys.*, 117, 1290–5.

Uniacke, T. and Fox, P. F. (2011). Milk of primates. In J. W. Fuquay, P. F. Fox and P. L. H. McSweeney (eds.), *Encyclopedia of Dairy Sciences, Volume 3*, 2nd edn., San Diego, CA: Academic Press, pp. 613–31.

Vaia, B., Smiddy, M. A., Kelly, A. L. and Huppertz, T. (2006). Solvent-mediated disruption of bovine casein micelles at alkaline pH. *J. Agr. Food Chem.* 54, 8288–93.

van Calcar, S. C., MacLeod, E. L., Gleason, S. T., Etzel, M. R., Clayton, M. K., Wolff, J. A., and Ney, D. M. (2009). Improved nutritional management of phenylketonuria by using a diet containing glycomacropeptide compared with amino acids. *Am. J. Clin. Nutr.*, 89, 1068–77.

van Calcar, S. C., and Ney, D. M. (2012). Food products made with glycomacropeptide, a low-phenylalanine whey protein, provide a new alternative to amino acid-based medical foods for nutrition management of phenylketonuria. *J. Acad. Nutr. Diet.*, 112, 1201–10.

van Hekken, D. L. and Thompson, M. P. (1992). Application of PhastSystem to the resolution of bovine milk proteins on urea-polyacrylamide gel electrophoresis. *J. Dairy Sci.*, 75, 1204–10.

Vanderghem, D., Danthine, S., Blecker, C. and Deroanne, C. (2007). Effect of proteose-peptone addition on some physico-chemical characteristics of recombined dairy creams. *Int. Dairy J.*, 17, 889–95.

Villumsen, N. S., Jensen, H. B., Le, T. T. T., Møller, H. S., Nordvang, R. T., Hammershøj, M., Nielsen, L., R., Nielsen, S. B., Søresen, J., Hammershøj, M. and Larsen, L. B. (2015). Self-assembly of caseinomacropeptide as a potential key mechanism in the formation of visible storage induced aggregates in acidic whey protein isolate dispersions. *Int. Dairy J.*, 49, 8–15.

Visser, H. (1992). A new casein micelle model and its consequences for pH and temperature effects on the properties of milk. In H. Visser (ed.), *Protein Interactions*, Weinheim, Germany: VCH, pp. 135–65.

Visser, S., Noorman, H. J., Slangen, C. J. and Rollema, H. S. (1989a). Action of plasmin on bovine β-casein in a membrane reactor. *J. Dairy Res.*, 56, 323–33.

Visser, S., Slangen, C. J., Alting, A. C. and Vreeman, H. J. (1989b). Specificity of bovine plasmin in its action on α_{s2}-casein. *Milchwissenschaft*, 44, 335–9.

Wake, R. G. and Baldwin, R. L. (1961). Analysis of casein fractions by zone electrophoresis in concentrated urea. *Biochim. Biophys. Acta.*, 47, 225–39.

Walstra, P. (1990). On the stability of casein micelles. *J. Dairy Sci.*, 73, 1965–79.

Walstra, P. (1999). Casein sub-micelles: do they exist? *Int. Dairy J.*, 9, 189–92.

Walstra, P. and Jenness, R. (1984). *Dairy Chemistry and Physics*. New York: John Wiley and Sons.

Wang, P., Liu, H., Wen, P., Zhang, H., Guo, H. and Ren, F. (2013). The composition, size and hydration of yak casein micelles. *Int. Dairy J.*, 31, 107–10.

Wang, T. and Lucey, J. A. (2003). Use of multi-angle laser light scattering and size-exclusion chromatography to characterise the molecular weight and types of aggregates present in commercial whey protein products. *J. Dairy Sci.*, 86, 3090–101.

Ward, R. E., German, J. B. and Corredig, M. (2006). Composition, applications, fractionation, technological and nutritional significance of milk fat globule material. In P. F. Fox and P. L. H. McSweeney (eds.), *Advanced Dairy Chemistry, Volume 2*, 3rd edn. New York: Springer, pp. 213–44.

Waugh, D. F., Creamer, L. K., Slattery, C. W. and Dresdner, G. W. (1970). Core polymers of casein micelles. *Biochemistry*, 9, 786–95.

Waugh, D. F., and von Hippel, P. H. (1956). κ-Casein and the stabilization of casein micelles. *J. Am. Chem. Soc.*, 78, 4576–82.

Webb, G. W., Dean, M. M., Humes, R. A. and Heywood, J. S. (2009). A comparison of the ability of three commercially available diluents to maintain the motility of cold stored stallion semen. *J. Equine Vet. Sci.*, 29, 229–32.

White, J. C. D. and Davies, D. T. (1966). The stability of milk protein to heat: III. Objective measurement of heat stability of milk. *J. Dairy Res.*, 33, 93–102.

Wu, C. (2001). Whey treatment process for achieving high concentration of α-lactalbumin. Patent, US6312755B1.

Wynn, P. C. and Sheehy, P. A. (2013). Milk proteins, including growth factors. In P. L. H. McSweeney and P. F. Fox (eds.), *Advanced Dairy Chemistry, Volume 1 A: Proteins: Basic Aspects*, 4th edn. New York: Springer, pp. 317–35.

Wynn, P. C., Morgan, A. J. and Sheehy, P. A. (2011). Minor proteins, bovine serum albumin, vitamin-binding proteins. In J. W. Fuquay, P. F. Fox and P. L. H. McSweeney (eds.), *Encyclopedia of Dairy Sciences, Volume 3*, 2nd edn. Oxford: Academic Press, pp. 795–800.

Yamada, M., Murakami, K., Wallingford, J. C. and Yuki, Y. (2002). Identification of low-abundance proteins of bovine colostral and mature milk using two-dimensional electrophoresis followed by microsequencing and mass spectrometry. *Electrophoresis*, 23, 1153–60.

Yazdi, S. Y., Corredig, M. and Dalgleish, D. G. (2014). Studying the structure of β-casein-depleted bovine casein micelles using electron microscopy and fluorescent polyphenols. *Food Hydrocoll.*, 42, 171–7.

Zhu, D. and Damodaran, S. (2012). Short communication: Annatto in Cheddar cheese-derived whey protein concentrate is primarily associated with milk fat globule membrane. *J. Dairy Sci.*, 95, 614–17.

Chapter 2

Bioactive components in cow's milk

Young W. Park, Fort Valley State University, USA

1 Introduction

Milk is a complex, multicomponent lacteal fluid secreted from the mammary glands. In recent decades, understanding of the role of milk has expanded beyond viewing it as a source of nutrients to neonates of mammalian species, as well as for growth of children and nourishment of adult humans (Park, 2009a). Apart from nutritional values of milk, milk-borne biologically active compounds such as proteins, peptides, lactoferrin (LF), enzymes, lipids and carbohydrates have been shown to be increasingly important for physiological and biochemical functions that have crucial impacts on human metabolism and health (Schanbacher et al., 1998; Korhonen and Pihlanto- Leppälä, 2004; Gobbetti et al., 2007; Park, 2009a; Phelan and Kerins, 2011; Sah et al., 2015). Various recent studies have demonstrated that milk furnishes a broad range of biologically active compounds that guard neonates and adults against pathogens and illnesses, such as immunoglobulins (Igs), antibacterial peptides, antimicrobial proteins, oligosaccharides and lipids, besides many other components at low concentrations, so-called 'minor' components, but with considerable potential benefits (Park, 2009a; Phelan and Kerins, 2011; Korhonen and Marnila, 2013; Elfahri et al., 2016).

Numerous studies have shown that milk contains an array of bioactivities that extend the range of influence of mother over young beyond nutrition (Gobbetti et al., 2007). Milk contains a wide range of proteins that provide protection against enteropathogens in the gut, as well as being essential for the manufacture and characteristic nature of certain dairy products (Korhonen and Pihlanto-Leppälä, 2004; Korhonen and Marnila, 2013). Milk

http://dx.doi.org/10.19103/AS.2016.0005.04

peptides are in latent or inactive state within protein molecules, but they can be released during enzymatic hydrolysis by digestion and/or fermentation, and become physiologically and biologically active (Park, 2009a; Korhonen and Marnila, 2013; Sah et al., 2015; Elfahri et al., 2016). A simple schematic representation of major bioactive compounds derived from milk is presented in Fig. 1.

In recent decades, major progress has been made in the science, technology and commercial applications of the many bioactive components particularly in bovine milk and colostrum (Korhonen and Marnila, 2013). Chromatographic and membrane separation techniques have been developed to fractionate and purify these components on an industrial scale (Smithers et al., 1996; Kulozik et al., 2003; Pouliot et al., 2006; Korhonen and Pihlanto, 2007). Production of bioactive milk ingredients by fractionation has emerged as a lucrative new sector for the dairy industry. Many of these bioactive compounds are currently being exploited for both dairy and non-dairy food formulations as well as pharmaceuticals (Korhonen et al., 1998; German et al., 2002; Playne et al., 2003; Krissansen, 2007). The dairy industry has played a pivotal role in the development of functional foods and has already commercialized bioactive products that enhance the immune system, reduce elevated blood pressure, combat gastrointestinal infections, exert anticancer and antioxidant activities, help control body weight and prevent osteoporosis (Zemel, 2004; FitzGerald et al., 2004; Cashman, 2006; Hartmann and Meisel, 2007; Elfahri et al., 2016).

This chapter reviews the current knowledge on bioactive components existing in cow milk and colostrum, their biological and nutritional functionalities, and exploitation of these biologically active compounds for human physiological metabolism and health. It will cover all the main bioactive components in bovine milk with potential to improve human health.

2 Bioactive proteins

2.1 Caseins

In recent years, scientific and commercial interest in the biological properties of milk proteins has increased significantly. Intact protein molecules of both milk proteins (caseins (CNs) and whey proteins) exhibit distinct physiological functions *in vivo* (Steijns, 2001; Walzem et al., 2002; Pihlanto and Korhonen, 2003; Madureira et al., 2007; Zimecki and Kruzel, 2007; Korhonen and Marnila, 2013). The major bioactive proteins found in bovine colostrum and milk and their concentration, molecular weight and suggested biological functions are shown in Table 1. Colostrum contains much greater concentrations of many bioactive whey proteins, notably Igs, LF and growth factors, than mature milk, reflecting its importance to the health of the newborn calf (Pakkanen and Aalto, 1997; Scammel, 2001).

Major milk components and precursors, corresponding bioactive compounds and their biological functionalities are listed in Table 2.

Cow (*Bos taurus*) milk contains approximately 3.5% protein. CN represents about 80% and whey proteins about 20% of the protein fraction. Recent research suggests that peptides derived from CN are not only nutritious but also a source of low-molecular-weight peptides with various biological activities. These peptides are generated and become

Table 1 Major bioactive protein components of bovine colostrum and milk: Concentration, molecular weight and potential biological functions[a]

Protein	Concentration (g/L)		Molecular weight	Biological activity
	Colostrum	Milk	Daltons	
Caseins (αs_1, αs_2, β and κ)	26	28	14 000–22 000	Ion carrier (Ca, PO_4, Fe, Zn, Cu), precursor for bioactive peptides immunomodulatory, anticarcinogenic
β-Lactoglobulin	8.0	3.3	18 400	Vitamin carrier, potential antioxidant, precursor for bioactive peptides, fatty acid binding
α-Lactalbumin	3.0	1.2	14 200	Effector of lactose synthesis in mammary gland, calcium carrier, immunomodulatory, precursor for bioactive peptides, potentially anticarcinogenic
Immunoglobulins	20–150	0.5–1.0	150 000–1 000 000	Specific immune protection through antibodies and complement system, potential precursor for bioactive peptides
Glycomacro-peptide	2.5	1.2	8000	Antimicrobial, antithrombotic, prebiotic, gastric hormone regulator
Lactoferrin	1.5	0.1	80 000	Antimicrobial, antioxidative, anticarcinogenic, anti-inflammatory, iron transport, cell growth regulation, precursor for bioactive peptides, immunomodulatory, stimulation of osteoblast proliferation
Lactoperoxidase	0.02	0.03	78 000	Antimicrobial, synergistic effects with immunoglobulins, lactoferrin and lysozyme
Lysozyme	0.0004	0.0004	14 000	Antimicrobial, synergistic effects with immunoglobulins, lactoferrin and lactoperoxidase
Serum albumin	1.3	0.3	66 300	Precursor for bioactive peptides
Milk basic protein	NA	NA	10 000–17 000	Stimulation of osteoblast proliferation and suppression of bone resorption
Growth factors	50 μg–40 mg/L	<1 μg–2 mg/L	6400–30 000	Stimulation of cell growth, intestinal cell protection and repair, regulation of immune system

[a] Compiled and adapted from Pihlanto and Korhonen (2003), Korhonen and Pihlanto (2007), Korhonen (2009) and NA = not announced.

Table 2 Major biologically active milk components and their functions[a]

Milk precursors or components	Bioactive compounds	Bioactivities observed
α-, β-Caseins	Casokinins	ACE inhibitory (increase blood flow to intestinal epithelium)
α-, β-Caseins	Casomorphins	Opioid agonist (decrease gut mobility, gastric emptying rate; increase amino acids and electrolytes uptake)
α-, β-Caseins	Immunopeptides casomorphins	Immunomodulatory (increase immune response and phagocytic activity)
	Casokinins	
α-, β-Caseins	Phosphopeptides	Mineral binding (Ca binding; increase mineral absorption, i.e. Ca, P, Zn)
α_{s1}-Casein	Isracidin	Antimicrobial
α_{s1}-Casein	Caseinophosphopeptide	Calcium binding and transport
α_{s2}-Casein	Casocidin	Antimicrobial
κ-Casein	Casoxins	Opioid antagonist
κ-Casein	Casoplatelins	Antithrombotic
κ-Casein	κ-casein glyco-macropeptide	Probiotic (growth of bifidobacteria in GI tract)
α-Lactalbumin (α-La) β-Lactoglobulin (β-La)	Lactorphins	Opioid agonist
Serum albumin	Serorphin	Opioid agonist
α-La, β-La and serum albumin	Lactokinins	ACE inhibitory
Immunoglobulins	IgG, IgA	Immunomodulatory (passive immunity)

Lactoferrin	Lactoferrin	Immunomodulatory (increase natural killer cell activity, humoral immune response, thymocyte trafficking immunological development and interleukins-6; decrease tumour necrosis factor-α)
		Antimicrobial (increase bacteriostatic inhibition of Fe-dependent bacteria; decrease viral attachment to and infections of cells)
		Probiotic activity (increase growth of bifidobacteria in GI tract)
Lactoferrin	Lactoferroxins	Opioid antagonist
Oligosaccharides	Oligosaccharides	Probiotic (increase growth of bifidobacteria in GI tract)
Glycolipids Oligosaccharides	Glycolipids Oligosaccharides	Antimicrobial (decrease bacterial and viral attachment to intestinal epithelial cells)
Prolactin	Prolactin	Immunomodulatory (increase lymphocyte and thymocyte trafficking, and immune development)
Cytokines	Interleukins-1,2, 6 and 10 Tumour necrosis factor-α	Immunomodulatory (lympocyte trafficking, immune development)
	Interferon-γ	
	Transforming growth	
	Factors-α, β; leukotriene B_4	
	Prostaglandin E_2, Fn	
Growth factors	IGF-1, TGF-α, EGF, TGF-β	Organ development and functions, stimulation of cell growth, intestinal cell protection and repair
Parathormone-P	PTHrP	Increase Ca^{+2} metabolism and uptake

[a]Adapted from Schanbacher et al. (1998), Meisel (1998), Clare and Swaisgood (2000) and Park (2009a).

active after digestion by proteolytic enzymes or during the fermentation and maturation processes of cheese and yogurt (Korhonen and Pihlanto, 2007). The production of such peptides has been reported in caseinates of different species milk, such as cow (López-Expósito et al., 2007; Phelan et al., 2009), goat (Park, 2009b), sheep (Recio et al., 2009a) and buffalo (Pandya and Haenlein, 2009).

Digested or catabolized CNs produce a variety of bioactive peptides. These include antihypertensive and immuno-stimulating peptides generated from bovine β-CN as well as caprine β-CN (Geerlings et al., 2006; Silva et al., 2006). The functional and biological properties of individual CN fractions differ due to differences in their amino acid composition (Akuzawa et al., 2009).

The principal CN fractions are divided into α_{s1}-, α_{s2}, β- and κ-CN, and their amino acid compositions are influenced by genetic polymorphisms in different species milk (Park et al., 2007). CNs have molecular weights between 19 and 24 kilodaltons (kDa) and are all assembled to form a complex micelle structure which comprises α_{s1}-CN and β-CN as the dominant fractions (each about 38–40%), followed by α_{s2}-CN and κ-CN, both presenting 10–12% of total CN (Park et al., 2007; Korhonen and Marnila, 2013). CN fractions exist as micelles mainly in an aggregate colloidal form. These are discrete spherical particles of about 200 nm diameter and encapsulate insoluble minerals (calcium and phosphate) and small amounts of other minerals (Fox and Brodkorb, 2008).

The large-scale fractionation of whole CN from milk by current technologies is based either on acidic coagulation at an isoelectric point of pH 4.6 or on enzymatic hydrolysis of CNs with rennet (Korhonen and Marnila, 2013). Intact whole CN and electrophoretic CN fractions can exhibit different biological activities, such as immunomodulation (Bennett et al., 2005; Gauthier et al., 2006b). Examples include regulating proliferation of lymphocytes and production of antibodies and cytokines (Yousefi et al., 2009; Bonomi et al., 2011). CN hydrolysates can become a potential source of highly functional ingredients for different food applications. Examples of such products are fermented milk products based on hypotensive tripeptides Val-Pro-Pro and Ile-Pro-Pro derived from both β-CN and κ-CN (Korhonen and Marnila, 2013).

As shown in Table 1, many bioactive compounds are generated from different CN fractions, including casomorphins, casokinins, phosphopeptides, immunopeptides, isracidin, casocidin, casoxins and casoplatelins, which exert different bioactive functionalities such as opioid agonist, opioid antagonist, angiotensin-I converting enzyme (ACE) inhibitory, mineral binding, immunomodulatory, antimicrobial and antithrombotic activities. β-casomorphin-7 (BCM7), a peptide sequence present in β-CN, has been associated with an increased risk for certain non-communicable diseases, such as autism, cardiovascular diseases and type I diabetes (McLachlan, 2001), although this was not confirmed by some other researchers (Clemens, 2011). Enzymatic hydrolysis can release BCM7 from β-CN, but the release is determined by the proteins' genetic variants, thus being associated with the specific breed (Korhonen and Marnila, 2013).

2.2 Whey proteins

Whey proteins represent about 20% of total milk proteins. They are made up of 50% β-lactoglobulin (β-lg), 12% α-lactalbumin (α-la), 10% Igs, 5% serum albumin, 0.23% proteose peptones, LF and lactoperoxidase (LP) (Horton, 1995; Siso, 1996). Individual whey protein components and their hydrolyzed peptide fragments exhibit various bioactivities, including opioid agonist, antimicrobial and antiviral actions, immune system stimulation,

Table 3 Bioactive peptides derived from whey proteins

Precursor protein	Fragment	Peptide sequence	Name	Function
α-Lactalbumin	50–53	Tyr-Gly-Leu-Phe	α-Lactorphin	Opioid agonist ACE inhibition
α-Lactoglobulin	102–105	Tyr-Leu-Leu-Phe	β-Lactorphin	Non-opioid stimulatory effect on ileum
	142–148	Ala-Leu-Pro-Met-His-Ile-Arg	–	ACE inhibition
	146–149	His-Ile-Arg-Leu	β-Lactotensin	Ileum contraction
Bovine serum albumin	399–404	Tyr-Gly-Phe-Gln-Asp-Ala	Serorphin	Opioid
	208–216	Ala-Leu-Lys-Ala-Trp-Ser-Val-Ala-Arg	Albutensin A	Ileum contraction, ACE inhibition
Lactoferrin	17–42	Lys-Cys-Arg-Arg-Trp-Glu-Trp-Arg-Met-Lys-Lys-Leu-Gly-Ala-Pro-Ser-Ile-Thr-Cys-Val-Arg-Arg-Ala-Phe	Lactoferricin	Antimicrobial

Adapted from Korhonen et al. (1998).

anticarcinogenic activity and other metabolic functions. Bioactive peptides derived from whey proteins are listed in Table 3.

2.2.1 α-Lactalbumin

α-La in cow milk accounts for 2–5% of the total protein. In cow colostrum, the α-la content varies from 1.4 to 2.6 gL^{-1}, while mature milk contains about 1 gL^{-1} (Korhonen and Marnila, 2013). The α-la content in different species varies: human milk is 3–4 gL^{-1}, buffalo milk 1.4 gL^{-1}, yak milk 2 gL^{-1}, goat milk 1–2.5 gL^{-1}, donkey milk 1.5–2 gL^{-1}, mare milk 3 gL^{-1}, lama milk 3.4 gL^{-1} and camel milk about 0.9 gL^{-1}, respectively (El-Hatmi et al., 2006; Pandya and Haenlein, 2009; Pecka et al., 2012). Cow α-la has the molecular weight of about 14 kDa, and it is synthesized in the mammary gland, and it acts as coenzyme for lactose biosynthesis (Korhonen, 2009). In human milk, α-la is the predominant whey protein constituting about 10–20% of the total protein in mature milk.

α-La from cow milk and its hydrolysates can be used as a supplement in infant formulae owing to the high degree (74%) of amino acid homology to human α-la (Lönnerdal, 2011), and a few α-la enriched formulae have been commercialized. However, intact α-la is also one of the major allergens in cow milk. Binding of calcium influences spatial conformation and stability of α-la, which also functions as a calcium carrier.

Bovine α-la hydrolysates and their specific peptides have been shown to have antihypertensive, antimicrobial, anticarcinogenic, immunomodulatory, opioid and prebiotic properties (Pihlanto and Korhonen, 2003; Chatterton et al., 2006). α-La has a high tryptophan content (5%). It is a good source of both the essential amino acids tryptophan and cystein, which are precursors of serotonin and glutathion, respectively (Korhonen, 2009). An α-la enriched diet has been found to reduce stress and depression (Markus et al., 2000). Markus et al. (2002) observed that α-la improved cognitive functions in stress-vulnerable subjects by increased brain tryptophan and serotonin activity. α-La may also protect against induced gastric mucosal injury in a way comparable to a typical antiulcer drug (Matsumoto et al., 2001; Ushida et al., 2003).

Cow α-la can induce apoptosis in tumour and immature cells without affecting healthy differentiated cells (Barbana et al., 2011). Pettersson et al. (2006) reported that purified porcine, equine and caprine α-la forms also express similar apoptotic activities. Sulphuric amino acid cysteine is rich in α-la (6%), and an intake of cysteine-rich whey protein diet has been shown to improve glycemic control, alleviate sucrose-induced oxidative stress and development of insulin resistance (Blouet et al., 2007). A high-Ca diet (1.8% $CaCO_3$) with α-la (18% of energy) significantly increased weight loss during an energy restriction diet.

2.2.2 β-Lactoglobulin

The concentration of β-lg (MW = 18.3 kDa) in cow milk is 3.0–3.5 gL^{-1}, which accounts for about 50% of whey proteins (Table 1). This major whey protein is also present in buffalo, mare, donkey, yak and goat milks, but is not found in human and camel milk (Chatterton et al., 2006). β-Lg exists as a reversible dimer at the native pH of milk and binds a range of small hydrophobic and amphipathic molecules, such as fatty acids, retinol and vitamin D2 (Korhonen and Marnila, 2013). Using chromatographic and membrane filtration techniques, β-lg can be isolated on an industrial scale from cheese whey. Because β-lg has excellent heat-set gelation properties, it is used for products requiring good water-binding and texturizing properties (Korhonen and Marnila, 2013).

In terms of bioactive functions, β-lg from cow, goat and sheep milk has proven to be an excellent source of peptides with a wide range of bioactivities, such as antihypertensive, antimicrobial, antioxidative, anticarcinogenic, immunomodulatory, opioid, hypocholesterolemic and other metabolic effects (Hernández-Ledesma et al., 2008; Park, 2009b; Recio et al., 2009a; Korhonen and Marnila, 2013) (Tables 1 and 2). Colloidal interactions of β-lg may play a role in protein and fat metabolism and could be exploited in optimizing nutrition (Mackie and Macierzanka, 2010). Glycation of cow milk β-lg with galactooligosaccharides (GOSs) via the Maillard reaction reportedly forms stable glycated peptides in subsequent simulated digestion. These glycated peptides appeared to exhibit similar bifidogenic activity when compared to unconjugated GOS (Hernández-Hernández et al., 2011). This research observation could open up new applications for Maillard reaction products as prebiotic compounds.

In a study with rats, the peptide β-lactosin B [Ala-Leu-Pro-Met; f(142–145)] has been shown to have antihypertensive properties (Murakami et al., 2004). The tryptic peptide [Ile-Ile-Ala-Glu-Lys; f(71–75)] has hypocholesterolemic properties (Nagaoka et al., 2001). In addition, an opioid peptide β-lactorphin [Tyr-Leu-Leu-Phe; f(102–105)] has been shown to improve arterial function (Sipola et al., 2002). Further *in vivo* studies are needed to substantiate animal studies. As an example, β-lg-derived ACE-inhibitory peptide [ALPMHIR; f(142–148)] was rapidly degraded upon *in vitro* incubation with human serum and, after ingestion, could not be detected in sera of human subjects (Roufik et al., 2006).

2.2.3 Lactoferrin

LF is an iron-binding glycoprotein which is present in colostrum, milk, other body secretions, cells of most mammalian species and neutrophilic leukocytes (Korhonen, 2009; Korhonen and Marnila, 2013). As shown in Tables 1 and 2, the main known *in vivo* activities of cow milk LF and LF-derived peptides include defending the udder and gastrointestinal tract against microbial pathogens, iron sequestering and transport, immune system regulating activities, antineoplastic activity and mitogenic and trophic activities in gastrointestinal tract and associated lymphoid tissue (GALT) (Marnila and Korhonen, 2009).

Bovine LF is comprised of a single-chain polypeptide sequence (689 amino acids, sequence homology with human 69%). Depending on the degree of glycosylation, MW of cow LF is 80–84 kDa (Korhonen and Marnila, 2013). Under pH 7–8 of physiological conditions, LF is positively charged and binds strongly polyanions, such as heparin, bacterial lipopolysaccharides (LPS), lysozyme (LZM), Igs (especially IgA), CN and DNA. With high affinity (Kd~10^{-30}), each LF molecule can bind two ferric ions (Fe^{3+}) with concomitant incorporation of bicarbonate (HCO_3^-) or carbonate ($CO_3^=$) ion. Bovine LF has an iron saturation rate of 20–30%. Iron-depleted LF (5% < Fe saturation) is called apo-LF, while the saturated form is known as holo-LF (Korhonen and Marnila, 2013).

The synthesis of LF occurs in the epithelial cells of the mammary gland and also exists at high concentrations in neutrophils. The LF content of cow colostrum is on average 1.5 gL^{-1} ranging from 0.2 to 5 gL^{-1} and decreases to around 0.1 (0.07–1.2) gL^{-1} in mature milk. The LF contents of other species are human milk is 0.7–1.5 gL^{-1}, buffalo milk 0.32 gL^{-1}, yak milk 0.67 gL^{-1}, donkey milk 0.3 gL^{-1} and camel milk 0.2–0.3 gL^{-1}, respectively (El-Hatmi et al., 2006; Pandya and Haenlein, 2009; Medhammar et al., 2012). The standard pasteurization regimes (72°C for 15 s) used in the dairy industry usually do not affect the structure, antibacterial activity and bacterial interaction of LF. Nevertheless, UHT treatment abolishes the ability of holo-LF to bind to bacteria as well as the bacteriostatic activity of apo-LF. A marginal loss of LF activity was observed in spray-drying of milk. In heat treatments, apo-LF denatures faster than holo-LF (Marnila and Korhonen, 2009). When microbial invasion such as mastitis, occurs, LF is delivered by degranulating neutrophils into the infected sites in various concentrations.

LF is regarded as an important host defence molecule and is known to confer many biological activities, including antimicrobial, antioxidative, anti-inflammatory, anticancer and immune regulatory properties (Wakabayashi et al., 2006; Zimecki and Kruzel, 2007). By the action of digestive enzyme pepsin, several antimicrobial peptides, such as lactoferricin B f(18–36) and lactoferrampin f(268–284), can be cleaved from LF. In addition, LF is known to play an important role in the body's innate defence system against microbial infections and degenerative processes induced by free oxygen radicals. The role of LF was initially limited to antimicrobial activity alone, whereas the LF has been well recognized by its multifunctionality now. Three mechanisms account for the antimicrobial activity of LF: (a) iron binding from the medium leading to inhibition of bacterial growth; (b) direct binding of LF to the microbial membrane, especially to LPS in Gram-negative bacteria, causing fatal structural damage to outer membranes and inhibition of viral replication and (c) prevention of microbial attachment to epithelial cells or enterocytes (Korhonen, 2009).

The bactericidal effect of LF can be augmented by the action of LZM or antibodies (Pan et al., 2007). LF can also elevate susceptibility of bacteria to certain antibiotics, such as vancomycin, penicillin and cephalosporins. Antimicrobial activity of LF and the derivatives *in vitro* system have been demonstrated against a wide range of pathogenic microbes, including enteropathogenic *Escherichia coli, Clostridium perfringens, Candida albicans, Haemophilus influenzae, Helicobacter pylori, Listeria monocytogenes, Pseudomonas aeruginosa, Salmonella typhimurium, S. enteriditis, Staphylococcus aureus, Streptoccccus mutans, Vibrio cholerae,* hepatitis C, G and B virus, HIV-1, cytomegalovirus, poliovirus, rotavirus and herpes simplex virus (Farnaud and Evans, 2003; Pan et al., 2007).

LF has also been studied intensively on the antitumour activity over the last decade, and several mechanisms have been suggested, such as iron chelation-related antioxidative property, immunoregulatory and anti-inflammatory functions. In *in vitro* experiments, LF has been shown to regulate both cellular and humoral immune systems by (a) stimulation of proliferation of lymphocytes; (b) activation of macrophages, monocytes, natural killer cells and neutrophils; (c) induction of cytokine and nitric oxide production and (d) stimulation of intestinal and peripheral antibody response (Pan et al., 2007).

Orally administered LF and related compounds in mice or rats have shown to suppress the overgrowth and translocation of certain intestinal bacteria, such as *E. coli, Streptococcus* and *Clostridium* strains, but did not affect intestinal bifidobacteria (Teraguchi et al., 2004; Wakabayashi et al., 2006; Pan et al., 2007). Moreover, oral administration of LF and lactoferricin reduced the infection rate of *H. pylori, Toxoplasma gondii,* candidiasis and tinea pedis as well as prevented clinical symptoms of influenza virus infection. Orally ingested LF and related compounds also improved nutritional status by reducing iron-deficient anaemia and drug-induced intestinal inflammation, colitis, arthritis and decreased mortality caused by endotoxin shock (Teraguchi et al., 2004).

LF has been shown to be transported into the cell nucleus, where it can bind DNA to regulate the phenotypic traits of the host. Animal model studies suggest that LF and its derived peptides participate in the defence against development and progression of tumours. LF from cow milk had a synergistic effect with black tea polyphenols in the prevention of chemically induced genotoxicity and development of oral carcinomas in hamster buccal pouch (Mohan et al., 2008). Oral LF administration to mice regulated the bone cell activity and increased bone formation (Cornish et al., 2004). In further animal studies, they demonstrated the beneficial effects of LF ingestion on inhibition of carcinogen-induced tumours in colon, oesophagus, lung, tongue, bladder and liver. In clinical infant trials with bovine LF preparations, oral administration increased the number of bifidobacteria in faecal flora and the serum ferritin level, while the ratios of Enterobacteriaceae, *Streptococcus* and *Clostridium* decreased. King et al. (2007) observed that LF supplementation to healthy infants for 12 months was associated with less respiratory tract illnesses and higher haematocrits, as compared to the control group fed with regular infant formula.

In other human studies, LF increased the eradication rate of *H. pylori* gastritis, decreased the incidence of bacteremia and severity of infection in neutropenic patients, alleviated symptoms of hepatitis C virus infection, reduced small intestine permeability in drug-induced intestinal injury and promoted the cure of tinea pedis (Korhonen, 2009). Bovine LF administration could also be beneficial in stress-related neurodegenerative disorders and treatment of certain cancer types (Zimecki and Kruzel, 2007). Due to these positive biological functionalities, bovine LF has attracted increasing commercial interest and many products containing added LF have already been launched in the market particularly in Asian countries. Industrial production of LF has been launched by many companies worldwide, and its use as an ingredient in functional foods and pharmaceutical preparations will increase drastically in the near future (Tamura, 2004). In Japan and Taiwan, commercial applications of LF in yogurt products have already been made and marketed. Furthermore, LF supplemented baby foods and infant formulae have been marketed in South Korea, Japan and China. LF also has been applied in different dietary supplements, which combine LF and bovine colostrum or probiotic bacteria. Owing to potential synergistic actions, LF has been incorporated together with LZM and LP into human oral health care products, such as toothpastes, mouth rinses, moisturizing gels and chewing gums (Korhonen, 2009).

LF provides part of the regulatory feedback loop in the inflammatory process. Kruzel et al. (2007) observed that LF can control the cytokine induced pro-inflammatory cascade during the development of a systemic inflammatory response in tissues. If infection occurs by Gram-negative bacteria, LF binds to LPS and inhibits the progressive inflammatory cascade (Kruzel et al., 2007). LF can thus reduce oxidative damage in inflamed tissues.

2.2.4 Immunoglobulins

Igs in colostrum and milk provide the newborn young or calf with passive immunological protection (antibodies) against invading environmental pathogens and toxins. Igs are essential for the newborn for establishing its immunity during the development of its own immune system as well as for the protection of the udder against infections (Butler, 1994). Igs account for 70–80% of total protein in cow colostrum during 1–3 days lactation, whereas in milk they represent only 1–2% of total protein content. Ig concentrations in cow milk vary according to the breed, age, health status and stage of lactation. The IgG contents in the first colostrum vary from 20 to 180 gL^{-1}, and then decline rapidly to

Table 4 Immunoglobulins in cow colostrum and milk: concentrations, physicochemical and immunological properties

	IgG_1	IgG_2	IgA	SIgA	I gM
Concentration (g/L)					
Colostrum*	60 (15–180)	13	Traces	3.5 (1–6)	5 (3–9)
Milk	0.35 (0.3–0.6)	0.02–0.12	Traces	0.05–0.14	0.04–1.0
Physicochemical properties					
MW (kDA)	146–163	146–154	160	385–430	900
Structure	Monomer	Monomer	Monomer	Dimer	Pentamer
Carbohydrates %	2.8–3.1	2.6–3.0	6–10		10–12
Immunological functionality					
Opsonization	+++	+	0	0	+++
Complement fixation	++	+	0	0	+++
Agglutination	+	+	++	++	+++

* = first milking *postpartum*, 0 = no activity, + = low activity, ++ = moderate activity, +++ = strong activity. Data compiled from Mehra et al. (2006) and Korhonen and Marnila (2009), and adapted from Korhonen and Marnila (2013).

0.3–1 gL^{-1} in mature milk. IgG contents of different species are reportedly similar such as in goat, camel, donkey, mare and yak milk (Levieux et al., 2002; Konuspayeva et al., 2007; Park et al., 2007; Medhammar et al., 2012; Pecka et al., 2012). As shown in Table 4, the Ig classes present in bovine milk are IgG_1, IgG_2, IgM and IgA; the major class being IgG_1 (Park et al., 2007; Korhonen and Marnila, 2009).

There are three major classes in bovine and human lacteal secretions, which are IgG, IgM and IgA. All Igs have a similar basic structure and are composed of two identical light chains and two identical heavy chains, and joined together with disulphide bonds. The complete basic Ig molecule has a Y-shaped structure and has a MW of about 160 kDa (Korhonen et al., 2000). Colostrum-based products for farm animals are commercially available, and have also been growing in the worldwide market as dietary supplements for humans (Scammel, 2001; Tripathi and Vashishtha, 2006; Struff and Sprotte, 2007). Specific Igs augment the recognition and phagocytosis of microbes by phagocytic leukocytes (opsonization). Igs can prevent the adhesion of microbes, inhibit bacterial metabolism, agglutinate bacteria, opsonization, complement fixation, augment phagocytosis of bacteria, kill bacteria through activation of complement-mediated bacteriolytic reactions and neutralize toxins and viruses (Mehra et al., 2006). Igs are associated with various parts of the cellular and humoral immune system. Pentameric IgM antibodies are considerably more effective in most of aforementioned activities than IgGs, especially in complement fixation. SIgA does not fixate complement but is effective in neutralizing viruses and toxins and in agglutinating bacteria. The Fab part binds to antigen and other parts (mostly the Fc-region) interact with other elements such as leukocyte receptor (Korhonen and Marnila, 2013).

Bioseparation techniques have been advanced and it made possible to fractionate and enrich these antibodies and formulate so-called immune milk preparations (Mehra et al., 2006). Such immune milk preparations have been studied by many researchers, and the preparations can be effective in the prevention of human and animal diseases caused by different pathogenic microbes, such as rotavirus, *E. coli, Candida albicans, Clostridium difficile, Shigella flexneri, Streptococcus mutans, Cryptosporidium parvum* and *H. pylori*. However, the therapeutic efficacy of these preparations has been very limited (Weiner et al., 1999; Korhonen et al., 2000; Korhonen and Marnila, 2006). Thus, some countries have a few commercial immune milk products on the market, while the unclear regulatory status of these products has emerged as a constraint for global commercialization in many countries (Hoerr and Bostwick, 2002; Mehra et al., 2006). Considering the globally increasing problem of antibiotic-resistant strains causing endemic hospital infections, the development and marketing of appropriate Ig containing milk products to combat these infections poses as a great challenge for future research.

Most Ig products are prepared from colostrum of cheese whey by removing the fat, followed by microfiltration/ultrafiltration or pasteurization. Heating at 60°C for 120 min does not affect the neutralizing activity of IgG, whereas the biological activities can be lost rapidly at temperatures above 65°C (McMartin et al., 2006). Typical Ig products are in the form of lyophilized or spray-dried powders, though some are produced as filtered colostral whey liquids or concentrates (Korhonen and Marnila, 2013).

The most important protective mechanism of milk Igs is preventing adhesion of pathogenic microbes to surfaces like epithelial cell lining or mucous membrane, such adhesion being the critical first step in colonization and infection (Korhonen and Marnila, 2009). The Igs agglutinating ability reduces the capability of bacteria to adhere to surfaces. Colostrum and milk contain agglutinating Igs for a large variety of microorganisms.

Microfold cells in intestinal epithelial lining have specific receptors for SIgA, and they can bind and transcytose antigen-SIgA-receptor complexes into aggregated lymphoid nodules (Peyer's patches) for antigen presentation and induction of mucosal immune responses (Korhonen and Marnila, 2013). Igs may also contribute to fighting pathogenic microbes by activating the milk complement system, though this function is limited to the immune defence of the udder. The milk Igs have also synergistic effects with some non-specific antimicrobial factors in milk, such as LZ, LF and LP–thiocyanate–hydrogen peroxide system (Loimaranta et al., 1998; Bostwick et al., 2000).

2.2.5 Glycomacropeptide

Glycomacropeptide (GMP) is a C-terminal glycopeptide f(106–169) released from the κ-CN molecule at ^{105}Phe-^{106}Met by the action of chymosin enzyme. GMP is hydrophilic and is the most abundant protein fraction in cheese whey amounting to 20–25% of the proteins (Abd El-Salam et al., 1996). GMP remains in the whey fraction in the cheese manufacturing process, while the remaining part of κ-CN, called para-κ-CN, precipitates into the cheese curd. The molecular weight of GMP is about 8000 Daltons, and it contains a significant amount of carbohydrate fraction (50–60% of the total GMP). GMP is composed of galactose, *N*-acetyl-galactosamine and *N*-neuraminic acid. Pure GMP can be recovered in large quantities from cheese whey by chromatographic or ultrafiltration techniques (Thomä-Worringer et al., 2006). GMP is postulated to exhibit many biological properties, while clinical evidences are limited in its potential health benefits. Through several *in vitro* studies, GMP has been demonstrated to inactivate microbial toxins of *E. coli* and *V. cholerae*, inhibit adhesion of cariogenic bacteria and influenza virus, modulate immune system responses, promote growth of bifidobacteria, suppress gastric hormone activities and regulate blood circulation through antihypertensive and antithrombotic activity (Manso and López-Fandiño, 2004).

Caseinomacropeptide (CMP) is the non-glycosylated form of GMP. As precursors of κ-CN, CMPs are heterogeneous mixture, because of genetic variations and differently glycosylated and phosphorylated forms (Park et al., 2007). The amino acid sequences of CMP from different species are well established, and they are characterized by the absence of aromatic amino acids (Phe, Trp and Tyr) and Arg, and by high levels of acidic and hydroxyl amino acids (Mercier et al., 1976; Stewart et al., 1984; Park et al., 2007). An excellent review on the physicochemical, technological, biological and nutritional aspects of κ-CN macropeptides, including CMPs from sheep and goat milk, has been published (Manso and López-Fandiño, 2004). In the perspective of bioactivity, the content of *N*-acetyl neuraminic and glycolyl neuraminic (sialic acid) in CMP is interesting. Large amounts of this carbohydrate are found in the brain and in the central nervous system in the form of gangliosides and glycoproteins, contributing to the functioning of cell membranes, membrane receptors and to normal brain development (Park et al., 2007). CMPs from several different species milk have been reported as a good source of antithrombotic peptides. CMP and its fractions in an *in vitro* study have also shown to have a protective effect against dental erosion, when applied in acidic drinks for example (Setarehnejad et al., 2010).

GMP is rich in branched-chain amino acids but low in methionine, which makes it a useful ingredient in diets for patients suffering from hepatic diseases (Korhonen and Marnila, 2013). In addition, GMP does not contain phenylalanine, which makes it suitable for patients suffering from phenylketonuria. The high sialic acid content of GMP may deliver beneficial effects for brain development and improvement of learning ability, which has been shown by an animal study (Wang et al., 2007). Some *in vitro* animal studies have demonstrated that GMP inhibits gastric secretions and slows down stomach motility by means of stimulating the release of cholecystokinin (CCK) (Burton-Freeman, 2008; Guilloteau et al., 2010), and the satiety hormone involved in controlling food intake and digestion in the duodenum of animals and humans (Yvon et al., 1994). This hormone is able to control appetite and food intake, as well as digestion in the duodenum of animals and humans.

Both GMP and a whey protein isolate (WPI) decreased weight gain and altered body composition in male Wistar rats (Royle et al., 2008). Although the mechanism is not known, GMP has a significant additional effect on reduction of fat accumulation when combined with WPI. Mouse model and human studies have shown that bifidobacteria residing in the gut may affect weight gain. Recio et al. (2009b) suggested that GMP may have a beneficial role in modulation of gut microflora, due to the promoting activity of GMP in the growth of bifidobacteria. For the purpose of appetite control and weight management, commercial products and supplements containing GMP are now available in many countries. However, the efficacy of these products has yet to be proved by human clinical trials. Chabance et al. (1998) reported that GMP can be absorbed intact and partially digested forms into the blood circulation of adult humans after ingestion of milk or yogurt.

For the purpose of appetite control and weight management, GMP or CMP containing commercial products have been launched in the market (Korhonen and Marnila, 2013). On the other hand, Gustafson et al. (2001) reported that CMP had no effect on food energy intake or subjective indicators of satiety in human adults for a short time period. Nevertheless, GMP has a good potential to play a beneficial role in the modulation of gut microflora, because GMP has been shown to promote the growth of bifidobacteria, probably with its carbohydrate content (Manso and López-Fandiño, 2004). Multiple physicochemical properties of GMP make it a potential ingredient for many food applications (Thomä-Worringer et al., 2006). Pure GMP can be produced in large quantities from cheese whey by chromatographic or ultrafiltration techniques.

2.3 Enzymes

2.3.1 Lactoperoxidase (LP)

LP is a glycoprotein that occurs naturally in colostrum and milk (Korhonen and Marnila, 2013). LP has a molecular weight of about 78 kDa, and it is the most abundant enzyme in cow milk (about 30 mgL^{-1}). It can be obtained from whey by using chromatographic techniques. Cow milk LP is relatively heat-resistant, and about 50% of its original activity is retained after HTST pasteurization (Kussendrager and van Hooijdonk, 2000).

LP catalyses peroxidation of thiocyanate anion and some halides in the presence of hydrogen peroxide (H_2O_2) to generate short-lived oxidation products of the thiocyanate anion (SCN^-), primarily hypothiocyanate (OSCN-), which kill or inhibit the growth of a wide range of microorganisms, including bacteria, viruses, fungi, moulds and protozoa (Reiter and Perraudin, 1991; Korhonen and Marnila, 2013; Seifu et al., 2005). The inhibition by the LP system is bacteriostatic in the case of Gram-positive bacteria, including *Listeria* spp., *Staphylococcus* spp. and *Streptococcus* spp., whereas the microbes are killed in the case of Gram-negative bacteria, such as *E. coli*, *Salmonella* spp., *Pseudomonas* spp., *Campylobacter* spp. and *H. pylori*.

The LP system is now regarded as an important part of the natural host defence system in mammals, where the protective function may be mediated by several mechanisms (Boots and Floris, 2006). LP is resistant to acidic pH as low as pH 3, and to human gastric juice. The thiocyanate ion is known as widely distributed in animal tissues and secretions. The presence of LP in human airway epithelia may actively combat respiratory infections caused by microbial invaders (Gerson et al., 2000). Seifu et al. (2005) showed that the LP system is active against a wide range of microorganisms, including bacteria, viruses, fungi, moulds and protozoa. In addition, this mechanism is inhibitory to *Candida* spp. and the protozoan *Plasmodium falciparum*, and also inactivate *in vitro* the HIV type 1 and polio virus. The LP antibacterial system is particularly effective against cariogenic bacteria *Streptococcus mutans* and periodontitis-associated bacteria, such as *Actinobacillus actinomycetemcomitans*, *Porphyromonas gingivalis* and *Fusobacterium nucleatum* (Korhonen and Marnila, 2013). Gastritis caused by *H. pylori* can be another potential new application, since this pathogen is killed *in vitro* by the LP system (Shin et al., 2002). Further applications can be envisaged for preservation of different products, such as meat, fish, vegetables, fruits and flowers (Boots and Floris, 2006).

The LP system has an important function in preserving raw milk (Korhonen and Marnila, 2013). The natural concentrations of thiocyanate and H_2O_2 are critical in milk and the system usually requires activation by addition of a source of these components. The effectiveness of the activated LP system has been demonstrated in many pilot studies and field trials (Reiter and Härnulv, 1984; Seifu et al., 2005). Therefore, the LP system has been approved by the Codex Alimentarius Committee for preservation of raw milk under specified conditions since 1991. The method is now being practised in a number of countries, especially where raw milk-cooling facilities are unavailable or inadequate.

The antimicrobial function of the LP system been demonstrated for species such as cow, buffalo, camel and goat (FAO/WHO, 2006; Pandya and Haenlein, 2009). The LP system has been shown to reduce diarrhoea in newborn calves (Reiter and Perraudin, 1991). It has been shown to have a synergistic antimicrobial action with other natural defence components, such as LF, LZM and antibodies (Floris et al., 2003). Shin et al. (2005) had shown that oral administration of LP in combination with LF had attenuated symptoms of pneumonia in influenza virus-infected mice.

2.3.2 Lysozyme (LZM)

LZM is a potent antibacterial enzyme, whose MW is about 15 kDa. It is also known as muramidase (peptidoglycan *N*-acetylmuramoyl hydrolase; EC 3.2.1.17), and catalyses the hydrolysis of the β-(1–4) linkage between *N*-acetylmuramic acid and *N*-acetylglucosamine of bacterial peptidoglycan (Floris et al., 2003; Korhonen and Marnila, 2013). Since LZM was discovered by Alexander Fleming in the early 1920s, it has been known as an effective antimicrobial agent. This enzyme component is especially rich in the cell wall of Gram-positive bacteria. LZM is widely distributed in various biological fluids and tissues, including avian egg, plants, bacteria and animal secretions, but its concentration varies considerably in different sources (Korhonen and Marnila, 2013). In addition, significant amounts of LZM are present in tears and saliva, while its concentration in cow and buffalo colostrum and milk is relatively low (0.4 mgL^{-1}) (Floris et al., 2003; Pandya and Haenlein, 2009).

The concentration of LZM in mammalian milks varies widely from 13 μg/100 mL in buffalo milk to 79 mg/100 mL in mare milk (El-Agamy et al., 1998). The rich sources of LZM are chicken egg white (1–3 gL^{-1}), human milk (up to 0.4 gL^{-1}), camel milk (0.23–0.50 gL^{-1}), mare's milk (0.4–1 gL^{-1}) and donkey's milk (1–2 gL¯1) (Korhonen and Marnila, 2013). El-Agamy and Nawar (2000), in a comparative study of LZM concentration in milk of different species, showed that camel milk had a considerably high concentration of LZM than cow, buffalo, sheep and goat milk, whereas LZM in camel milk was lower than those in human, donkey and mare milks. The equivalent concentration of LZM in camel milk was 11, 18, 10 and 8 times that of cow, buffalo, sheep and goat milk, respectively. LZM concentrations in milk vary according to some factors such as lactation period and health status of the animal (El-Agamy, 2009).

LZM exhibits a high activity against mesophilic and thermophilic spore-forming bacteria, such as bacilli and clostridia, while it shows some activity against other spoilage and pathogenic microorganisms, for example, *Listeria monocytogenes*, *Staphylococcus aureus*, *Candida albicans*, and *Streptococcus mutans* (Johnson, 1994). LZM, LF and antibodies have a synergistic bactericidal effect on many microorganisms (Pan et al., 2007). LZM from different sources has some variability towards their antibacterial spectrum and specificity towards different types of mucopolysaccharides.

In terms of heat stability, the entire activity of LFs was lost at 85°C for 30 min in cow, buffalo and camel milks. However, at this level of temperature, the activity losses of LZM were 56, 74 and 81.7% for camel, cow and buffalo milk, respectively (El-Agamy, 2009). It was generally concluded that protective proteins of camel milk are significantly ($p \leq 0.01$) more heat-resistant than cow and buffalo milk proteins. Among the protective proteins, the order of heat resistance found was LZM > LF > IgG_S (El-Agamy, 2009).

Because LZM is fairly resistant to digestive enzymes, it can be effective even after oral administration (Sava, 1996). LZM has been shown to be effective by oral and topical applications in prevention and control of several viral skin infections, including herpes simplex and chicken pox (Sava, 1996). In the application of the food industry, egg white LZM is being utilized as a preservative in fermented cheese varieties to prevent the late fermentation defect caused by contaminating butyric acid bacteria.

Aside from antibacterial activity, LZM exerts many other functions, including inactivation of certain viruses, enhancement of antibiotic effects, anti-inflammatory and antihistamine actions, activation of immune cells and antitumour activity (Floris et al., 2003). However, the potential physiological significance of natural milk LZM has not been fully explored or elucidated.

2.4 Bioactive peptides

There are more than two hundred biologically and functionally active peptides that exist in milk and dairy products. Biologically and physiologically active peptides are encrypted as inactive within the amino acid sequence of their parent protein molecule (Korhonen, 2009). Proteins are comprised of 20 different amino acids that are arranged diversely into several conformations, giving each protein molecule its own unique structure and function. Bioactive peptides can be liberated by (a) hydrolysis by digestive enzymes, (b) fermentation of milk with proteolytic starter cultures and (c) proteolysis by microbial or plant-derived enzymes (Korhonen and Pihlanto, 2007; Park, 2009b; Vercruysse et al., 2009; Korhonen and Marnila, 2013). Amino acid bonds can be cleaved at specific sites on the proteins to release peptide sequences of different sizes. Bioactive peptides are considered protein

Table 5 Bioactive peptides identified in fermented dairy products[a]

Product	Examples of identified bioactive peptides	Bioactivity	Reference
Cheese type			
Cheddar	α_{s1}- and β-casein fragments	Several phosphopeptides	Singh et al. (1997)
Italian variety:	β-cn f(58–72)	ACE inhibitory	Smacchi and Gobbetti (1998)
Mozzarella, Crescenza, Italico, Gorgonzola			
Gouda	α_{s1}-cn f(1–9), β-cn f(60–68)	ACE inhibitory	Saito et al. (2000)
Festivo	α_{s1}-cn f(1–9), f(1–7), f(1–6)	ACE inhibitory	Ryhänen et al. (2001)
Emmental	α_{s1}- and β-casein fragments	Immunostimulatory, several phosphopeptides, antimicrobial	Gagnaire et al. (2001)
Manchego (sheep milk)	Sheep α_{s1}-, α_{s2}- and β-casein fragments	ACE inhibitory	Gomez-Ruiz et al. (2002)
Cheddar	α_{s1}-cn f(1–6), f(1–7), f(1–9), f(24–32), f(102–110), β-cn f(47–52), f(193–209)	ACE inhibitory	Ong et al. (2007)
Fermented milks			
Sour milk	β-cn f(74–76, f(84–86) κ-cn f(108–111)	Antihypertensive	Nakamura et al. (1995)
Dahi	Ser-Lys-Val-Tyr-Pro	ACE inhibitory	Ashar and Chand (2004)
Yogurt (sheep milk)	Active peptides not identified	ACE inhibitory	Chobert et al. (2005)
Kefir (goat milk)	PYVRYL, LVYPFTGPIPN[b]	ACE inhibitory	Quiros et al. (2005)
Fermented milk (probiotic and dairy strains)	Active peptides not identified	ACE inhibitory	Donkor et al. (2007)

Abbreviations: α_{s1}-cn = α_{s1}-casein, β-cn = β-casein, κ-cn = κ-casein.
[a]Adapted from Korhonen (2009).
[b]One-letter amino acid code.

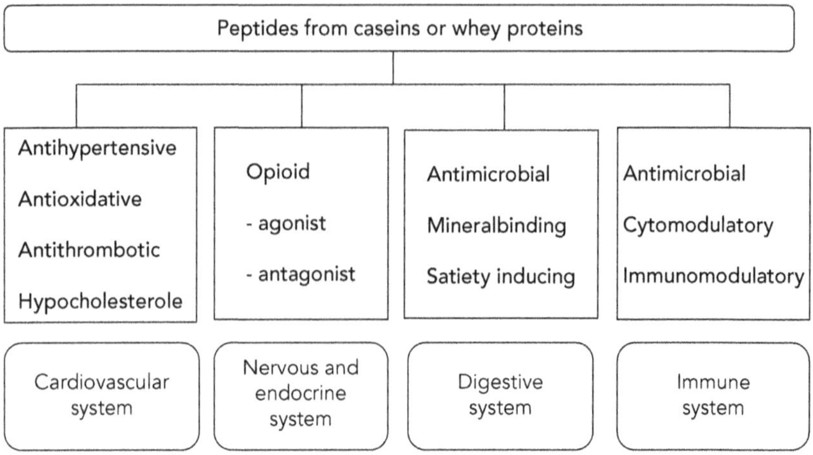

Figure 1 Physiological functionality of milk-derived bioactive peptides (Korhonen, 2009).

fragments that, upon hydrolysis by proteolytic enzymes or fermentation, exhibit positive functions or conditions that have positive effects on human health (Kitts and Weiler, 2003). The activities of peptides exhibit different functionalities on the basis of their inherent amino acid composition and sequence. The size of active sequences may vary from two to twenty amino acid residues, and many peptides are known to display multifunctional properties (Park et al., 2007; Korhonen, 2009).

Bioactive peptides affect functions in the body such as gastrointestinal, cardiovascular, endocrine, immune and nervous systems (Korhonen, 2009; Park, 2009b). They demonstrate antimicrobial, antihypertensive, antithrombotic, antioxidative, opioid and immunomodulatory properties (FitzGerald and Meisel, 2003; Park et al., 2007; Sah et al., 2015). Some demonstrate a range of immunostimulatory, opioid and ACE-inhibitory activities (Korhonen and Pihlanto, 2007; Nagpal et al., 2011).

Pepsin, trypsin and chymotrypsin have been shown to produce a number of antihypertensive, antibacterial, antioxidative, immunomodulatory and opioid peptides, and caseinphosphopeptides (CPPs) from CNs and whey proteins α-la, β-lg and GMP (López-Expósito et al., 2007; del Mar Contreras et al., 2009). Commercial proteolytic enzymes, including alcalase, flavourzyme, thermolysin and subtilisin and other proteases, have been utilized to produce various bioactive peptides, both from CNs and whey proteins (Pihlanto-Leppälä et al., 2000; Otte et al., 2007; Ortiz-Chao et al., 2009). The major functions of bioactive peptides are illustrated in Fig. 1.

2.4.1 Antihypertensive peptides

These peptides possess the capacity of lowering blood pressure and are studied more than any other milk protein-derived peptides. Currently, more than 150 antihypertensive peptides have been identified from different milk proteins from cow, buffalo, goat and sheep (Korhonen and Marnila, 2013). Angiotensin is one of two polypeptide hormones and a powerful vasoconstrictor controlling arterial blood pressure (Gobbetti et al., 2007; Park, 2009). The ACE causes elevation of blood pressure by converting angiotensin-I to the potent vasoconstrictor, angiotensin-II and by degrading bradykinin, a vasodilatory peptide and enkephalins (Petrillo and Ondetti, 1982). ACE (peptidyl-peptide hydrolases; EC 3.4.15.1) has been regarded as a multifunctional ectoenzyme located in different body tissues, including plasma, lung, kidney, heart, skeletal muscle, pancreas, arteries and brain, and plays a key physiological role in regulating peripheral blood pressure, and also in the renin–angiotensin, kallikrein–kinin and immune systems (Gobbetti et al., 2007; Korhonen and Pihlanto, 2007; Park, 2009).

The ACE-inhibitory or antihypertensive peptides have been isolated from the enzymatic digest of various food proteins and they are recently the most investigated type of bioactive peptides (Korhonen and Pihlanto, 2007). Exogenous ACE inhibitors having an antihypertensive effect *in vivo* were first discovered in snake venom (Ondetti et al., 1977). As shown in Table 5, several ACE-inhibitory peptides were identified by *in vitro* enzymatic digestion of milk proteins or chemical synthesis of peptide analogs (Gobbetti et al., 2004; Korhonen, 2009). The ACE inhibitors generated from milk proteins are attributed to different fragments of CN, known as casokinins (Meisel and Schlimme, 1994), or whey proteins, named lactokinins (FitzGerald and Meisel, 2000).

Meisel et al. (1997) found low MW ACE-inhibitory peptides in several ripened cheeses, and observed that the ACE-inhibitory activity was increased as proteolysis developed, whereas the ACE-inhibitory effect was reduced when the cheese maturation exceeded a certain level during proteolysis. The hydroylsates of caprine β-lg digested with thermolysin had shown to contain four novel ACE-inhibitory peptides: those are f46–53, f58–61, f103–105 and f122–125 (Table 5; Gobbetti et al., 2007). ACE-inhibitory peptides with very high potency were found, whereby these were corresponding to casokinins such as α_{s1}-CN f23–27 and f1–9, β-CN f60–68 and f177–183, and α_{s2}-CN f174–181 and f174–179, having IC_{50} values lower than 20 μmol/L (Saito et al., 2000; Meisel, 2001; Park, 2009b).

The antihypertensive efficacy of ACE-inhibitory milk peptides has been studied in *in vivo* and *in vitro* experiments, including rat model studies (Murray and FitzGerald, 2007; Gobbetti et al., 2007; Jäkälä and Vapaatalo, 2010; Ricci et al., 2010). In these studies, the tripeptides VPP and IPP have proven to be the most effective ones (Nakamura et al., 1995; Jauhiainen et al., 2005; De Leeuw et al., 2009). In human studies, after consumption of fermented dairy products or tablets containing these peptides, moderate or significant reduction of blood pressure has been observed in mildly hypertensive subjects (Hirota et al., 2007; Boelsma and Kloek, 2009; Usinger et al., 2010). Comparing with the placebo group, the ACE-inhibitory peptides group had reductions of 1.5–14.0 mm Hg for systolic blood pressure (SBP) and 0.5–6.8 mm Hg for diastolic blood pressure (DBP). Effective dosages of lactotripeptides (VPP and IPP) range from 3.07 to 52.5 mg/d. Blood pressure-lowering effects of lactotripeptides have typically been observed after 4–6 weeks of treatment (Korhonen and Marnila, 2013).

Nakamura et al. (1995) purified ACE-inhibitory peptides from a Japanese soft drink manufactured from skim milk fermented by *Lactobacillus helveticus* and *S. cerevisiae*. Milk innoculated with *Lb. helveticus* has shown to release Val-Pro-Pro and Ile-Pro-Pro peptides from α_{s1}- and β-CN (Yamamoto et al., 1994). It was observed that the hypertensive patients had significant decreases in blood pressure after 4 and 8 weeks of daily ingestion of 95 mL sour milk, which contained the two tripeptides. This is an equivalent ingestion of 1.2–1.6 mg/day ACE-inhibitory peptides in a placebo-controlled study (Hata et al., 1996). After active treatment for 8–12 weeks, maximum blood pressure reductions of approximately 13 mmHg of SBP and 8 mmHg of DBP, respectively, have also been observed (Phelan and Kerins, 2011). The absorption from the gastrointestinal tract into circulation and a dose-dependent antihypertensive effect *in vivo* of VPP and IPP have been established in rat model and human studies (Hata et al., 1996; Jauhiainen et al., 2007). A placebo-controlled, full crossover intervention study showed that IPP was absorbed intact from a fermented milk drink into the circulation of healthy human subjects (Foltz et al., 2007). However, several other studies were not able to establish any significant effect of drinks containing ACE-inhibitory peptides (Lee et al., 2007).

2.4.2 Antioxidative peptides

Like other bioactive peptides, antioxidative peptides are inactive within the sequence of the parent protein but can be released during enzyme hydrolysis. Once released, these antioxidative peptides have been shown to possess radical scavenging, metal ion chelation properties and the ability to inhibit lipid peroxidation (Park, 2009b; Power et al., 2013).

Pihlanto (2006) reported that milk-derived antioxidative peptides are composed of 5–11 amino acids including hydrophobic amino acids, proline, histidine, tyrosine or tryptophan in the sequence. However, the structure–activity relationship or the antioxidant mechanism of peptides is not fully understood. Antioxidative peptides can be released from different sources of proteins such as CNs, soya bean and gelatine in hydrolysis by the action of proteolytic enzymes (Korhonen and Pihlanto, 2003; Sah et al., 2015; Elfahri et al., 2016).

Researchers have demonstrated that peptides derived from α_s-CN have free radical-scavenging activity and inhibit enzymatic and non-enzymatic lipid peroxidation (Suetsuna et al., 2000; Rival et al., 2001). The low temperature-processed whey protein has been shown to contain high levels of specific dipeptides (glutamylcysteine). These dipeptides can promote the synthesis of glutathione, which is an important antioxidant involved with cellular protection and repair processes (Bounous and Gold, 1991).

2.4.3 Antithrombotic peptides

Antithrombotic peptides exist in milk. Clare and Swaisgood (2000) found that undecapeptide (residues 106–116) from bovine κ-CN have clotting properties. Casoplatelin, a peptide derived from κ-CN, affected platelet function and inhibited both the aggregation of ADP-activated platelets and the binding of human fibrinogen λ-chain to its receptor region on the platelets surface (Jolles et al., 1986). CMP is a peptide separated from κ-CN during milk coagulation by rennin. CMP has peptide sequences that inhibit the aggregation of blood platelets and the binding of the human fibrinogen γ-chain to platelet surface fibrinogen receptors (Fiat et al., 1993). Bovine κ-caseinoglycopeptides, an antithrombotic peptide derived from κ-CN, have been detected in infants fed with cow milk-based formula (Chabance et al., 1995, 1998).

Qian et al. (1995a) observed that the C-terminal residues of sheep κ-CN or κ-caseinoglycopeptide (106–171) decreased thrombin- and collagen-induced platelet aggregation in a dose-dependent manner. They also found that thrombin-induced platelet aggregation was inhibited with pepsin digests of sheep and human LF. A single peptide peak possessing of this thrombin-induced platelet aggregation activity was obtained by reverse-phase chromatography of the hydrolysate (Qian et al., 1995b).

2.4.4 Hypocholesterolemic peptides

Hypocholesterolemic or serum cholesterol-lowering activity has been found in a tryptic hydrolysate of β-lg (Nagata et al., 1982; Nagaoka, 2001). Cholesterol is rendered soluble in bile salt-mixed micelles and then reabsorbed (Wilson and Rudel, 1994; Park, 2009b). It has been suggested that an orally administered peptide possesses high bile acid-binding capacity, could inhibit the reabsorption of bile acid in the ileum and thus reduce blood cholesterol levels (Iwami et al., 1986).

A hypocholesterolemic peptide (Ile-Ile-Ala-Glu-Lys) containing tryptic hydrolysate of β-lg has been shown to demonstrate hypocholesterolemic activity (Nagaoka et al., 2001). The bioactive peptides that were identified were β-lg f9–14, f41–60, f71–75 and f142–146. However, the mechanism underlining the hypocholesterolemic effect of these peptides has not been explained (Korhonen and Pihlanto, 2007).

2.4.5 Opioid peptides

Opioid peptides such as enkephalins have opiate-like effects (Gobbetti et al., 2007) (Tables 2 and 4). The major opioid peptides are fragments of β-CN called β-casomorphins (Branti et al., 1981; Yoshikawa et al., 1986). Opioid peptides have also been obtained from pepsin hydrolysis of bovine α_{s1}-CN fractions (Paroli, 1988; Meisel and Schlimme, 1990). Similar peptides have been observed from human β-CN fraction (Yoshikawa et al., 1984), and the common bovine β-casomorphin, Y-P-F sequence, was also found in the primary structure of human β-CN (Clare and Swaisgood, 2000).

Opioid peptides are opioid receptor ligands with agonistic or antagonistic activities. Endogenously produced opioid peptides (endorphins) are derived from proenkephalin, propiomelanocortin and prodynorphin and exhibit a definite N-terminal sequence Y-G-G-F (Chiba and Yoshikawa, 1986; Clare and Swaisgood, 2000). Opioid peptides produced from food proteins (exorphins) have been noticed in milk protein and wheat gluten hydrolysates (Teschemacher, 2003; Meisel and Schlimme, 1990; Schanbacher et al., 1998). Milk-derived peptides, generated by hydrolysis of various precursor proteins such as α- and β-CN, α-la, and β-lg are called 'atypical' exomorphic, agonist peptides and exhibit morphine-like activity (Zioudrou et al., 1979).

The α_{s1}-CN-exorphin (α_{s1}-CN f90–96), β-casomorphins-7 and -5 (β-CN f60–66 and f60–64, respectively), and lactorphins (α-la f50–53 and β-lg f102–105) act as opioid agonists, whereas casoxins (i.e. k-CN f35–42, f58–61 and f25–34) act as opioid antagonists (Meisel and FitzGerald, 2000; Gobbetti et al., 2007). β-Casomorphins have also been found in cow and human β-CN (Meisel and Schlimme, 1996). The hydrolysis of *Lactobacillus* GG fermented UHT milk by pepsin/trypsin released several opioid peptides derived from α_{s1}- and β-CN, and α-la (Rokka et al., 1997). Proteolysis of α-la with pepsin produces α-lactorphin, while digestion of β-lg with pepsin and then trypsin produces β-lactorphin (Gobbetti et al., 2007).

Casomorphines are known to affect mood (Panskeep et al., 1984; Paroli, 1988), increase analgesic behaviour (Matthies et al., 1984; Paroli, 1988), prolong gastrointestinal transient time (Tome et al., 1987), modulate amino acid transport, exert antisecretory (antidiarrhoeal) action (Daniel et al., 1990) and stimulate endocrine responses such as the secretion of insulin and somatostatin (Meisel and Schlimme, 1990). Milk-derived opioid-like peptides also regulate appetite by modifying pancreatic endocrine activity, which results in an increase of insulin output (Nieter et al., 1981). Casoxins A and B have been synthesized.

Gobbetti et al. (2007) has shown that the common structural feature in most endogenous and exogenous opioid peptides is a Tyr residue at the amino terminal end and another aromatic residue, Phe or Tyr, in the third or fourth position. This Tyr residue appears to be essential for opioid activity (Chang et al., 1981). The Pro residue in the second position is also crucial to maintain the proper orientation of the Tyr and Phe side chains for opioid activity (Mierke et al., 1990).

2.4.6 Mineral-binding peptides

Mineral-binding peptides include phosphopeptides, CPPs and calcium-binding phosphopeptides (CCPs). By forming soluble organophosphate salts, they function as carriers for different minerals (Meisel and Olieman, 1998). These include Ca^{2+}, Zn^{2+}, Mn^{2+} and Fe^{2+} (Korhonen and Pihlanto- Leppälä, 2004). CPPs lead to increased Ca absorption by limiting the precipitation of Ca in the distal ileum (Korhonen and Pihlanto, 2007). Gobbetti et al. (2007) found that most CPPs include a sequence of three phosphoseryl, followed by two glutamic acid residues. The negatively charged side chains of these amino acids provide binding sites for minerals, particularly phosphates. Factors affecting binding have been reviewed by Berrocal et al. (1989), Sato et al. (1983) and Schlimme and Meisel (1995). CN can reduce CPPs via enzymatic digestion using pancreatic endoproteinases such as trypsin, as well as chymotrypsin, pancreatin, papain, pepsin, thermolysin and pronase (FitzGerald, 1998).

CPPs have been shown to increase Ca^{2+} and Zn^{2+} absorption (Hansen, 1995). CPP supplementation enhanced calcium absorption in rats (Saito et al., 1998). However, enhancement of calcium bioavailability may be limited (Meisel, 1997; Meisel and Fitzgerald, 2003). CPPs also have anticariogenic effects, leading to some commercial applications (Morgan et al., 2008). Mice fed with a CCP preparation had a higher level of serum and intestinal antigen-specific IgA than those fed with the control diet (Otani et al., 2000).

2.4.7 Anti-appetizing peptides

The total whey protein in the diet has shown to be associated with a lowering of LDL cholesterol and to increased release of an appetite-suppressing hormone, cholecystokinin (Zhang and Beynen, 1993). The bioactivity for total whey protein in suppressing appetite may be linked to the combination of active whey protein fractions or amino acid sequences.

Schaafsma (2006) reported that dietary proteins can help to reduce energy intake and promote a healthy body composition with less body fat due to their positive effects on satiation/satiety. Replacement of either fat or carbohydrates by whey proteins can be helpful in reduction of energy intake. They expounded that the diet-induced thermogenesis increased post-prandial concentration of plasma amino acids, and their effects on gut hormones may play a role in the brain gut axis. This physiological role of whey protein implies a great potential for processed whey products in developing new and lucrative health food markets as functional food ingredients (Regester et al., 1997).

Luhovyy et al. (2007) stressed that whey protein reduces short-term food intake compared to placebo, carbohydrate and other proteins ingested groups, and reiterated that whey protein has a potential for the regulation of body weight by providing satiety signals that affect both short-term and long-term food intake (i.e. anti-appetizing) regulation. They further postulated that whey protein affects satiation and satiety by the actions of (1) whey protein fractions per se, (2) bioactive peptides, (3) amino acids released after digestion and (4) combined action of whey protein and/or peptides and/or amino acids with other milk constituents. Whey protein is insulinotropic, and whey-borne peptides affect the renin–angiotensin system. Therefore, whey protein has potential as physiologically functional food component for persons with obesity and its co-morbidities (hypertension, type II diabetes, hyper- and dislipidemia). However, the favourable effects of whey on food intake, subjective satiety and intake-regulatory mechanisms in humans are not clear.

2.4.8 Antimicrobial peptides

Lactenin was probably the first antibacterial factor found in milk (Jones and Simms, 1930). Antimicrobial milk proteins, such as LF, were reported in the 1970s (Bullen et al., 1972; Lahov et al., 1971). Antimicrobial peptides are able to modulate inflammatory responses in addition to killing microorganisms (Devine and Hancock, 2002).

Peptides demonstrating antimicrobial activities have been purified from bovine milk protein hydrolysates as well as other foods (Clare et al., 2003; Floris et al., 2003; Pellegrini, 2003; Gobbetti et al., 2004). The overall antibacterial effect in milk is greater than the individual contributions of LF, LP and lysozyme, proteins or peptides (Gobbetti et al., 2007). This is due to the synergistic activity of naturally occurring proteins and peptides, in addition to peptides generated from inactive protein precursors (Clare and Swaisgood, 2000).

Lactoferricins are the most studied antimicrobial peptides derived from bovine and human LF (Kitts and Weiler, 2003; Wakabayashi et al., 2003). Lactoferricins exhibit antimicrobial activity against various Gram-positive and -negative bacteria, yeasts and filamentous fungi (Korhonen and Pihlanto, 2007; Gobbetti et al., 2007).

Casecidins are a group of basic, glycosylated and high-molecular-weight (about 5 kDa) polypeptides. Casecidin was among the first purified antimicrobial peptides, demonstrating activity against *Staphylococcus, Sarcina, Bacillus subtilis, Diplococcus pneumoniae* and *Streptococcus pyogenes* (Lahov and Regelson, 1996). Bovine milk casocidin-I, a cationic α_{s2}-CN-derived peptide, inhibits growth of *E. coli* and *Staphylococcus carnosus* (Zucht et al., 1995; Clare and Swaisgood, 2000).

Isracidin is another antibacterial peptide derived from α_{s1}-CN treated with chymosin (Hill et al., 1974). Isracidin was shown to exert inhibitory effect on the *in vitro* growth of lactobacilli and other Gram-positive bacteria, including *S. aureus, Streptococcus pyogenes* and *Listeria monocytogenes*.

Milk also contains peptides that exert antifungal properties in combination with azole antifungal agents, antifungal activity of LF or its peptides (i.e. lactoferrincin B) has been demonstrated with *Candida albicans* (Wakabayashi et al., 1996). Bellamy et al. (1994) found that several filamentous fungi, including agents of skin disease (dermatophytes), were susceptible to the mixture of lactoferricin B and azole antifungal agents.

2.4.9 Immunomodulatory peptides

Peptides derived from milk CNs and major whey proteins as well as protein hydrolysates have immunomodulatory effects exhibiting immune cell functions, such as lymphocyte proliferation, antibody synthesis and cytokine regulation (Gill et al., 2000; Gauthier et al., 2006). Intact whey proteins α-la and β-lg have been reported to prime human neutrophils to inflammatory cells seen as increased chemotaxis, degranulation and oxygen radical production due to secondary stimuli (Rusu et al., 2010). The whey proteins also stimulate synthesis of pro-inflammatory cytokines in epithelial-like $CaCO_2$ cells (Ustunol and Wong, 2010).

Milk-derived immunomodulatory peptides include α_{s1}-CN f194–199 (α_{s1}-immunocasokinin) and β-CN f193–202, f63–68, f191–193 (immunopeptides), which are synthesized by hydrolysis with pepsin–chymosin as well as by fermentation (Kayser and Meisel, 1996). These peptides have been found to modulate the proliferation of human lymphocytes, to down-regulate the production of certain cytokines, and to stimulate the phagocytic activities of macrophages (Korhonen and Pihlanto, 2003a,b, 2007; Matar et al., 2003). Because of immune cell functions, these peptides have been of special interest to food researchers and to the food processing industry (Korhonen and Pihlanto, 2013; Sah et al., 2015; Elfahri et al., 2016).

Kayser and Meisel (1996) found that β-casomorphin-7 and β-CN immunopeptides suppressed the proliferation of human peripheral blood lymphocytes at low concentrations ($<10^{-7}$ mol/L), but these peptides stimulated the proliferation at higher concentration. β-CN-derived several peptides enhanced the IgG production in mouse spleen cell cultures (Gobbetti et al., 2007). The proliferation of human colonic lamina propria lymphocytes was inhibited by β-casomorphin-7, where the antiproliferative effect of micromolar concentrations was reversed by the opiate receptor antagonist naloxone (Elitsur and Luk, 1991). Peptides containing glutamine may substitute for the free amino acid glutamine, which is required for lymphocyte proliferation and utilized at a high rate by immunocompetent cells (Calder, 1994).

β-Casomorphin-7 and β-casokinin-10 are CN-derived peptides, which may have either suppressive or stimulatory effects on proliferation of human lymphocytes, depending on their concentrations (Meisel, 2004; Gauthier et al., 2006). GMP was reported to have anti-inflammatory activity (López-Posadas et al., 2010) as has interleukin (IL)-6 (Korhonen and Pihlanto, 2013). Milks fermented with *L. helveticus* have been shown to increase immune response against subcutaneous fibrosarcomas (LeBlanc et al., 2002) as well as show anti-inflammatory properties (Vinderola et al., 2007). Peptide fractions generated from fermented milk by *L. helveticus* were found to inhibit IL-6 and TNF-α production and respiratory burst activity of LPS-primed human THP-1 promonocytes (Tompa et al., 2011).

2.4.10 Cytomodulatory peptides

Milk-derived peptides may trigger susceptibility of cancer cells to cytomodulation (Gobbetti et al., 2007). Feeding rats primed with the procarcinogen dimethylhydrazine CN or whey protein resulted in significantly fewer tumours per treatment group compared to a red meat or soya bean meals (McIntosh et al., 1995). Purified peptides that are equivalent to sequences of CN, also showed the modulation of cell viability such as proliferation and apoptosis in different human cell culture models (Hartmann et al., 2000). Anti-colon cancer and antioxidant activities of the bovine skim milk fermented by *Lactobacillus helveticus* were also recently observed (Sah et al., 2015; Elfahri et al., 2016).

Cytomodulatory effects were also observed in CPPs. Cytomodulatory peptides derived from CN fractions inhibited cancer cell growth or stimulated the activity of immunocompetent cells and neonatal intestinal cells (Meisel and FitzGerald, 2003). Proliferation of leukaemia cells was inhibited by the peptides from a lyophilized extract of Gouda cheese. Cancer cell lines were more reactive to peptide-induced apoptotic stimulation than non-malignant cells (Gobbetti et al., 2007). Bacterial hydrolysis of CN using commercial yogurt starter cultures can yield bioactive peptides that affect colon cell Caco-2 kinetics *in vitro* (McDonald et al., 1994). Bovine skimmed milk digested with cell-free extract of the yeast *Saccharomyces cerevisiae* had also been shown to have antiproliferative activity towards leukaemia cells (Roy et al., 1999).

Whey protein-derived cysteine and glutathione (GSH) are important substrates for the immune response (White et al., 1994). Glutathione supplementation in vitro to peripheral blood mononuclear cells (PBMCs) from young and old subjects enhanced the interleukin-2 production and the T-cell-mediated mitogenic response (Wu et al., 1994). A whey protein concentrate increased the proliferation *in vitro* of PBMCs and also the intracellular GSH concentration. However, tumour cell lines were inhibited in cell proliferation, with a concomitant decrease in cellular GSH (Baruchel and Viau, 1996).

3 Bioactive lipids

3.1 Conjugated linoleic acid (CLA)

Conjugated linoleic acid (CLA) has been shown to have several bioactive functionalities on human health, including anticarcinogenic, antiatherogenic, antidiabetic, immune-stimulating, growth-promoting and body fat reducing activities (Pariza et al., 1996; Dhiman et al., 1999; Park, 2006, 2009b). CLA refers to a collection of positional and geometrical isomers of *cis*-9, *cis*-12-octadecadienoic acid (C18:2) with a conjugated double-bond configuration. The major CLA isomer in milk fat is 9-*cis*, 11-*trans*, also called rumenic acid (Collomb et al., 2006; Korhonen, 2009), which accounts for more than 82% of the total CLA isomers in dairy products (Dhiman et al., 1999; Park, 2006). CLA is formed partially by bioconversion of polyunsaturated fatty acids in the rumen by anaerobic bacteria, such as *Butyrivibrio fibrisolvens*, and primarily endogenously by Δ9-desaturation of vaccenic acid in the mammary gland of lactating cows, goats and sheep (Griinari et al., 2000; Korhonen, 2009). In milk fat, the *cis*-9, *trans*-11 isomer amounts to 75–90% of the total CLA. Average daily intake of CLA varies from 95 to 440 mg (Korhonen, 2009). A daily intake of 3.0 to 3.5 g may be required to provide anticarcinogenic response in humans (Collomb et al., 2006). Jensen (2002) described that most of these fatty acids are esterified to the glycerol molecule backbone and appear in milk fat mostly as triacylglycerols.

Milk fat contains the highest amount of both CLA and vaccenic acid (the precursor of CLA) (Park, 2009b). Factors affecting CLA concentration include breed and feeding regimen which account for variations in CLA contents of milk fat from 2 to 53.7 mg/g fat (Collomb et al., 2006; Jahreis et al., 1999). Pasture feeding and grass composition can increase CLA concentrations significantly (Mir et al., 1999; Collomb et al., 2001; Park, 2009b). Seed/oil supplemented diets rich in PUFA can significantly increase CLA in the milk of cows and other breeds (Stanton et al., 2003; Chilliard and Ferlay, 2004; Park, 2009b). Linoleic acid (rapeseed, soya bean, sunflower)-rich concentrates have a better impact on increase in CLA than other polyunsaturated plant oils (peanut, linseed). Fish oils combined with plant oils significantly increase CLA content (Shingfield et al., 2006; Korhonen, 2009).

The concentrations of CLA in milk and dairy products affect product characteristics. CLA-enriched milk resulted in softer butter compared to the butter made from ordinary milk. Only minor effects were observed on the CLA content of final cheese products during processing (Chamba et al., 2006; Bisig et al., 2007). In Cheddar and Edam type cheeses made with CLA-enriched milk, the texture of cheeses was found to be softer than in control cheese, whereas no major differences were found in the organoleptic properties (Avramis et al., 2003; Ryhänen et al., 2005; Korhonen, 2009). CLA level has been shown to be higher in organic milk and organic dairy products in comparison with conventionally produced milk and products (Bisig et al., 2007; Korhonen, 2009). Although starter cultures, such as propionibacteria, lactobacilli and bifidobacteria can convert linoleic acid into CLA in culture media, results for during yogurt and cheese production are variable. Production of CLA in special culture medium and addition of the isolated CLA into dairy products have been suggested as a technological solution (Sieber et al., 2004; Bisig et al., 2007).

Bioactive functionalities of CLA have been studied in animal models. Park et al. (1999) have shown that *trans*-10, *cis*-12 isomer is responsible for inducing reduction of body fat in mice, whereas *cis*-9, *trans*-11 isomer enhances growth in young animal models (Pariza et al., 2001). Strong evidence from animal trials supports an influence of CLA intake on lowering of body weight and fat mass, and increase in lean body mass. A great number of *in vitro* experiments and animal trials have been conducted on anticarcinogenic effects of synthetic CLA and rumenic acid-enriched milk fat (Ip et al., 2003; Parodi, 2004; Lee and Lee, 2005). A majority of these studies support the role for dietary CLA in protection against various types of cancer, for example, breast, prostate and colon cancer, but the underlying mechanisms warrant further studies.

As far as human studies are considered, the CLA intake did not support body weight and body fat losses, while an epidemiological study has suggested an inverse association between dietary and serum CLA and risk of breast cancer in postmenopausal women (Aro et al., 2000). Another cohort study reported the high intake of CLA containing dairy foods may reduce the risk of colorectal cancer and increased lean body mass (Larsson et al., 2005).

3.2 Phospholipids

Phospholipids are mainly located in the milk fat globule membrane (MFGM). Phospholipids are known to be essential components of cell membranes in human, animal and plant tissues. They are involved in the function of cell membranes and have the ability to interact with metabolites, ions, hormones, antibodies and other cells (Weihrauch and Son, 1983; Park, 2009b). Phospholipids are also major constituents of the brain, nerve tissue, heart muscle, liver and sperm (Renner et al., 1989). The major fractions of phospholipid include phosphatidylethanolamine (PE), phosphatidylcholine (PC) and lesser amounts of phosphatidylserine (PS) and phosphatidylinositol (PI). The major sphingolipid fraction is sphingomyelin (SP) with smaller portions of ceramides and gangliosides (Jensen, 2002). These polar lipid compounds are secondary messengers involved in transmembrane signal transduction and regulation, growth, proliferation, differentiation and apoptosis of cells. In addition, polar lipids play a role in neuronal signalling, are linked to age-related diseases, blood coagulation, immunity and inflammatory responses (Pettus et al., 2004; Korhonen, 2009).

Phospholipids are made up approximately 1.6% of the total lipids. Of the polar lipid fraction, glycolipids make up 16% in goat milk, as compared to the 6% in cow milk (Morrison et al., 1965). Phospholipid fractions of bound lipids of bovine milk contain 35.0% PE, 2.0% PS, 5.0% PI, 30.0% PC (lecithin) and 24.0% SP (Cerbulis et al., 1982; Renner et al., 1989). Species differences in levels of phospholipid fractions are minimal, while goat milk contains slightly higher PE, SP and PS than cow milk (Table 6). Human milk contains

Table 6 Distribution of phospholipid sub-classes in goat, cow and human milks

	Percent of total phospholipids		
Phospholipid fraction	Goat milk	Cow milk	Human milk
Phosphatidyl ethanolamine	35.4	35	32
Phosphatidyl choline	28.2	30	29
Sphingomyelin	29.2	24	29
Phosphatidyl inositol	4.0	5	5
Phosphatidyl serine	3.2	2	4

Data from Cerbulis et al. (1982) and Renner et al. (1989).

more PS, PC and SP than goat milk. The MFGM consists approximately 60% of proteins and 40% of lipids that are mainly composed of triglycerides, cholesterol, phospholipids and sphingolipids (Park, 2009b). The level of polar lipid in raw milk is reported to range between 9.4 and 35.5 mg per 100 g of milk. In processing of milk into different dairy products, the polar lipids are preferentially enriched in the aqueous phases like skimmed milk, buttermilk and butter serum (Korhonen, 2009).

Being amphipathic molecules, polar lipids contribute to a rapid absorption of fat by forming a membrane around fat globules, thereby facilitating their emulsification into smaller fat droplets (Park, 2009b). Polar lipids aid fat transport from the liver through their lipotropic activity. Imaizumi et al. (1983) found that dietary supplementation of PE decreased serum cholesterol. Galli et al. (1985) found that PC (lecithin) administration could reduce platelet lipid and cholesterol.

Ingestion of lecithin improved learning and memory in animals and humans, although the phospholipids may not be considered as the essential nutrient because the body itself can synthesize PC (Weihrauch and Son, 1983). Sphingolipids and their derivatives are particularly considered highly bioactive components possessing anticancer, cholesterol-lowering and antibacterial activities (Rombaut and Dewettinck, 2006; Park, 2009b). Furthermore, butyric acid and butyrate have been shown to inhibit development of colon and mammary tumours (Parodi, 2003). Sphingolipid-rich foods or supplements could be beneficial in the prevention of breast and colon cancers and bowel-related diseases (Park, 2009b). These positive results from cell culture and animal model studies may need further confirmation and human clinical investigations for bioactive functions of polar lipids.

3.3 Cholesterol and minor lipids

Cholesterol is mainly synthesized in the liver from acetic acid via acetyl coenzyme A. Although high blood cholesterol is linked to atherosclerosis, stroke and coronary heart disease, it is essential for normal cell function and body metabolism, as a structural component of cellular and subcellular membranes, plasma lipoproteins and nerve cells (Innis, 1985; Renner et al., 1989). Cholesterol is a metabolic precursor of bile acids and steroid hormones including vitamin D, and it is required for the metabolic systems involved in DNA synthesis and cell division, as well as playing a role in lipid transport (Innis, 1985).

Sterols are a minor fraction of total lipids in milk, where the main sterol is cholesterol (300 mg/100 g fat, equivalent to 10 mg/100 mL bovine milk) (Park et al., 2007, 2009b). The MFGM is the main source of milk cholesterol, which represents 0.4–3.5% of the membrane lipids (Renner et al., 1989). An 85–90% milk cholesterol exists as free cholesterol, while a minor portion presents as esterified form, usually bind to long-chain fatty acids (Renner et al., 1989). The body synthesizes 1–4 g cholesterol daily, and the total amount of cholesterol in the human body is 100–150 g, while 10–12 g is constantly present in blood (Renner et al., 1989; Park, 2009b).

Milk minor lipids are considered as bioactive components. These include gangliosides, glycolipids, glycosphingolipids and cerebrosides. Since these compounds are ubiquitously present in mammalian tissues, studies on the minor lipids in bovine and human milk have been reported. The functionalities of these minor lipids are important in cell-to-cell interaction, cell differentiation, proliferation and immune recognition. In addition, the receptor functions in relation to protein hormones, interferon, fibronectin and bacterial toxins are significant (Renner et al., 1989; Park, 2009b). Extracted a ganglioside from human milk was shown to inhibit *E. coli* heat-labile enterotoxin and cholera toxin (Laegreid and Kolsto Otnaess, 1985). Bovine MFGM contains ceramide glucoside and ceramide dihexoside which are the major neutral glycosphingolipids and gangliosides, where ceramide galactoside and ceramide dihexoside were identified in pooled human milk (Takamizawa et al., 1986). In addition, these minor milk compounds contain long-chain fatty acids (C22–C24), which are essential for the synthesis of glycosphingolipids in the constitution of the nervous system (Bouhours et al., 1984).

Alkylglycerol is another milk minor lipid, which occurs as non-esterified or esterified with fatty acids and/or phospholipids in milk (Ahrne et al., 1983). Alkylglycerol is a highly potent substance at nanomolar concentrations and is identified as a platelet-activating factor (Ahrne et al., 1983; Bjorck, 1985). The total amounts of neutral alkylglycerols are 0.1–0.2 mg/g of milk fat with higher amounts in colostrum. Several therapeutic functions, such as tuberculostatic and anti-inflammatory effects, have been reported from studies using crude mixtures of alkylglylcerols (Renner et al., 1989; Park, 2009b).

4 Bioactive carbohydrates

4.1 Lactose

Lactose is the major carbohydrate in milk with contents of 4.1, 4.7 and 6.9 g/100 mL for goat, cow and human milks, respectively (Park, 2006, 2009b; Hernández-Ledesma et al., 2011). Lactose is a disaccharide of valuable nutrient, since it favours the intestinal absorption of Ca, Mg and P, and the utilization of vitamin C (Hernández-Ledesma et al., 2011). The high level of lactose in human milk may explain for the stimulation of the development of a *bifidus* flora, which is associated with a decrease in the luminal pH and an increased resistance against the colonization of pathogens in the human intestine (Schulze and Zunft, 1991; Park, 2009b).

Calcium absorption occurs mainly in the ileum, where this segment of the intestine has a limited capacity of active calcium absorption (Lee et al., 1991). Lactose was shown to stimulate the vitamin D-independent component of the intestinal calcium transport system (Schaafsma et al., 1988). Studies also have demonstrated that lactose has a lower glycemic index than sucrose or glucose, indicating that lactose can be regarded as suitable in the diet of diabetics for stabilizing blood glucose levels (Wolever et al., 1985; Park, 2009b). Lactose is also reportedly less cariogenic for dental health compared to other major sugars, including glucose, fructose and maltose (Edgar, 1993). A part of lactose can also escape digestion and absorption in the small intestine, and serves as a substrate for the colonic *bifidus* microflora.

4.2 Lactose-derived compounds

Lactulose, lactitol, lactobionic acid and GOSs are known as the major lactose-derived products. Lactulose is derived from lactose during heat processing of milk, and it is a disaccharide of galactose and fructose (Park, 2009b). Lactulose was first acknowledged as a bifidus growth promoter and has been used in therapeutic-enhancing products as well as an ingredient of infant milk formulae (Regester et al., 1997). Lactulose content in pasteurized milk was reported to range from 4 to 200 mg/100 mL (Andrews, 1989).

Lactulose and lactitol are lactose-derived compounds possessing good bioactivities. They are non-digestible, but serve as a source of soluble fibre. They are widely used in the treatment of constipation and chronic hepatic encephalopathy, where they act in a similar way as substrates for the intestinal microflora (Camma et al., 1993; Blanc et al., 1992; Park, 2009b). Research had shown that calcium absorption in postmenopausal women was enhanced by a daily intake of 10 g lactulose (Van den Heuvel et al., 1999).

Lactobionic acid increases mineral absorption including Ca by the formation of soluble complexes with minerals, due to its prebiotic properties (Schaafsma and Steijns, 2000). Lactobionic acid is indigestible in the small intestine, but it can be fermented by the intestinal microbial flora. The GOSs are also non-digestible, which may have prebiotic properties owing to their selective stimulation of bifidobacteria in the intestine, and these lactose-derived compounds also have the bioactivity of increasing calcium absorption in the intestine (Chonan and Watanuki, 1995).

4.3 Oligosaccharides

Oligosaccharides are a class of carbohydrates that comprise from 2 to 10 monosaccharide units (Park, 2009b). More than 50 oligosaccharides have been identified in human milk and more than 30 from bovine and caprine milks (Saito et al., 1984). Oligosaccharides in milk belong to the group of bifidus factor, promoting the growth of *Lactobacillus bifidus* in the intestinal tract. Human milk oligosaccharides are beneficial for infants in relation to their prebiotic and anti-infective properties (Martinez-Ferez et al., 2006).

Besides lactose, milk contains minor amounts of glucose and galactose that are free as well as bound to lipids, proteins or phosphate. Among these compounds, oligosaccharides may be the most important compound, which contain fucose, *N*-acetylglucosamine and *N*-acetylneuraminic acid (NANA) in different proportions (Cheetham and Dube, 1983; Renner et al., 1989; Park, 2009b). Bovine milk contains 10–15 mg/100 mL of these oligosaccharides associated with glucose and galactose. Martinez-Ferez et al. (2006) characterized and quantified the neutral and sialylated lactose-derived oligosaccharides in mature caprine milk and compared those in ovine, bovine and human milks. They found that a large amount and variety of acidic and neutral oligosaccharides were present in goat milk relative to cow and sheep milk. In addition, these authors were able to identify 15 new oligosaccharide structures in caprine milk.

Caprine milk is uniquely different in milk carbohydrate patterns compared to bovine milk. Moreover, caprine milk and human milk have been shown to have ten times greater oligosaccharides than those in bovine milk. This condition may be of special interest to infant nutrition since caprine milk oligosaccharides may have higher bioactive functionalities on infant growth and human nutrition (Park, 2009b). Human milk oligosaccharides are thought to be beneficial for the infant in relation to their prebiotic and anti-infective properties, which would promote gut health (Martinez-Ferez et al., 2006; Park, 2009b).

Using a rat model, Lara-Villoslada et al. (2006) studied the effect of goat milk oligosaccharides (GMO) on dextran sodium sulphate (DSS)-induced colitis, and found that DSS caused body weight loss, while the rats fed the GMO did not have the loss. They also observed that GMO rats exhibited less severe colonic lesions and a more favourable intestinal microbiota, indicating that GMO can reduce intestinal inflammation and contribute to the recovery of damaged colonic mucosa.

Lacto-oligosaccharides can be produced by enzymatic inversion of lactose from milk. Oligosaccharides are produced at a capacity of approximately 2000 tonnes per year from milk lactose, and were used in ice cream, chocolate and frozen desserts in the 1990s (Regester et al., 1997).

5 Bioactive other compounds in milk

5.1 Growth factors

Growth-promoting or growth-inhibitory factors for different cell types in human colostrum and milk were first reported during the 1980s and then shown in bovine colostrum, milk and whey (Pakkanen and Aalto, 1997; Pouliot and Gauthier, 2006; Korhonen, 2009). The main biological functions of growth factors of milk have been reviewed by many researchers (Pouliot and Gauthier, 2006; Gauthier et al., 2006; Tripathi and Vashishtha, 2006; Korhonen, 2009). In bovine milk, a number of growth factors have been identified and reported as follows: EGF (epidermal growth factor), IGF-I and IGF-II (insulin-like growth factor), FGF1 and FGF2 (fibroblast growth factor), TGF-β1 and TGF-β2 (transforming growth factor), BTC (β-cellulin) and PDGF (platelet-derived growth factor).

The colostrum during the first hours after calving contains the highest concentrations of all known growth factors, and then the levels decrease substantially thereafter. Growth factors are basically polypeptides, and the range of their molecular masses is between 6000 and 30000 daltons, with amino acid residues varying from 53 (EGF) to about 425 (TGF-β2), respectively (Korhonen, 2009). Among growth factors, the most abundant ones in bovine milk are EGF (2–155 ng/mL), IGF-I (2–101 μg/L), IGF-II (2–107 μg/Ll) and TGF-β2 (13–71 μg/L), whereas the concentrations of the other known growth factors remain below 4 μg/L (Pouliot and Gauthier, 2006; Korhonen, 2009).

EGF and BTC belong to the epidermal growth factor family, and their low concentrations in milk are sufficient to induce physiological responses. Members of the EGF family stimulate the growth of epithelial cells, the predominant function being to be readily available to stimulate repair at sites of gastrointestinal damage (Korhonen and Marnila, 2013). EGF inhibits the secretion of gastric acid and modulates the synthesis of a number of hormones (Ghosh and Playford, 2006). The concentrations of EGF vary between 4–320 μg/L and 2–155 μg/L in colostrum and mature bovine milk, respectively. BTC is present in cow colostrum (2.30 μg/L) and in cheese whey (2.59 μg/L) (Korhonen and Marnila, 2013). It has been observed that the growth factors in milk seemed to withstand and retain relatively well in pasteurization and even UHT heat treatment of milk (Gauthier et al., 2006).

EGF and BTC essentially stimulate the proliferation of epidermal, epithelial and embryonic cells. In addition, they inhibit the secretion of gastric acid and promote wound healing and bone resorption (Korhonen, 2009). The transforming growth factor (TGF-β family) plays an important role in the development of embryo, tissue repair, formation of bone and cartilage, and in regulation of the immune system (Pouliot and Gauthier, 2006). Both TGF-β1 and TGF-β2 are shown to stimulate proliferation of connective tissue cells and inhibit proliferation of lymphocytes and epithelial cells. Both forms of insulin-like growth factor (IGF) are also known to stimulate proliferation of many cell types and regulate some metabolic functions, such as glucose uptake and synthesis of glycogen (Pouliot and Gauthier, 2006; Korhonen, 2009).

IGF-I and IGF-II have been shown to promote cell proliferation and differentiation, and may contribute to development of the neonate (Korhonen and Marnila, 2013). The IGF peptide is released from binding proteins by acid treatment. Bovine colostrum contains 32–800 μg/L of IGF-I and mature bovine milk 4–27 μg/L (Gauthier et al., 2006; Ghosh and Playford, 2006). Platelet-derived growth factor (PDGF) is an acid-stable molecule synthesized and secreted by platelets and macrophages. PDGF promotes ulcer healing (Gauthier et al., 2006).

TGF-β_2 is the predominant isoform in the TGF-β family and is present at high concentrations in cow colostrum and milk (150–1150 and 13–71 μg/L, respectively). TGF-β_1 also occurs in cow colostrum and milk (12–43 and 0.8–4 μg/L, respectively) (Korhonen and Marnila, 2013). The TGF-βs maintain normal epithelial function in the non-damaged mucosa (Gauthier et al., 2006; Ghosh and Playford, 2006). Fibroblast growth factor-I and -II (FGF-I and FGF-II) play an important role in proliferation, differentiation and survival of a variety of different cell types. FGF-II has angiogenic properties involved in wound healing and hematopoiesis (Korhonen and Marnila, 2013). FGF-I has a concentration of 0.006 μg/L in cow milk, whereas FGF-II is detected at 0.02 μg/L in milk and at 0.5–1 μg/L in colostrum (Gauthier et al., 2006).

Caprine milk has a much higher level of growth factor activity than bovine milk does (Wu and Elsasser, 1995), thereby it was suggested that goat milk can be a feasible nutraceutical for gastrointestinal disorders (Wu et al., 2006; Park, 2009b). The existence of EGF in goat milk was found by using a human EGF (hEGF) polyclonal antibody (Denhard et al., 2000). Caprine milk also possesses the ability to reduce heat-induced gastrointestinal hyperpermeability (Prosser et al., 2004). Growth factors were sustainable in the harsh conditions of gastric acid exposure, and can be absorbed through the GI tract in neonates to act on other tissues (Gauthier et al., 2006; Park, 2009b), while some GFs are digested and may not be absorbed in adults. Playford et al. (2000) suggested that milk-derived growth factors may be useful as an orally administered treatment for a wide variety of gastrointestinal disorders.

5.2 Cytokines

Cytokines include chemokines, interferons, interleukins, lymphokines and tumour necrosis factor but generally not hormones or growth factors. A broad range of cells produce cytokines, including immune cells such as macrophages, B lymphocytes, T lymphocytes and mast cells, as well as endothelial cells, fibroblasts and various stromal cells (Stedman's

Medical Dictionary, 2006; Ibelgaufts, 2013). Cytokines are particularly important in the immune system. They modulate the balance between humoral and cell-based immune responses, and regulate the growth, maturation, repair of different cell types and responsiveness of specific cell populations. Some cytokines enhance or inhibit the action of other cytokines in complex ways (Ibelgaufts, 2013; Korhonen and Marnila, 2013).

Cytokines alter cellular metabolism and trigger acute cellular response during inflammation (Ghosh and Playford, 2006). Bovine colostrum and mastitic milk have shown to contain a number of cytokines derived from leukocytes, epithelial cells or lymphoid tissues of the udder. Cytokines, such as IL-1β, TNF-α, interferon-γ and soluble IL-1 receptor agonist, stimulate the immune system (Korhonen and Marnila, 2013).

Macrophages, T cells and epithelial cells secrete a pleiotropic cytokine called osteopontin (OPN) which affects immune and anti-inflammatory responses (Lönnerdal, 2011; Korhonen and Marnila, 2013). OPN is associated with the generation of T helper type 1 (Th1) and Th17 cells that inhibit some autoimmune diseases (Uede, 2011; Korhonen and Marnila, 2013). OPN has been associated with regulating immune response in infants (Lönnerdal, 2011).

5.3 Milk hormones

Hormones are substances that are released into the extracellular medium by the cells of a given tissue, and are any member of a class of signalling molecules produced by glands in multicellular organisms that are transported by the circulatory system to target distant organs to regulate physiology and behaviour as endocrine function (Rodriques, 2013). Hormones in milk originate from the blood and are secreted in milk through an active transport within the mammary gland. Also, some hormones can be synthesized by the mammary gland and excreted to milk (Grosvenor et al., 1993; Rodriques, 2013). In general, the hormones occur in very small concentrations (picograms or nanograms per millilitre) and the highest quantities are usually found in colostrum, declining thereafter drastically at the onset of the main lactation period (Korhonen and Marnila, 2013).

There are several main categories of hormones in milk: gonadal hormones (oestrogens, progesterone, androgens), adrenal (glucocorticoids), pituitary (prolactin, growth hormone) and hypothalamic hormones (gonadotropin-releasing hormone, luteinizing-hormone-releasing hormone, thyrotropin-releasing hormone and somatostatin) (Korhonen and Marnila, 2013; Rodriques, 2013). Other hormones of interest are bombesin, calcitonin, insulin, melatonin and parathyroid hormone (Korhonen and Marnila, 2013). Bovine colostrum and mature milk contain a large number of hormones of either steroidic or peptidic origin (Grosvenor et al., 1993; Jouan et al., 2006). The concentration of prolactin, for instance, found in colostrum varies between 500 and 800 ng/mL compared to 6–8 ng/mL in milk (Jouan et al., 2006). The hormones occurring in mammary secretions are considered important both in the regulation of specific functions of the mammary gland and in the growth of the newborn, including development and maturation of its gastrointestinal and immune systems. Bernt and Walker (1999) also elucidated that hormones in colostrum could temporarily regulate the activity of some endocrine glands until the newborn's hormonal system reaches maturity. In addition, oral ingestion of melatonin-enriched milk was observed to improve sleep and diurnal activity in animal model and human studies (Valtonen et al., 2005).

The category of gonadal hormone includes oestrogens, progesterone and androgens, where the androgens are the least studied. Because the amount of hormones in milk and milk products is very low, their accurate quantification remains a challenge (Rodriques, 2013). The concentrations of oestrogens, mainly 17β-estradiol, estrone and estriol in milk and several dairy products have been reported (Wolford and Argoudelis, 1979). It was found that the bovine milk fat fraction contains 65% of 17β-estradiol and 80% of estrone. Progesterone contents in cow milk determined by gas chromatography are reported as around 0.0003–0.0004 ng/mL (Darling et al., 1974). Ginther et al. (1976) reported progesterone levels of several cow dairy products, such as 11.4 ng/mL in whole milk, 4.7 ng/mL in skim milk and 58.8 ng/mL in cream.

Adrenal gland hormones, mainly glucocorticoids, have been identified in bovine milk, and their concentrations were found to be between 0.7 and 1.4 ng/mL^{-1} (Gwazdauskas et al., 1977). No remarkable differences have been found between whole and skim milk. Cortisol and corticosterone are the main glucocorticoids in blood plasma of cows (Tucker and Schwalm, 1977). The milk glucocorticoid levels during lactation represent only 4% of the blood plasma concentrations, indicating that only a small amount of glucocorticoid is transferred to the milk (Rodriques, 2013).

In pituitary hormones, two main hormones have been identified, namely prolactin and the growth hormone. Prolactin was detected by radioimmunoassay in bovine milk, with concentrations ranging from 5–200 ng mL^{-1} (Malven and McMurtry, 1974; Rodriques, 2013). The somatotropin or growth hormone had been detected in milk at lower than 1 ng mL^{-1} (Torkelson, 1987). The category of hypothalamic hormones includes: gonadotropin-releasing hormone, luteinizing hormone-releasing hormone, thyrotropin-releasing hormone and somatostatin. All these pituitary hormones in bovine milk have been detected and quantified by radioimmunoassay. Gonadotropin-releasing hormone level in milk was shown to be 5–6 times greater than in blood plasma (Baram et al., 1977).

Although less known and characterized, other types of hormones, such as proteins related to the parathyroid hormone, insulin, calcitonin, bombesin, erythropoietin and melatonin, have also been identified in milk (Rodriques, 2013). The existence of parathyroid hormone-related protein in bovine milk has been reported by many researchers (Ratcliffe et al., 1990; Rodriques, 2013). Bovine colostrum contains parathyroid hormone-related protein ranging from 0.67 to 5.0 nM, which is 100-fold greater than that in the blood plasma (Ballard et al., 1982). Calcitonin has been found to inhibit the liberation of prolactin, and calcitonin concentrations in human milk have been reported as 700 ng/mL (Koldovsky and Thornburg, 1987). Moreover, bombesin is known to influence the gastric hormonal secretions following ingestion (Lazarus et al., 1986).

5.4 Nucleosides and nucleotides

Nucleotides, nucleosides and nucleobases are non-protein-nitrogen (NPN) fraction of milk, where NPN contains other compounds such as urea, uric acid and ammonia. Schlimme et al. (2000) suggested that these nucleobase milk components can act as pleiotrophic factors in the development of brain functions. Nucleotides and nucleobases are the preferred forms for absorption in the intestine (Michaelidou, 2008). These minor nucleobase compounds exhibit species-specific patterns in the milk of different species, and have specific physiological influence on the early life of different mammals. The average nucleotide compositions of different species are shown in Table 7.

Schlimme et al. (2000) performed a detailed review on the compositional and chemical aspects of nucleosides and nucleotides in bovine milk and colostrum. The significance

Table 7 Average nucleotide[a] concentration in milk of various species (μmol/L)

Time postpartum	CMP	UMP	GMP	AMP	Reference
Human					
2 days	55.1	17.7	3.3	33.4	
15 days	26.4	7.0	–	26.0	Gil and Sanchez-Medina (1982)
Mature milk	18.3	9.3	–	15.1	
Mature milk	66.0	11.0	1.5	5.7	Thorell et al. (1996)
2 days	55.1	17.7	3.3	33.4	Boza (1998)
15 days	26.4	7.0	–	26.0	
Colostrum	23.0	6.8	1.0	1.4	Duchen and Thorell (1999)
Mature milk	61.5	6.4	1.0	1.9	
Bovine					
1–2 days	36.8	394.9	–	53.8	
5 days	30.2	28.7	8.3	31.5	Gil and Sanchez-Medina (1981)
15 days	49.0	–	–	29.1	
Mature milk	2.9	–	1.8	–	Tiemeyer et al. (1984)
Mature milk	26.6	Traces	Traces	Traces	Ferreira et al. (2001)
Ovine					
1–2 days	362.0	925.6	39.6	286.8	
3 days	104.3	1451.5	–	146.3	Gil and Sanchez-Medina (1981)
15 days	71.7	200.7	–	118.7	
Mature milk	48.6	110.7	Traces	54.1	Ferreira et al. (2001)
1–3 days	28.3	378.4	9.3	17.2	
4–5 days	31.4	300.9	9.7	16.3	Plakantara et al. (2007)
15 days	21.6	250.3	4.1	18.3	
Caprine					
1–2 days	39.4	558.6	–	23.1	
5 days	80.7	123.7	–	110.0	Gil and Sanchez-Medina (1981)
15 days	22.8	160.8	9.9	27.9	
Mature milk	72.5	227.2	Tracess	85.6	Ferreira et al. (2001)
1–3 days	20.2	292.0	8.8	20.3	
4–5 days	15.9	269.2	8.7	11.1	Plakantara et al. (2007)
15 days	8.7	145.4	6.6	5.5	

[a]UMP, uridyl-5′-monophosphate; CMP, cytidyl-5′-monophosphate; GMP, guanosyl-5′-monophosphate; AMP, adenosyl-5′-monophosphate.

Source: Michaelidou (2008).

of various actions of nucleotides especially to preterm and gestational age infants has long been recognized, and this premise has been suggested to apply in different clinical situations. Modified nucleosides may have effectiveness in human cell model systems, and may inhibit cell proliferation and activate apoptosis (Michaelidou, 2008). Food-derived inducers of apoptosis may be significant as exogenous anticarcinogens in the control of malignant cell proliferation, and the intestinal tract could be the primary target site for a possible selective apoptotic stimulant against malignant cells (Schlimme et al., 2000).

Owing to the bio- and trophochemical properties of dietary nucleosides and nucleotides, some infant and follow-on formulae have been supplemented with specific ribonucleotide salts. The European Commission has permitted the supplementation of specific ribonucleotide salts in the manufacture of those formulae. The increased interest in the bioactive role of nucleotides in infant nutrition has led to various studies and resulted in many scholarly publications on this research (Cosgrove, 1998; Michaelidou et al., 1998; Yu, 2002; Alles et al., 2004).

Dietary nucleotides have further effects on biosynthetic processes and modulation of gene expression, at least on those genes involved in nucleotide metabolism (Sanchez-Pozo and Gil, 2002). The impact of dietary nucleotides is exerted at the mucosal barrier, specifically through a sensing mechanism involved in purinergic signalling (Grimble and Westwood, 2001). Supplementation of nucleosides and nucleotide may be beneficial to the functions of the brain, while the influence on the gut may be dependent on the type of damage (Yamamoto et al., 1997). Caprine and ovine milk could be of interest in this regard due to significantly higher levels of certain nucleotides, compared to those of bovine and human milk.

5.5 Polyamines

A polyamine is an organic compound having two or more primary amino groups as cations.

Low-molecular-weight linear polyamines perform essential functions in all living cells. Primary examples of polyamines are putrescine, cadaverine, spermidine and spermine, which are found in milk. Polyamins are involved in DNA, RNA and protein synthesis, and they are also intimately engaged in the control of cell growth. However, the most important function of polyamines is to mediate the actions of all known hormones and growth factors, indicating that every cell in the body requires polyamines for its proper function (Bardocz et al., 1999). Moreover, the importance of polyamines in cell function is reflected in a strict regulatory control of their intracellular levels.

Polyamines are flexible polycations, fully charged under physiological pH conditions, essential for cell growth and proliferation, and exhibit various roles in cellular metabolism (Bardocz and White, 1999; Eliassen et al., 2002; Gugliucci, 2004; Larque et al., 2007). Considerable quantitative interspecies and interbreed variations were reported in the polyamine pattern in mammalian milk. As the milking time advances, considerable changes occur in polyamine concentrations, which may be attributed to the changes in the needs of the animals with age. Mature caprine milk has substantially higher polyamine contents, especially in spermidine and putrescine, than those in bovine and ovine milk. Human milk contains even much lower contents than bovine and ovine milks (Buts et al., 1993; Ploszaj et al., 1997).

The exogenous polyamines from the food sources are needed if polyamine requirements cannot be met by biosynthesis in the body (Michaelidou, 2008). Dietary polyamine intake can spare the organism the cost of de novo synthesis and may optimize tissue function (Jeevanandam et al., 1997). Thus, the importance of dietary polyamines depends on the physiological and pathological state of the individual. Polyamine requirements are higher in both animals and human for the periods of rapid growth. Michaelidou (2008) reported that the gastrointestinal tract, pancreas and spleen, which have a high cell turnover rate, are particularly dependent on dietary polyamines.

5.6 Organic acids

Organic acids are organic compounds with acidic properties. Examples of organic acids in milk include lactic acid, citric acid, pyruvic acid, uric acid, orotic acid, nucleic acid and neuraminic acid. They are, in general, weak acids and do not dissociate completely in water. Lower molecular mass organic acids such as formic and lactic acids are miscible in water, but higher molecular mass organic acids, such as benzoic acid, are insoluble, while most organic acids are very soluble in organic solvents (Park, 2006; Wikipedia, 2016).

Citric acid constitutes approximately 90% of the organic acids in milk (Park, 2009b). Average concentration of citrate in bovine milk is 1.7 g/L, ranging from 0.9 to 2.3 g/L (Renner, 1983). Citrate functions as a part of the buffer system of milk, contributes to the stability of the calcium caseinate complex and is the starting material for flavour substances in cultured milk products. It is a carboxylic acid, synthesized in the mammary gland from pyruvic acid (Renner, 1983; Park, 2009b).

Pyruvic acid is an important organic acid as a key intermediate in the intermediary metabolism of carbohydrates, amino acids and citrate by many organisms. Pyruvate is excreted into the milk during the catabolism of lactose and the oxidative deamination of alanine by microorganisms, and the initial content is 1 mg/L (Marshall et al., 1982; Park, 2009b). Pyruvic acid can be converted to a variety of end products, as a transitional substance in bacterial metabolism. Neuraminic or sialic acid plays a role in the stability of the CN complex, as well as functions in inhibition of the growth of *E. coli* and staphylococci bacteria (Park, 2009b). Neuraminic acid in milk presents in acetylated form as *N*-acetyl neuraminic acid, where the mean sialic acid level in bovine milk is approximately 150 mg/L, ranging 80–1000 mg/L (Renner et al., 1989).

Approximately 80% of the nucleotides in bovine milk exist in the form of orotic acid, which is almost absent in human milk (Counotte, 1983). Orotic acid is an intermediate compound in pyrimidine biosynthesis, and it is considered as a component of all cells. Ruminant milks contain high concentrations of orotic acid. Cow milk was shown to contain about 60 mg/L with a range between 10 and 120 mg/L (about 400–600 μmole/L), whereas goat and sheep milks have lower levels, about 120 and 30 μmole/L, respectively (Tiemeyer et al., 1984).

Nucleic acids in milk exist as ribonucleic acid (RNA), deoxyribonucleic acid (DNA) and nucleotides, and are constituents of all cells. Bovine milk and human milk have similar DNA contents as 1.2 and 1.5 mg/100 mL, respectively. However, human milk contains higher RNA than cow milk (11.5 vs. 5.4 mg/100 mL) (Renner, 1983). Increased intake of nucleic acids can lead to the formation of uric acid, which may result in urinary calculi and gout. Uric acid is a final product of the purine metabolism and is known to exert antioxidative activity. The uric acid content in bovine milk is around 200 μmole/L of protein and fat-free milk extract (Renner, 1983; Park, 2009b).

6 Bioactive minerals and vitamins

6.1 Minerals

Milk is a rich source of major and trace minerals, which exhibit a number of bioactive functionalities in the body. Milk as dietary source of minerals plays an important role in human health, nutrition and metabolisms. There is an important relationship between dietary minerals and the occurrence of specific diseases such as hypertension, osteoporosis, cancer and cardiovascular disease (Park, 2009b).

Underwood (1977) reported that six major and eight trace elements were recognized as essential minerals for growth, metabolism and development of pathology up to 1960. The six essential macro minerals include sodium, potassium, calcium, phosphorus, magnesium and chloride, and the eight essential trace minerals are iron, iodine, copper, manganese, zinc, cobalt, selenium and chromium. Differences in concentrations of major and trace minerals among three species milk (human, cow and goats) are compared in Table 8. Goat milk has been known to contain higher calcium, phosphorus, potassium, magnesium and chlorine, and lower sodium and sulphur contents than cow milk (Haenlein and Caccese, 1984; Park, 2006, 2009b). All essential minerals are involved in various types of milk enzymes and metabolic functions of the body.

6.1.1 Macro minerals

Among major minerals, calcium, phosphorus and potassium are present in substantially high concentrations in ruminant milk, compared to other minerals. Calcium is an important bioactive nutrient involved in the growth, metabolism and health of bone (Kanis, 1993; Park, 2006, 2009b). Calcium as a bioactive mineral is demonstrated widely in a range of calcium-fortified foods, including modified milk and beverages. Milk and dairy products provide approximately 70% of the recommended daily intake for calcium in Western diets. The US Food and Drug Administration has advised men and women over 50 years age to increase their calcium intakes towards 1200 mg/day. Calcium is integral for the development and maintenance of skeletal integrity and for the prevention of osteoporosis (Schaafsma et al., 1988).

Calcium plays an important role as the protective factor in the aetiology of colon cancer (Sorenson et al., 1988). Calcium is also associated with binding and removal of carcinogenic agents (bile salts, etc.) along the gastrointestinal tract (Regester et al., 1997), and even involved in resistance against infections of pathogenic bacteria (Bovee-Oudenhoven et al., 1997). Low calcium intake has been related to hypertension, whereby calcium supplementation reduced blood pressure in hypertensive patients (Grobbee and Hofman, 1986). People ingesting diets low in sodium and high in potassium, magnesium and calcium have been found to have less hypertension and cardiovascular disease (Morgan et al., 1986).

Caprine milk contains about 134 mg Ca and 121 mg P/100g (Park and Chukwu, 1988; Park, 2006), while human milk contains only one-fourth to one-sixth of these minerals. Phosphorus exhibits several important bioactive metabolic functions in the body, such as bone mineralization, energy metabolism (i.e. ATP and chemical energy), fat and carbohydrate metabolisms, body buffer system (acid–base balance and pH of the body) and formation and transport of nucleic acids and phospholipids across cell membranes for

Table 8 Mineral and vitamin contents of three species of milk

	Goat	Cow	Human
Constituents	Amount in 100 g		
Mineral			
Ca (mg)	134	122	33
P (mg)	121	119	43
Mg (mg)	16	12	4
K (mg)	181	152	55
Na (mg)	41	58	15
Cl (mg)	150	100	60
S (mg)	2.89	–	–
Fe (mg)	0.07	0.08	0.20
Cu (mg)	0.05	0.06	0.06
Mn (mg)	0.032	0.02	0.07
Zn (mg)	0.56	0.53	0.38
I (mg)	0.022	0.021	0.007
Se (μg)	1.33	0.96	1.52
Vitamin			
Vitamin A (I.U.)	185	126	190
Vitamin D (I.U.)	2.3	2.0	1.4
Thiamine (mg)	0.068	0.045	0.017
Riboflavin (mg)	0.21	0.16	0.02
Niacin (mg)	0.27	0.08	0.17
Pantothenic acid (mg)	0.31	0.32	0.20
Vitamin B_6 (mg)	0.046	0.042	0.011
Folic acid (μg)	1.0	5.0	5.5
Biotin (μg)	1.5	2.0	0.4
Vitamin B_{12} (μg)	0.065	0.357	0.03
Vitamin C (mg)	1.29	0.94	5.00

Data from Posati and Orr (1976), Park and Chukwu (1988), Jenness (1980), Haenlein and Caccese (1984), Park (2006, 2009b).

body cell functioning. Due to lack of availability of cow milk and limited meat consumption in under-developed countries, caprine milk has been an important daily food source of animal protein, phosphate and calcium (Haenlein and Caccese, 1984; Park, 1991, 2009b).

6.1.2 Trace minerals

There are important trace minerals in milk giving bioactive functions, including iron, zinc, iodine, selenium and manganese. Several other trace minerals exhibit active bioactivities in the body metabolisms, including Mo, Cr, Co, Cu, F, As, Sn and V. Unlike major minerals, concentrations of trace minerals are affected by diet, breed, individual animals and stages of lactation (Park and Chukwu, 1988; Park, 2006). A large proportion of copper, zinc and manganese are bound to milk CN. Iron and manganese are partly bound to LF, which is a bacteriostatic whey protein (0.2 mg/mL) (Lönnerdal et al., 1983, 1985).

Iron occurs in milk in combination with several proteins, such as LF and transferrin. Iron also occurs in blood as haemoglobin and transferrin in the plasma in the ratio of 1000:1, and a small quantity of ferritin is present in erythrocytes. Iron deficiency causes anaemia, impaired growth and lipid metabolism (Underwood, 1977). In a comparative Fe bioavailability study, Park et al. (1986) found that goat milk had a greater haemoglobin regeneration efficiency and iron bioavailability than cow milk. Rats fed with goat milk diet had greater iron and copper apparent digestibility coefficient and bioavailability in different animal organs, compared to those fed with cow milk diet, especially those animals with malabsorption syndrome (Barrionuevo et al., 2002).

Goat and cow milks contain greater zinc content than in human milk (Park and Chukwu, 1988). Zinc deficiency causes skin lesions, disturbed immune function, growth retardation and impaired wound healing (Underwood, 1977; Park, 2006). Rats fed on goat milk diet exhibited a greater bioavailability of zinc and selenium, compared to those fed on cow milk diet (Alferez et al., 2003). Iodine contents of caprine and bovine milk are significantly higher than human milk, which would be important for human nutrition, since iodine and thyroid hormone are closely related to the metabolic rate of physiological body functions (Underwood, 1977). Selenium contents of goat and human milk are significantly higher than in cow milk (Table 8). Less than 3% of the total selenium is related to the lipid fraction of milk. The selenium-dependent enzyme, glutathione peroxidase, is higher in goat milk than in human and cow milk. Total peroxidase activity of goat milk was 65%, as opposed to 29% for human and 27% for cow milk (Debski et al., 1987; Park, 2009b).

6.2 Vitamins

Vitamin is an organic compound that is essential for maintaining normal physiological function of the body. Vitamins are physiologically, biochemically and metabolically bioactive compounds that occur in milk and colostrum. Vitamins are divided into two categories: water-soluble and fat-soluble vitamins, and all known vitamins are contained in milk (Park, 2009b). Bovine milk is the rich source of human dietary requirements of vitamins, particularly riboflavin (B_2) and vitamin B_{12} (Park, 2006, 2009b).

Extensive researches have shown that subtle deficiencies in B vitamins may be risk factors for vascular and neurological diseases and cancers (Brachet et al., 2004). A combined deficiency of folate and vitamin B_{12} is associated with neuropsychiatric disorders among the elderly, development of dementia and Alzheimer's disease (Seshadri et al., 2002). Caprine milk has higher amounts of vitamin A than bovine milk. Caprine milk supplies adequate amounts of vitamin A and niacin, and excesses of thiamin, riboflavin and pantothenate for a human infant (Table 8; Parkash and Jenness, 1968; Park, 2006). If a human infant is fed solely on goat milk, it would oversupply with protein, Ca, P, vitamin A, thiamin, riboflavin, niacin and pantothenate in relation to the FAO-WHO requirements (Jenness, 1980; Park, 2009b). Compared to cow milk, goat milk has a significant drawback of deficiencies in folic acid and vitamin B_{12} (Davidson and Townley, 1977; Jenness, 1980; Park et al., 1986; Park, 2009b). Cow milk has five times more folate and vitamin B_{12} than goat milk, where folate is necessary for the synthesis of haemoglobin (Collins, 1962; Davidson and Townley, 1977). Folate and vitamin B_{12} deficiency has been reportedly implicated in 'goat milk anaemia', which is a megaloblastic anaemia in infants (Parkash and Jenness, 1968; Park, 2006).

Vitamin C has many bioactivities, including antioxidant, collagen biosynthesis, enzyme co-factors and tissue repairs. Vitamin C contents of cow, camel, buffalo, sheep, goat, human, donkey and mare milks are 27, 52, 22, 29, 16, 35, 49 and 61 mg/L, respectively (El-Agamy and Nawar, 2000). The levels of fat-soluble vitamins (A, D, E, K) in cow milk are influenced by different factors such as breed, parity, lactation period production level and health status (Baldi, 2005). Tocotrienols are a member of the vitamin E family, which possess powerful neuroprotective, antioxidant, anticancer and cholesterol-lowering properties (Sen et al., 2006). Vitamin K may have protective actions against osteoporosis, atherosclerosis and hepatocarcinoma (Kaneki et al., 2006).

7 Conclusions

Milk is a highly complex, and multicomponents containing mammary gland secretion, and is a most important source of essential nutrients and bioactive components for proper nutrition and health of all mammalian species, including human. Research in the identification of milk bioactive components and their functions has progressed tremendously in recent years for cow milk, while this line of research is in its beginning stage for other dairy species milk. Numerous components in milk modulate crucial physiological functions and regulatory processes, which include hormone secretion (casomorphins), immune defence (casokinins, casomorphins, immunopeptides), nutrient uptake (phosphopeptides and casomorphins), neurological transmission (casokinins), antioxidant, antimicrobial, antihypertensive, opioid and anticancer activities.

The development of health-promoting functional foods promoted by the current global interest of health foods will provide a timely opportunity to tap the myriad of innate bioactive milk components for inclusion in formulations of various health foods, infant formulae and pharmaceutical agents. The processing techniques for industrial or semi-industrial scale have been available for fractionation and isolation of major proteins from colostrum and milk. The advances in processing technology, including ultrafiltration, microfiltration, reverse osmosis or ion exchange among others, have resulted in the presence of a multitude of whey products as bioactive ingredients in the market. Health-promoting products such as CNs and whey-derived native proteins, peptides, growth factors and lipid fractions have already been manufactured commercially.

In the near future, it is expected that several breakthrough products based on these bioactive compounds from milk and colostrum will be launched on foods and pharmaceutical markets worldwide. The development of these nutrition and health-promoting products may be targeted to infants, the elderly, as well as people with chronic diseases, allergies and immune deficiencies. Furthermore, there is a need to study and exploit the health potential of other types of minor bioactive components naturally occurring in colostrum and milk. More scientific studies are desired in the future to investigate the hidden health-promoting potentials in the wide range of traditional dairy products consumed worldwide. In addition, more industrial and commercial production of milk-derived functional ingredients may be produced from milks of non-bovine dairy species, such as goats, buffalo, sheep, camels, mares, yaks and other domesticated mammals.

8 Where to look for further information

Further information may be found in the following two publications and elsewhere:

Bioactive Components in Milk and Dairy Products. 2009. In Y. W. Park (ed.), Wiley-Blackwell Publishers, Ames, IA and Oxford, England, p. 440. ISBN 978-0-8138-1982-2.

Milk and Dairy Products in Human Nutrition. 2013. In Y. W. Park and G. F. W. Haenlein (eds), Wiley-Blackwell Publishers, Ames, IA and Oxford, England, p. 728. ISBN 978-0-470-67418-5

9 References

Abd El-Salam, M. H., El-Shibini, S. and Buchheim, W. 1996. Characteristics and potential uses of the casein macropeptide. *Int. Dairy J.* 6: 327–41.

Ahrne, L., Bjorck, L. and Claesson, O. 1983. Alkylglycerols in bovine colostrum and milk in relation to precursor levels. *J. Dairy Res.* 50: 201.

Akuzawa, R., Miura, T. and Kawakami, H. 2009. Bioactive components in caseins, caseinates and cheeses. In *Bioactive Components in Milk and Dairy Products*, edited by Y. W. Park, 217–33. Wiley-Blackwell, Iowa.

Alferez, M. J., Lopez-Aliage, I., Barrionuevo, M. and Campos, M. S. 2003. Effects of dietary inclusion of goat milk on the bioavailability of zinc and selenium in rats. *J. Dairy. Res.* 70: 181–7.

Alles, M. S., Scholtens, P. A. M. J. and Bindels, J. G. 2004. Current trends in the composition of infant milk formulas. *Curr. Pediatr.* 14: 51–63.

Andrews, G. 1989. Lactulose in heated milk. *Bull. Int. Dairy Fed.* 238: 46–52.

Aro, A., Männistö, S., Salminen, I., Ovaskainen, M.-L., Kataja, V. and Uusitupa, M. 2000. Inverse association between dietary and serum conjugated linoleic acid and risk of breast cancer in postmenopausal women. *Nutr. Cancer* 38: 151–7.

Ashar, M. N. and Chand, R. 2004. Fermented milk containing ACE-inhibitory peptides reduces blood pressure in middle aged hypertensive subjects. *Milchwissenschaft* 59: 363–6.

Avramis, C. A., Wang, H., McBride, B. W., Wright, T. C. and Hill, A. R. 2003. Physical and processing properties of milk, butter, and Cheddar cheese from cows fed supplemental fish meal. *J. Dairy Sci.* 86: 2568–76.

Baldi, A. 2005. Vitamin E in dairy cows. *Livestock Prod. Sci.* 98: 117–22.

Ballard, F., Nield, M., Francis, G., Dahlenburg, G. and Wallace, J. 1982. The relationship between the insulin content and inhibitory effects of bovine colostrum on protein break down in cultured cells. *J. Cell Biol.* 110: 249–54.

Baram, T., Koch, Y., Hazum, E. and Frikin, M. 1977. Gonadotropin releasing hormone in milk. *Science* 198: 300–2.

Barbana, C., Sánchez, L. and Pérez, M. D. 2011. Bioactivity of α-lactalbumin related to its interaction with fatty acids: a review. *Crit. Rev. Food Sci. Nutr.* 51(8): 783–94.

Bardocz, S., Grant, G., Hughes, E. L., Brown, D. S. and Pusztai, A. 1999. Uptake,inter-organ distribution and metabolism of dietary polyamines in the rat. In *Polyamines in Health and Nutrition*, edited by S. Bardocz and A. White, 241–58. Kluwer Academic Publishers, Massachusetts.

Bardocz, S. and White, A. 1999. Estimaton of the polyamine boby pool: contribution by de novo biosynthesis, diet and luminal bacteria. In *Polyamines in Health and Nutrition*, edited by S. Bardocz and A. White, 117–22. Kluwer Academic Publishers, Massachusetts.

Barrionuevo, M., Alferez, M. J. M., Lopez Aliaga, I., Sanz Sampelayo, M. R. and Campos, M. S. 2002. Beneficial effect of goat milk on nutritive utilization of iron and copper in malabsorption syndrome. *J. Dairy Sci.* 85: 657–64.

Baruchel, S. and Viau, G. 1996. In vitro selective modulation of cellular glutathione by a humanized native milk protein isolate in normal cells and rat mammary carcinoma model. *Anti. Res.* 16: 1095–100.

Bellamy, W. R., Yamaguchi, K., Wakabayashi, H., Takase, M., Shimamura, S. and Tomita, M. 1994. Antifungal properties of lactoferricin, a peptide derived from the N-terminal region of bovine lactoferrin. *Lett. Appl. Microbiol.* 18: 230–3.

Bennett, L. E., Crittenden, R., Khoo, E. and Forsyth, S. 2005. Evaluation of immune-modulatory dairy peptide fractions. *Aus. J. Dairy Technol.* 60: 106–9.

Bernt, K. M. and Walker, W. A. 1999. Human milk as carrier of biochemical messages. *Acta Pediatr.* 430(Suppl.): 27–41.

Berrocal, R., Chanton, S., Juillerat, M. A., Paavillard, B., Scherz, J.-C. and Jost, R. 1989. Tryptic phosphopeptides from whole casein: II. Physiochemical properties related to the solubilization of calcium. *J. Dairy. Res.* 56: 335–41.

Bisig, W., Eberhard, P., Collomb, M. and Rehberger, B. 2007. Influence of processing on the fatty acid composition and the content of conjugated linoleic acid in organic and conventional dairy products- a review. *Lait* 87: 1–19.

Bjorck, L. 1985. *Alkylglycerols in Milk-Fat.* Lipidforum Symposium, Goteborg, p. 23.

Blanc, P., Daures, J. P. and Rouillon, J. M. 1992. Lactitol of lactulose in the treatment of chronic hepatic encephalopathy: results of a meta analysis. *Hepatology* 15: 222–8.

Blouet, C., Mariotti, F., Mikogami, T., Tome, D. and Huneau, J. 2007. Meal cysteine improves postprandial glucose control in rats fed a high-sucrose meal. *J. Nutr. Biochem.* 18(8): 519–24.

Boelsma, E. and Kloek, J. 2009. Lactotripeptides and antihypertensive effects: a critical review. *Br. J. Nutr.* 101(6): 776–86.

Bonomi, F., Brandt, R., Favalli, S., Ferranti, P., Fierro, O., Frökier, H., Ragg, E. and Lametti, S. 2011. Structural determinants of the immunomodulatory properties of the C-terminal region of β-casein. *Int. Dairy J.* 21: 770–6.

Boots, J.-W. and Floris, R. 2006. Lactoperoxidase: from catalytic mechanism to practical applications. *Int. Dairy J.* 16: 1272–6.

Bostwick, E. F., Stejins, J. and Braun, S. 2000. Lactoglobulins. In *Natural Food Antimicrobial Systems*, edited by A. S. Naidu, 133–58. CRC Press, Boca Raton, FL, USA.

Bouhours, J. F., Grousson, D., Dorier, A. and Baques, C. 1984. Glycolipids in the milk of women, cows, and rabbits. *Reprod. Nutr. Devel.* 24: 698.

Bounous, G. and Gold, P. 1991. The biological activity of undenatured dietary whey proteins: role of glutathione. *Clin. Invest. Med.* 14: 296–309.

Bovee-Oudenhoven, I., Termont, D., Weerkamp, A. H., Faassen-Peeters, M. A. W. and van der Meer, R. 1997. Dietary calcium inhibits the intestinal colonization and translocation of *Salmonella* in rats. *Gestroenterology* 113: 550–7.

Boza, J. 1998. Nucleotides in infant nutrition. *Monatsschr. Kinderheilkunde* 146: S39–48.

Brachet, P., Chanson, A., Demigne, C., Batifoulier, F., AlexandreGouabau, M. C. and Tyssandier, V. 2004. Age-associated B vitamin deficiency as a determinant of chronic diseases. *Nutr. Res. Rev.* 17: 55–68.

Branti, V., Teschemacher H., Blasig J., HenschenA. and Lottspeich F. 1981. Opioid activities of β-casomorphins. *Life Sci.* 28: 1903–9.

Bullen, J. J., Rogers, H. J. and Leigh, L. 1972. Iron binding proteins in milk and resistance to *E. coli* infections in infants. *Br. Med. J.* 1: 69–75.
Burton-Freeman, B. M. 2008. Glycomacropeptide (GMP) is not critical to whey-induced satiety, but may have a unique role in energy intake regulation through cholecystokinin (CCK). *Physiol. Behav.* 93: 379–87.
Butler, J. E. 1994. Passive immunity and immunoglobulin diversity. *Int. Dairy Fed. Spec. Issue* 9404: 14–50.
Buts, J. P., De Keyser, N., Kolanowski, J., Sokal, E. and Van Hoof, F. 1993. Maturation of villus and crypt cell functions in rat small intestine role of dietary polyamines. *Dig. Dis. Sci.* 38: 1091–8.
Calder, P. C. 1994. Glutamine and the immune system. *Clin. Nutr.* 13: 2–8.
Camma, C., Fiorello, F., Tine, F., Marchesini, G., Fabbri, A. and Pagliaro, L. 1993. Lactitol in the treatment of chronic hepatic encephalopathy: a meta-analysis. *Dig. Dis. Sci.* 38: 916–22.
Cashman, K. D. 2006. Milk minerals (including trace elements) and bone health. *Int. Dairy J.* 16: 1389–98.
Cerbulis, J., Parks, O. W. and Farrell, H. M. 1982. Composition and distribution of lipids of goats milk. *J. Dairy Sci.* 65: 2301.
Chabance, B., Jolles, P., Izquierdo', C., Mazoyer, E., Francoual, C., Drouet, L. and Fiat, A.-M. 1995. Characterization of an antithrombotic peptide from K-casein in newborn plasma after milk ingestion. *Br. J. Nutr.* 73: 583–90.
Chabance, B., Marteau, P., Rambaud, J. C., Migliore-Samour, D., Boynard, M., Perrotin. P., Guillet, R., Jollés, P. and Fiat, A.-M. 1998. Casein peptide release and passage to the blood in humans during digestion of milk or yogurt. *Biochimie* 80: 155–65.
Chamba, J.-F., Chardigny, J.-M., Gnädig, S., Perreard, E., Chappaz, S., Rickert, R., Steinhart, H. and Sébédio, J.-L. 2006. Conjugated linoleic acid (CLA) content of French Emmental cheese: effect of season, region of production, processing and culinary preparation. *Lait* 86: 407–80.
Chang, K. J., Killian, A., Hazum, E. and Cuatrecasas, P. 1981. Morphiceptin (NH_4-Tyr-Pro-Phe-Pro-$CONH_2$): a potent and specific agonist for the morphine (μ) receptors. *Science*, Washington, DC, 212: 75–7.
Chatterton, D. E. W., Smithers, G., Roupas, P. and Brodkorb, A. 2006. Bioactivity of β-lactoglobulin and α-lactalbumin- technological implications for processing. *Int. Dairy J.* 16: 1290–40.
Cheetham, N. W. H. and Dube, F. E. 1983. Preparation of lacto-n-neotetraose from human-milk by high- performance liquid-chromatography. *J. Chromatog.* 262: 426.
Chiba, H. and Yoshikawa, M. 1986. Biologically functional peptides from food proteins. In *Protein Tailoring for Food and Medical Uses*, edited by R. E. Feeney and J. R. Whitaker, 123–53. Marcel Dekker, New York.
Chilliard, Y. and Ferlay, A. 2004. Dietary lipids and forages interactions on cow and goat milk fatty acid composition and sensory properties. *Reprod. Nutr. Dev.* 44: 467–92.
Chobert, J.-M., El-Zahar, K., Sitohy, M., Dalgalarrondo, M., Métro, F., Choiset, Y. and Haertle, T. 2005. Angiotensin I-converting-enzyme (ACE)- inhibitory activity of tryptic peptides of ovine β-lactoglobulin and of milk yoghurts obtained by using different starters. *Lait* 85: 141–52.
Chonan, O. and Watanuki, M. 1995. Effect of galactooligosaccharides on calcium absorption in rats. *J. Nutr. Sci. Vitaminol.* 41: 95–104.
Clare, D. A., Catignani, G. L. and Swaisgood, H. E. 2003. Biodefense properties of milk: the role of antimicrobial proteins and peptides. *Curr. Pharm. Des.* 9: 1239–55.
Clare, D. A. and Swaisgood, H. E. 2000. Bioactive milk peptides: a prospectus. *J. Dairy Sci.* 83: 1187–95.
Clemens, R. A. 2011. Milk A1 and A2 peptides and diabetes. *Nestle Nutr. Workshop Ser. Pediatr. Program.* 67: 187–95. doi: 10.1159/000325584.
Collins, R. A. 1962. Goat's milk anemia in retrospect. *Am. J. Clin. Nutr.* 11: 169.
Collomb, M., Bütikofer, U., Sieber, R., Bosset, J. O. and Jeangros, B. 2001. Conjugated linoleic acid and *trans* fatty acid compositionof cow's milk fat produced in lowlands and highlands. *J. Dairy Res.* 68: 519–23.

Collomb, M., Schmid, A., Sieber, R., Wechsler, D. and Ryhänen, E.-L. 2006. Conjugated linoleic acids in milk fat: variation and physiological effects. *Int. Dairy J.* 16: 1347–61.

Cornish, J., Callon, K. E., Naot, D., Palmano, K. P., Banovic, T., Bava, U., Watson, M., Lin, J. M., Tong, P. C., Chen, Q., Chan, V. A., Reid, H. E., Fazzalari, N., Baker, H. M., Baker, E. N., Haggarty, N. W., Grey, A. B. and Reid, I. R. 2004. Lactoferrin is a potent regulator of bone cell activity and increases bone formation *in vivo*. *Endocrinol.* 145: 4366–74.

Cosgrove, M. 1998. Nucleotides. *Nutrition* 14: 748–51.

Counotte, G. H. M. 1983. Determination of orotate in ruminant milk by high-performance liquid-chromatography. *J. Chromatogr.* 276(2): 423–6.

Daniel, H., Vohwinkel, M. and Rehner, G. 1990. Effect of casein and beta-casomorphins on gastrointestinal motility in rats. *J. Nutr.* 3: 252–7.

Darling, J., Laing, A. and Harkness, R. 1974. A survey of the steroids in cow's milk. *J. Endocrinol.* 62: 291–7.

Davidson, G. P. and Townley, R. R. W. 1977. Structural and functional abnormalities of the small intestine due to nutritional folic acid deficiency in infancy. *J. Pediat.* 90: 590.

Debski, B., Picciano, M. F. and Milner, J. A. 1987. Selenium content and distribution of human, cow and goat milk. *J. Nutr.* 117.

De Leeuw, P. W., Van der Zander, K., Kroon, A. A., Rennenberg, R. M. and Koning, M. M. 2009. Dose-dependent lowering of blood pressure by dairy peptides in mildly hypertensive subjects. *Blood Press.* 18(1–2): 44–50.

del Mar Contreras, M., Carron, R., Montero, M. J., Ramos, M. and Recio, I. 2009. Novel casein-derived peptides with antihypertensive activity. *Int. Dairy J.* 10: 566–73.

Denhard, M., Claus, R., Munz, O. and Weiler, U. 2000. Course of epidermal growth factor (EGF) and insulin-like growth factor (IFG-I) in mammary secretions of the goat during end-pregnancy and early lactation. *J. Vet. Med. Ser. A* 47: 533–40.

Devine, D. A. and Hancock, R. E. W. 2002. Cationic peptides distribution and mechanism of resistance. *Curr. Pharm. Design.* 8: 703–14.

Dhiman, T. R., Helmink, E. D., McMahon, D. J., Fife, R. L. and Pariza, M. W. 1999. Conjugated linoleic acid content of milk and cheese from cows fed extruded oilseeds. *J. Dairy Sci.* 82: 412–19.

Donkor, O., Henriksson, A., Vasiljevic, T. and Shah, N.P. 2007. Proteolytic activity of dairy lactic acid bacteria and probiotics as determinant of growth and in vitro angiotensin-converting enzyme inhibitory activity in fermented milk. *Lait* 86: 21–38.

Duchen, K. and Thorell, L. 1999. Nucleotide and polyamine in colostrum and mature milk in relation to maternal atopy and atopic development in chollдres. *Acta Paediatr.* 88: 1338–43.

Edgar, W. M. 1993. Extrinsic and intrinsic sugars: a review of recent UK recommendations on diet and caries. *Caries Res.* 27 (Suppl 1): 64–7.

El-Agamy, E. I. 2009. Bioactive components of camel milk. In *Bioactive Components in Milk and Dairy Products*, edited by Y. W. Park, 159–94. Wiley-Blackwell Publishers, Ames, Iowa and Oxford, England.

El-Agamy, E. I., Abou Shloue, Z. I. and Abdel-Kader, Y. I. 1998. Antimicrobial factors and nutritive value of human milk and other species. *J. Agr. Sci. Mansoura Univ.* 23(1): 245–54.

El-Agamy, E. I. and Nawar, M. A. 2000. Nutritive and immunological values of camel milk: a comparative study with milk of other species. In 2nd International Camelied Conference. Agroeconomics of camelied farming, Almaty, Kazakhstan, 8–12 September.

Elfahri, K. R., Vasiljevic, T., Yeager, T. and Donkor, O. N. 2016. Anti-colon cancer and antioxidant activities of bovine skim milk fermented by selected *Lactobacullus helveticus* stranis. *J. Dairy Sci.* 99: 31–40.

El-Hatmi, H., Levieux, A. and Levieux, D. 2006. Camel (*Camelus drome*darius) immunoglobulin G, alpha-lactalbumin, serum albumin and lactoferrin in colostrum and milk during the early post partum period. *J. Dairy Res.* 73(3): 288–93.

Eliassen, K. A., Reistad, R., Risoen, U. and Ronning, H. F. 2002. Dietary polyamines. *Food Chem.* 78: 273–80.

Elitsur, Y. and Luk, G. D. 1991. β-casomorphin (BCM) and human colonic lamina propria lymphocyte proliferation. *Clin. Exp. Immunol.* 85: 493–7.

FAO/WHO. 2006. Benefits and potential risks of the lactoperoxidase system of raw milk preservation. *Report of an FAO/WHO technical meeting.* FAO Headquarters, Rome, 28 November to 2 December 2005.

Farnaud, S. and Evans, R. W. 2003. Lactoferrin-amultifunctional protein with antimicrobial properties. *Mol. Immunol.* 40: 395–405.

Ferreira, I. M. P. L. V. O., Mendes, E., Gomes, A. M. P., Faria M. A. and Ferreira, M. A. 2001. The determination and distribution of nucleotides in Dairy products using HPLC and diode arrayDetection. *Food Chem.* 74: 239–44.

Fiat, A. M., Miglilore-Samour, D., Jolles, P., Crouet, L., Collier, C. and Caen, J. 1993. Biologically active peptides from milk proteins with emphasis on two example concerning antithrombotic and immuno-modulating activities. *J. Dariy Sci.* 76: 301–10.

FitzGerald, R. J. 1998. Potential uses of caseinophosphopeptides. *Int. Dairy J.* 8: 451–7.

FitzGerald, R. J. and Meisel, H. 2000. Milk protein derived peptide inhibitors of angiotensin-I converting enzyme. *Br. J. Nurt.* 84(Suppl. I): S33–7.

FitzGerald, R. J. and Meisel, H. 2003. Milk protein hydrolysates and bioactive peptides. In *Advances in Dairy Chemistry*, edited by P. F. Fox and P. L. H. McSweeney, 675–98, 3rd ed, Kluwer Academic/ Plenum Publishers, New York.

FitzGerald, R. J., Murray, B. A. and Walsh, D. J. 2004. Hypotensive peptides from milk proteins. *J. Nutr.* 134: 980S–8S.

Floris, R., Recio, I., Berkhout, B. and Visser, S. 2003. Antibacterial and antiviral effects of milk proteins and derivatives there of. *Curr. Pharm. Des.* 9: 1257–75.

Foltz, M., Meynen, E. E., Bianco, V., van Platerink, C., Koning, T. M. and Klock, J. 2007. Angiotensin converting enzyme inhibitoy peptides from a lactotripeptide – enriched milk beverage are absorbed intact into the circulation. *J. Nutri.* 137: 953–8.

Fox, P. F. and Brodkorb, A. 2008. The casein micelle: historical aspects, current concepts and significance. *Int. Dairy J.* 18: 677–84.

Gagnaire, V., Molle, D., Herrouin, M. and Leonil, J. 2001. Peptides identified during Emmental cheese ripening: origin and proteolytic systems involved. *J. Agric. Food Chem.* 49: 4402–13.

Galli, C., Tremoli, D., Giani, D., Maderna, P., Bianfranceschi, G. and Sirtori, C. R. 1985. Oral poly-unsaturated phosphatidylcholine reduces platelet lipid and cholesterol contents in healthy volunteers. *Lipids* 20: 561.

Gauthier, F. G., Pouliot, Y. and Saint-Sauveur, D. 2006. Immunomodulatory peptides obtained by the enzymatic hydrolysis of whey proteins. *Int. Dairy J.* 16: 1315–23.

Geerlings, A., Villar, I. C., Hidalga Zarco, F., Sanchez, M., Vera, R., Zafra Gomes, A., Boza, J. and Duarte, J. 2006. Identification and characterization of novel angiotensin-converting enzyme inhibitors obtained from goat milk. *J. Dairy Sci.* 89: 3326–35.

German, J. B., Dillard, C. J. and Ward, R. E. 2002. Bioactive components in milk. *Curr. Opin. Clin. Metab. Care.* 5: 653–8.

Gerson, C., Sabater, J., Scuri, M., Torbati, A., Coffey, R., Abraham, J., et al. 2000. The lactoperoxidase system functions in bacterial clearance of airways. *Am. J. Respir. Cell Mol. Biol.* 22: 665–71.

Ghosh, S. and Playford, R. J. 2006. Relevance of growth factors for the gastrointestinal tract and other organs. chapter 11. In *Nutraceutical Proteins and Peptides in Health and Disease*, edited by Y. Mine and F. Shahidi, 217–41. CRC Press, Taylor & Francis Group, LLC, Boca Raton, USA.

Gill, H. S., Coull, F., Rutherfurd, K. J. and Cross, M. L. 2000. Immunoregulatory peptides in bovine milk. *Br. J. Nutr.* 84: S111–17.

Gil, A. and Sanchez-Medina, F. 1981. Acid-soluble nucleotides of cow's, goat's and sheep's milks at different stages of lactation. *J. Dairy Res.* 48: 35–44.

Gil, A. and Sanchez-Medina, F. 1982. Acid-soluble nucleotides of human milk at different stages of lactation. *J. Dairy Res.* 49: 301–7.

Ginther, O., Nuti, L., Garcia, M., Wentworth, B. and Tyler, W. 1976. Factors affecting progesterone concentration in cow's milk and dairy products. *J. Anim. Sci.* 42: 155–9.

Gobbetti, M., Minervini, F. and Rizzello, C. G. 2004. Angiotensin I-converting-enzyme-inhibitory and antimicrobial bioactive peptides. *Int. J. Dairy Technol.* 57: 172–88.

Gobbetti, M., Minervini, F. and Rizzello, C. G. 2007. Bioactive peptides in dairy products. In *Handbook of Food Products Manufacturing*, edited by Y. H. Hui, 489–517. John Wiley & Sons, Inc., New York.

Gomez-Ruiz, J. A., Ramos, M. and Recio, I. 2002. Angiotensin-converting enzyme-inhibitory peptides in Manchego cheeses manufactured with different starter cultures. *International Dairy Journal* 12: 697–706.

Griinari, J. M., Corl, B. A., Lacy, S. H., Choinard, P. Y., Nurmela, K. V. and Bauman, D. E. 2000. Conjugated linoleic acid is synthesized endogenously in lactating dairy cows by Δ9-desaturase. *J. Nutr.* 130: 2285–91.

Grimble, R. F. and Westwood, O. M. 2001. Nucleotides as immunomodulators in clinical nutrition. *Curr. Opin. Clin. Nutr. Metab. Care* 4: 57–64.

Grobbee, D. E. and Hofman, A. 1986. Effect of calcium supplementation on diastolic blood-pressure in young-people with mild hypertension. *Lancet* 2(8509): 703–7.

Grosvenor, C. E., Picciano, M. F. and Baumrucker, C. R. 1993. Hormones and growth factors in milk. *Endocr. Rev.* 14: 710–28.

Gugliucci, A. 2004. Polyamines as clinical laboratory tools. *Clin. Chim. Acta* 344: 23–35.

Guilloteau, P., Romé, V., Delaby, L., Mendy, F., Roger, L. and Chayvialle, J. A. 2010. A new role of phosphopeptides as bioactive peptides released during milk casein digestion in the young mammal: regulation of gastric secretion. *Peptides* 30: 2221–7.

Gustafson, D. R., McMahon, D. J., Morrey, J. and Nan, R. 2001. Appetite is not influenced by a unique milk peptide: caseinomacropeptide (CMP). *Appetite* 36: 157–63.

Gwazdauskas, F., Pappe, M. and Mc Gilliard, M. 1977. Milk and plasma glucocorticoid alterations after injections of hydrocortisone and adrenocorticotropin. *Proc. Soc. Exp. Biol. Med.* 154: 543–5.

Haenlein, G. F. W. and Caccese, R. 1984. Goat milk versus cow milk. In *Extension Goat Handbook*, edited by G. F. W. Haenlein and D. L. Ace, E–1: 1–4. USDA Publication, Washington, DC.

Hansen, M. 1995. The effect of casein phosphopeptides on zinc and calcium absorption evaluated in experimental models and humans. PhD thesis. The Royal Veterinary and Agricultural University, Denmark.

Hartmann, R., Gunther, S., Martin, D., Meisel, H., Pentzien, A. K., Schlimme, E. and Scholz, N. 2000. Cytochemical model systems for the detection and characterization of potentially bioactive milk components. *Kieler Milchwirtschaftliche Forschungsberichte* 52: 61–85.

Hartmann, R. and Meisel, H. 2007. Food-derived peptides with biological activity: from research to food applications. *Curr. Opin. Biotechnol.* 18: 1–7.

Hata, Y., Yamamoto, M., Ohni, H., Nakajima, K., Nakamura, Y. and Takano, T. 1996. A placebo-controlled study of the effect of sour milk on blood pressure in hypertensive subjects. *Am. J. Clin. Nutr.* 64: 767–71.

Hernández-Hernández, O., Moreno, F. J., Kolida, S., Rastall, R. A. and Sanz, M. L. 2011. Effect of glycation of bovine β-lactoglobulin with galactooligosaccharides on the growth of human faecal bacteria. *Int. Dairy J.* 21(12): 949–52.

Hernández-Ledesma, B., Ramos, M. and Gomez-Ruiz, J. A. 2011. Bioactive components of ovine and caprine cheese whey. *Small Rum. Res.* 101: 196–204.

Hernández-Ledesma, B., Recio, I. and Amigo, L. 2008. beta-Lactoglobulin as source of bioactive peptides. *Amino Acids* 35: 257–65.

Hill, R. D., Lahov, E. and Givol, D. 1974. A rennin-sensitive bond in alpha and beta casein. *J. Dairy Res.* 41: 147–53.

Hirota, T., Ohki, K., Kawagishi, R., Kajimoto, Y., Mizuno, S. and Nakamura, Y. 2007. Casein hydrolysate containing the antihypertensive tripeptides Val-Pro-Pro and Ile-Pro-Pro improves vascular endothelial function independent of blood pressure–lowering effects: contribution of the inhibitory action of angiotensin-converting enzyme. *Hypertens. Res.* 30: 489–96.

Hoerr, R. A. and Bostwick, E. F. 2002. Commercializing colostrum-based products: a case study of GalaGen. *Bull. Int. Dairy Fed.* 375: 33–46.

Horton, B. S. 1995. Commercial utilization of minor milk components in the health and food industries. *J. Dairy Sci.* 78: 2584–9.

Ibelgaufts, H. 2013. Cytokines. In *Cytokines & Cells Online Pathfinder Encyclopedia Version 31.4* Spring/Summer 2013 Edition.

Imaizumi, K., Mawatari, K., Murata, I. and Sugno, M. J. 1983. The contrasting effect of dietary phosphatidylethanolamine and phosphatidylcholine on serum lipoproteins and liver lipids in rats. *Nutrition* 113: 2403.

Innis, S. M. 1985. The role of diet during development on the regulation of adult cholesterol homeostasis. *Can. J. Physiol Pharmacol.* 63: 557.

Ip, M. M., Masso-Welch, P. A. and Ip, C. 2003. Prevention of mammary cancer with conjugated linoleic acid: Role of the stroma and the epithelium. *Journal of Mammary Gland and Neoplasia* 8: 103–18.

Iwami, K., Sakakibara, K. and Ibuki, F. 1986. Involvement of post-digestion hydrophobic peptides in plasma cholesterol-lowering effect of dietary plant protein. *Agri. Bio. Chem.* 50: 1217–22.

Jahreis, G., Fritsche, J. and Kraft, J. 1999. Species dependent, seasonal, and dietary variation of conjugated linoleic acid in milk. In *Advances in Conjugated Linoleic Acid*, Volume 1, edited by M. P. Yurawecz, M. M. Mossoba, J. K. G. Kramer, M. W. Pariza and G. J. Nelson, 215–25, American Oil Chemists Society, Champaign, IL.

Jäkälä, P. and Vapaatalo, H. 2010. Antihypertensive peptides from milk proteins. *Pharmaceuticals* 3: 251–72.

Jauhiainen, T., Rönnback, M., Vapaatalo, H., Wuolle, K., Kautiainen, H. and Korpela, R. 2007. *Lactobacillus helveticus* fermented milk reduces arterial stiffness in hypertensive subjects. *In. Dairy J.* 17: 1209–11.

Jauhiainen, T., Vapaatalo, H., Poussa, T., Kyrönpalo, S., Rasmussen, M. and Korpela, R. 2005. *Lactobacillus helveticus* fermented milk lowers blood pressure in hypertensive subjects in 24-h ambulatory blood pressure measurement. *Am. J. Hypertens.* 18: 1600–5.

Jeevanandam, M., Holaday, N. J., Begay, C. K. and Petersen, S. R. 1997. Nutritional efficacy of a spermidine supplemented diet. *Nutrition* 13: 788–94.

Jenness, R. 1980. Composition and characteristics of goat milk: review 1968–1979. *J. Dairy Sci.* 63: 1605–30.

Jensen, R. G. 2002. The composition of bovine milk lipids: January 1995–December 2000. *J. Dairy Sci.* 85: 295–350.

Johnson, E. A. 1994. Egg white lysozyme as a preservative for use in foods. In *Egg uses and Processing Technologies, New Developments*, edited by J. S. Sim and S. Nakai, 177–91. CAB International, Oxfordshire.

Jolles, P., Levy-Toledano, S., Fiat, A. M., Soria, C., Gillesen, D., Thomaidis, A., Dunn, F. W. and Caen, J. 1986. Analogy between fibrinogen and casein: effect of an undecapeptide isolated from k-casein on platelet function. *Eur. J. Biochem.* 158: 379–82.

Jones, F. S. and Simms, H. S. 1930. The bacterial growth inhibitor (lactenin) of milk. *J. Exp. Med.* 51: 327–39.

Jouan, P.-N., Pouliot, Y., Gauthier, S. F. and Laforest, J.-P. 2006. Hormones in milk and milk products: a survey. *Int. Dairy J.* 16: 1408–14.

Kaneki, M., Hosoi, T., Ouchi, Y. and Orimo, H. 2006. Pleiotropic actions of vitamin K: protector of bone health and beyond?. *Nutrition* 22: 845–52.

Kanis, J. A. 1993. Factors affecting bone metabolism and osteoporosis: calcium requirements for optimal skeletal health in women. In *Dairy Products in Human Health and Nutrition*, edited by Serrano Rios et al., 415–22. Balkema, Rotterdam.

Kayser, H. and Meisel, H. 1996. Stimulation of human peripheral blood lymphocytes by bioactive peptides derived from bovine milk proteins. *FEBS Lett.* 383: 18–20.

King, Jr., J. C., Cummings, G. E., Guo, N., Trivedi, L., Readmond, B. X., Keane, V., Feigelman, S. and de Waard, R. 2007. A double-blind, placebo-controlled, pilot study of bovine lactoferrin supplementation in bottle-fed infants. *J. Pediatr. Gastroenterol. Nutr.* 44: 245–51.

Kitts, D. D. and Weiler, K. 2003. Bioactive proteins and peptides from food sources. Applications of bioprocesses used in isolation and recovery. *Curr. Pharm. Des.* 9: 1309–23.

Koldovsky, O. and Thornburg, W. 1987 Hormones in milk. *J. Ped. Gastroenterol. Nutr.* 6: 172–96.

Konuspayeva, G., Faye, B., Loiseau, G. and Levieux, D. 2007. Lactoferrin and immunoglobulin contents in camel's milk (*Camelus bactrianus, Camelus dromedarius,* and Hybrids) from Kazakhstan. *J. Dairy Sci.* 90(1): 38–46.

Korhonen, H. 2009. Bioactive components in bovine milk. In *Bioactive Components in Milk and Dairy Products,* edited by Y. W. Park, 15–42. Wiley-Blackwell, Iowa.

Korhonen, H. and Marnila, P. 2006. Bovine milk antibodies for health. In *Proteins and Peptides in Health and Disease,* edited by Y. Mine and S. Shahidi, 137–59. Taylor & Francis Group, Boca Raton, FL, USA.

Korhonen, H. and Marnila, P. 2009. Bovine milk immunoglobulins against microbial human diseases. In *Dairy-derived Ingredients-Food and Nutraceutical Uses,* edited by M. Corredig, 269–89. Woodhead Publishing, Cambridge, UK.

Korhonen, H. and Marnila, P. 2013. Milk bioactive proteins and peptides. Chapter 8. In *Milk and Dairy Products in Human Nutrition,* edited by Y. W. Park and G. F. W. Haenlein, 148–71. Wiley-Blackwell Publishers, Oxford, UK.

Korhonen, H., Marnila, P. and Gill, H. 2000. Bovine milk antibodies for health: a review. *Br. J. Nutr.* 84(Suppl. 1): 135–46.

Korhonen, H. and Pihlanto, A. 2003a. Bioactive peptides: new challenge and opportunities for the dairy industry. *Austr. J. Dairy Tech.* 58(2): 129–34.

Korhonen, H. and Pihlanto, A. 2003b. Food-derived bioactive peptides – opportunities for designing future foods. *Curr. Pharm. Des.* 9: 1297–308.

Korhonen, H. and Pihlanto, A. 2007. Technological options for the production of health- promoting proteins and peptides derived from milk and colostrum. *Curr. Pharm. Des.* 13: 829–43.

Korhonen, H. and Pihlanto-Leppala, A. 2004. Milk-derived bioactive peptides: formation and prospects for health promotion. In *Handbook of Functional Dairy Products,* edited by C. Shortt and J. O'Brien, 109–24. CRC Press, Boca Raton, FL, USA.

Korhonen, H., Pihlanto- Leppälä, A., Rantamaki, P. and Tupasela, T. 1998. The functional and biological properties of whey proteins: prospects for the development of functional foods. *Agric. Food Sci. Finl.* 7: 283–96.

Krissansen, G. W. 2007. Emerging health properties of whey proteins and their clinical implications. *J. Am. Coll. Nutr.* 26: 713S–23S.

Kruzel, M. L., Actor, J. K., Boldogh, I. and Zimecki, M. 2007. Lactoferrin in health and disease. *Postepy Higieny i medycyny doświadczalnej* 61: 261–7.

Kulozik, U., Tolkach, A., Bulca, S. and Hinrichs, J. 2003. The role of processing and matrix design in development and control of microstructures in dairy food production- a survey. *Int. Dairy J.* 13: 621–30.

Kussendrager, K. D. and van Hooijdonk, A. C. M. 2000. Lactoperoxidase: physico-chemical properties, occurrence, mechanism of action and applications. *Br. J. Nutr.* 84(Suppl. 1): 19–25.

Laegreid, A. and Kolsto Otnaess, A. B. 1985. The identification of a human milk ganglioside which inhibits E. coli and V. cholerae enterotoxins. *Protides Biol. Fluids.* 32: 231.

Lahov, E. and Regelson, W. 1996. Antibacterial and immunostimulating casein-derived substances from milk: casecidin, isracidin peptides. *Food Chem. Toxicol.* 34: 131–45.

Lahov, E., Edelsten, D., Sode-Mogensen, M. T. and Sofer, E. 1971. Properties of basic glycopeptides from cow milk protein by heat. *Milchwissenschaft* 26: 489–95.

Lara-Villoslada, F., Debras, E., Neito, A., Concha, A., Galvez, J., Logez-Huertas, E. and Boza, J. J. 2006. Oligosaccharides isolated from goat milk reduce intestinal inflammation in a rat model of dextran sodium sulfate-induced colitis. *Clin. Nutr.* 25: 477–88.

Larque, E., Sabater Molina, M. and Zamora, S. 2007. Biological significance of dietary polyamines. *Nutrition* 23: 87–95.

Larsson, S. C., Bergkvist, L. and Wolk, A. 2005. High-fat dairy food and conjugated linoleic acid intakes in relation to colorectal cancer incidence in the Swedish mammography cohort. *American Journal Clinical Nutrition* 82: 894–900.

Lazarus, L. H., Gaunido, G., Wilson, W. E. and Erspamer, V. 1986. An immunoreactive peptide in milk contains bombesin-like bioactivity. *Experientia* 42: 822–3.

LeBlanc, J. G., Matar, C., Valdez, J. C., LeBlanc, J. and Perdigon, G. 2002. Immunomodulating effects of peptidic fractions issued from milk fermented with *Lactobacillus helveticus*. *J. Dairy Sci.* 85(11): 2733–42.

Lee, D. B. N., Hu, M. S., Kayne, L. H., Nakhoul, F. and Jamgotchian, N. 1991. The importance of non-vitamin D-mediated calcium absorption. *Contrib. Nephrol.* 91: 14–20.

Lee, K. W. and Lee, H. J. 2005. Role of the conjugated linoleic acid in the prevention cancer. *Critical Reviews in Food Science and Nutrition*. 45: 135–44.

Lee, Y.-M., Skurk, T., Hennig, M. and Hauner, H. 2007. Effect of a milk drink supplemented with whey peptides on blood pressure in patients with mild hypertension. *Eur. J. Nutr.* 46: 21–7.

Levieux, D., Morgan, F., Geneix, N., Masle, I. and Bouvier, F. 2002. Caprine immunoglobulin G, beta-lactoglobulin, alpha-lactalbumin and serum albumin in colostrum and milk during the early post partum period. *J. Dairy Res.* 69(3): 391–9.

Loimaranta, V., Tenovuo, J. and Korhonen, H. 1998. Combined inhibitory effect of bovine immune whey and peroxidase-generated hypothiocyanite against glucose uptake by Streptococcus mutans. *Oral Microbiol. Immunol.* 13(6): 378–81.

Lönnerdal, B. 2011. Biological effects of novel bovine milk fractions. Nestle Nutrition Workshop Series. Pediatric Programme, Nestec Ltd., Vevey/S.Karger AG, Basel, 67: 41–54.

Lönnerdal, B., Jiang, R. and Du, X. 2011. Bovine lactoferrin can be taken up by the human intestinal lactoferrin receptor and exert bioactivities. *J. Pediatr. Gastroenterol. Nutr.* 53(6): 606–14.

Lönnerdal, B., Keen, C. L. and Hurley, L. S. 1983. Manganese binding in human-milk and cows milk – an effect on bioavailability. *Fed. Proc.* 42: 926.

Lönnerdal, B., Keen, C. L. and Hurley, L. S. 1985. Manganese binding-proteins in human and cows milk. *Am. J. Clin. Nutr.* 41: 550–9.

López-Expósito, I., Quiros, A., Amigo, L. and Recio, I. 2007. Casein hydrolysates as a source of antimicrobial, antioxydant and antihypertensive peptides. *Lait* 87: 241–9.

López-Posadas, R., Requena, P., González, R., Suárez, M. D., Zarzuelo, A., Sánchez de Medina, F. and Martínez-Augustin, O. 2010. Bovine glycomacropeptide has intestinal antiinflammatory effects in rats with dextran sulfate-induced colitis. *J. Nutr.* 140(11): 2014–19.

Luhovyy, B. L., Akhavan, T. and Anderson, G. H. 2007. Whey proteins in the regulation of food intake and satiety. *J. Am. Coll. Nutr.* 26: 704S–12S.

Mackie, A. and Macierzanka, A. 2010. Colloidal aspects of protein digestion. *Curr. Opin. Colloid Interface Sci.* 15: 102–8.

Madureira, A. R, Pereira, C. I., Gomes, A. M. P., Pintado, M. E. and Malcata, F. X. 2007. Bovine whey proteins- Overview on their main biological propwerties. *Food Res. Int.* 40: 1197–211.

Malven, P. and McMurtry, J. 1974. Measurement of prolactin in milk by radio immunoassay. *J. Dairy Sci.* 57: 411–15.

Manso, M. A. and López-Fandiño, R. 2004. κ-Casein Macropeptides from cheese whey: physicochemical, biological, nutritional and technological features for possible uses. *Food Rev. Int.* 20: 329–55.

Markus, C. R., Olivier, B. and de Haan, E. H. 2002. Whey protein rich in α-lactalbumin increases the ratio of plasma tryptophan to the sum of the large neutral amino acids and improves cognitive performance in stress-vulnerable subjects. *Am. J. Clin. Nutr.* 75: 1051–6.

Markus, C. R., Olivier, B., Pamhuysen, G. E., Van der Gugten, J., Alles, M.-S., Tuiten, A., et al. 2000. The bovine protein α-lactalbumin increases the plasma ratio of tryptophan to the other large neutral amino acids, and in vulnerable subjects raises brain serotonin activity, reduces cortisol concentration and improves mood under stress. *American Journal of Clinical Nutrition* 75: 1051–6.

Marnila, P. and Korhonen, H. J. 2009. Lactoferrin for human health. In *Dairy derived Ingredients-Food and Nutraceutical Uses*, edited by M. Corredig, 290–307. Woodhead Publishing Ltd., Cambridge, UK.

Marshall, R. T., Obrien, B. L., Lee, Y. H. and Moats, W. A. 1982. Pyruvate in producer and commingled manufacturing grade milk. *J. Food Prot.* 45(13): 1236–41.

Martinez-Ferez, A., Rudloff, S., Guadix, A., Henkel, C. A., Pohlentz, G., Boza, J. J., Gaudix, E. M. and Kunz, C. 2006. Goats' milk as a natural source of lactose-derived oligosaccharides: isolation by membrane technology. *Int. Dairy J.* 16: 173–81.

Matar, C., LeBlanc, J. G., Martin, L. and Perdigon, G. 2003. Active peptides released in fermented milk: role and functions. In *Handbook of Fermented Functional Foods*. Functional Foods and Nutraceuticals series, edited by E. R. Farnworth, 177–201, CRC Press, Boca Raton

Matsumoto, H., Shimokawa, Y., Ushida, Y., Toida, T. and Hayasawa, H. 2001. New biological function of bovine α-lactalbumin: protective effect against ethanol- and stress-induced gastric mucosal injury in rats. *Biosci. Biotechnol. Biochem.* 65: 1104–11.

Matthies, H., Stark, H., Hartrodt, B., Ruethrich, H. L., Spieler, H. T., Barth, A. and Neubert, K. 1984. Derivatives of β-casomorphins with high analgesic potency. *Peptides* 5: 463–70.

McDonald, R. S., Thornton, W. H. and Marshall, R. T. 1994. A cell culture model to identify biologically active peptides generated by bacterial hydrolysis of casein. *J. Dairy Sci.* 77: 1167–75.

McIntosh, G. H., Register, G. O., Le Leu, R. K., Royle, P. J. and Smithers, G. W. 1995. Dairy proteins protect against dimethylhydrazine-induced intestinal cancers in rats. *J. Nutr.* 125: 809–16.

McLachlan, C. N. S. 2001. β-casein A1, ischaemic heart disease mortality, and other illnesses. *Med. Hypotheses* 56(2): 262–72.

McMartin, S., Godden, S., Metzger, L., Freitag, J., Bey, R., Stabel, J., Goyal, S., Fetrow, J., Wells, S. and Chester-Jones, H. 2006. Heat treatment of bovine colostrum. I: effects of temperature on viscosity and immunoglobulin G level. *J. Dairy Sci.* 89(6): 2110–18.

Medhammar, E., Wijesinha-Bettoni, R., Stadlmayr, B., Nilsson, E., Charrondiere, U. R. and Burlingame, B. 2012. Composition of milk from minor dairy animals and buffalo breeds: a biodiversity perspective. *J. Sci. Food Agric.* 92(3): 445–74. doi: 10.1002/jsfa.4690.

Mehra, R., Marnila, P. and Korhonen, H. 2006. Milk immunoglobulins for health promotion. *Int. Dairy J.* 16: 1262–71.

Meisel, H. 1997. Biochemical properties of regulatory peptides derived from milk proteins. *Biopolymers* 43(2): 119–28.

Meisel, H. 1998. Overview on milk protein-derived peptides. *Int. Dairy J.* 8: 363–73.

Meisel, H. 2001. Bioactive peptides from milk proteins: a perspective for consumers and producers. *Aust. J. Dairy. Technol.* 56: 83–91.

Meisel, H. 2004. Multifunctional peptides encrypted in milk proteins. *BioFactors* 21: 55–61.

Meisel, H. and FitzGerald, R. J. 2000. Opioid peptides encrypted in intact milk protein sequences. *Br. J. Nutr.* 84: 27–31.

Meisel, H. and FitzGerald, R. J. 2003. Biofunctional peptides from milk proteins: mineral binding and cytomodulatory effects. *Curr. Pharm. Des.* 9: 1289–95.

Meisel, H., Goepfert, A. and Gunther, S. 1997. ACE inhibitory activites in milk products. *Milchwissenschaft* 52: 307–11.

Meisel, H. and Olieman, C. 1998. Estimation of calcium-binding constants of casein phosphopeptides by capillary zone electrophoresis. *Anal. Chim. Acta* 372: 291–7.

Meisel, H. and Schlimme, E. 1990. Milk proteins: precursors of bioactive peptides. *Trends Food Sci. Technol.* 1: 41–3.

Meisel, H. and Schlimme, E. 1994. Inhibitors of angiotensin-converting enzyme derived from bovine casein (casokinins). In *β-casomorphins and Related Peptides: Recent Developments*, edited by V. Brantl and H. Teschemacher, 27–33. VCH, Weinheim.

Meisel, H. and Schlimme, E. 1996. Bioacive peptides derived from milk proteins: ingredients for functional foods. *Kieler Milchwirtschaftliche Forschungsberichte* 48: 343–57.

Mercier, J. C., Chobert, J. M. and Addeo, F. 1976. Comparative structure of the caseinmacropeptides from seven species. *FEBS Lett.* 7: 208–14.

Michaelidou, A. 2008. Factors influencing nutritional and health profile of milk and milk products. *Small Ruminant Res.* 79: 42–50.

Michaelidou, A., Alichanidis, E., Urlaub, H., Polychroniadou, A. and Zerfidis, G. K. 1998. Isolation and identification of some major water soluble peptides in feta cheese. *J. Dairy Sci.* 81: 3109–16.

Mierke, D. F., Nobner, G., Shiller, P. W. and Goodman, M. 1990. Morphiceptin analogs containing 2-aminocyclopentane carboxylic acid as a peptidomimetic for proline. *Int. J. Peptide Res.* 35: 34–45.

Mir, Z., Goonewardene, L. A., Okine, E., Jaegar, S. and Scheer, H. D. 1999. Effect of feeding carola oil on constituents, conjugated linoliec acid (CLA) and long chain fatty acids in goats milk. *Small Rum. Res.* 33: 137–43.

Mohan, K. V., Letchoumy, P. V., Hara, Y. and Nagini, S. 2008. Combination hemoprevention of hamster buccal pouch carcinogenesis by bovine milk lactoferrin and black tea polyphenols. *Cancer Invest.* 26(2): 193–201.

Morgan, T., Nowson, C., Snowden, R., Teow, B. H., Hadji, E., Hodgson, M., Anderson, A., Wilson, D. and Adam, W. 1986. The effect of sodium, potassium, calcium, and magnesium on blood pressure. In *Recent Advances in Clinical Nutrition*, edited by M. L. Wahlqvist and A. S. Truswell, 9–19. John Libbey, London.

Morgan, M. V., Adams, G. G., Bailey, D. L., Tsao, C. E., Fischman, S. L. and Reynolds, E. C. 2008. The anticariogenic effect of sugar-free chewing gum containing CPP-ACP nancomplexes on approximal caries determined using digital bitewing radiography. *Caries Research* 42: 171–84.

Morrison, W. R., Jack, E. L. and Smith, L. M. 1965. Fatty acids of bovine milk glycolipids and phospholipids and their specific distribution in the diacylglycerophospholipids. *J. Am. Oil Chem. Soc.* 42: 1142.

Murakami, M., Tonouchi, H., Takahashi, R., Kitazawa, H., Kawai, Y., Negishi, H., et al. 2004. Structural analysis of a new anti-hypertensive peptide (β-lactosin B) isolated from a commercial whey product. *J. Dairy Sci.* 87: 1967–74.

Murray, B. A. and FitzGerald, R. J. 2007. Angiotensin converting enzyme inhibitory peptides derived from food proteins: biochemistry, bioactivity and production. *Curr. Pharm. Des.* 13: 773–91.

Nagaoka, S., Futamura, Y., Miwa, K., Awano, T., Yamauchi, K., Kanamaru, Y., Tadashi, K. and Kuwata, T. 2001. Identification of novel hypocholesterolemic peptides derived from bovine milk β-lactoglobulin. *Biochem. Biophys. Res. Commun.* 218: 11–17.

Nagata, Y., Ishiwaki, M. and Sugan, M. 1982. Studies on the mechanism of antihypercholesterolemic action of soy protein and soy protein-type amino acid mixtures in relation to the casein counterparts in rats. *J. Nutr.* 112: 1614–25.

Nagpal, R., Behare, P., Rana, R., Kumar, A., Kumar, M., Arora, S., Morotta, F., Jain, S. and Yadav, H. 2011. Bioactive peptides derived from milk proteins and their health beneficial potentials: an update. *Food Funct.* 2: 18–27.

Nakamura, Y., Yamamoto, M., Sakai, K., Okubo, A., Yamazaki, S. and Takano, T. 1995. Purification and characterization of angiotensin I-converting enzyme inhibitors from sour milk. *J. Dairy Sci.* 78: 777–83.

Nieter, M., Teschemacher, H. J. and Schatz, H. 1981. Insulotropic actions of endorphins and β-casomorphin, an opioid-like fragment of milk casein. *Diabetologia* 21: 309–15.

Ondetti, M. A., Rubin, B. and Cushman, D. W. 1977. Design of Specific inhibitors of angiotensin-converting enzyme: new class of orally active antihypertensive agents. *Science* 196: 441–4.

Ong, L., Henriksson, A. and Shah, N. P. 2007. Angiotensin converting enzyme-inhibitory activity in Cheddar cheeses made with the addition of probiotic *Lactobacillus casei* sp. *Lait* 87: 149–65.
Ortiz-Chao, P., Gomez-Ruiz, J. A., Rastall, R. A., Mills, D., Cramer, R., Pihlanto, A., Korhonen, H. and Jauregi, P. 2009. Production of novel ACE inhibitory peptides from β-lactoglobulin using Protease N Amano. *Int. Dairy J.* 19: 69–76.
Otani, H., Kihara, Y. and Park, M. 2000. The immunoenhancing property of a dietary casein phosphopeptide preparation in mice. *Food Agr. Immunol.* 12: 165–73.
Otte, J., Shalaby, S. M., Zakora, M., Pripp, A. H. and El-Shabrawy, S. A. 2007. Angiotensin-converting enzyme inhibitory activity of milk protein hydrolysates: effect of substrate, enzyme and time of hydrolysis. *Int. Dairy J.* 17: 488–503.
Pakkanen, R. and Aalto, J. 1997. Growth factors and antimicrobial factors of bovine colostrum. *Int. Dairy J.* 7: 285–97.
Pan, Y., Rowney, M., Guo, P. and Hobman, P. 2007. Biological properties of lactoferrin: an overview. *Aust. J. Dairy Technol.* 62: 31–42.
Pandya, A. J. and Haenlein, G. F. W. 2009. Bioactive components in buffalo milk. In *Bioactive Components in Milk and Dairy Products,* edited by Y. W. Park, 105–57. Wiley-Blackwell, Iowa.
Panskeep, J., Normansell, L., Siviy, S., Rossi, J. and Zolovick, A. J. 1984. Casomorphins reduce separation distress in chicks. *Peptides* 4: 829–31.
Pariza, M. W., Park, Y. and Cook, M. E. 2001. The biological active isomers of conjugated linoleic acid. *Progress in Lipid Research* 40: 283–98.
Pariza, M. W., Park, Y., Cook, M., Albright, K. and Liu, W. 1996. Conjugated linoleic acid (CLA) reduces body fat. *FASEB J.* 10: 3227.
Park, Y. W. 1991. Relative buffering capacity of goat milk, cow milk, soy-based infant formulas, and commercial non-prescription antacid drugs. *J. Dairy Sci.* 74: 3326–33.
Park, Y. W. 2006. Goat milk – Chemistry and Nutrition. In *Handbook of Milk of Non-Bovine Mammals,* edited by Y. W. Park and G. F. W. Haenlein, 34–58. Blackwell Publishers, Ames, Iowa and Oxford, England.
Park, Y. W. 2009a. Overview of bioactive components in milk and dairy products. In *Bioactive Components in Milk and Dairy Products*, edited by Y. W. Park, 3–14. Wiley-Blackwell Publishers, Ames, Iowa and Oxford, England.
Park, Y. W. 2009b. Bioactive components of goat milk. In *Bioactive Components in Milk and Dairy Products*, edited by Y. W. Park, 43–82. Wiley-Blackwell Publishers, Ames, Iowa and Oxford, England.
Park, Y. W. and Chukwu, H. I. 1988. Macro-mineral concentrations in milk of two goat breeds at different stages of lactation. *Small Rumin. Res.* 1: 157–66.
Park, Y. W., Juarez, M., Ramos, M. and Haenlein, G. F. W. 2007. Physico-chemical characteristics of goat and sheep milk. *Small Rumin. Res.* 68: 88–113.
Park, Y. W., Mahoney, A. W. and Hendricks, D. G. 1986. Bioavailability of iron in goat milk compared with cow milk fed anemic rats. *J. Dairy Sci.* 69: 2608–15.
Park, Y., Storkson, J. M., Albright, K. J., Liu, W. and Pariza, M. W. 1999. Evidence that the *trans-10,cis-12* isomer of conjugated linoleic acxid induces body composition changes in mice. *Lipids* 34: 235–41.
Parkash, S. and Jenness, R. 1968. The composition and characteristics of goat's milk: a review. *Dairy Sci. Abstr.* 30: 67.
Parodi, P. W. 2003. Anti-cancer agents in milkfat. *Aust. J. Dairy Technol.* 58: 114–18.
Parodi, P. W. 2004. Milk fat in human nutrition. *Australian Journal of Dairy Technology* 59: 3–59.
Paroli, E. 1988. Opioid peptides from food (the exorphins). *World Rev. Nutr. Diet.* 55: 58–97.
Pecka, E., Dobrzański, Z., Zachwieja, A., Szulc, T. and Czyż, K. 2012. Studies of composition and major protein level in milk and colostrum of mares. *Anim. Sci. J.* 83: 162–8.
Pellegrini, A. 2003. Antimicrobial peptides from food proteins. *Curr. Pharm. Des.* 9: 1225–38.
Petrillo, Jr E. W. and Ondetti, M. A. 1982. Angiotensin converting enzyme inhibitors: medicinal chemistry and biological actions. *Med. Res. Rev.* 2: 1–41.

Pettersson, J., Mossberg, A. K. and Svanborg, C. 2006. alpha-Lactalbumin species variation, HAMLET formation, and tumor cell death. *Biochem. Biophys. Res. Commun.* 345(1): 260–70.

Pettus, B. J., Chalfant, C. E. and Hannun, Y. A. 2004. Shingolipids in inflammation: roles and implications. *Curr. Moleccular Med.* 4: 405–18.

Phelan, M., Aherne, A., Fitzgerald, R. J. and O'Brien, N. M. 2009.Casein-derived bioactive peptides: Biological effects, industrial uses, safety aspects and regulatory status. *Ine. Dairy J.* 19: 643–54.

Phelan, M. and Kerins, D. 2011. The potential role of milk-derived peptides in cardiovascular disease. *Food Funct.* 2: 153–67.

Pihlanto, A. 2006. Antioxidative peptides derived from milk proteins: review. *Int. Dairy J.* 16: 1306–14.

Pihlanto, A. and Korhonen, H. 2003. Bioactive peptides and proteins. In *Advances in Food and Nutrition Research*, edited by S. L.Taylor, 175–276. Elsevier Inc., San Diego, USA.

Pihlanto-Leppälä, A., Koskinen, P., Piilola, K., Tupasela, T. and Korhonen, H. 2000. Angiotensin I-converting enzyme inhibitory properties of whey protein digests: concentration and characterization of active peptides. *J. Dairy Res.* 67: 53–64.

Playford, R. J., Macdonald, C. E. and Johnson, W. S. 2000. Colostrum and milk-derived peptide growth factors for the treatment of gastrointestinal disorders. *Am. J. Clin. Nutr.* 72: 5–14.

Playne, M. J., Bennett, L. E. and Smithers, G. W. 2003. Functional dairy foods and ingredients. *Aust. J. Dairy Technol.* 58: 242–64.

Ploszaj, T., Ryniewicz, Z. and Motyl, T. 1997. Polyamines in goat's colostrum and milk. *Comp. Biochem. Physiol. B: Biochem. Mol. Biol.* 118: 45–52.

Posati, L. P. and Orr, M. L. 1976. Composition of foods. Agric. Handbook No. 8-1. ARS, USDA, Washington, DC.

Pouliot, Y. and Gauthier, S. F. 2006. Milk growth factors as health products: some technological aspects. *Int. Dairy J.* 16: 1415–20.

Pouliot, Y., Gauthier, S. F. and Groleau, P. E. 2006. Membrane-based fractionation and purification strategies for bioactive peptides. In *Nutraceutical Proteins and Peptides in Health and Disease*, edited by Y. Mine and F. Shahidi, 639–58. CRC, Taylor & Francis Group, Boca Raton, FL, USA.

Power, O., Jakeman, P. and FitzGerald, R. J. 2013. Antioxidative peptides: enzymatic production, in vitro and in vivo antioxidant activity and potential applications of milk-derived antioxidative peptides. *Amino Acids.* 44(3): 797–820.

Prosser, C., Stelwagen, K., Cummins, R., Guerin, P., Gill, N. and Milne, C. 2004. Reduction of heat induced gastrointestinal hyperpermeability in rats by bovine colostrum and goat milk powders. *J. Appl. Physiol.* 96: 650–4.

Qian, Z. Y., Jolles, P., Migliore-Samour, D. and Fiat, A. M. 1995a. Isolation and characterization of sheep lactoferrin, an inhibitor of platelet aggregation and comparison with human lactoferrin. *Biochim. Biophys. Acta* 1243: 25–32.

Qian, Z. Y., Jolles, P., Migliore-Samour, D., Schoentgen, F. and Fiat, A. M. 1995b. Sheep kappa-casein peptides inhibit platelet aggregation. *Biochim. Biophys. Acta* 1244: 411–17.

Quiros, A., Hernandez-Ledesma, B., Ramos, M., Amigo, L. and Recio, I. 2005. Angiotensin-converting enzyme inhibitory activity of peptides derived from caprine kefir. *J. Dairy Sci.* 88: 3480–7.

Ratcliffe, W., Green, E., Emly, J., Norbury, S., Lindsay, M., Heath, D. and Ratcliffe, J. G. 1990. Identification and partial characterization of parathyroid hormone-related protein in human and bovine milk. *J. Endocrinol.* 127: 167–76.

Recio, I., de la Fuente, M. A., Juárez, M. and Ramos, M. 2009a. Bioactive components in sheep milk. In *Bioactive Components in Milk and Dairy Products,* edited by Y. W. Park, 83–104. Wiley-Blackwell, Iowa.

Recio, I., Moreno, E. J. and López-Fandino, R. 2009b. Glycosylated dairy components: their roles in nature and ways to make use of their functionality in dairy products. In *Dairy-derived Ingredients-Food and Nutraceutical Uses*, edited by M. Corredig, 170–211. Woodhead Publishing Ltd, Cambridge, UK.

Regester, G. O., Smithers, G. W., Mitchell, I. R., McIntosh, G. H. and Dionysius, D. A. 1997. Bioactive factors in milk: natural and induced. In *Milk Composition, Production and Biotechnology,* edited by R. Welch, D. Burns, S. Davis, A. Popay and C. Prosser, 119–32. CAB International, Oxfordshire.

Reiter, B. and Härnulv, G. 1984. Lactoperoxidase antibacterial system: natural occurrence, biological functions and practical applications. *J. Food Prot.* 47: 724–32.

Reiter, B. and Perraudin, J.-P. 1991. Lactoperoxidase: biological functions. In *Peroxidases in Chemistry and Biology,* edited by J. Everse, K. E. Everse and M. B. Grisham, 143–80. CRC Press, Boca Raton, FL.

Renner, E. 1983. *Milk and Dairy Products in Human Nutrition*, 4th ed. München: Volkswirtschaftlicher Verlag.

Renner, E., Schaafsma. G. and Scott, K. J. 1989. Micronutrients in milk. In *Micronutrients in Milk and Milk-based Products*, edited by E. Renner, 1–70. Elsevier Applied Science, New York.

Ricci, I., Artacho, R. and Olalla, M. 2010. Milk protein peptides with angiotensin I-converting enzyme inhibitory (ACEI) activity. *Critical Reviews in Food Science and Nutrition* 2010: 390–402.

Rival, S. G., Boeriu, C. G. and Wichers, H. J. 2001. Caseins and casein hydrolysates. 2. Antioxidative properties and relevance to lipoxygenase inhibition. *J. Agr. Food Chem.* 4: 295–302.

Rodriques, L. R. 2013. Milk minor constituents, enzymes, hormones, growth factors, and organic acids. In *Milk and Dairy Products in Human Nutrition*, edited by Y. W. Park and G. F. W. Haenlein, 220–45. Wiley-Blackwell Publishers, Oxford, England.

Rokka, T., Syvoja, E. L., Tuominen, J. and Korhonen, H. 1997. Release of bioactive peptides by enzymatic proteolysis of *Lactobacillus* GG fermented UHT milk. *Milchwissenschaft* 52: 675–8.

Rombaut, R. and Dewettinck, K. 2006. Properties, analysis and purification of milk polar lipids. *Int. Dairy J.* 16: 1362–73.

Roufik, S., Gauthier, S. F. and Turgeon, S. L. 2006. *In vitro* digestibility of bioactive peptides derived from bovine β-lactoglobulin. *Int. Dairy J.* 16: 294–302.

Roy, M. K., Watanabe, Y. and Tamai, Y. 1999. Induction of apoptosis in HL-60 cells by skimmed milk digested with a proteolytic enzyme from the yeast *Saccharomyces cerevisiae. J. Biosci. Bioeng.* 88: 426–32.

Royle, P. J., McIntosh, G. H. and Clifton, P. M. 2008. Whey protein isolate and glycomacropeptide decreases weight gain and alters body composition in male Wistar rats. *Br. J. Nutr.* 100: 88–93.

Rusu, D., Drouin, R., Pouliot, Y., Gauthier, S. and Poubelle, P. E. 2010. A bovine whey protein extract stimulates human neutrophils to generate bioactive IL-1Ra through a NF-kappaB- and MAPK-dependent mechanism. *J. Nutr.* 140(2): 382–91.

Ryhänen, E.-L., Pihlanto-Leppälä, A. and Pahkala, E. 2001. A new type of ripened; low-fat cheese with bioactive properties. *Int. Dairy J.* 11: 441–7.

Ryhänen, E.-L., Tallavaara, K., Griinari, J.M., Jaakkola, S., Mantere-Alhonen, S. and Shingfield, K. 2005. Production of conjugated linoleic acid enriched milk and dairy products from cows receiving grass silage supplemented with a cereal-based concentrate containing rapeseed oil. *Int. Dairy J.* 15: 207–17.

Sah, B. N., Vasiljevic, P. T., NcKechnie, S. and Donkor, O. N. 2015. Identification of anticancer peptides from bovine milk proteins and their potential roles in management of cancer: a critical review. *Comp. Rev. Food Sci. Food Saf.* 14: 123–38.

Saito, T., Itoh, T. and Adachi, S. 1984. Presence of two neutral disaccharides containing n-acetylhexosamine in bovine colostrum as free forms. *Biochim. Biophys. Acta* 801: 147.

Saito, T., Nakamura, T., Kitazawa, H., Kawai, Y. and Itoh, T. 2000. Isolation and structural analysis of antihypertensive peptides that exist naturally in Gouda cheese. *J. Dairy Sci.* 83: 1434–40.

Saito, T., Lee, Y. S. and Kimura, S. 1998. Minimum effective dose of casein phosphopeptides (CPP) for enhancement of calcium absorption in growing rats. *Int. J. Vitam. Nutr. Res.* 68: 335–40.

Sanchez-Pozo, A. and Gil, A. 2002. Nucleotides as semi essential nutritional components. *Br. J. Nutr.* 87: S135–7.

Sato, R., Noguchi, T. and Naito, H. 1983. The necessity for the phosphate portion of casein molecules to enhance Ca absorption from the small intestine. *Agri. Bio. Chem.* 47: 2415–17.

Sava, G. 1996. Pharmacological aspects and therapeutic applications of lysozymes. *EXS* 75: 433–49.

Scammel, A. W. 2001. Production and uses of colostrum. *Aust. J. Dairy Technol.* 56: 74–82.

Schaafsma, G. 2006. Health of issues of whey proteins: 2. Weight management. *Curr. Top. Nutraceutical Res.* 4: 123–6.

Schaafsma, G. and Steijns, J. M. 2000. Dairy ingredients as source of functional foods. In *Essentials of Functional Foods*, edited by M. K. Schmidl and T. P. Labuza, 181–204. Aspen Publishers, Gaithersbur.

Schaafsma, G., Visser, W. J., Dekker, P. R. and van Schaik, M. 1988. Effect of dietary calcium supplementation with lactose on bone in vitamin D-deficient rats. *Bone.* 8: 357–62.

Schanbacher, F. L., Talhouk, R. S., Murray, F. A., Gherman, L. I. and Willet, L. B. 1998. Milk-born bioactive peptides. *Int. Dairy J.* 8: 393–403.

Schlimme, E., Martin, D. and Meisel, H. 2000. Nucleosides and nucleotides and natural bioactive substances in milk and colostrum. *Br. J. Nutr.* 84: S59–68.

Schlimme, E. and Meisel, H. 1995. Bioactive peptides derived from milk proteins. Structural, physiological, and analytical aspects. *Die Nahrung* 39: 1–20.

Schulze, J. and Zunft, H. J. 1991. Lactose, a potential dietary fiber: on the regulation of its microecological efficiency in the intestinal tract. Part 1. Problems, state of knowledge, methodology [in German] *Nahrung.* 45: 849–66.

Seifu, E., Buys, E. M. and Donkin E. F. 2005. Significance of the lactoperoxidase system in the dairy industry and its potential applications: a review. *Trends Food Sci. Technol.* 16: 137–54.

Sen, C. K., Khanna, S. and Roy, S. 2006. Tocotrienols: Vitamin E beyond tocopherols. *Life Sci.* 78: 2088–98.

Seshadri, S., Beiser, A., Selhub, J., Jacques, P. F., Rosenberg, I. H. and D'Agostino, R. B. 2002. Plasma homocysteine as a risk factor for dementia and Alzheimer's disease. *New England J. Med.* 346: 476–83.

Setarehnejad, A., Kanekanian, A., Tatham, A. and Abedi, A. H. 2010. The protective effect of caseinomacropeptide against dental erosion using hydroxyapatite as a model system. *Int. Dairy J.* 20: 652–6.

Shin, K., Wakabayashi, H., Yamauchi, K., Teraguchi, S., Tamura, Y., Kurokawa, M. and Shiraki, K. 2005. Effects of orally administered bovine lactoferrin and lactoperoxidase on influenza virus infection in mice. *Journal of Medical Microbiology* 54: 717–23.

Shin, K., Yamauchi, K., Teraguchi, S., Hayasawa, H. and Imoto, I. 2002. Susceptibility of *Helicobacter pylori* and its urease activity to the peroxidase-hydrogen peroxide-thiocyanate antimicrobial system. *J. Med. Microbiol.* 51: 231–7.

Shingfield, K. J., Reynolds, C. K., Hervas, G., Griinari, J. M., Grandison, A. S. and Beever, D. E. 2006. Examination f the persistency of milk fatty acid composition responses to fish oil and sunflower oil in the diet of dairy cows. *J. Dairy Sci.* 89: 714–32.

Sieber, R., Collomb, M., Aeschlimann, A., Jelen, P. and Eyer, H. 2004. Impact of microbial cultures on conjugated linoleic acid in dairy products-a review. *Int. Dairy J.* 14: 1–15.

Silva, S. V., Pihlanto, A. and Malcata, F. X. 2006. Bioactive peptides in ovine and caprine cheeselike systems prepared with proteases from Cyanara cardunculus. *J. Dairy Sci.* 89: 3336–44.

Singh, T. K, Fox, P. F. and Healy, A. 1997. Isolation and identification of further peptides in the diafiltration retentate of the water-soluble fraction of Cheddar cheese. *J. Dairy Res.* 64: 433–43.

Sipola, M., Finckenberg, P., Korpela, R., Vapaatalo, H. and Nurminen, M.-L. 2002. Effect of long-term intake of milk products on blood pressure in hypertensive rats. *J. Dairy Res.* 69: 103–11.

Siso, M. I. G. 1996. The biotechnological utilization of cheese whey: a review. *Bioresour. Technol.* 57: 1–11.

Smacchi, E. and Gobbetti, M. 1998. Peptides from several Italian cheeses inhibitory to proteolytic enzymes of lactic acid bacteria; *Pseudomonas fluorescens* ATCC 948 and to the angiotensin I-converting enzyme. *Enzyme Microbiology and Technology* 22: 687–94.

Smithers, G. W., Ballard, F. J., Copeland, A. D., De Silva, K. J., Dionysius, D. A., Francis, G. L., Goddard, C., Grieve, P. A., McIntosh, G. H., Mitchell, I. R., Pearce, R. J. and Regester, G. O. 1996. New opportunities from the isolation and utilization of whey proteins. *J. Dairy Sci.* 79: 1454–9.

Sorenson, W., Slattery, M. S. and Ford, M. 1988. Calcium and colon cancer: a review. *Nutr. Cancer* 11: 135–45.

Stanton, C., Murphy, J., McGrath, E. and Devery, R. 2003. Animal feeding strategies for conjugated Linoleic Acid Enrichment of Milk. In *Advances in Conjugated Linoleic Acid Research*, Volume 2, edited by J. L. Sébédio, W. W. Christie and R. Adlof, 123–45, AOCS Press, Champaign. IL.

Stedman's Medical Dictionary. 2006. *Cytokines*. 28th ed. Wolters Kluwer Health, Lippincott, Williams & Wilkins.

Steijns, J. M. 2001. Milk ingredients as nutraceuticals. *Int. J. Dairy Technol.* 54: 81–8.

Stewart, A. F., Willis, I. M. and McKinlay, A. G. 1984. Nucleotide sequence of bovine $\alpha_{s1}$and κ-casein cDNAs. *Nucleic Acid Res.* 12: 3895–907.

Struff, W. G. and Sprotte, G. 2007. Bovine colostrum as a biologic in clinical medicine: a review. *Int. J. Clin. Pharm. Ther.* 45: 193–202.

Suetsuna, R., Ukeda, H. and Ochi, H. 2000. Isolation and characterization of free radical scavenging activities peptides derived from casein. *J. Nutr. Biochem.* 11: 128–31.

Takamizawa, K., Iwanori, M., Mutai, M. and Nagai, Y. 1986. Selective changes in gangliosides of human-milk during Lactation – a molecular indicator for the period of lactation. *Biochim. Biophys. Acta.* 879: 73.

Tamura, Y. 2004. Production and application of bovine lactoferrin. *Bull. Int. Dairy Fed.* 389: 64–8.

Teraguchi, S., Wakabayashi, H., Kuwata, H., Yamauchi, K. and Tamura, Y. 2004. Protection against infections by oral lactoferrin: evaluation in animal models. *Biometals* 17: 231–4.

Teschemacher, H. 2003. Opoid receptor ligands derived from food proteins. *Curr. Pharm. Des.* 9: 1331–44.

Tiemeyer, W., Stohrer, M. and Giesecke, D. 1984. Metabolites of nucleic acids in bovine milk. *J. Dairy Sci.* 67: 723–8.

Thomä-Worringer, C., Sörensen, J. and López-Fandino, R. 2006. Health effects and technological features of caseinomacropeptide. *Int. Dairy J.* 16: 1324–33.

Thorell, L., Sjoberg, L. B. and Hernell, O. 1996. Nucleotide in human milk: sources and metabolism by the newborn infant. *Pediatr. Res.* 40: 845–52.

Tome, D., Ben Mansour, A., Hautefeville, M. and Desjeux, J. F. 1987. Opiate activity and transepithelial passage of intact β-casomorphins in rabbit iluem. *Am. J. Physiol.* 253: G737–44.

Tompa, G., Laine, A., Pihlanto, A., Korhonen, H., Rogelj, I. and Marnila, P. 2011. Chemiluminescence of non-differentiated THP-1 promonocytes: developing an assay for screening anti-inflammatory milk proteins and peptides. *Luminescence* 26(4): 251–8.

Torkelson, A. 1987 Radioimmunoassay of somatotropin in milk from cows administered recombinant bovine somatotropin. *Proc. Am. Dairy Sci. Assoc.* 70: 146.

Tripathi, V. and Vashishtha, B. 2006. Bioactive compounds of colostrum and its application. *Food Rev. Int.* 22: 225–44.

Tucker, H. and Schwalm, J. 1977. Glucocorticoids in mammary tissue and milk. *J. Anim. Sci.* 45: 627–34.

Uede, T. 2011. Osteopontin, intrinsic tissue regulator of intractable inflammatory diseases. *Pathol. Int.* 61: 265–80.

Underwood, E. J. 1977. Chapter 1. Introduction. In *Trace Elements in Human and Animal Nutrition*, 1–12, 4th ed. Academic Press, New York.

Ushida, Y., Shimokawa, Y., Matsumoto, H., Toida, T. and Hayasawa, H. 2003. Effects of bovine α-lactalbumin on gastric defense mechanisms in naive rats. *Biosci. Biotechnol. Biochem.* 67: 577–83.

Usinger, L., Jensen, L. T., Flambard, B., Linneberg, A. and Ibsen, H. 2010. The antihypertensive effect of fermented milk in individuals with prehypertension or borderline hypertension. *J. Hum. Hypertens.* 24: 678–83.

Ustunol, Z. and Wong, C. 2010. Effect of nonfat dairy milk and major whey components on interleukin-6 and interleukin-8 production in human intestinal epithelial-like Caco-2 cells. *J. Dairy Sci.* 93: 2311–14.

Valtonen, M., Niskanen, L., Kangas, A. P. and Koskinen, T. 2005. Effect of melatonin-rich night-time milk on sleep and activity in elderly institutionalized subjects. *Nord. J. Psychriatry* 59: 217–21.

Van den Heuvel, E. G., Muijs, T. H., van Dokkum, W. and Schaafsma, G. 1999. Lactulose stimulates calcium absorption in post-menopausal women. *J. Bone Miner. Res.* 14: 1211–16.

Vercruysse, L., Van Camp, J., Dewettinck, K. and Smagghe, G. 2009. Production and enrichment of bioactive peptides derived from milk proteins. In *Dairy-derived Ingredients-Food and Nutraceutical Uses*, edited by M. Corredig, 51–67. Woodhead Publishing Ltd., Cambridge, UK.

Vinderola, G., Matar, C. and Perdigon, G. 2007. Milk fermentation products of *L. helveticus* R389 activate calcineurin as a signal to promote gut mucosal immunity. *BMC Immunol.* 8: 19. doi:10.1186/1471-2172-8-19.

Wakabayashi, H., Abe, S., Okutomi, T., Tansho, S., Kawase, K. and Yamaguchi, H. 1996. Cooperative anti-*Candida* effects of lactoferrin or its peptides in combination with azole antifungal agents. *Microbiol. Immunol.* 40: 821–5.

Wakabayashi, H., Takase, M. and Tomita, M. 2003. Lactoferricin derived from milk protein lactoferrin. *Curr. Pharm. Des.* 9: 1277–87.

Wakabayashi, H., Yamauchi, K. and Takase, M. 2006. Lactoferrin research, technology and applications. *Int. Dairy J.* 16: 1241–51.

Walzem, R. L., Dillard, C. J. and German. J. B. 2002. Whey components: millennia of evolution create functionalities for mammalian nutrition: what we know and what we may be overlooking. *Crit. Rev. Food Sci. Nutr.* 42: 353–75.

Wang, B., Yu, B., Karim, M., Hu, H. H., Sun , Y., McGreevy, P., Petocz, P., Held, S. and Brandmiller, J. 2007. Dietary sialic acid supplementation improves learning and memory in piglets. *American J. Clin. Nutri.* 85: 561–9.

Weihrauch, J. L. and Son, Y. S. 1983. The phospholipid content of foods. *Am. Oil Chem. Soc.* 60: 1971.

Weiner, C., Pan, Q., Hurtig, M., Boren, T., Bostwick, E. and Hammarström, L. 1999. Passive immunity against human pathogens using bovine antibodies. *Clin. Exp. Immunol.* 116: 193–205.

White, A. C., Thannickal, J. J. and Fanburg, B. L. 1994. Glutathione deficiency in human disease. *J. Nutr. Biochem.* 5: 218–26.

Wikipedia. 2016. Hormone. Free Encyclopidia.

Wilson, M. D. and Rudel, L. L. 1994. Review of cholesterol absorption with emphasis on dietary and biliary cholesterol. *J. Lipid Res.* 35: 943–55.

Wolever, T. M. S., Wong, G. S. and Kenshole, A. 1985. Lactose in the diabetic diet: a comparison with other carbohydrates. *Nutr. Res.* 5: 1334–5.

Wolford, S. and Argoudelis, C. 1979. Measurement of estrogen in cow's milk, human milk and dietary products. *J. Dairy Sci.* 62: 1458–63.

Wu, F. Y. and Elsasser, T. H. 1995. Studies on cell growth promoting activity in goat milk. *J. Chin. Agric. Chem. Soc.* 33: 326–32.

Wu, D., Meydani, S. N., Sastre, J., Hayek, M. and Meydani, M. 1994. In vitro glutathione supplementation enhances interleukin-2 production and mitogenic response of peripheral blood mononuclear cells from young and old subjects. *J. Nutr.* 124: 655–63.

Wu, F. Y., Tsao, P. H., Wang, D. C., Lin, S., Wu, J. S. and Cheng, Y. K. 2006. Factors affecting growth factor activity in goat milk. *J. Dairy Sci.* 89: 1951–5.

Yamamoto, N., Akino, A. and Takano, T. 1994. Antihypertensive effect of different kinds of fermented milk in spontaneously hypertensive rats. *Biosci. Biotechnol. Biochem.* 58: 776–8.

Yamamoto, S., Wang, M. F., Adjei, A. A. and Ameho, C. K. 1997. Role of nucleosides and nucleotides in the immune system, gut reparation after injury and brain function. *Nutrition* 13: 372–4.

Yoshikawa, M., Tani, F., Yoshimura, T. and Chiba, H. 1986. Opioid peptides from milk proteins. *Agric. Biol. Chem.* 50: 2419–21.

Yoshikawa, M., Yoshimura, T. and Chiba, H. 1984. Opioid peptides from human β-casein. *Agric. Biol. Chem.* 48: 3185–7.

Yousefi, R., Gaudin, J. C., Chobert, J. M., Poupak, Z., Moin, M., Moosavi-Movahedi, A. A. and Haertle, T. 2009. Micellization and immunoreactivities of dimeric β-caseins. *Biochim. et Biophys. Acta* 1794: 1775–83.

Yu, V. 2002. Scientific rationale and benefits of nucleotides supplementation of infant formula. *J. Paediatr. Child Health* 38: 543–9.

Yvon, M., Beucher, S., Guilloteau, P., Le Huerou-Luron, I. and Corring, T. 1994: Effects of caseinomacropeptide (CMP) on digestion regulation. *Reprod. Nutr. Dev.* 34: 527–37.

Zemel, M. B. 2004. Role of calcium and dairy products in energy portioning and weight management. *Am. J. Clin. Nutr.* 79(Suppl. 1): 907S–12S.

Zhang, X. and Beynen, A. 1993. Lowering effect of dietary milk-whey protein v. casein on plasma and liver cholesterol concentrations in rats. *Br. J. Nutr.* 70: 139–46.

Zimecki, M. and Kruzel, M. L. 2007. Milk-derived proteins and peptides of potential therapeutic and nutritive value. *J. Exp. Ther. Oncol.* 6: 89–106.

Zioudrou, C., Streaty, R. A. and Klee, W. A. 1979. Opioid peptides derived from food. The exorphins. *J. Biol. Chem.* 254: 2446–9.

Zucht, H. D., Raida, M., Adermann, K., Magert, H. J. and Forssman, W. G. 1995. Casocidin-I: a casein-alpha s2 derived peptide exhibits antibacterial activity. *FEBS Lett.* 372: 185–8.

Chapter 3

Ingredients from milk for use in food and non-food products: from commodity to value-added ingredients

Thom Huppertz and Inge Gazi, NIZO food research, The Netherlands

1 Introduction

Milk provides complete nutrition for the neonate and is, therefore, often described as nature's ultimate food. Not only does it provide energy in the form of lipids, carbohydrates and proteins, but it also provides essential nutrients and micronutrients in the form of minerals, trace elements, vitamins, nucleotides, etc. The proteins are not only a source of amino acids, but can also confer immunity and are a carrier of calcium phosphate, which is essential for bone growth. In addition, some of the milk proteins contain bioactive sequences that may be released upon hydrolysis during digestion. The proteins in the milk fat globule membrane (MFGM) are also known to have antimicrobial and antiviral properties. In addition to the main carbohydrate lactose, milk also contains smaller amounts of oligosaccharides, which are known to aid the development of the intestinal flora of the neonate, which provides important anti-infection properties and is an important factor stimulating post-natal development. The levels and ratios of the constituents of milk differ between species and with stage of lactation and are typically tailored to the demands of the neonate. This ensures that optimal nutrition is provided.

http://dx.doi.org/10.19103/AS.2016.0005.40

The high nutritional value of milk for the neonate was recognized early and has also led to the introduction of milk and dairy products from other species as an important part of the human diet in many parts of the world. This includes the consumption of liquid milk products, whose shelf-life is typically extended by heat treatment and homogenization, but also a wide variety of dairy products, such as cheese, yogurt, butter and related products. In addition, with the emergence of technologies such as membrane filtration and chromatography, as well as enzymatic modifications for widespread industrial application, milk has been used as a base or an ingredient in the preparation of a wide variety of food products. The application of milk in a wide range of products has also resulted in the development of a wide range of dairy ingredients. These ingredients include commodity ingredients such as whole, skim and fat-filled milk powder, buttermilk powder and whey powder, but also fractions isolated from these dairy streams. The milk proteins present the highest economic value and as such have received extensive attention with respect to preparing functional ingredients. Desired functionality may be either from a physical perspective or from a nutritional perspective. Techniques used include selective precipitation, membrane filtration and chromatography and all yield products with specific purity and functionality. In addition, enzymatic hydrolysis of proteins may also be used to either improve physical or nutritional functionality.

In addition, the carbohydrate fraction of milk has also been a rich source of ingredients. Lactose is a widely used carbohydrate in food products, but it is also used as an excipient in pharmaceutical products. In addition, lactose can also be converted into functional ingredients such as lactulose, lactitol and lactobionic acid. Furthermore, prebiotic galactooligosaccharides (GOS) can be produced from lactose and have found wide application, particularly for infant nutrition products. This chapter focuses on dairy-derived ingredients and their physical and nutritional functionalities and application of the products. Focus will be primarily on ingredients that are produced industrially and how scientific and technical innovations have allowed their production and optimization.

2 Commodity dairy ingredients

Commodity dairy ingredients include various milk powders and whey powders which are, in general, prepared from liquid dairy streams without further fractionation. Main products in this class are skim milk powder, whole milk powder, buttermilk powder, whey powder and fat-filled milk powder, which is compositionally similar to whole milk powder but

Table 1 Typical composition (%, m/m) of whole milk powder and skim milk powder

	Whole milk powder	Skim milk powder
Fat	25–29	0.5–1.5
Protein	25–28	34–37
Lactose	36–39	48–52
Ash	6–7	7–8
Moisture	2–4	3–5

contains vegetable fat rather than milk fat (Table 1). The preparation of skim milk powder typically contains the following steps (Kelly et al., 2003; Schuck, 2011a,b; Pisecky, 2012):

- Skimming
- Pre-heating
- Evaporation
- Spray-drying

Skimming of the milk is typically done centrifugally, wherein the milk is separated into a cream phase and a skim milk phase. The skim milk is subsequently pre-heated, the conditions of which vary widely and depend on the heat class of the milk powder to be produced, that is, low-heat, medium-heat or high-heat skim milk powder. Classifications are based on the extent of whey protein denaturation using the so-called whey protein nitrogen index (WPNI), which should be ≥6, 1.5–6 or ≤1.5 mg undenatured whey protein per gram of milk powder for low-heat, medium-heat and high-heat skim milk powder, respectively (Table 2). After pre-heating, the milk is evaporated to 45–50% dry matter using falling film evaporators after which the concentrate is spray-dried to a powder containing ≤4% moisture. Skim milk powder finds wide use in the dairy industry, where it is used in the production of yogurt, ice cream, processed cheese and recombined milk, but also in infant formula, bakery products and confectionery. Other applications include meat products, and dry mixes such as soups, cocoa mixes and sauces. Functional properties contributed by milk powder in these applications include emulsification, foaming, browning, flavour, water-binding and gelling (Table 3).

For whole milk powder, the typical processing steps include the following (Kelly et al., 2003; Schuck, 2011a,b; Pisecky, 2012):

- Standardization
- Pre-heating
- Evaporation
- Homogenization
- Spray-drying

Milk is first standardized to the desired fat-in-dry matter (FIDM) content (Table 1), after which the milk is pre-heated. Typically, for whole milk powder, pre-heating is in the medium- or high-heat range. Homogenization is typically used to reduce fat globule size and improve stability of the fat globules. This may be applied to the milk prior to evaporation or to the

Table 2 Overview of different heat classes of skim milk powder

Heat class	WPNI[a]	Typical pre-heat treatment	Potential applications
Low heat	>6.0	15–60 s at 70–80°C	Recombined milk, recombined cheese
Medium heat	1.5–6.0	30–60 s at 90–100°C	Ice cream confectionery, recombined sweetened condensed milk
High heat	<1.5	30 s–5 min at 115–125°C	Bakery products, recombined evaporated milk

[a] WPNI = whey protein nitrogen index, expressed as mg undenatured whey protein nitrogen/g powder.

Table 3 Functional properties of different constituents of milk powders

Constituent	Functional property	Application
Protein	Water-binding/thickening	Meat products, bakery products, confectionery, soups and sauces, chocolate
	Emulsification	Coffee whiteners, soups, sauces, ice cream confectionery, meat products
	Foaming/whipping	Ice cream, desserts, whipped toppings
	Gelation	Cheese, yogurt
	Colour/flavour development	Chocolate, confectionery
	Heat stability	Recombined milk, infant and clinical nutrition, soups and sauces
	Solubility	Recombined milk, infant and clinical nutrition
Fat	Texture/mouthfeel	Cheese, yogurt, confectionery, chocolate, soups, sauces, coffee whiteners
	Foaming/whipping	Ice cream, desserts, whipped toppings
	Colour (whitening)	Confectionery, chocolate, coffee creamer, soups and sauces
	Flavour	Beverages, ice cream, chocolate
Lactose	Colour development	Confectionery, chocolate
	Mouthfeel	Coffee whiteners, confectionery
	Humidity buffering	Bakery products

concentrate after evaporation. Evaporation is typically performed to solid contents in the range 50–55% dry matter, which is slightly higher than for skim milk powder, because the viscosity of whole milk concentrates is lower than that of skim milk concentrates. Following evaporation, the concentrate is spray-dried to a powder containing ≤4% moisture. Due to the presence of considerable amounts of fat on the powder particle surface, whole milk powder can be susceptible to lipid oxidation, which can limit shelf-life. Process control to minimize surface fat levels and gas flushing are required to ensure sufficient shelf-life of whole milk powder (Vignolles et al., 2007; Lloyd et al., 2009). Whole milk powder is used in a range of products, including milk beverages, infant formula, confectionery, ice cream, soups and sauces, and bakery products. Functional properties contributed by milk powder in these applications include emulsification, foaming, browning, flavour, mouthfeel, water-binding and gelling (Table 3). Whole milk powder is often used in chocolate manufacture, but for this purpose, it is often roller-dried, rather than spray-dried. Roller-dried whole milk powder typically has a higher level of surface fat on the powder particles, and unlike for other applications, this is actually desirable for chocolate manufacture (Liang and Hartel, 2004).

An alternative to whole milk powder in several applications is fat-filled milk powder, which is prepared from skim milk and a vegetable fat or blend thereof. Particularly, palm fat and coconut fat, and blends thereof, are used regularly, but other fat sources may also be used. The fat can be emulsified into the skim milk either before or after the evaporation process, after which the product is dried. Application areas for fat-filled milk powder are broadly similar to those of whole milk powder. Fat-filled milk powder is often economically more attractive, but can lack the typical flavour characteristics contributed by the milk fat in whole milk powder (Kelly et al., 2002; Pisecky, 2012).

Most whey powder is typically prepared by drying so-called sweet whey from cheese production, but some is also prepared from so-called acid whey, from the production of acid-coagulated products (Table 4). In this process, the following steps are typically involved:

- Pre-treatment of whey
- Concentration
- Pre-crystallization
- Spray-drying

Before further processing, the whey is typically pre-treated, which involves the centrifugal removal of fat and the removal of curd fines, which can be done centrifugally or by filtration. Heat treatment is also required in most cases to inactivate residual chymosin and starter bacteria. An optional pre-treatment also includes the demineralization, in case demineralized whey powders are produced. Demineralization may be performed by nanofiltration (Van der Horst et al., 1995) or electrodialysis (Houldsworth, 1980). Subsequently, the whey is concentrated to typically >60% dry matter. High solids contents can be reached due to the low protein content of the material. Concentration may either be carried out solely by evaporation or by reverse osmosis followed by evaporation. After concentration, a pre-crystallization step is carried out by controlled cooling of the concentrate. Because the concentrate is supersaturated with respect to lactose at this point, the cooling will result in crystallization of part of the lactose. This enables the subsequent drying of whey powder without excessive issues regarding stickiness, as well as the production of a low hygroscopic powder, which remains rather stable during storage (Pisecky, 2012). Whey powder is typically a comparatively low-cost ingredient, which is used in a wide range of areas, including bakery products, dry mixes, ice cream, confectionery, snack foods and beverages, but also in the meat industry. Demineralized whey powders are also used in infant formula manufacture and other nutritional products. Typical functionalities contributed by whey powder in these applications include emulsification, foaming, browning, flavour, water-binding and gelling.

Table 4 Composition (in %, m/m) of sweet, acid and demineralized whey powder

	Sweet whey powder	Acid whey powder	Demineralized whey powder
Protein	11–15	11–14	11–15
Lactose	63–75	61–70	70–80
Fat	0–2	0–2	0–2
Ash	8–10	10–13	1–7
Moisture	3–5	3–5	3–5

3 Caseins and caseinates

Caseins and caseinates are a class of milk protein ingredients that have been produced industrially for nearly a century. Initially, they were primarily used as high-quality ingredients in glues, adhesives, plastics and synthetic fibres, but over time, their use in food products has become the predominant application of caseins and caseinates. Rennet casein and acid casein are the two different types of casein ingredient classes available. Caseinates are classified on the basis of cation of the alkali used in their production, that is, sodium, potassium, calcium or magnesium caseinates. All these ingredients contain >90% protein in dry matter, with the remainder primarily being minerals; lactose and fat are found only in very small amounts in these products (Table 5).

Skim milk is the starting material for the production of all casein and caseinate ingredients. Acid casein manufacture is based on the acid-induced destabilization of casein micelles; that is, the pH is lowered sufficiently to induce coagulation of the casein fraction. While this can be done through lactic acid production by lactic acid bacteria, nowadays, mineral acids such as hydrochloric acid and sulphuric acid are mostly used. Once a pH ~4.6 is reached, the caseins form a coagulum. This coagulum is subsequently cooked to 50–55°C to remove whey and further washing steps are carried out to ensure that all residual lactose, whey protein and minerals are removed. Following washing, the acid casein curd can be dried to obtain acid casein powder. For caseinate production, the acid casein curd is milled and the suspension is subsequently neutralized by the addition of base. Sodium, potassium, calcium and magnesium hydroxide are mostly used for this purpose and yield the caseinates of the respective cations. The neutralized suspension is then heated and dried. The manufacture of rennet casein differs from that of acid casein and caseinates, in that rennet-induced coagulation, rather than acid-induced coagulation, is used. Following rennet-induced coagulation, the coagulum is again cooked, dewheyed, washed and dried (Mulvihill and Ennis, 2003; Carr and Golding, 2016).

Caseins and caseinates are used in an extremely wide range of products, but the choice of the type of casein is highly application dependent (Singh, 2011; Carr and Golding, 2016). Acid casein and rennet casein are virtually insoluble in water, whereas sodium

Table 5 Typical compositions (in %, m/m) of acid casein, sodium caseinate, calcium caseinate and rennet casein

	Acid casein	Sodium caseinate	Calcium caseinate	Rennet casein
Moisture	8–12	3–5	3–5	8–12
Protein	84–90	90–94	90–94	80–85
Ash	1–2	3–6	3–5	6–9
Lactose	0–1	0–1	0–1	0–1
Fat	0–1	0–1	0–1	0–1
Sodium		1–2		
Calcium			1–2	2–3
pH	4.6–5.2	6.5–7.0	6.5–7.0	7.0–7.5
Solubility in water	0	>98	>95	0

caseinate and calcium caseinate are highly water-soluble. A suspension of sodium or potassium caseinate consists of rather small associations of molecules, reaching a size of approximately 20–40 nm. As a result, sodium and potassium caseinate suspensions are rather translucent. Sodium and potassium caseinate are excellent emulsifiers and foamers, and also have high heat stability, strong water-binding functionality and excellent nutritional properties. They are, therefore, widely applied in coffee creamers and other high fat products, cream liqueurs, bakery products, whipped toppings, soups, sauces, ice cream, meat products and infant and clinical nutrition (Singh, 2011). Calcium and magnesium caseinates form turbid milky-white suspensions in water, containing particles up to several hundred nanometres (Moughal et al., 2000). The main advantage of using calcium and magnesium caseinates over sodium and potassium caseinates is that suspensions are generally lower in viscosity. As such, calcium caseinates often find application in clinical nutrition and other nutritional products, in which the high viscosity arising from sodium caseinate is of concern. Calcium caseinate generally has poorer heat stability than sodium caseinates, so sodium caseinate is also favoured when severe heating is applied.

Rennet casein and acid casein are virtually insoluble in water. This was not a major concern for some of the initial non-food applications, but successful application in many food products requires the formation of a workable suspension of the product. A suspension of acid casein can be achieved by neutralization, thus effectively producing sodium and calcium caseinate suspensions. Advantages of the use of acid casein, rather than sodium caseinate or calcium caseinate, are that the user of the acid casein can control the degree of neutralization and hydration that is required for the specific application. In contrast to acid casein, rennet casein cannot be suspended by pH adjustment, but requires the chelation of micellar calcium phosphate. In the primary area of application of rennet casein, that is, processed cheese products and cheese analogs, the suspension of rennet casein is achieved by the addition of so-called melting salts, for example, citrates and (poly)phosphates, which chelate calcium (Ennis et al., 1998). By creating such suspensions, the rennet casein is suspended, the fat can be efficiently emulsified in the matrix and a gelled texture will form on cooling.

4 Whey protein ingredients

4.1 Whey protein concentrates and isolates

Whey protein concentrates (WPCs) are prepared by concentrating the protein fraction of whey by ultrafiltration (UF). Typically, membranes with a molecular weight cut-off of ≤10 kDa are used, which retain the proteins in the retentate but allow lactose, minerals and other small constituents to permeate through the membrane. In addition, diafiltration with water can be applied to further increase the proportion of proteins in the product (Morr and Ha, 1993; Mulvihill and Ennis, 2003; Westergaard, 2004; Jelen, 2009). The protein content of WPCs typically ranges from 35 to 80%, with WPC35, WPC60, WPC75 and WPC80 being the most commonly produced variants (Table 6). With increasing protein content, fat content also increases, whereas ash and lactose contents decrease proportionally.

WPC35 is very similar to SMP in terms of gross composition, and is as such often applied as a cost-effective partial or full skim milk replacer, although it does provide a different protein and mineral composition. Typical applications of WPC35 include dairy beverages,

Table 6 Typical composition (in %, m/m) of whey protein concentrates

	WPC35	WPC60	WPC75	WPC80
Protein	34–36	60–62	75–77	80–82
Lactose	48–52	25–30	10–15	4–8
Fat	3–5	3–7	4–8	4–8
Ash	6–8	4–6	4–6	3–4
Moisture	3–5	3–5	3–5	3–5

ice cream and cultured products, such as yogurt and fresh cheese. WPC35 also finds application in the bakery industry, soups, snacks and nutritional products. WPC60, WPC75 and WPC80 provide more concentrated sources of highly nutritional protein for protein supplementation for a wide variety of applications. The proteins are highly soluble, also under acidic conditions, and have good emulsifying properties and excellent water-binding, gelling and thickening properties. These products are widely employed as value-added ingredients in infant formula, clinical nutrition, nutritional bars, beverages and mixes, as well as processed cheese, yogurts, desserts and processed meat and fish products.

Whey protein isolate (WPI) typically contains 90–92% protein, a maximum of 1% of fat and lactose, 2–3% ash and 4–5% moisture. WPI can be prepared by membrane filtration, whereby the whey is first subjected to microfiltration (MF) to reduce the fat content of the whey. Subsequently, the MF permeate is subjected to UF and diafiltration, to achieve the desired proportion of protein in the dry matter of the product (Rowan, 1998). WPI can also be prepared through ion exchange (IEX) chromatography, whereby the whey proteins are selectively absorbed onto the IEX resin, whereas the fat, lactose and minerals are eluted. The whey proteins are subsequently eluted from the resin by increasing ionic strength (Mulvihill and Ennis, 2003). These different processing technologies result in differences in protein and mineral composition between WPI prepared by membrane filtration and by IEX chromatography. WPI prepared using IEX chromatography generally contains no glycomacropeptide, lactoferrin (LF) and peptide fragments because these are not adsorbed by the resin. Thus, levels of α-lactalbumin (α-LA), β-lactoglobulin (β-LG) and bovine serum albumin (BSA) are higher in WPI prepared by IEX chromatography than in WPI prepared by membrane filtration (Affertsholt and Nielsen, 2007). The minerals in WPI prepared by membrane filtration occur at ratios similar to those observed in WPC, whereas in WPI prepared by IEX chromatography, the minerals naturally present in the whey are largely removed, and the monovalent cation of the mineral used to achieve elution of the proteins from the resin predominates (Affertsholt and Nielsen, 2007).

4.2 Fractionated whey protein ingredients

In addition to WPCs and WPIs, as described in the previous section, a number of fractionated whey protein ingredients, which are strongly enriched in a single whey protein, are being prepared, that is, α-LA, β-LG and LF. Preparation of these products requires additional fractionation and processing steps. Since α-LA is the major whey protein in human milk, which is devoid of β-LG, purified α-LA and α-LA-enriched whey protein products are of interest in the infant formula industry. Rather pure α-LA preparations (>95% protein, of which >90% is α-LA), can be isolated from sweet whey using chromatographic processes (Affertsholt and Nielsen, 2007). However, the comparatively high price for the production of these ingredients limits widespread use and shifts interest to α-LA-enriched fractions. Most processes for the isolation of α-LA-enriched fractions from whey exploit the low thermal stability of the calcium-depleted, apo-form of α-LA, which can be formed by reducing pH, or treatment with a cation exchange resin or calcium-chelating agent. Apo-α-LA denatures and aggregates on heating at approximately 35–55°C. The exact temperature required to induce denaturation depends on pH, ionic strength and the type of calcium chelator added. The denatured α-LA can subsequently be separated from the remaining soluble whey proteins using various centrifugation or filtration techniques. The separated apo-α-LA can subsequently be resolubilized in water, neutralized, concentrated and spray-dried (Pearce, 1983, 1987; de Wit and Bronts, 1994; Bramaud et al., 1997; Gesan-Guiziou et al., 1999). These techniques can yield products where >70% of the proteinaceous fraction consists of α-LA and can be used in infant nutrition.

β-LG is the most abundant protein in whey. Like α-LA, it can be isolated from milk chromatographically, but the high prices prevent widespread application. More economically feasible are β-LG-enriched fractions, which are the by-products of the aforementioned procedure for the isolation of α-LA (Pearce, 1983, 1987; de Wit and Bronts, 1984; Bramaud et al., 1997; Gesan-Guiziou et al., 1999). Further removal of other whey proteins, for example, immunoglobulins (Igs) and BSA, can be achieved by membrane filtration, and β-LG preparations containing >80% protein, of which >75% of protein is β-LG, can be prepared. Primary applications for β-LG concentrates are technofunctional, rather than nutritional, since the protein has excellent foaming, emulsification, heat-gelation, water-binding properties and solubility at low pH. As such, β-LG can be used as an egg white replacer in bakery products, a fat replacer in processed meat products, a source of protein fortification in drinks and a stabilizer in desserts (de Wit, 1998).

LF plays an important role in the human cellular immune system response and protects the body against infections, as well as provides primary defence against pathogens, stimulation of the immune system and regulation of the iron status in the body. Isolation of LF from dairy streams has been commercialized and the product is used in infant formula, meat preservation, dietary supplements, pharmaceutical products and cosmetics, and oral hygiene products. LF can be isolated from skim milk, but is most commonly isolated from whey by cation exchange chromatography (Paul et al., 1980; Prieels and Peiffer, 1986; Martin-Hernandez et al., 1990; Chui and Etzel, 1997). The high isoelectric point (~8.7) and thus positive charge at neutral pH of the protein is exploited herein, by using a cationic exchange resin. LF preparations typically contain 93–96% protein, of which >95% is LF. Lactoperoxidase, which is also positively charged at neutral pH, is co-adsorbed to the cation exchange resin in the isolation of LF. Subsequent separation of the two can be achieved by using elution gradients, membrane filtration or size exclusion chromatography (Prieels and Peiffer, 1986; Burling, 1989; Chui and Etzel, 1997).

5 Milk protein concentrates

Among the different milk protein ingredients, the milk protein concentrates (MPCs) are rapidly gaining popularity. MPCs containing ~40–90% protein in dry matter are being produced. Typical compositions of MPCs (Table 7) show that MPCs are enriched in protein

Table 7 Typical composition (in %, m/m) of skim milk powder and different milk protein concentrates

	SMP	MPC42	MPC56	MPC70	MPC80	MPC85
Protein	34.0	42.0	56.0	70.0	80	85
Fat	0.8	1	1.3	1.4	1.8	1.8
Lactose	53.5	45.7	31.2	17	5.5	1.2
Ash	7.9	7.8	7.7	7.2	7.4	7.4
Moisture	3.8	3.5	3.8	4.4	4.5	4.6

and depleted in lactose compared to milk powder, whereas the ash, fat and moisture contents are reasonably constant. Most high-quality MPCs are prepared by UF of skim milk, and typically the process consists of the following steps:

- Skimming
- Pasteurization
- Ultrafiltration and diafiltration
- Evaporation
- Drying

The skim milk is heated in the low-heat region (e.g. 10−20 s at 70−75°C), and is subsequently concentrated by UF using membranes with a pore size of ≤10 kDa. During this step, caseins, whey proteins, micellar salts and residual fat are concentrated in the retentate, whereas lactose, soluble salts and non-protein nitrogen are removed with the permeate (Babella, 1989; Bastian et al., 1991). For high-protein MPCs, for example, MPC70–MPC90, UF alone is insufficient to achieve the required protein:solids ratio in the retentate, and diafiltration is applied (Singh, 2007). The maximum protein content achievable is limited to ~90% protein in dry matter because of the retention of micellar calcium phosphate and residual fat (Kelly, 2011). After the desired protein:solids ratio has been achieved, the retentate is evaporated and dried. Because of the considerably higher protein:solids ratio in the retentate, evaporation cannot achieve a solids content for MPCs that is similar to that of SMP.

Because MPCs are the only milk protein ingredients that contain the caseins and whey proteins in their natural ratio and in their native state, they are a popular ingredient for standardization of cheese milk, protein fortification of yogurt, or in ice cream mixes and clinical and infant nutrition products. Because micellar calcium phosphate is largely retained in the micelles during the UF process, MPCs contain high levels of encapsulated bioavailable calcium, thus making them interesting ingredients for nutritional products.

One of the issues for the high-protein MPCs in particular is that they can suffer loss of solubility during storage, particularly on storage at above ambient temperatures (Anema et al., 2006; Fang et al., 2011; Gazi and Huppertz, 2015) and at high moisture content and water activity (Baldwin and Truong, 2007). Solubility is often better at higher temperature (McKenna, 2000; Mimouni et al., 2009) or with the application of shear (McKenna, 2000), but this is not feasible in cases where whey protein denaturation or damage to emulsion droplets is undesirable, for example, in standardization of cheese milk. The poor solubility of high-protein MPCs has been related to casein interactions on the surface of the powder particle (Anema et al., 2006; Havea, 2006; Gazi and Huppertz, 2015). MPCs have been incorporated into a variety of common dairy products, most notably yogurt, ice cream, and particularly cheese. For yogurt, it has been shown that MPCs can be used efficiently as replacements for traditional skim milk ingredients, such as SMP and skim milk concentrate, which are added in many instances to increase the protein content and improve the texture and stability of the product.

6 Milk protein hydrolysates

Milk protein hydrolysates are a class of milk protein ingredients that have been gaining more and more interest in the last few decades. Hydrolysis of milk proteins can be achieved either chemically in alkaline or acidic conditions, or enzymatically. The most commonly used enzymes for this purpose are digestive proteases and peptidases. The composition of the hydrolysates resulting from chemical hydrolysis is difficult to control. Due to the higher specificity of enzymes, enzymatic hydrolysis is preferred in the manufacture of milk protein hydrolysates.

Milk protein hydrolysates can be divided into three main categories on the basis of their designated application:

- hydrolysates or specific peptides with biological activity;
- hydrolysates for consumers with specific nutritional needs;
- hydrolysates for improved protein functionality.

6.1 Hydrolysates with biological activity

The hydrolysates with biological activity contain peptides that resemble endogenous structures with functions of hormones, signal peptides or neurotransmitters (Hernández-Ledesma et al., 2014). The biological activity of milk protein hydrolysates has been extensively studied *in vitro*, many of these findings having also been confirmed in studies performed *in vivo* on animals and/or on humans in clinical studies.

Antihypertensive peptides are one of the most well-known categories of milk peptides with biological activity. One of the causes of hypertension is the activity of the angiotensin-converting enzyme (ACE), which elevates the blood pressure by causing the blood vessels to constrict (Erdmann et al., 2008). The antihypertensive milk peptides act as ACE inhibitors, thus lowering the blood pressure. Peptide sequences with antihypertensive activity have been identified from both caseins and whey proteins. The biological activity of the antihypertensive peptides has been confirmed *in vivo* by measuring the reduction of systolic blood pressure in spontaneously hypertensive lab rats (Fitzgerald et al., 2004; Hernández-Ledesma et al., 2011; Martínez-Maqueda et al., 2012b). Production of these peptides was achieved through two main ways: enzymatic hydrolysis of milk protein ingredients and fermentation by proteolytic bacteria in yogurt and cheese type of products.

Alongside hypertension, chronic inflammation, oxidative stress and plasma lipid profiles are also factors responsible for cardiovascular diseases. *In vivo* studies on healthy humans have confirmed whey-derived peptide NOP-47 to show anti-inflammatory and antioxidant properties (Ballard et al., 2009). *In vitro* studies have shown that other whey- and casein-derived peptides show the potential to reduce inflammation, through regulating the secretion of pro-inflammatory cytokines (Mao et al., 2011; Tompa et al., 2010). These findings have not yet been confirmed through clinical studies. The consumption of fermented milk products in animal and human trials was found to counteract effects of oxidative stress, improving cardiovascular health (Kullisaar et al., 2003; Zommara et al., 1998). The compounds with antioxidative activity in the fermented milk products are believed to be peptides resulting from the proteolytic activity of lactic acid bacteria. Peptides from tryptic (Nagaoka et al., 2001) and chymotryptic (Yamauchi et al., 2003) hydrolysates of β-lg were shown to have hypocholesterolaemic activity when administered to rats. The consumption of these hydrolysates caused a decrease in total cholesterol and low-density lipoprotein cholesterol, and an increase in high-density lipoprotein cholesterol.

Hydrolysates from both caseins and whey proteins were found to have a protective effect on gut mucosa and intestinal health. The surface of the intestine is covered by a mucus gel, the main component of which are high-molecular-weight mucins, secreted by goblet cells (Corfield et al., 2000; Linden et al., 2008). The protective effect of milk protein peptides was found to come from regulating the number and expression of goblet cells (Plaisancié et al., 2013), as well as from regulating mucin secretion by the goblet cells (Claustre et al., 2002; Trompette et al., 2003; Martínez-Maqueda 2012a, 2013a,b). A protective effect on intestinal health can also be exercised by milk protein hydrolysates with antimicrobial activity. There are several mechanisms through which peptides can show antimicrobial activity. The most common antimicrobial peptides are positively charged and have an amphiphilic structure (López-Expósito and Recio, 2008). It is hypothesized that due to the positive charge, they could attach to the membranes of the negatively charged bacteria, preventing them from attaching to the intestinal lining. The amphiphilic character of the peptides could contribute to disruption of the bacterial membrane and inactivation of the bacteria. Antimicrobial activity against pathogenic bacteria could also be the indirect result of a prebiotic effect of milk protein hydrolysates. The most commonly known peptides with antimicrobial activity are derived from LF (Bellamy et al., 1992; Bolscher et al., 2012; Puknun et al., 2013; Recio and Visser, 1999; Van der Kraan et al., 2004). Nevertheless, tryptic, chymotryptic (Pellegrini et al., 1999, 2001; Demers-Matthieu et al., 2013) and peptic (Theolier et al., 2013) hydrolysates of α-la, β-lg, WPI and caseins were also shown to have antimicrobial activity. Next to antimicrobial activity, the protective effects of milk protein hydrolysates against infection also come from immunomodulatory activity (Gauthier et al., 2006; Gill et al., 2000). Ingestion of whey proteins, such as α-la or LF, whey protein-derived peptides or casein phosphopeptides in mice or humans led to increases in the levels of serum, bile, intestine or faeces Ig A (Kitamura and Otani, 2002; Miyauchi et al., 1997; Otani et al., 2000, 2003; Saint-Sauveur et al., 2009).

At the intestinal level, caseinophosphopeptides (CPPs) can modulate absorption of minerals (Meisel and Fitzgerald, 2003). It is hypothesized that CPPs can increase the solubility of calcium in the alkaline pH of the intestine, potentially leading to better absorption of calcium. *In vivo* studies have found increased concentrations of soluble calcium in the intestine attributed to CPP released from ingested casein. Calcium bound to CPPs was also found to help remineralize teeth, as well as show anti-cariogenic potential (Luo and Wong, 2004; Reynolds et al., 2003). Like calcium, the bioavailability of iron was also increased in the presence of CPPs in studies performed *in vivo* on rats. Mineral absorption modulatory effects of CPPs have not been clearly proven in human studies. It is hypothesized that the composition and physical properties of food may influence the effects of CPPs.

Positively charged peptides derived from LF, that is, lactoferricins, were found to also have antiproliferative effect on cancer cells. Similar to the antimicrobial activity, the cationic peptides are believed to interact with and destabilize the negatively charged membranes of cancer cells (Hoskin and Ramamoorthy, 2008). Casein-derived β-casomorphins were found to show antiproliferative activity on breast cancer cells. The antiproliferative effects of β-casomorphins and other milk-derived opioid peptides could be the result of interaction of the peptides with opioid receptors.

Consumption of whey protein and whey protein hydrolysates was found to regulate insulin secretion and blood glucose levels in both healthy and diabetic individuals (Sousa et al., 2012). The biological activity is believed to be the result of a homology between whey protein-derived peptides and hormones such as glucagon-like peptide-1 (GLP-1), that regulate insulin secretion and absorption of glucose from the blood stream.

Consumption of protein in general, but specifically whey protein, is known to increase the sensation of satiety (Anderson et al., 2004; Fromentin et al., 2012). The mechanism through which this occurs has not yet been fully understood, but several possible explanations have been presented in literature. Satiety can be caused by physical properties of the ingested food, such as viscosity, aggregation behaviour in the stomach and effect on the speed of stomach emptying. The feeling of fullness is also influenced by the release of satiety hormones, for example, cholecystokinin, GLP-1, ghrelin or peptide YY. The effect of whey proteins and whey protein hydrolysates on satiety can come either from influencing the secretion of these hormones or from structural homology with the satiety hormones.

Specific milk protein-derived peptides also exert biological activity on the nervous system. Peptides derived from tryptic hydrolysis of α_{s1}-CN were found to have a relaxing effect, decreasing elevated blood pressure caused by stress (Cakir-Kiefer et al., 2011; Miclo et al., 2001). LF consumption has also been shown to have anti-stress effects, as well as analgesic activity, and potentiation of the analgesic activity of morphine.

6.2 Hydrolysates for consumers with specific nutritional needs

There are three main categories of milk protein hydrolysates designated to address specific nutritional needs (Nongonierma et al., 2016):

- Milk protein hydrolysates used as ingredients in hypoallergenic infant milk formulas developed for infants that suffer from cow's milk protein allergies;
- Low-phenylalanine hydrolysates for consumers who suffer from phenylketonuria;
- Mildly hydrolysed milk proteins for easier digestion developed either for infants or for the elderly.

Cow's milk protein allergies are common in infants who are introduced to cow's milk early in their diets. The main allergens in cow's milk are the caseins, as well as β-lg, which is absent from human milk. Milk protein hydrolysates for hypoallergenic infant formulas usually have a high degree of hydrolysis. This is done in order to ensure removal of the allergenic epitopes of the milk proteins. The infant milk formulas with the highest degree of hydrolysis are the elemental, amino-acid-based formulas.

Sufferers of phenylketonuria show a deficiency in phenylalanine (Phe) hydrolase, which normally converts Phe into tyrosine, resulting in accumulation of Phe in the body. As a result, the intake of Phe must be limited. Milk protein hydrolysates depleted in Phe have specifically been developed for this nutritional category. Phe is enzymatically released from the proteins, followed by treatment with active carbon or other adsorbents in order to remove the Phe from the hydrolysate mixture.

Bioavailability of protein decreases in the elderly population, with often an increase in protein intake being necessary. The decrease in protein bioavailability can lead to loss of muscle mass, which, in turn, leads to decreased mobility and health decline. Mild hydrolysis of the milk proteins facilitates their digestion and increases the bioavailability and conversion of ingested protein to muscle mass. Similarly, protein digestion in infants is improved when infants are fed with comfort-type infant milk formulas containing mildly hydrolysed milk proteins. Milk protein hydrolysates with faster absorption are also used in sports nutrition, enteral formulas, nutritional bars and nutritional drinks.

6.3 Hydrolysates for improved protein functionality

Hydrolysis of whey proteins can increase their solubility over a broader pH range and decrease the viscosity of the solutions (Bansal and Bhandari, 2016). Mild hydrolysis can improve surface activity, emulsifying and foaming ability, as well as increase thermal stability.

7 Lactose and lactose derivatives

7.1 Lactose

Lactose is the main carbohydrate in mammalian milk and is a reducing carbohydrate composed of galactose and glucose linked by a $\beta 1 \rightarrow 4$ glycosidic bond (Fox, 2009). Lactose is mostly isolated from whey permeates (e.g. from WPC production) and is used in a wide variety of products (Paterson, 2009). One main market for lactose is the infant formula market. Furthermore, lactose is also used in baking and confectionery application, owing to the fact that lactose is less sweet than sucrose. For baking applications, the fact that lactose is not metabolized by yeast and therefore remains available as a reducing carbohydrate for Maillard-induced browning during the baking steps makes lactose a popular ingredient (Nickerson, 1976). Lactose is also used in a variety of other food products (Nickerson, 1976), as well as in the pharmaceutical industry, where it is used as an excipient for making tablets and as a carrier in dry powder inhalers (Durham et al., 2004; Paterson, 2009). Different grades of lactose are available for different purposes; for example,

- Crude lactose (95–98% purity)
- Edible lactose (95–98% purity)
- Refined edible lactose (99–99.5% purity)
- Pharmaceutical lactose (99–99.5% purity; strict requirements on particle size)

Production of edible-grade lactose products, which are in the α-lactose monohydrate form, typically use the permeate from UF of whey as a source material. Typically, the permeate will contain 5–6% dry matter, of which at least 80% is lactose, and this must be concentrated to ~110 g lactose per 100 g water, that is, ~58% solids (Paterson, 2009;

Wong and Hartel, 2014). Concentration can be achieved by reverse osmosis, followed by evaporation, or solely by evaporation. In general, the higher the dry matter content achieved, the higher the yield, although the downside that should be taken into account is the fouling of evaporators that occurs at high solids content and should be minimized for an efficient process (Paterson, 2009; Wong and Hartel, 2014). Following concentration, the lactose needs to be crystallized, which should be done in a controlled manner to ensure that high yields and the desired particle sizes are reached. For this purpose, seeding crystals may also be added. After evaporation, flash cooling to 30–40°C may be used to initiate nucleation and subsequently, controlled cooling at a rate of 2–3°C per hour to 10–15°C is generally carried out (Nickerson and Moore, 1974; Paterson, 2009; Schuck, 2011; Wong and Hartel, 2014). The crystallization stage is the key stage in lactose production and requires careful control in terms of temperature, stirring speeds and flow patterns (Paterson, 2009; Schuck, 2011; Wong and Hartel, 2014). Following crystallization, which can take up to 24 h or more, the crystals produced are removed by using centrifuges or decanters, followed by a washing step where residual mother liquor is removed. Subsequently, the remaining free moisture is removed in flash drier, followed by a fluid bed drier, which also cools the product (Paterson, 2009). Failure to cool the product sufficiently results in a tendency in the product to cake in the bags during storage (Bronlund and Paterson, 2004).

For pharmaceutical-grade lactose, different forms are available and used for specific applications. Milled α-lactose monohydrate is typically used for wet granulation, whereas anhydrous lactose is the product of choice for dry granulation and spray-dried lactose. Anhydrous lactose can also be used for direct compression applications, for example, tablets. Pharmaceutical α-lactose monohydrate can be produced from edible-grade lactose described above by redissolving the powder and removing impurities (riboflavin, minerals, some proteins, lactose phosphate) by a combination of adsorption and filtration processes, followed by re-crystallization and drying. Alternatively, the adsorption and filtration processes can be included in the first production cycle stage, thus eliminating the need for a second cycle. To produce anhydrous pharmaceutical lactose, it is important that lactose is crystallized at temperatures >93.5°C to ensure that primarily β-lactose is formed. Typically, this is achieved by spraying a cleaned lactose solution as outlined above on a hot roller drier, resulting in a fine cake of crystals typically containing up to 80% anhydrous β-lactose crystals and the remainder anyhydrous α-lactose crystals. The third form of pharmaceutical-grade lactose, spray-dried lactose, is prepared by spray-drying suspensions of α-lactose monohydrate, resulting in spherical agglomerates of lactose crystals. This spray-dried lactose has excellent flowability and is highly suitable for tableting (Durham et al., 2004; Paterson, 2009).

7.2 Galactooligosaccharides

While lactose is by far the most abundant carbohydrate in the milk of the main dairying species, they also contain oligosaccharides (Urashima et al., 2011). Because of their prebiotic function, they are extremely valuable ingredients, particularly from a nutritional perspective. However, as concentrations in milk are generally low and difficult to isolate, routes for producing oligosaccharides from lactose have been researched and have led to the commercial production of GOS.

GOS refers to β-linked galactose moieties with galactose or glucose at the reducing end and typically having a degree of polymerization of 3–8. A large number of human studies have consistently shown that GOS has a number of health benefits; for example, it is resistant to hydrolysis in the human digestive system and thus has a low caloric value. Because it is non-digestible in the digestive tract, it enters the colon intact, where it selectively stimulates the growth of bifidobacteria and is fermented to short-chain fatty acids. Furthermore, GOS has also been shown to aid regulation of bowel function and stool production and improve stool consistency. Besides, there are also indications that GOS can aid absorption of Ca and Mg in the gastrointestinal tract and inhibit adhesion of enteropathogenic *E. coli* (Playne and Crittenden, 2009; Gosling et al., 2010; Torres et al., 2010).

Industrial production of GOS uses lactose as a substrate and is an enzymatic reaction. Glycosyltransferases are ideal for preparing GOS, but are not commercially available. Hence, industrial GOS production is typically carried out with β-galactosidases as enzymes, which are more commonly applied to hydrolyse lactose into glucose and galactose in products. In this reaction, water molecules typically act as the galactosyl acceptor molecule. However, in systems where water availability is limiting, other molecules, that is, glucose, galactose or GOS, may also act as acceptor molecules (Playne and Crittenden, 2009; Gosling et al., 2010; Torres et al., 2010). Therefore, GOS production should be carried out in concentrated systems, containing at least 30% (m/m) lactose, irrespective of the enzyme used (Torres et al., 2010). Due to the comparatively low solubility of lactose at room temperature (~20%, m/m), the reactions need to be performed at high temperature, where lactose solubility is higher. Typically, GOS production is carried out at temperatures in the range of 40–70°C, depending primarily on enzyme activity and stability. GOS yield typically strongly increases with increasing lactose content up to ~30%, above which only limited further increases are observed. Enzyme cost is a main contributor to overall cost in GOS production and may be reduced by using immobilized enzymes (Playne and Crittenden, 2009; Gosling et al., 2010; Torres et al., 2010). After the required degree of conversion of lactose to GOS has been achieved, the solution is heated to inactivate the enzyme, as prolonged incubation typically leads to hydrolysis of the oligosaccharides by β-galactosidase. Further treatment with membranes and absorbing resins to remove impurities (proteins, riboflavin, minerals) may be conducted prior to concentration to a syrup or spray-drying to a powder.

7.3 Lactose derivatives

Lactose is also the precursor for a number of other lactose derivatives, that is, the ion sequestrant lactobionic acid, the laxative lactulose and the sweeteners, lactosucrose, lactitol and D-tagatose. All have specific advantages and application, and are described further below.

Lactobionic acid (β-4′galactosylglucuronic acid) is produced commercially by the chemical oxidation of lactose at high pH over a noble metal catalyst, whereas enzymatic and electrolytic methods have also been researched. In the oxidation reaction, the aldehyde group of the glucose moiety in lactose is oxidized to a carboxyl group (Harju, 1993; Playne and Crittenden, 2009; Gutiérrez et al., 2012). The mineral binding capacity of lactobionic acid is exploited in its use in the Wisconsin transplantation solution to reduce oxidative damage by metal ions to tissues and organs during storage. Furthermore, lactobionic acid is used in the pharmaceutical industry for intravenous delivery of erythromycin and also in calcium supplementation. It is also used in chlorohexidine-based disinfectants, where it suppresses tissue damage facilitating wound healing. Furthermore, lactobionic acid has applications in the skincare industry as well as used as calcium carrier and sweet acidifier in food products (Playne and Crittenden, 2009; Gutiérrez et al., 2012).

The disaccharide lactulose (galactosyl-β-1→4-fructose) is a lactose isomer that is formed in small quantities on heating milk, for example, during UHT treatment or sterilization (Olano et al., 1989; Van Boekel, 1998). However, industrial production of lactulose involves synthetic isomerization reactions at alkaline pH (Aider and de Halleux, 2007; Playne and Crittenden, 2009). Most lactulose is used in the pharmaceutical industry in the treatment of constipation and hepatic encephalopathy. However, it has also been approved for use as a prebiotic in several countries and applications as a prebiotic ingredient in functional food products are also emerging. Furthermore, it has also been suggested as a low-calorie sweetener (Tamura et al., 1993; Aider and de Halleux, 2007; Playne and Crittenden, 2009; Panesar and Kumari, 2011).

Lactosucrose can be produced enzymatically by transferring the fructosyl moiety of sucrose to a lactose molecule using a β-fructofuranosidase in a transfructosylation reaction (Mussatto and Mancilha, 2007; Fujita et al., 2009). Similar to β-galactosidases in GOS production, the enzymes also have hydrolytic activity, so conditions have to be chosen such that the transfructosylation reaction predominates (Crittenden and Playne, 1996; Playne and Crittenden, 2009). Lactosucrose is primarily used as a prebiotic and sweetener in a range of dairy and non-dairy food products, including beverages, bakery products, confectionery, yogurt and desserts (Fujita et al., 2009; Playne and Crittenden, 2009).

Lactitol is a sugar alcohol prepared from lactose by reduction of the glucose moiety (Harju, 1993; Playne and Crittenden, 2009). The main use of lactitol is as a low-calorie sweetener in food products, particularly bakery products, because it does not participate in the Maillard reaction. It is also used in chocolate, confectionery and ice cream. In pharmaceutical applications, lactitol may serve as an alternative to lactulose, whereas it may also serve as a cryoprotectant, for example, for surimi. Prebiotic effects of lactitol have also been described (Van Velthuijsen, 1979; Playne and Crittenden, 2009).

D-tagatose is produced by isomerization of galactose, which must first be released from lactose enzymatically using β-galactosidase. Following removal of glucose, the galactose can be isomerized under alkaline conditions to form D-tagatose. The main market for tagatose is as a sweetener in diabetic foods, as well as weight-loss products (Levin, 2002; Kim, 2004; Playne and Crittenden, 2009).

8 Milk fat globule membrane material

In addition to the milk proteins and carbohydrates, the MFGM also forms an interesting source for both functional and nutritional dairy ingredients (Jiménez-Flores and Brisson, 2008; Dewettinck et al., 2008). The MFGM is a trilayer structure naturally surrounding the milk fat globules and consists of polar lipids, neutral lipids and proteins. The polar lipids and proteins of the MFGM, in particular, have desired functional properties, for example, emulsification and stabilization, bactericidal effects and prevention of pathogen. Buttermilk is a natural source of enriched MFGM material due to the fact that the membrane is damaged during the churning process in butter manufacture. With the emergence of the NIZO butter process in the 1960, sweet rather than acidified buttermilk became available for processing, giving opportunities for further processing to functional ingredients. An even richer source of MFGM material is butter serum, the aqueous phase by-product of anhydrous milk fat manufacture. Both buttermilk and butter serum still contain high levels of skim milk constituents (proteins, lactose, minerals) which can be removed by precipitation or membrane filtration, whereas removal of neutral lipids to supercritical CO_2 extraction has been described as a useful technique for removing polar lipids (Astaire et al., 2003). From a physical perspective, MFGM material has desired properties for emulsion stabilization and controlling protein interactions and as such has been used in ice cream, evaporated milk, cheese and processed cheese and nutritional products. Nutritional benefits include antibacterial and antiviral activities and the inhibition of cell growth (Jiménez-Flores and Brisson, 2008; Dewettinck et al., 2008).

9 Conclusions and future trends

Similar to the past three decades, the dairy ingredient sector is expected to remain one of continuous development, improvement and innovation in the future. Technological developments will undoubtedly facilitate the development of ingredients with higher levels of purity or functionality, but ultimate uptake hereof will depend on several other factors. Firstly, of course, demand for dairy ingredients for specific applications and market segments. For instance, demand for dairy ingredients for nutritional products for infants and elderly people, but also for performance nutrition and nutritional snacks will become an even bigger driver. Secondly, competition of non-dairy ingredients, particularly proteins, will also become a driving factor, but will also allow to further establish the uniqueness of dairy ingredients. Finally, any developments in dairy ingredients should, of course, fit in a global sustainability perspective and allow for the use of all, and not just some, fractions of milk as functional ingredients, thereby attaining maximum physical and nutritional functionality out of milk.

10 Where to look for further information

A standard introduction to the subject is provided in *Dairy Science and Technology* by Walstra et al. (2005) and the *Encyclopedia of Dairy Sciences* by Fuquay et al. (2011). More detailed information on dairy constituents and their conversion to functional ingredients can be found in the *Advanced Dairy Chemistry* Series (Fox and McSweeney, 2006; McSweeney and Fox, 2009, 2013; McSweeney and O'Mahoney, 2016). Furthermore, resources provided by International Dairy Journal (http://www.fil-idf.org/) and the US Dairy Export Council (www.usdec.org) provide valuable resources.

Fuquay, J. W., Fox, P. F. and McSweeney, P. L. (2011), *Encyclopedia of Dairy Sciences 2nd Edition*. Academic Press.
McSweeney, P. L. H. and Fox, P. F. (2006), *Advanced Dairy Chemistry. Volume 2: Lipids*, 3rd Edition, Springer.
McSweeney, P. L. H. and Fox, P. F. (2009), *Advanced Dairy Chemistry. Volume 3: Lactose, Water, Salts and Minor Constituents*, 3rd Edition, Springer.

McSweeney, P. L. H. and Fox, P. F. (2013), *Advanced Dairy Chemistry. Volume 1A: Proteins: Basic Aspects*, 4th Edition, Springer.
McSweeney, P. L. H. and O'Mahony, J. A. (2016), *Advanced Dairy Chemistry. Volume 1B: Proteins: Applied Aspects*, 4th Edition, Springer.
Walstra, P., Wouters, J. T. and Geurts, T. J. (2005), *Dairy Science and Technology*, CRC press.

11 References

Affertsholt, T. and Nielsen, M. D. (2007), *The World Market for Whey and Lactose Products*, pp. 2006–10. Aarhus: 3A Business Consulting.
Aider, M. and de Halleux, D. (2007), 'Isomerization of lactose and lactulose production: Review', *Trends Food Sci. Technol.*, *18*, 356–64.
Anderson, G. H., Tecimer, S. N., Shah, D. and Zafar, T. A. (2004), 'Protein source, quantity, and time of consumption determine the effect of proteins on short-term food intake in young men', *J. Nutr.*, *134*, 3011–15.
Anema, S. G., Pinder, D. N., Hunter, R. J. and Hemar, Y. (2006), 'Effects of storage temperature on the solubility of milk protein concentrate (MPC85)', *Food Hydrocolloids*, *20*, 386–93.
Astaire, J. C., Ward, R., German, J. B. and Jiménez-Flores, R. (2003), 'Concentration of polar MFGM lipids from buttermilk by microfiltration and supercritical fluid extraction', *J. Dairy Sci.*, 86, 2297–2307.
Babella, G. (1989), 'Scientific and practical results with use of ultrafiltration in Hungary', *Bull. Int. Dairy Fed.*, *244*, 7–25.
Baldwin, A. and Pearce, D. (2005), 'Milk powder', in *Encapsulated and Powdered Foods*, ed. C. Onwulata, pp. 387–433. Boca Raton, FL: CRC Press.
Baldwin, A. J. and Truong, G. N. T. (2007), 'Development of insolubility in dehydration of dairy milk powders', *Food Bioprod. Process.*, *85*, 202–8.
Ballard, K. D., Bruno, R. S., Seip, R. L., Quann, E. E., Volk, B. M., Freidenreich, D. J., Kawiecki, D. M., Kupchak, B. R., Chung, M. Y., Kraemer, W. J. and Volek, J. S. (2009), 'Acute ingestion of a novel whey-derived peptide improves vascular endothelial responses in healthy individuals: A randomized, placebo controlled trial', *Nutr. J.*, *8*, Article No 34.
Bastian, E. D., Collinge, S. K. and Ernstrom, C. A. (1991), 'Ultrafiltration: Partitioning of milk constituents into permeate and retentate', *J. Dairy Sci.*, *74*, 2423–34.
Bellamy, W., Takase, M., Yamauchi, K., Wakabayashi, H., Kawase, K. and Tomita, M. (1992), 'Identification of the bactericidal domain of lactoferrin', *Biochim. Biophys. Acta*, *1121*, 130–6.
Bansal, N. and Bhandari, B. (2016), 'Functional milk proteins: Production and utilization – whey-based ingredients', in *Advanced Dairy Chemistry 1B: Proteins: Applied Aspects*, eds. P. L. H. McSweeney and J. A. O'Mahoney, pp. 67–98. New York: Springer.
Bolscher, J., Nazmi, K., van Marle, J., van't Hof, W. and Veerman, E. (2012), 'Chimerization of lactoferrin and lactoferrampin peptides strongly potentiates the killing activity against *Candida albicans*', *Biochem. Cell Biol.*, *90*, 378–88.
Bramaud, C., Aimar, P. and Daufin, G. (1997), 'Whey protein fractionation: Isoelectric precipitation of α-lactalbumin under gentle heat', *Biotechnol. Bioeng.*, *56*, 391–7.
Bronlund, J. and Paterson, T. (2004), 'Moisture sorption isotherms for crystalline, amorphous and predominantly crystalline lactose powders', *Int. Dairy J.*, *14*, 247–54.
Burling, H. (1989), 'Process for extracting pure fractions of lactoperoxidase and lactoferrin from milk serum', International Patent Application WO 89/04608 A1.
Cakir-Kiefer, C., Le Roux, Y., Balandras, F., Trabalon, M., Dary, A., Laurent, F., Gaillard, J. L. and Miclo, L. (2011), '*In vitro* digestibility of α-casozepine, a benzodiazepine-like peptide from bovine casein, and biological activity of its main proteolytic fragment', *J. Agric. Food Chem.*, *59*, 4464–72.

Carr, A. and Golding, M. (2016), 'Functional milk protein production and utilization: Casein-based ingredients', in *Advanced Dairy Chemistry 1B: Proteins: Applied Aspects*, eds. P. L. H. McSweeney and J. A. O'Mahoney, pp. 35–67. New York: Springer.

Chui, C. K. and Etzel, M. R. (1997), 'Fractionation of lactoferrin and lactoperoxidase from bovine whey using a cation exchange membrane', *J. Food Sci.*, 62, 996–1000.

Claustre, J., Toumi, F., Trompette, A., Jourdan, G., Guignard, H., Chayvialle, J. A. and Plaisancié, P. (2002), 'Effects of peptides derived from dietary proteins on mucus secretion in rat jejunum', *Am. J. Physiol. Gastrointest. Liver Physiol.*, *283*, G521–8.

Corfield, A. P., Myerscough, N., Longman, R., Sylvester, P., Arul, S. and Pignatelli, M. (2000), 'Mucins and mucosal protection in the gastrointestinal tract: New prospects for mucins in the pathology of gastrointestinal disease', *Gut*, *47*, 589–94.

Crittenden, R. G. and Playne, M. (1996), 'Production, properties and applications of food-grade oligosaccharides', *Trends Food Sci. Technol.*, *7*, 353–61.

Demers-Matthieu, V., Gauthier, S. F., Britten, M., Fliss, I., Roobitaille, G. and Jean, J. (2013), 'Antibacterial activity of peptides extracted from tryptic hydrolysate of whey protein by nanofiltration', *Int. Dairy J.*, *28*, 94–101.

Dewettinck, K., Rombaut, R., Thienpont, N., Le, T. T., Messens, K. and Van Camp, J. (2008), 'Nutritional and technological aspects of milk fat globule membrane material', *Int. Dairy J.*, *18*, 436–57.

de Wit, J. N. (1998), 'Nutritional and functional characteristics of whey proteins in food products', *J. Dairy Sci.*, 81, 597 – 608.

de Wit, J. N. and Bronts, H. (1994), 'Process for the recovery of alpha-lactalbumin and beta-lactoglobulin from a whey product', European Patent Application 0 604 864.

Durham, R. J., Sleigh, R. W. and Hourigan, J. A. (2004), 'Pharmaceutical lactose: A new whey with no waste', *Aust. J. Dairy Technol.*, *59*, 138.

Erdmann, K., Cheung, B. W. Y. and Schröder, H. (2008), 'The possible roles of food-derived bioactive peptides in reducing the risk of cardiovascular disease', *J. Nutr. Biochem.*, *19*, 643–54.

Ennis, M. P., O'Sullivan, M. M. and Mulvihill, D. M. (1998), 'The hydration behaviour of rennet caseins in calcium chelating salt solution as determined using a rheological approach', *Food Hydrocolloids*, *12*, 451–7.

Fang, Y., Selomulya, C., Ainsworth, S., Palmer, M. and Chen, X. D. (2011), 'On quantifying the dissolution behaviour of milk protein concentrate', *Food Hydrocolloids*, *25*, 503–10.

Fitzgerald, R. J., Murray, B. A. and Walsh, D. J. (2004), 'Hypotensive peptides from milk proteins', *J. Nutr.*, *134*, S980–8.

Fox, P. F. (2009), 'Lactose: Chemistry and properties', in *Advanced Dairy Chemistry* 3, eds. P. L. H. McSweeney and P. F. Fox, pp. 1–15. New York: Springer.

Fromentin, G., Darcel, N., Chaumontet, C., Marsset-Baglieri, A., Nadkarni, N. and Tomé, D. (2012), 'Peripheral and central mechanisms involved in the control of food intake by dietary amino acids and proteins', *Nutr. Res. Rev.*, *25*, 29–39.

Fujita, K., Ito, T. and Kishino, E. (2009), 'Characteristics and applications of lactosucrose', *Proc. Res. Society of Japan Sugar Refineries', Technologists*, 57, 13–21.

Gauthier, S. F., Pouliot, Y. and Saint-Sauveur, D. (2006), 'Immunomodullatory peptides obtained by enzymatic hydrolysis of whey proteins', *Int. Dairy J.*, *16*, 1315–23.

Gill, H. S., Doull, F., Rutherfurd, K. J. and Cross, M. L. (2000), 'Immunoregulatory peptides in bovine milk', *Br. J. Nutr.*, *84*, S111–17.

Gazi, I. and Huppertz, T. (2015), 'Influence of protein content and storage conditions on the solubility of caseins and whey proteins in milk protein concentrates', *Int. Dairy J.*, *46*, 22–30.

Gesan-Guiziou, G., Daufin, G., Timmer, M., Allersma, D. and van der Horst, C. (1999), 'Process steps for the preparation of purified fractions of α-lactalbumin and β-lactoglobulin from whey protein concentrate', *J. Dairy Res.*, 66, 225–36.

Gosling, A., Stevens, G. W., Barber, A. R., Kentish, S. E. and Gras, S. L. (2010), 'Recent advances refining galactooligosaccharide production from lactose', *Food Chem.*, *121*, 307–18.

Gutiérrez, L. F., Hamoudi, S. and Belkacemi, K. (2012), 'Lactobionic acid: A high value-added lactose derivative for food and pharmaceutical applications', *Int. Dairy J.*, *26*, 103–111.

Harju, M. (1993), 'Production and properties of lactulose, lactitol and lactobionic acid', *Bul. Int. Dairy Fed.*, *289*, 27–30.

Havea, P. (2006), 'Protein interactions in milk protein concentrate powders', *Int. Dairy J.*, *16*, 415–22.

Hernández-Ledesma, B., Contreras, M. M. and Recio, I. (2011), 'Antihypertensive peptides: Production, bioavailability and incorporation into foods', *Adv. Colloid Interface Sci.*, *165*, 23–35.

Hernández-Ledesma, B., García-Nebot, M. J., Fernández-Tomé, S., Amigo, L. and Recio, I. (2014), 'Dairy protein hydrolysates: Peptides for health benefits', *Int. Dairy J.*, *38*, 82–100.

Houldsworth, D. W. (1980), 'Demineralization of whey by means of ion exchange and electrodialysis', *Int. J. Dairy Technol.*, *33*, 45–51.

Hoskin, D. W. and Ramamoorthy, A. (2008), 'Studies on anticancer activities of antimicrobial peptides', *Biochim. Biophys. Acta*, *1778*, 357–85.

Jelen, P. (2009), 'Dried whey, whey proteins, lactose and lactose derivative products', in *Dairy Powders and Concentrated Products*, ed. A. Y. Tamime, pp. 255–67. Chichester: Blackwell Publishing.

Jiménez-Flores, R. and Brisson, G. (2008), 'The milk fat globule membrane as an ingredient: Why, how, when?', *Dairy Sci. Technol.*, *88*, 5–18.

Kelly, P. M. (2011), 'Milk protein concentrate', in *Encyclopedia of Dairy Sciences*, 2nd edn, eds. J. W. Fuquay, P. F. Fox and P. L. H. McSweeney, pp. 848–54. San Diego, CA: Academic Press.

Kelly, A. L., O'Connell, J. E. and Fox, P. F. (2003), 'Manufacture and properties of milk powders', in *Advanced Dairy Chemistry—1 Proteins*, eds. P. F. Fox and P. L. H. McSweeney, pp. 1027–61. New York: Springer.

Kelly, J., Kelly, P. M. and Harrington, D. (2002), 'Influence of processing variables on the physicochemical properties of spray dried fat-based milk powders', *Lait*, *82*, 401–12.

Kim, P. (2004), 'Current studies on biological tagatose production using L-arabinose isomerase: A review and future perspective', *Appl. Microbiol. Biotechnol.*, *65*, 243–9.

Kitamura, H. and Otani, H. (2002), 'Fecal IgA levels in healthy persons who ingested cakes with or without bovine casein phosphopeptides', *Milchwissenschaft*, *57*, 11–12.

Kullisaar, T., Songisepp, E., Mikelsaar, M., Zilmer, K., Vihalemm, T. and Zilmer, M. (2003), 'Antioxidative probiotic fermented goats' milk decreases oxidative stress-mediated atherogenicity in human subjects', *Br. J. Nutr.*, *90*, 449–56.

Levin, G. V. (2002), 'Tagatose, the new GRAS sweetener and health product', *J. Medicinal Food*, *5*, 23–36.

Liang, B. and Hartel, R. W. (2004), 'Effects of milk powders in milk chocolate', *J. Dairy Sci.*, *87*, 20–31.

Linden, S. K., Sutton, P., Karlsson, N. G., Korolik, V. and McGuckin, M. A. (2008), 'Mucins in the mucosal barrier to infection', *Mucosal Immunol.*, *1*, 183–97.

Lloyd, M. A., Hess, S. J. and Drake, M. A. (2009), 'Effect of nitrogen flushing and storage temperature on flavor and shelf-life of whole milk powder', *J. Dairy Sci.*, *92*, 2409–22.

López-Expósito, I. and Recio, I. (2008), 'Protective effect of milk peptides: Antibacterial and antitumor properties', *Adv. Exp. Med. Biol.*, *606*, 271–93.

Luo, S. J. and Wong, L. L. (2004), 'Oral care confections and method of using', US Patent, 6733818.

Mao, X.-Y., Cheng, X., Wang, X. and Wu, S.-J. (2011), 'Free-radical-scavenging and anti-inflammatory effect of yak milk casein before and after enzymatic hydrolysis', *Food Chem.*, *126*, 484–90.

Martin-Hernandez, M. C., van Markwijk, B. W. and Vreeman, H. J. (1990), 'Isolation and properties of lactoperoxidase from milk', *Neth. Milk Dairy J.*, 44, 213–31.

Martínez-Maqueda, D., Miralles, B., Cruz-huerta, E. and Recio, I. (2013a), 'Casein hydrolysate and derived peptides stimulate mucin secretion and gene expression in human intestinal cells', *Int. Dairy J.*, *32*, 13–19.

Martínez-Maqueda, D., Miralles, B., de Pascual-teresa, S., Reverón, I., Muñoz, R. and Recio, I. (2012a), 'Food-derived peptides stimulate mucin secretion and gene expression in intestinal cells', *J. Agric. Food Chem.*, *60*, 8600–5.

Martínez-Maqueda, D., Miralles, B., Ramos, M. and Recio, I. (2013b), 'Effect of β-lactoglobulin hydrolysate and β-lactorphin on intestinal mucin secretion and gene expression in human goblet cells', *Food Res. Int.*, *54*, 1287–91.

Martínez-Maqueda, D., Miralles, B., Recio, I. and Hernández-Ledesma, B. (2012b), 'Antihypertensive peptides from food proteins: A review', *Food Funct.*, *3*, 350–61.

McKenna, A. B. (2000), 'Effect of processing and storage on the reconstitution properties of whole milk and ultrafiltered skim milk powders', PhD diss., Massey University, Palmerston North, New Zealand.

Meisel, H. and Fitzgerald, R. J. (2003), 'Biofunctional peptides from milk proteins: Mineral binding and cytomodulatory effects', *Curr. Pharm. Des.*, *9*, 1289–95.

Miclo, L., Perrin, E., Driou, A., Papadopoulos, V., Boujrad, N., Vanderesse, R., Boudier, J. F., Desor, D., Linden, G. and Gaillard, J. L. (2001), 'Characterization of α-casozepine, a tryptic peptide from bovine α_{s1}-casein with benzodiazepine-like activity', *FASEB J.*, *15*, 1780–2.

Mimouni, A., Deeth, H. C., Whittaker, A. K., Gidley M. J. and Bhandari, B. R. (2009), 'Rehydration process of milk protein concentrate powder monitored by static light scattering', *Food Hydrocolloids*, *23*, 1958 – 65.

Miyauchi, H., Kaino, A., Shinoda, I., Fukuwatari, Y. and Hayasawa, H. (1997), 'Immunomodulatory effect of bovine lactoferrin pepsin hydrolysate on murine splenocytes and Peyer's patch cells', *J. Dairy Sci.*, *80*, 2330–9.

Morr, C. V. and Ha, E. Y. W. (1993), 'Whey protein concentrates and isolates: Processing and functional properties', *Crit. Rev. Food Sci. Nutr.*, *33*, 431–76.

Moughal, K. I., Munro, P. A. and Singh, H. (2000), 'Suspension stability and size distribution of particles in reconstituted, commercial calcium caseinates', *Int. Dairy J.*, *10*, 683–90.

Mulvihill, D. M. and Ennis, M. P. (2003), 'Functional milk proteins: Production and utilization', in *Advanced Dairy Chemistry—1 Proteins*, eds. P. F. Fox and P. L. H. McSWeeney, pp. 1175–1228. New York: Springer.

Mussatto, S. I. and Mancilha, I. M. (2007), 'Non-digestible oligosaccharides: A review', *Carbohydr. Polym.*, *68*, 587–97.

Nagaoka, S., Futamura, Y., Miwa, K., Awano, T., Yamauchi, K., Kanamaru, Y., Tadashi, K. and Kuwata, T. (2001), 'Identification of novel hypocholesterolemic peptides derived from bovine milk β-lactoglobulin', *Biochem. Biophys. Res. Commun.*, *281*, 11–17.

Nongonierma, A. B., O'Keeffe, M. B. and FitzGerald, R. J. (2016), 'Milk protein hydrolysates and bioactive peptides', in *Advanced Dairy Chemistry 1B: Proteins: Applied Aspects*, eds. P. L. H. McSweeney and J. A. O'Mahoney, pp. 417–82. New York: Springer.

Nickerson, T. A. (1976), 'Use of milk derivative, lactose, in other foods', *J. Dairy Sci.*, *59*, 581–7.

Nickerson, T. A. and Moore, E. E. (1974), 'Factors influencing lactose crystallization', *J. Dairy Sci.*, *57*, 1315–19.

Olano, A., Calvo, M. M. and Corzo, N. (1989), 'Changes in the carbohydrate fraction of milk during heating processes', *Food Chem.*, *31*, 259–65.

Otani, H., Kihara, Y. and Park, M. (2000), 'The immunoenhancing property of a dietary casein phosphopeptide preparation in mice', *Food Agric. Immunol.*, *12*, 165–73.

Otani, H., Nakano, K. and Kawahara, T. (2003), 'Stimulatory effect of a dietary casein phosphopeptide preparation on the mucosal Iga response of mice to orally ingested lipopolysaccharide from *Salmonella typhimurium*', *Biosci. Biotechnol. Biochem.*, *67*, 729–35.

Panesar, P. S. and Kumari, S. (2011), 'Lactulose: Production, purification and potential applications', *Biotechnol. Adv.*, *29*, 940–8.

Paterson, A. H. J. (2009), 'Production and uses of lactose', in *Advanced Dairy Chemistry* 3, eds. P. L. H. McSweeney and P. F. Fox, pp. 105–20. New York: Springer.

Paul, K. G., Ohlsson, P. I. and Hendrikson, A. (1980), 'The isolation and some liganding properties of lactoperoxidase', *FEBS Lett.*, 110, 200–4.

Pearce, R. J. (1983), 'Thermal separation of β-lactoglobulin and α-lactalbumin in bovine Cheddar cheese whey', *Aust. J. Dairy Technol.*, *38*, 144–9.

Pearce, R. J. (1987), 'Franctionation of whey proteins', *Int. Dairy Fed. Bul., 212*, 150–3.
Pellegrini, A., Dettling, C., Thomas, U. and Hunziker, P. (2001), 'Isolation and characterization of four bactericidal domains in the bovine β-lactoglobulin', *Biochim. Biophys. Acta, 1526*, 131–40.
Pellegrini, A., Thomas, U., Bramaz, N., Hunziker, P. and Von Fellenberg, R. (1999), 'Isolation and identification of three bactericidal domains in the bovine α-lactalbumin molecule', *Biochim. Biophys. Acta, 1426*, 439–48.
Pisecky, J. (2012), '*Handbook of milk powder manufacture*', GEA Process Engineering A/S.
Plaisancié, P., Claustre, J., Estienne, M., Henry, G., Boutrou, R., Paquet, A. and Léonil, J. (2013), 'A novel bioactive peptide from yoghurts modulates expression of the gel-forming MUC2 mucin as well as population of goblet cells and Paneth cells along the small intestine', *J. Nutri. Biochem., 24*, 213–21.
Playne, M. J. and Crittenden, R. G. (2009), 'Galacto-oligosaccharides and other products derived from lactose', in *Advanced Dairy Chemistry* 3, eds. P. L. H. McSweeney and P. F. Fox, pp. 121–201. New York: Springer.
Prieels, J. P. and Peiffer, R. (1986), 'Process for the purification of proteins from a liquid such as milk', UK Patent Application GB2 171 102 A1.
Puknun, A., Bolscher, J. G. M., Nazmi, K., Veerman, E. C. I., Tungpradabkul, S., Wongratanacheewin, S., Kanthawong, S. and Taweechaisupapong, S. (2013), 'A heterodimer comprised of two bovine lactoferrin antimicrobial peptides exhibits powerful bactericidal activity against *Burkholderia pseudomallei*', *World J. Micro. Biot., 29*, 1217–24.
Recio, I. and Visser, S. (1999), 'Two ion-exchange chromatography methods for the isolation of antibacterial peptides from lactoferrin. In situ enzymatic hydrolysis on an ion-exchange membrane', *J. Chromatogr. A, 831*, 191–201.
Reynolds, E. C., Cai, F., Shen, P. and Walker, G. D. (2003), 'Retention in plaque and remineralization of enamel lesions by various forms of calcium in a mouthrinse or sugar-free chewing gum', *J. Dent. Res., 82*, 206–11.
Rowan, C. (1998), 'The whey ahead', *Food Manuf., 73*, 19–20.
Saint-Sauveur, D., Gauthier, S. F., Boutin, Y., Montoni, A. and Fliss, I. (2009), 'Effect of feeding whey peptide fractions on the immune response in the healthy and *Escherichia coli* infected mice', *Int. Dairy J., 19*, 537–44.
Schuck, P. (2011a), 'Milk powder: Types and manufacture', *Encyclopedia of Dairy Sciences*, 108–16.
Schuck, P. (2011b), 'Physical and functional properties of milk powders', *Encyclopedia of Dairy Sciences*, 117–24.
Singh, H. (2007), 'Interactions of milk proteins during the manufacture of milk powders', *Lait, 87*, 413–23.
Singh, H. (2011), 'Functional properties of milk proteins', *Encyclopedia of Dairy Sciences*, 887–93.
Sousa, G. T., Lira, F. S., Rosa, J. C., de Oliveira, E. P., Oyama, L. M., Santos, R. V. and Pimentel, G. D. (2012), 'Dietary whey protein lessens several risk factors for metabolic diseases: A review', *Lipids Health Dis., 11*, 67–75.
Tamura, Y., Mizota, T., Shimamura, S. and Tomita, M. (1993), 'Lactulose and its application to the food and pharmaceutical industries', *Bul. Int. Dairy Fed., 289*, 43–43.
Theolier, J., Hammami, R., labelle, P., Fliss, I. and Jean, J. (2013), 'Isolation and identification of antimicrobial peptides derived by peptic cleavage of whey protein isolate', *J. Funct. Foods, 5*, 706–14.
Tompa, G., Laine, A., Pihlanto, A., Korhonen, H., Rogelj, I. and Marnilab, P. (2010), 'Chemiluminescence of non-differentiated THP-1 promonocytes: Developing an assay for screening anti-inflammatory milk proteins and peptides', *Luminescence, 26*, 251–8.
Torres, D. P., Gonçalves, M. D. P. F., Teixeira, J. A. and Rodrigues, L. R. (2010), Galacto-oligosaccharides: Production, properties, applications, and significance as prebiotics', *Compr. Rev. Food Sci. Food Safe., 9*, 438–54.
Trompette, A., Claustre, J., Caillon, F., Jourdan, G., Chayvialle, J. A. and Plaisancié, P. (2003), 'Milk bioactive peptides and β-casomorphins induce mucus release in rat jejunum', *J. Nutr., 133*, 3499–503.

Urashima, T., Asakuma, S., Kitaoka, M. and Messer, M. (2011), 'Indigenous oligosaccharides in milk', in *Encyclopedia of Dairy Science*, 2nd edn, 3rd vol., pp. 241–73. Amsterdam: Elsevier.

Van Boekel, M. A. J. S. (1998), 'Effect of heating on Maillard reactions in milk', *Food Chem., 62*, 403–14.

Van der Horst, H. C., Timmer, J. M. K., Robbertsen, T. and Leenders, J. (1995), 'Use of nanofiltration for concentration and demineralization in the dairy industry: Model for mass transport', *J. Membrane Sci., 104*, 205–18.

Van der Kraan, M. I. A., Groenink, J., Nazmi, K., Veerman, E. C. I., Bolscher, J. G. M. and Nieuw Amerongen, A. V. (2004), 'Lactoferrampin: A novel antimicrobial peptide in the N1-domain of bovine lactoferrin', *Peptides, 25*, 177–83.

Van Velthuijsen, J. A. (1979), 'Food additives derived from lactose: Lactitol and lactitol palmitate', *J. Agric. Food Chem., 27*, 680–6.

Vignolles, M. L., Jeantet, R., Lopez, C. and Schuck, P. (2007), 'Free fat, surface fat and dairy powders: Interactions between process and product. A review', *Lait, 87*, 187–236.

Westergaard, V. (2004), *Milk Powder Technology: Evaporation and Spray Drying*, 5th edn. Copenhagen, Denmark: Niro A/S.

Wong, S. Y. and Hartel, R. W. (2014), 'Crystallization in lactose refining: A review', *J. Food Sci., 79*, R257–72.

Yamauchi, R., Ohinata, K. and Yoshikawa, M. (2003), 'β-Lactotensin and neurotensin rapidly reduce serum cholesterol via NT_2 receptor', *Peptides, 24*, 1955–61.

Zommara, M., Tougo, H., Sakanao, M. and Imaizumi, K. (1998), 'Prevention of peroxidative stress in rats fed on a low vitamin E containing diet by supplementing with a fermented bovine milk whey preparation: Effect of lactic acid and β-lactoglobulin on the antiperoxidative action', *Biosci. Biotechnol. Biochem., 62*, 710–17.

Chapter 4

Understanding and preventing spoilage of cow's milk

G. LaPointe, University of Guelph, Canada

1 Introduction

Milk provides water, sugar, proteins and minerals that are suitable for the growth of microorganisms. While some microorganisms can be pathogenic for animals and humans, others render the milk unsuitable for consumption. Processing reduces or delays milk spoilage and provides the means for diversifying the forms in which we consume milk. The key issues include understanding and controlling the sources of contamination, storage conditions and processing treatments. This chapter will focus on the types and methods of transmission of spoilage bacteria affecting milk before and after heat treatment.

Over a lactation period of 305 days, cows produce an average of 9780 kg of milk, with an average content of 3.85% fat and 3.22% protein, providing nutrients that can be used by microbes for growth. Improvements on the farm have led to better, longer-lasting milk products through enhanced feeding and milking techniques, faster cooling and transport. However, intensified processing and the need for high-quality ingredients introduce economic motivation to manage spore counts and spoilage bacteria. Microbial counts can vary from farm to farm, and even among individual cows. Thermoduric and thermophilic bacteria reduce product shelf life and are especially problematic for concentrates and powders. A spore count of 50 per mL of raw milk would give a whole milk powder with 2.6 log10 spores per g. Contamination with spore-formers has been traced all across the value chain, from on-farm to plant equipment. Rotating sanitation processes

http://dx.doi.org/10.19103/AS.2016.0005.06

and managing product lines in order to first treat products with lower spore counts are some of the strategies for reducing the impact of spore-formers in the processing plant. More research is needed to determine critical points in the origin of spore-formers on the farm as well as in the processing plant and to introduce control strategies. As spore counts are not correlated with somatic cell counts in milk, appropriate biomarkers must be identified.

2 Causes of milk spoilage

Spoilage is defined as becoming unfit for human consumption due to changes in the food's texture, colour, odour or flavour. These changes can occur through chemical or microbiological mechanisms. Milk components such as proteins, fat and carbohydrates can be degraded by microbes or by enzymes released from cells and microbes. Microorganisms that can be found in milk are psychrotrophs (cold-loving microorganisms), lactic acid bacteria (LAB), coliforms, fungi (yeasts and moulds) and spore-forming bacteria. The spore-forming bacteria belong to two general classes: *Bacilli* (aerobic, or requiring oxygen for growth) and *Clostridia* (anaerobic, or requiring the absence of oxygen for growth). Thermophilic microorganisms have optimal growth at high temperatures while thermoduric refers more to their capacity to resist heat treatment. Thus, psychrotrophic, thermophilic or thermoduric and spore-forming microorganisms (PTS) can contaminate milk, grow in chilled bulk tanks and survive heat treatments to reduce shelf life. They also produce thermotolerant lipolytic and proteolytic enzymes that can survive the pasteurization process. Persistent contamination of dairy-processing plants can be due to biofilm formation by *Bacillus* spp. and *Geobacillus stearothermophilus* on stainless steel equipment surfaces, gasket seals and interfaces. *Paenibacillus* sp. could pose a particular problem in processing, as it is a facultative anaerobe (can grow with or without oxygen) and a psychrotrophic spore-former.

Somatic cells (also called leucocytes) in milk can release heat-stable proteases and lipases that cause spoilage. Microorganisms produce a number of hydrolases that can degrade lactose, milk proteins and lipids. Determinants of enzymatic spoilage include the initial quantity of cells, their growth rate, enzyme production, and time and temperature of storage. Conditions leading to lipolysed flavour are high numbers of lipase-producing bacteria, the thermal stability of the enzymes, storage time, temperature, pH and water activity.

Enzymatic activity leading to spoilage can include proteolysis, lipolysis and carbohydrate degradation. Rancidity develops when lipoprotein lipase (LPL) produces short-chain fatty acids (e.g. butyric acid) released from fat globule membranes during agitation, foaming and pumping of milk. Other factors that may increase enzymatic activity are late lactation, mastitis (increased somatic cell count), as well as hay and grain feed ratio.

Lactose fermentation leads to acid/sour defects in milk. Fruity off-flavours are caused by enzymes from *Pseudomonas fragi*, a psychrotrophic spoilage organism. Protease action can lead to bitter/putrid as well as malty flavour and aromas. Ropy defect is caused by exopolysaccharides produced by microorganisms.

Methods for detecting and measuring milk quality are explained in Chapter 9. Regulatory standards for indicator organisms are reviewed in Ledenbach and Marshall (2009).

3 Origins of spoilage microbes

The sources of contamination of milk by spoilage microorganisms are essentially the farm environment, cows, milking equipment, tanker and transport chain during milk production (Fig. 1). During processing, raw materials, inadequate sanitation of equipment and post-process contamination are sources of spoilage microorganisms. Chapter 18 describes the routes of contamination by pathogens.

Spoilage microbes can come from the mammary glands, the external surface of udder and teats as well as from the environment, including dust, feed components, forage, skin, hair of cows and contaminated silage (e.g. *C. butyricum*), in addition to wash water. Recontamination with raw milk should also be considered a possible entry point for spoilage microbes.

Even though the cells in mammary glands are considered sterile, the glands may become infected over the lactation period. The diameter of the teat canal increases from the early to late lactation stages, allowing microorganisms to enter the canal and contaminate the milk along

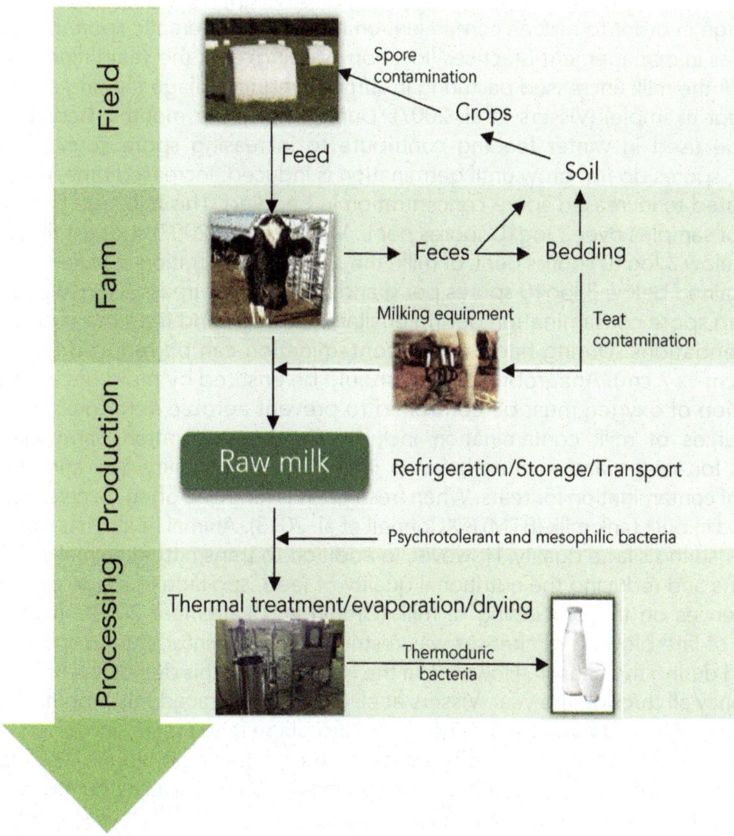

Figure 1 Pathways for entry and growth of spoilage microorganisms in milk.

the way. At the beginning of lactation, microbial numbers are low, but as milking progresses over the season, microorganisms can accumulate to the upper reaches of the teat. This is why forestripping has been associated with reduced microbial counts in bulk tanks. Teat cleaning and drying procedures are critical to reducing the microbial load of milk. However, pre-milking disinfectants are less effective at reducing thermoduric bacteria counts. Refrigeration actually facilitates the growth of psychrotrophs, which compete with other microbial species in milk.

Contaminating microorganisms come from the environment (soil, bedding material, feed and silage) or from personnel and milking equipment (transfer). Bedding material can introduce spore-formers, thermodurics, psychrotrophs and coliforms, while inadequate cleaning and sanitation regimes allow the build-up of thermoduric bacteria in milking equipment. When cleaning and maintenance are not adequately carried out, the combination of residual liquid and milk residues along with bacteria can lead to the formation of biofilms in equipment and pipes and even on stainless steel surfaces. Biofilms can entrap microorganisms, which multiply and, subsequently, release spores into the milk flow.

Silage quality can suffer from anaerobic degradation by germination of clostridial spores or growth of butyric acid bacteria (BAB), due to inadequate decline in pH or from inadequate aerobic stability (oxygen penetration; yeast consume organic acids such as lactate to raise pH in pockets, which allows clostridial spores to develop). Some management practices avoid silage in order to reduce contamination levels of problematic spore-formers.

Changes in management practices (location, feeding) over the year influence the spore content of the milk (increased pasturing in summer reduces silage feeding and time spent in barn, for example) (Vissers et al. 2007). During the winter months, bedding materials and silage used in winter feeding contribute to increasing spore counts in raw milk. However, spores do not grow until germination is induced. Increased time in the barn has been related to increased spore concentration in bedding. This milk also had the highest fraction of samples over 3 log10 spores per L. Vissers et al. (2007) suggest that in order to remain below 3 log10 spores per L of milk, the spore concentration in mixed silage should be maintained below 3 log10 spores per g and should not surpass 5 log10 per g.

Limiting spore contamination during ensilage, storage and feed out is one of the key recommendations. During harvest, soil contamination can be reduced by higher cuts (e.g. 10 cm vs 7 cm). Anaerobic stability should be ensured by rapid decline in pH, and penetration of oxygen must be controlled to prevent aerobic deterioration of silage.

As sources of milk contamination include silage deterioration, farm management practices for forage and manure have an impact on milk quality. Soil and bedding are sources of contamination for teats. When fresh grass is renewed often, *B. cereus* population is reduced in bulk tank milk (BTM) (O'Connell et al. 2013). Animal health has been the main goal of ensuring silage quality. However, in addition to transmitting animal and/or human pathogens and reducing the nutritional quality of feed, spoilage of silage can have direct consequences on the processing of milk (Drouin and Lafrenière 2012). In the past, the problem of late blowing of cheeses was restricted to the winter, due to spore content of silage fed during this season. However, in the Netherlands, this defect has been increasing in frequency all through the year. Vissers et al. (2007) have traced this problem to changes in farming practices during the summer, including silage feeding and increased time in the barn. This example shows how important it is to integrate microbiological risk analysis into farm management practices that have such an impact on milk quality during processing.

Refrigerated shelf life was reduced for milk from farms with high levels of spores in milk (over 3 log10 cfu per mL), while milk from farms with under 3 log10 cfu per mL had a shelf life of at least 21 days (Masiello et al. 2014). Even when temperature is kept low, the initial microbial load of the milk determines the time milk can be held. Poor hygienic conditions lead to short storage life.

Silage-fed herds produce milk with higher counts of BAB spores, which cause late gas blowing defects when spores are allowed to germinate and proliferate by converting lactate to acetate, butyrate, CO_2 and H_2 (due to higher ripening temperatures). Gas blowing defect caused by facultative heterofermenting lactobacilli (salt tolerant and mesophilic lactobacilli) is encountered in Cheddar-type cheese as well as in Swiss and Dutch cheeses, although it is more frequent in raw milk cheeses (Sheehan 2013). This problem is due to the conversion of residual lactose to galactose and CO_2 by species such as *Lactobacillus brevis* and *L. fermentum*. Low starter activity and high salt-in-moisture levels are some of the factors leading to residual lactose. Some non-starter LAB may also be capable of producing biogenic amines (BA). If these bacteria survive pasteurization temperatures, they can continue their metabolic activity, even when injured, and generate defects during cheese ripening. Thermal inactivation is highly dependent on the initial load in the milk, and viable counts do not reveal the injured cells that are able to survive but unable to multiply.

Temperature, moisture and oxygen are three of the major determinants of microbial growth, and these factors can significantly impact the composition of the microbial ecosystem. On the basis of their heat resistance (thermoduric) as well as their preferred growth temperature, spoilage microorganisms can be classified as psychrotrophic, mesophilic or thermophilic. Thermodurics survive pasteurization; some, such as *Bacillus cereus* and *Paenibacillus*, grow in chilled milk, and they are hence considered psychrotrophic. These psychrotrophs are the main spoilage bacteria of extended shelf life (ESL) products. *Paenibacillus* has been found in soil, compost, faeces and even UHT-treated milk (Ivy et al. 2012), although their persistence in processing equipment is not known. Mesophilic spore-formers include *B. licheniformis*, *B. pumilus*, *B. sporothermodurans* (proteolytic organisms), *B. subtilis*, *C. tyrobutyricum* (gas forming organisms, butyric acid spoilage organisms) and *C. beijerinckii*. These species do not grow well at refrigeration temperatures. Highly heat-resistant spore-formers are the aerobic *B. cereus*, *B. licheniformis*, *B. subtilis*, *B. sporothermodurans* (thermophilic; heat resistant), *B. aleronius*, *B. siralis*, *Brevibacillus borstelensis* and *Aneurinibacillus* spp. (resists 100°C for 30 min). Thermophiles such as *Anoxybacillus flavithermus*, *Geobacillus* spp. (proteolysis, acid spoilage), *C. butyricum* and *C. sporogenes* may thrive in heating and evaporation equipment. The microbial profile thus reflects the cumulative history of chilling and heat treatment of the milk, as well as cow and equipment hygiene, selecting for thermoduric psychrotrophs.

Although total microbial load can reflect farm hygiene, it does not predict thermoduric load. The microbiology quality requirements for milk powder (besides pathogens and aerobic plate counts, APC, coliforms) require that sulphur-reducing clostridia and thermophilic spore-formers remain under 10 CFU (colony-forming units) per g. Of particular concern during milk powder production are thermophilic microorganisms that grow at 40–55°C and in high numbers in pre-heaters and evaporators. Thermoduric bacteria are considered an indicator of equipment sanitation, and can also predict the spoilage rate of the downstream product. This is tested by laboratory pasteurization count (LPC), where pasteurization selects for thermodurics, while in the subsequent agar-based enumeration step, incubation temperature distinguishes psychrotrophs (low temperature growth) from mesophiles (growth at 20–30°C) and thermophiles (growth at high temperature). Generally, psychrotrophs do not survive pasteurization; however, if the cell counts are high, heat-stable lipolytic and proteolytic enzymes will be released, causing problems during storage and thus reducing product quality. The preliminary incubation count (PIC) can help identify sources of contamination, selecting for microorganisms that can grow at lower temperatures. The milk is held for 18 h at 12.8°C (55°F) and then counted for SPC. PIC counts are generally higher than SPC, as there is an incubation step. Inadequate cleaning of the milking equipment or cows, long storage times of raw milk or problems with cooling, all contribute to raising PIC levels. When the PIC is only slightly higher than the SPC, contamination with mastitis organisms may be the problem. Mastitis organisms do not generally grow at the temperature used for PIC incubation.

The types of microorganisms found are representative of their environmental or disease origin as well as the cooling conditions.

1 Mastitis organisms. Mastitis microbes most frequently found in bulk tank milk counts are *Streptococcus* species, such as *S. agalactiae* and *S. uberis*. *Staphylococcus aureus* does not contribute significantly to counts in bulk tank milk. These organisms can come from the environment as well as from cows with mastitis.
2 Environmental contamination. Microbes in bedding material can contaminate teats and udders. They include streptococci, staphylococci, spore-formers, coliforms and other Gram-negative species. Thermoduric and psychrotrophic bacteria are recovered from teat surfaces.
3 Cleaning and sanitation. Build-up of thermoduric bacteria may take some time to show up as increases in total bulk tank counts. Although bulk tanks themselves are not difficult to keep clean, wear and tear on rubber parts has been associated with build-up of thermoduric bacteria. Cleaning failures can also lead to increases in Gram-negative bacteria (coliform and *Pseudomonas*), which are fast-growing, but less-resistant microorganisms. Sanitizers, when used effectively, reduce psychrotrophic bacteria.
4 Refrigeration. Elevated psychrotrophs indicate failures in sanitation of refrigerated bulk tanks, even if there is low initial load in the milk. Lafarge et al. (2004) have shown that molecular methods reveal an increase in psychrotrophs after only 24 h refrigeration, while classical microbiological methods show increase in psychrotrophs after 48 h of storage at 4°C. They also found that refrigeration reduces the presence of bacteria such as *Lactococcus*, *Streptococcus* and certain lactobacilli (*L. plantarum* and *L. pentosus*). The application of molecular methods such as qPCR has revealed how bacteria such as *S. aureus*, *Aerococcus viridans*, *Acinetobacter calcoaceticus*, *Corynebacterium variabile* and *S. uberis* can be stably maintained in milk stored at 4°C for 7 days (Rasolofo et al. 2010).

4 Controlling milk spoilage during production

Essentially, three processes must be controlled in order to maintain high-quality milk: contamination, cleaning and cooling. Routine testing can help strengthen the ability to diagnose and pre-empt potential problems.

4.1 Contamination

Management practices having repercussions on the bacterial content of milk are mastitis control, udder hygiene, milking routine, environmental sanitation (including feed and bedding), tank and truck sanitation, processing conditions and equipment sanitation.

Udder hygiene and teat preparation (cleaning and drying) are considered critical points. However, the concentration of spores in silage and feed during housing periods is now regarded as having significant impact on the spore load of milk (Vissers et al. 2007).

4.2 Silage stability and quality

The life cycle of silage can be divided into four phases: the initial aerobic phase at harvest, the fermentation phase, the anaerobic storage phase and the feed-out phase. There is a high risk of compromised silage preservation when dry matter content and the amount of soluble sugars are too low to allow formation of enough lactic acid to overcome the buffering capacity in order to reach target pH (Weinberg and Muck 1996). Solutions to control the anaerobic degradation of silage include the use of silage additives, ensuring rapid pH decline and inhibition of the proliferation and germination of anaerobic spore-forming bacteria (mainly *Clostridia* spp.). Aerobic deterioration of silage by yeasts and moulds leads to substrate availability for spore germination (aerobic spore-formers). However, high butyric spore counts have also been associated with aerobic deterioration (Vissers et al. 2007).

Critical points for aerobic stability are crop composition, packing density and additives. Packing density and porosity determines the rate of oxygen penetration and thus the growth of yeasts and moulds. Aerobic deterioration occurs because the lactic acid built up during fermentation can serve as substrate when oxygen begins to penetrate during the feed-out phase after the silo is opened. Acid-tolerant lactate-assimilating yeasts start the process of aerobic deterioration. Acids and carbohydrates are oxidized to CO_2 and water, accompanied by a raise in temperature. This depends on the dry matter content (more equals higher temperature rise). Temperature rises more at the periphery where more mould and yeast growth occurs, than at the centre. Higher instability occurs with higher mould and yeast counts at the time of harvest. As lactic acid is metabolized, the pH rises, allowing growth of moulds and clostridia, which consume both carbohydrates and protein. Paradoxically, aerobic instability is characterized by growth of anaerobic bacteria as well, because of favourable conditions in small, localized niches. Initial yeast and mould counts can increase when wilting or the first aerobic phases are prolonged, as with poor weather.

Soil introduction during harvest allows ingress of microorganisms (coliforms, clostridia and fungi). If they develop during the initial aerobic phase, they can produce acetic acid and ammonia, but do not contribute to lactic acid production, thus leading to suboptimal fermentation. However, contaminated silage may paradoxically be more stable when exposed to air, showing less aerobic deterioration due to yeasts and moulds. Aerobic instability is related to high amounts of lactic acid compared with other acids, as the dissociation constant of lactic acid is lower (3.08 pKa). During the anaerobic phase, high lactic acid contributes to stability. Heterofermentation (mixed acid, CO_2 and loss of dry matter, DM) is less efficient for acidification and preservation of nutrients than homofermentation.

Reduced silage stability is correlated with yeast counts over 5 log10 per g fresh weight, regardless of additive treatments (for maize silage). LAB improve fermentation quality by increasing lactic acid, decreasing acetic acid and residual water-soluble carbohydrates that could be used by yeasts and moulds. However, LAB-inoculated silage, particularly homofermentative LAB, can be a liability for aerobic stability. Heterofermentative inoculants such as *L. buchneri* improve aerobic stability by metabolizing lactic acid to acetic acid and other molecules that inhibit yeast. Other additives being used to inhibit yeasts and moulds are sorbate, propionate, sulphite and benzoate. However, biological control agents have more potential as sustainable resources than chemical additives.

Attention should be focused on protecting silage from lengthy exposure to air during the feed-out period by attending to silo design, to achieving the target silage density and to improving in-silo processes such that aerobic stability is achieved with minimal nutrient loss. Air penetrates the feed-out face by 1–2 m, so the optimal removal rate is 1–2 m per week, but the silo face is often too wide, especially for bunker silos. The preferred configuration is long and narrow (especially at high ambient air temperature). The rate of silage removal should match or exceed the depth of air penetration into the silo. The feed-out phase is even more critical for bale silage because the density is lower, and the surface-to-volume ratio is higher. Both baled and silo-stored silage require plastic barrier film. Ideally, planning should coordinate the entire silage-making and feeding operations. Improving aerobic stability will also reduce risk of outgrowth of anaerobic and aerobic spore-forming microorganisms in addition to yeast and moulds (Borreani et al. 2013).

Isolating natural products of fermentation, killer toxins, diacetyl, and yeast- and mould-growth-inhibiting LAB and implementing strategies for reducing spore germination are among the perspectives in the control of silage quality. Knowing the yeast and mould populations on the crops before ensiling would be ideal, but there is a limitation with the timescale of 3 days between sampling and obtaining results. However, analysis is impractical at present because of the short time interval and high variability in these counts. When conditions are not optimal at harvest, silo management is even more important to minimize aerobic deterioration through additives, proper compaction during filling, oxygen barriers and feed-out processes. Anaerobic instability has been controlled through lactic acid fermentation and work is in progress to increase aerobic stability as well. New microbiological approaches are required to control the development of yeasts and moulds and thus reduce aerobic deterioration of silage.

Bacterial counts in bedding are positively correlated with bacterial counts on teat ends and the rate of mastitis in lactating dairy cattle. In some studies, compost bedded pack (CPB) contains high numbers of microorganisms, including mastitis-causing species, but not in others (Barberg et al. 2007), so the quality is variable. In the large resting space, manure is incorporated into the bedding (sawdust or wood shavings) by aeration, generally twice per day.

Although the initial investment in anaerobic digestion and composting is high, this may be offset by the production of energy (biogas) while the liquid phase can be used for fertilizer. This strategy is economically feasible for larger operations. Recycling bedding through the use of good composting practices is among the future trends for sustainability on smaller farms.

More detailed information can be found in Part 1, 'Ensuring the safety and quality of milk on the farm', as well as in Part 2, on 'Sustainability' (Chapters 31 and 32 deal with water and manure management to improve safety and environmental impact).

4.3 Sanitary conditions: cleaning and cooling

In addition to initial microbial load, time and temperature are the primary factors that determine milk spoilage. With more frequent pick-ups and more rapid cooling times, as well as low wait times until processing, the progression of microbial growth can be controlled. Sanitation of equipment is crucial to preventing build-up of thermoduric microorganisms (Gleeson, O'Connell and Jordan 2013). Both cleaning systems and cow hygiene can be responsible when bacterial counts in bulk tanks reach or exceed regulatory limits. Guidelines from the National Mastitis Council are available to help dairy producers manage the interpretation of microbial diagnostics (see Section 7 for the NMC website). Diagnosing potential problems requires relative comparison of several types of microbes in the milk, each of which can be counted using one of three methods (Fig. 2). Standard plate count, direct bacterial count, and individual bacteria count by BactoScan™ indicate overall microbial load, but do not give clear clues to the origin of a problem. High counts can signify the following four: hygiene problems (cow, milking machine or bulk tank), mastitis, cooling fault or environmental contamination. Coliforms indicate environmental/faecal contamination, but high levels can occur for other reasons as well, such as temperature of the milk remaining above recommended thresholds (bacterial increase through incubation). LPC (or thermoduric count) can indicate cleaning failures in milking and storage equipment, but elevated LPC counts can occur for several other reasons. Mastitis organisms and coliforms do not survive the pasteurization temperature of the LPC. LPC should be lower than 100–200 cfu/mL, and a level of 10 cfu/mL indicates excellent equipment sanitation. The following examples show how bacterial counts can be used in source-tracking problems leading to milk contamination.

1 Cow hygiene: When SPC is moderately high (5000–20 000 CFU/mL) and coliform count is between 100 and 1000 cfu/mL, but LPC is less than the coliform count, the problems may reside with cow hygiene.

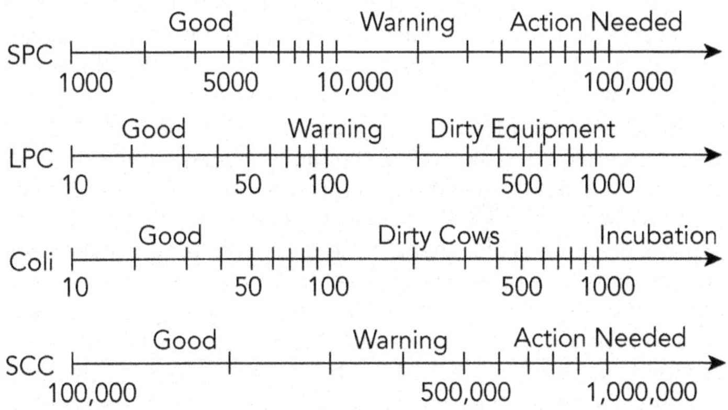

Figure 2 Diagnostic chart for bulk tank bacteria and somatic cell counts with thresholds (colony-forming units, cfu/mL) indicating potential problems and corrective action required. The standard plate count (SPC) of 50 000 cfu/mL would be equivalent to 121 000 IBC/mL (individual bacterial count by BactoScan™). LPC = Laboratory Pasteurization Count; Coli = Coliform Plate Count; SCC = Somatic Cell Count.

Source: Adapted from: Guterbok and Blackmer (1984).

2 Milking equipment: When SPC is moderately high (5000–20 000 cfu/mL), but coliforms are less than LPC (between 100 and 1000 cfu/mL), a persistent cleaning problem of the milking equipment may be at fault.
3 Milk handling and storage: When SPC is extremely elevated (greater than 50 000–100 000 cfu/mL, or too numerous to count, TNTC), coliforms are greater than 1000 (or TNTC), while LPC remains less than the coliform count, but greater than 100 cfu/mL (or TNTC), there could be incubation in the milk handling system, or there may be multiple sanitation problems. Short-term spikes in SPC or LPC are more probably due to environmental contamination than due to milking equipment sanitation, as microbes take some time to build-up in equipment. If the suspected origin of high bulk tank counts is the milk handling equipment, then the cause could be corrected by replacing worn or broken parts, proper cleaning frequency and procedures circulation of cleaning solutions for adequate contact time, removing milk residues and preventing build-up of mineral deposits and biofilms.

As bacteria grow exponentially, their growth data should be log transformed to convert counts to a normal distribution and then statistical methods can be applied. Long-term trends are difficult to discern due to the variability of individual data points, so it is useful to calculate moving averages over several successive data points in order to visualize these trends using a time series plot.

5 Controlling milk spoilage during processing

Pasteurization kills pathogens and most spoilage microbes. However, ESL requires higher heat treatments or other treatments such as bactofugation or microfiltration, which remove spores as well as bacterial cells. A critical step after pasteurization of fluid milk is to protect from recontamination by equipment sanitation before packaging in sealed containers. Lysozyme or nitrate, bactofugation or microfiltration are employed to control the spoilage of fluid milk, but these methods are not widely used for milk destined for cheese manufacturing.

High numbers of LAB are added to milk destined for cultured dairy products such as cheese. These starter microbes can control the growth of psychrotrophic bacteria. Spores can germinate during milk processing and can cause defects in cheese quality. Hard and semi-hard cheeses (Emmental, Gouda, Edamer, Comté) may provide favourable conditions for spore germination in terms of higher pH and lactic acid as germinating agent.

Salting the curd and using brine temperatures below 12°C can control gassing defects. Salt in moisture over 3% controls gas production from spore-former growth in ripened cheeses. Spore content does not interfere with processing of dry-salted cheeses, where the salt content prevents spore germination. When brining is used, the salt gradient allows spore germination in the core of the cheeses, resulting in slits or late-blowing defect. The major spore-forming species involved in late blowing are *Clostridium butyricum*, *C. sporogenes*, *C. beijerinckii* and *C. tyrobutyricum*. Other types of microorganisms causing late-blowing are *L. fermentum*, heterofermentative lactobacilli, *Propionibacteria* and *Eubacterium* sp. (facultative anaerobe). By contrast, early blowing is mainly due to coliforms and yeast.

Spoilage microbes are generally destroyed by the time and temperature treatments during cottage cheese making, but recontamination can occur during washing of curd if water quality is not maintained. Efforts should be made to limit recontamination of cream cheese and processed cheese as there is no competing microbial community. In dairy powder products, thermophilic spore-formers have been identified as the major problem, particularly in non-fat dry milk and sweet whey powders (Watterson et al. 2014). Increases of spore counts in final products can occur through concentration from liquid or contamination from equipment or the environment.

6 Summary and future trends

Good dairy farming practices contribute to reducing spoilage by relying on three steps: (1) avoiding the introduction of contaminants into milk by preparing animals for hygienic milking using suitable, well-maintained and clean equipment for milking and milk storage; (2) harvesting milk under hygienic conditions; and (3) practicing proper handling after milking to minimize spoilage by refrigerating and storing milk under hygienic conditions. Frequent cleaning of pens, changes in bedding types and milking routines, improvements in equipment cleaning and avoiding aerobic fermentation of silage are important management practices that contribute to reducing contamination.

New inline techniques are being developed for automated detection of problems during forestripping. Novel technologies are desirable in order to expand the tools available to control fungi in silage. More knowledge is needed on combining control technologies. Improved methods are necessary, especially for detecting slow-growing microorganisms and for finding niche environments in the processing plant. Whole-genome sequencing is being applied to tracing foodborne disease outbreaks, and this approach will also be useful for tracking the origin and dissemination of contaminants. Entire microbial communities can be profiled with molecular methods based on taxonomically conserved sequences, complementing culture-dependent techniques. Microbiological control is becoming more demanding all along the value chain from production to processing and distribution, so increased performance is required for methods monitoring product quality and controlling spoilage.

7 Where to look for further information

Key societies

National Mastitis Council (https://www.nmconline.org)
American Dairy Science Association
Dairy Farmers of Canada
Dairy Farmers of Ontario

Research and extension centers

Canada

Dairy at Guelph: University of Guelph, Guelph, Ontario, Canada
Canadian Bovine Mastitis and Milk Quality Research Network, Québec, Canada (http://www.medvet.umontreal.ca/reseau_mammite/en/index.php)

INAF/Centre STELA: Centre for Dairy Science and Technology, Québec, Québec, Canada
Maritime Quality Milk, Atlantic Veterinary College, UPEI, PE, Canada (http://www.milkquality.ca)

USA

Dairy Research Institute, Innovation Centre for U.S. Dairy (http://www.usdairy.com)
California Dairy Research Foundation (http://cdrf.org)
Food Safety Laboratory and Milk Quality Improvement Program, Department of Food Science, Cornell University (http://foodsafety.foodscience.cornell.edu)
Ohio State University, OARDC Mastitis Lab (http://www.oardc.ohio-state.edu/mastitis/t08_pageview2/Home.htm)
UW Milk Quality, Wisconsin (http://milkquality.wisc.edu)
Dairy Team, Milk Quality, Iowa State University (http://www.extension.iastate.edu/dairyteam/milk-quality-mastitis)
Murphy, S. C. and Boor, K. J. (2010), 'Sources and causes of high bacteria counts in raw milk: an abbreviated review', Extension article. http://articles.extension.org/pages/11811/sources-and-causes-of-high-bacteria-counts-in-raw-milk:-an-abbreviated-review (Accessed 4 November 2016).

8 References

Barberg, A. E., Endres, M. I., Salfer, J. A. and Reneau, J. K. (2007), 'Performance and welfare of dairy cows in an alternative housing system in Minnesota', *J. Dairy Sci.*, 90, 1575–83.
Borreani, G, Dolci, P., Tabacco, E. and Cocolin, L. (2013), 'Aerobic deterioration stimulates outgrowth of spore-forming *Paenibacillus* in corn silage stored under oxygen-barrier or polyethylene films', *J. Dairy Sci.*, 96, 5206–16.
Drouin, P. and Lafreniere, C. (2012), 'Clostridial spores in animal feeds and milk', *Milk Production – An Up-to-Date Overview of Animal Nutrition, Management and Health*. InTech, pp. 375–93.
Gleeson, D., O'Connell, A. and Jordan, K. (2013), 'Review of potential sources and control of thermoduric bacteria in bulk-tank milk', *Irish J. Agric. Food Res.*, 52, 217–27.
Guterbok, W. M. and Blackmer, P. E. (1984), 'Veterinary Interpretation of Bulk Tank Milk', *Vet. Clin. North America: Large Anim. Prac.*, 6(2), 257–68.
Ivy, R. A., Ranieri, M. L. and Martin, N. H. (2012), 'Identification and characterization of psychrotolerant sporeformers associated with fluid milk production and processing', *Appl. Environ. Microbiol.*, 78, 1853–64.
Lafarge, V., Ogier, J. C., Girard, V., Maladen, V., Leveau, J. Y., Gruss, A. and Delacroix-Buchet, A. (2004), 'Raw cow milk bacterial population shifts attributable to refrigeration', *Appl. Environ. Microbiol.*, 70, 5644–50.
Ledenbach, L. H. and Marshall, R. T. (2009), 'The microbiological spoilage of dairy products', In Sperber, W. H. and M. P. Doyle (eds), *Compendium of Microbiological Spoilage of Food and Beverages*, pp. 41–67. Food microbiology and food safety Series, Springer, New York.
Masiello, S. N., Martin, N. H., Watters, R. D., Galton, D. M., Schukken, Y. H., Wiedmann, M. and Boor, K. J. (2014), 'Identification of dairy farm management practices associated with the presence of psychrotolerant sporeformers in bulk tank milk', *J. Dairy Sci.*, 97, 4083–96.
Murphy, S. C. and Boor, K. J. (2000), 'Trouble-shooting sources and causes of high bacteria counts in raw milk', *Dairy, Food Environ. Sanitation*, 20(8), 606–11.
O'Connell, A., Ruegg, P. L. and Gleeson, D. (2013), 'Farm management factors associated with the *Bacillus cereus* count in bulk tank milk', *Irish J. Agric. Food Res.*, 52, 1–13.

Rasolofo, E. A., St-Gelais, D., LaPointe, G. and Roy, D. (2010), 'Molecular analysis of bacterial population structure and dynamics during cold storage of untreated and treated milk', *Int. J. Food Microbiol.*, 138, 108–18.

Sheehan, J. J. (2013), 'Milk quality and cheese diversification', *Irish J. Agric. Food Res.*, 52, 243–53.

Vissers, M. M. M., Driehuis, F., Te Giffel, M. C., De Jong, P. and Lankveld, J. M. G. (2007), 'Minimizing the level of butyric acid bacteria spores in farm tank milk', *J. Dairy Sci.*, 90, 3278–85.

Watterson, M. J., Kent, D. J., Boor, K. J., Wiedmann, M. and Martin, N. H. (2014), 'Evaluation of dairy powder products implicates thermophilic sporeformers as the primary organisms of interest', *J. Dairy Sci.*, 97(4), 2487–97.

Weinberg, Z. G. and Muck, R. E. (1996), 'New trends and opportunities in the development and use of inoculants for silage', *FEMS Microbiol. Rev.*, 19, 53–68.

Chapter 5

Sensory evaluation of cow's milk

Stephanie Clark, Iowa State University, USA

1 Introduction

Milk contains nine of the essential nutrients at good (at least 10% of the daily value) to excellent (at least 20% of the daily value) levels recommended for human nutrition, including protein, calcium, phosphorus, potassium, riboflavin, niacin and vitamins A, D and B12. It is well established and accepted that dairy products contribute to bone and dental health (Black et al. 2002; Rockell et al. 2005; Gao et al. 2006; Huncharek et al. 2008; Moschonis et al. 2010; Davoodi et al. 2013). Additionally, dairy products also contribute to overall health and even have protective effects against coronary heart disease, stroke, type 2 diabetes, certain types of cancer and other diseases (Elwood et al. 2005, 2007, 2008, 2010; Kleim and Givens 2011; Davoodi et al. 2013). Thus, ensuring consumption of milk and dairy products is nutritionally relevant for a vibrant population.

However, an abundant supply of milk matters little if poor quality prevents it from being consumed by those who need it. High-quality raw milk is essential, not only for the production of fluid milk but also for the production of all subsequent value-added dairy products made from milk. Because of high moisture (approximately 88% water) and nutrient composition (Table 1), milk is a highly perishable product. Although pasteurization ensures the safety of milk, the flavour quality of milk fresh from the cow cannot be improved. For instance, if milk is contaminated by microorganisms at any stage between cow and consumer, and those are allowed to proliferate (i.e., via temperature abuse), off-flavours can be produced that will persist (and sometimes become enhanced) with subsequent processing steps. Perhaps, the only exception includes the use of vacuum evaporation, which enables removal of absorbed volatile off-flavours (such as manure aroma) from milk.

http://dx.doi.org/10.19103/AS.2016.0005.08

Table 1 Typical cow milk composition (whole, 3.25%, without added vitamin A and D)*

Component	Unit	Amount per 100 g
Water	g	88.13
Energy	Kcal	61
Macronutrient		
Carbohydrate (predominantly lactose)	g	4.90
Fat (total lipid)	g	3.27
Fibre (total dietary)	g	0.0
Protein	g	2.15
Minerals		
Calcium, Ca	mg	113
Iron, Fe	mg	0.03
Magnesium, Mg	mg	10
Phosphorus, P	mg	84
Potassium, K	mg	132
Sodium, Na	mg	43
Zinc, Zn	mg	0.37
Vitamins		
Vitamin C (total ascorbic acid)	μg	3
Thiamine	μg	46
Riboflavin	μg	169
Niacin	μg	89
Vitamin B6	μg	36
Folate	μg	5
Vitamin B12	μg	0.45
Vitamin A	μg	46
Vitamin E	μg	70
Vitamin D (D2 + D3)	μg	0.1
Vitamin K	μg	0.3
Lipids		
Fatty acids, total saturated	g	1.87
Fatty acids, total monounsaturated	g	0.81
Fatty acids, total unsaturated	g	0.20
Cholesterol	mg	10

*Adapted from National Nutrient Database for Standard Reference (USDA 2015).

2 Milk evaluation processes

Milk may be evaluated in raw (unprocessed) form, immediately after processing or at any time during storage. The primary, though not exclusive, sense used in the evaluation of milk, is the smell. However, milk quality evaluation involves almost all of the senses: sight, smell, taste and tactile.

The sequence of milk evaluation begins from the cow, when the handler checks for the presence/absence of mastitis. The handler may use a combination of udder palpitation (to feel for lumps or abnormalities) and milk ejection onto a grate or into a strip cup (to observe for the presence of flocculation and/or blood). Cow handlers and milkers should be trained to understand the critical role they play in ensuring milk safety and quality. Proper use of teat pre- and post-dips, individual udder-drying cloths, cleaning and sanitizing inflations and all milk handling equipment, withholding milk from antibiotics-treated animals, attention to cow nutrition and health and denying strongly flavoured feed components (i.e., garlic, onion, mint) are all vital for ensuring high-quality milk. Miller and colleagues (2015) reported that management practices (including herd size, milking routine and bedding) play a significant role in controlling the levels of mesophilic and thermophilic spores in milk.

The milk hauler uses visual cues to evaluate whether cream or butter granules have risen to the top or if sediment has settled to the bottom of the bulk tank. Consumers also make visual evaluations. One of the reasons consumers like to purchase milk in translucent plastic high-density polyethylene (HDPE) packaging is that they like to see what they are buying; also, they can see how much milk is left in the container without having to pick it up. Consumers also look for sedimentation and flocculation on the sides of translucent containers, which are clear signs of spoilage. Another visual cue that must not be overlooked is the 'best by' or 'sell by' date printed on the carton. While some consumers simply grab the carton closest to the front of the refrigerated case at the market, many reach through the display case to procure the carton labelled furthest in the future. There are many reasons for this practice, including, but not limited to, the expectation that the milk will not be finished by the date printed, assumption that the milk will spoil by the printed date, or unfounded fear that the milk will not be safe after the printed date.

The second step in milk evaluation is generally by smelling. Milk receivers or haulers evaluate milk quality by turning off the agitator of the bulk tank, lifting the lid and sniffing the headspace of the bulk tank before pumping the milk from the tank to the truck. The hauler can reject the milk based solely on aroma quality. Cow's milk is composed of over 52 identified volatile compounds (Yue et al. 2015). A summary of aromatic compounds in milk is included in Table 2. Using a triple-channel comparative analysis combining the use of an electronic nose (E-Nose), gas chromatography mass spectrometry (GC-MS) and GC-olfactometry (GC-O), Ai et al. (2015) demonstrated that volatile flavour compounds affected consumer preference of raw whole and skim milk. Siefarth and Buettner (2014) identified 54 aromatic compounds in goats' milk, which were influenced by season and heat treatment.

Before one can identify an off-flavour in milk, recognition of what is 'normal' for milk flavour must be understood. Fresh from the cow, or freshly pasteurized, milk should have no offensive aroma. Aromatic compounds are more readily released from warm milk than from cold milk. Thus, to effectively smell milk that has been refrigerated, the evaluator

Table 2 Recognized volatile aroma compounds in cow and/or goat milk, along with descriptors*

Compound (also known as)	Odour descriptor
Acetic acid	Vinegar
Butan-2,3-dione (diacetyl)	Buttery
Butanoic acid (C4; butyric acid)	Green, cheesy, sweaty
Deca-2,4-dienal	Fatty, green, chicken broth-like
Decanal	Sweet, orange skin, flowery
δ-Decalactone	Coconut
γ-Deltalactone	Peach-like
Decanoic acid (C10; capric acid)	Tallow, fatty, goat-like, leather-like
Dimethyl sulphide (methylthiomethane)	Sweaty, fatty
δ-Dodecalactone	Flowery, daisy-like
γ-Dodecalactone	Soapy, fatty, flowery
Dodecanoic acid (C12; lauric acid)	Fatty, soapy, goat-like, faecal
3-Ethylphenol	Fatty, leather-like, medicinal
trans-4,5-epoxy-dec-2-enal	Metallic
2(5H)-Furanone (butenolide)	Unspecified
Hexanoic acid (C6; caproic acid)	Sour, urine-like, pungent
Hexadecanoic acid (C16; palmitic acid)	Fatty
3-Hydroxy-2-butanone	Fatty
2-Methylbutanoic acid	Sweaty, cheesy
3-Methylbutanoic acid (isovaleric acid)	Sweaty, cheesy
4-Methyloctanoic acid	Musty, urine-, stable-like
3-Methylphenol (*m*-cresol)	Leather-like, medicinal
3-Methylthiopropanal (methional)	Cooked potato-like
Nona-2,6-dienal	Fatty, cucumber-like
Nona-2,4-dienal	Fatty, green
δ-Nonalactone	Sweet, peach-like
γ-Nonalactone	Peach-like
Nonanal	Fatty
Non-4-enal	Green, fatty
Non-2-enal	Fatty, green, body odour
Nonanoic acid (pelargonic acid)	Fatty, faecal, male goat-like
γ-Octalactone	Coconut-like
Octanoic acid (C8; caprylic acid)	Rancid, medicinal, fatty, musty
1-Octanol	Fruity
Oct-2-enal	Green, fatty, plastic-like
Oct-1-en-3-one	Mushroom-like, metallic
Pentanoic acid (valeric acid)	Sweaty, cheesy
Phenylethan-2-ol (phenethyl alcohol)	Flowery, fruity, sweet
Tetradecanoic acid (C14; myristic acid)	Fatty

*Compiled from Ai et al. (2015) and Siefarth and Buettner (2014).

should place approximately 15 mL of milk into a waxed or plastic (not paper or foam) cup. The evaluator is advised to place one hand over the cup and the other hand around the bottom of the cup, so that the warmth of the lower hand can penetrate the cup, enabling volatile aromatic components to fill the headspace, trapped by the upper hand. Swirling of the cup facilitates aromatic release. The evaluator should place his or her nose close to the upper hand, lift the hand from the top and immediately sniff the headspace. Fresh milk, if full of fat, should have a clean aroma of milk fat. Any deviation from this suggests potential defects. Various off-flavours (namely acid or sour, barny, cooked, cowy, feed, fruity/fermented, garlic/onion, malty, oxidized, rancid) can be readily recognized just by sniffing the milk. Reading Section 3 may help the evaluator create a mental database of potential off-flavours, to help in future identification. Better yet, evaluators can utilize recipes, such as those included in Table 3 or in *The Sensory Evaluation of Dairy Products* (Costello and Clark 2009), and practice sniffing and tasting various intensities of the defects to aid in identification.

While off-flavours will be most readily noted by sniffing the headspace, off-flavours will be noted retronasally, if the evaluator takes a sip, swirls the milk around the mouth, expectorates and breathes out through the nose.

It is also very common for consumers to smell milk. Although not trained to identify aroma compounds, each consumer has his or her own definition of what is considered offensive; thresholds and tolerance levels vary among consumers. While some consumers are forgiving of an occasional offense, others may steer clear of a particular brand of milk if it offends, particularly if it spoils before the code date printed on the carton. Some consumers discard containers once they reach the printed 'best by' or 'sell by' date, without even sniffing, while others continue drinking milk beyond the date as long as it smells 'OK'. Thus, it is in the best interest of companies to conduct shelf-life tests and to monitor product sensory quality throughout shelf-life.

Taste refers to the basic sensations acquired from stimulation, on the tongue, by general classes of tastants, including sweet, sour, salty, bitter, umami (savoury) and fat (fatty). While the basic acid or sour and fatty tastes often have aromatic components associated with them, the basic tastes, sweet, salty, bitter and umami can only be recognized by tasting the milk. The basic tastes are commonly associated with locations on the tongue, but taste sensation varies among humans. Flavour is a broader term, referring to the combined sensation of taste and aroma. Fresh 'normal' milk should taste slightly sweet, umami and fatty, with no offensive flavour or aftertaste. Upon tasting, the first thing that an evaluator should notice is the natural sweetness of milk, since lactose is the primary solid component in milk (approximately 5%). The sweet sensation can be noted readily on many parts of the tongue and may be associated with a warming feeling. The basic taste of umami, which is associated with proteins, is often described as 'brothy'. Milk with less fat (low-fat (1%) or skim milk) will have a more pronounced umami taste than milk with more fat (reduced fat (2%) or whole milk (at least 3.25% fat)).

In many facilities, taste/flavour quality is evaluated throughout production runs. Staff take cartons off the processing line, to the laboratory, where a variety of quality and regulatory compliance tests are run throughout a shift. It is a great practice to have trained sensory personnel in milk quality control/quality assurance laboratories. Such personnel can identify potential quality issues early, and potentially circumvent customer complaints. Regulatory compliance tests, including, but not limited to, bacterial limits, temperature recording, proximate analysis, somatic cell count, drug residues, phosphatase, etc., are beyond the scope of this chapter. The reader is encouraged to

Table 3 Recipes* for production of samples for milk off-flavour identification

Off-flavour	Description/recipes/notes
No defect	Fresh, wholesome whole or 2% milk should have no offensive aroma, and should taste slightly sweet, fatty and umami. There should be no offensive flavour or aftertaste. When purchasing from a store, milk packaged in paperboard may have no defect, but milk in glass or HDPE will likely be oxidized. Skim milk and low-fat (1%) milk will lack the fatty taste and will have more pronounced sweet and umami tastes.
Acid/sour	Add 1–2 mL of cultured buttermilk to 1 L of milk. Alternatively, add 3 mL of a 10% lactic acid solution to 1 L of milk.
Bitter	Add 1–2 mL of a 0.25% quinine sulphate solution or a 0.25% caffeine solution to 1 L of milk.
Cooked	Purchase UP or UHT milk. Alternatively, boil 500 mL of milk and bring to 1 L volume with 'no defect' milk.
Feed	Add 5 mL 'alfalfa tea' (steep alfalfa or Timothy hay 5 min in boiling water) to 1 L of milk within 6 h of tasting (aroma dissipates over time).
Flat	Use 'no defect' low-fat (1%) or non-fat milk or combine 500 mL skim milk with 500 mL 2% milk. Alternatively, add up to 20% water to 1 L of milk.
Foreign	Anything that should not be in milk, but not otherwise identified by one of the other off-flavour categories, is foreign. To 1 L milk, add (1) 0.25 mL iodine sanitizer or 1 mL bleach solution (1 mL/1 L water) within 1 h of tasting (unstable); (2) 0.1 mL vanilla or other flavour; (3) 1 g sugar (or 0.25 g of a high-intensity sweetener); (4) 0.25 mL fish oil; (5) 0.25 g vitamin pre-mix.
Fruity/ fermented	Add up to 3 mL pineapple juice, apple juice or V8 Splash® Tropical Blend juice to 1 L of milk.
Garlic/onion	Add 0.25 mL of garlic or onion juice to 1 L of milk.
Lacks freshness	Dissolve 1–2 g of stale (stored) non-fat dry milk (NFDM) in 1 L of milk. Alternatively, cut a sanitized green pepper and place it into approximately 200 mL milk. Fish the pepper out of the milk after 30 min and bring to 1 L volume with 'no defect' milk. Another option is to taste milk stored 7 days past the code date, but other off-flavours may be noted (fruity/fermented, bitter).
Light oxidized	Buy milk in any plastic or glass container or place 'no defect' milk into a glass or plastic container and expose to fluorescent or UV light for 15 to 30 min.
Malty	Dissolve 1 g malted milk powder (Carnation® or other brand) in about 200 mL of milk that has been warmed in a microwave for 20 seconds; swirl or mix with a sanitized spoon; bring to 1 L volume.
Metal oxidized	Add a cleaned and sanitized copper penny to 1 L milk; allow oxidization for at least 4 h; filter out the penny before serving. Alternatively, add 1 mL 0.25% copper sulphate to 1 L of milk at least 4 h in advance of serving.
Rancid	Add 0.2 g kid lipase to 1 L milk and allow to hydrolyze for at least 4 h under refrigeration. Alternatively, add 10 g grated Romano cheese to 500 mL milk and refrigerate overnight. Filter cheese out of milk (4–6 layers of cheesecloth or 1 coffee filter) and bring to 1 L volume with 'no defect' milk.
Salty	Add approximately 0.25 g table salt to 1 L 'no defect' milk.
Unclean	Combine at least two of the following off-flavours and bring to 1 L volume: bitter, fruity/fermented, acid/sour, malty, rancid.

*Recipes can be adjusted up or down to train varying intensities of off-flavours.

reference the *Grade 'A' Pasteurized Milk Ordinance* (USDHHS et al. 2011) for details. Microbiological tests that may be conducted as indicators of milk quality, including, but not limited to, standard plate count, coliform count, laboratory pasteurization count, preliminary incubation count, direct microscopic count and psychrotrophic/psychrophilic or spore pasteurization count will also not be covered in this chapter. The reader is encouraged to reference the *Standard Methods for the Examination of Dairy Products* for detailed methodology (Wehr and Frank 2004). Some processors 'hold back' products at various stages of production runs, for storage in-house and for sensory and/or microbial evaluation throughout the stated shelf-life. Shelf-life testing will be elaborated upon in a later section of this chapter.

Tactile issues are least likely to be experienced in milk because of the likelihood that consumers will use their other senses. Visually, if milk has coagulated (by acidification or proteolysis), rejection will occur well before the milk is tasted. However, if coagulation is not visually noted (for instance, if someone drinks spoiled milk directly out of a paperboard carton), the milk will likely taste either sour (acidification) or fruity/fermented and/or bitter (proteolysis). Age gelation, resulting from the natural plasmin enzyme system in milk, can be an issue in ultrahigh temperature (UHT) processed milk stored beyond 6 months or at elevated temperatures (Chavan et al. 2011).

3 Off-flavours in milk: categories, causes and remedies

There are four basic categories that off-flavours in milk can be divided into, namely absorbed, bacterial, chemical and delinquency. Preventing absorbed off-flavours generally involves good cow nutrition (appropriate feeds, balanced rations) and management (ventilation, health monitoring, manure management) practices. Preventing bacterial off-flavours hinges on the good training of staff that prepare teats for milking and proper maintenance of equipment, temperature control, proper selection of application of cleaning and sanitizing chemicals, and prompt milk processing. Preventing chemical off-flavours involves keeping milk away from light and reactive metals, avoiding excessive agitation and using appropriate processing controls. Preventing delinquency off-flavours relies on attentive care by all who handle milk, from cow to consumer.

3.1 Absorbed

If cow paddocks or resting areas are not properly maintained, and if manure is not cleared away, the aroma of manure may be absorbed into the milk and be perceived by consumers. The barny off-flavour, described as 'faecal', is readily observed by aroma.

The cowy off-flavour is described as medicinal, acetone or ketone aroma/flavour. The off-flavour indicates that cows in the herd are suffering from acetonaemia or ketosis, a result of abnormal fat metabolism when energy demands exceed energy intake.

Because of the absorbent nature of milk fat, strong feed flavour compounds may be absorbed from the blood into milk. The off-flavour, which can be described with a variety of terms, including, but not limited to, 'grassy', 'alfalfa', 'green', 'tea', 'grain' or 'molasses', is not objectionable to all consumers.

Closely related to the feed off-flavour, garlic/onion off-flavours are absorbed into the milk if cows are exposed to wild onions or garlic. Garlic/onion off-flavour is readily noted by its aroma. While some may not even notice, an unpleasant aftertaste can linger for those who are sensitive to the off-flavour.

Lacks freshness is a general term to capture any off-flavour that is not well described by any of the other terms. However, the word 'stale' or 'taste like the smell of the refrigerator' have been used to describe lacks freshness. Lacks freshness may be of absorbed or bacterial origin. Milk may absorb aromas from strongly flavoured foods in the refrigerator. Psychrotrophic bacteria, including *Pseudomonas, Paenibacillus, Bacillus* and *Microbacterium* species, may produce off-flavours associated with lacks freshness. Lacks freshness can be smelled. The stale flavour may not clean up readily.

3.2 Bacterial

The acid/sour taste has been described as sharp, piercing and even cooling sensation; it is commonly experienced on the sides and top of the tongue. In dairy products, acid describes the taste of lactic acid, while sour includes volatile components associated with the lactic acid bacteria (LAB) that produce not only lactic acid but also volatile components such as diacetyl in such products as cultured buttermilk, sour cream, cream cheese, etc. 'Sour' is perhaps the most common specific sensory description for spoiled milk by consumers. However, acid/sour is a rare off-flavour in milk. The reason for this disconnect is likely that consumers do not have an established vocabulary to describe milk defects and the word 'sour' has been carried through generations. Souring is the result of fermentation of lactose into lactic acid by LAB, including, but not limited to, species of *Lactococcus, Lactobacillus, Streptococcus* and *Leuconostoc*. Depending on the genus, LAB are most active at temperatures ranging from as low as 25–35°C (mesophilic species) to as high as 35–45°C (thermophilic species), and most grow slowly at temperatures as low as 4°C. The U.S. Food and Drug Administration's *Grade 'A' Pasteurized Milk Ordinance* (USDHHS et al. 2011) for interstate milk shipment requires that milk be cooled to below 7°C within 2 h of the complete milking of the last cow. Since all fluid milk in the United States must meet Grade 'A' standards, when present in raw milk, LAB proliferation is slowed down by the cold temperatures required by the PMO. Additionally, LAB are killed by pasteurization, which is also required for fluid milk sold across state lines, and even within many states' borders. Although pasteurization kills LAB, it will not improve the flavour of milk if the defect (acid/sour) is already present. Because of these two practices, milk does not sour as readily today as it did in the early years of transporting milk to consumers. However, it should be noted that post-pasteurization contamination of milk with LAB enables formation of the sour/acid off-flavour in milk.

Fruity/fermented is one of the most common off-flavours in milk that has been stored beyond its code date. Fruity/fermented off-flavours result from ethyl ester formation by esterases from thermoduric psychrotrophic bacteria, especially *Pseudomonas*, including, but not limited to, *Pseudomonas fragi*. Other species that predominate in milk stored too long, and that produce fruity/fermented off-flavours, include Gram-positive psychrotrophic *Paenibacillus, Bacillus* and *Microbacterium* species (Fromm and Boor 2004). Fruity milk may smell like apples, pineapples, mangos, strawberries and so on, while fermented milk smells like fermented fruit or sauerkraut. Fruity/fermented milk may taste dirty and tends to leave an aftertaste.

Similar to acid/sour, the malty off-flavour results from temperature abuse of raw milk. The microorganism responsible for this off-flavour, *Lactococcus lactis* ssp. *maltigines*, grows well at temperatures above 18°C. *Lactococcus lactis* subsp. *lactis* biovar. *diacetylactis* has also been implicated in malty flavour compounds, including 2- and 3-methylbutanal and 2- and 3-methyl-1 1-butanol (Mutukumira et al. 2009). Although the microorganisms are killed by pasteurization, the malty flavour will persist once produced. The malty off-flavour can be readily sensed by sniffing. Although not terribly objectionable in aroma, it is a clear sign of bacterial degradation; counts will be in the millions per millilitre. Malty smells and tastes like the milk left over after eating a bowl of breakfast cereal (toasted barley, corn, oat or Grape Nuts® cereal).

Milk that is 'unclean' tastes dirty. It generally smells bad and leaves an unpleasant aftertaste. The unclean off-flavour results from activity by psychrotrophic and/or mesophilic bacteria. While outgrowth of psychrotrophic bacteria is expected at refrigeration temperatures, an outgrowth of mesophilic bacteria indicates temperature abuse. If left long enough (e.g. ~15 days beyond code), most milk will eventually turn unclean.

3.3 Bacterial/chemical

Bitterness typically has a slower onset than the other basic tastes, is noted at the back of the tongue and throat, can give a dull or throbbing sensation, and tends to linger longer than the other basic tastes. The sensitivity to or threshold for bitterness varies widely among humans; some people are 'bitter blind' while others are particularly offended by the off-taste. Fresh milk should never taste bitter, unless cows are fed certain feedstuffs. However, bitterness is one of the most common defects in milk that has been stored beyond its code date, because of proteolysis. Proteolysis may be initiated from two sources: native plasmin enzyme system and bacterial origins. The term 'plasmin enzyme system' refers to a family of proteins, specifically plasminogen (an inactive form, or precursor of plasmin), and two plasminogen activators (which enable conversion of plasminogen to plasmin), plasmin, plasmin inhibitors and plasmin activator inhibitors (Metwalli et al. 1998; Ismail and Nielsen 2010). Plasmin and plasminogen are highly heat stable, which is why bitterness and age gelation occur even in UHT milk. Because milk is rapidly cooled, and because many microorganisms that survive pasteurization are proteolytic, the dairy industry has naturally selected thermoduric spore-forming psychrotrophic microorganisms to dominate in dairy environments and dairy products over time.

Although enjoyed in some types of cheeses (Asiago, feta, blue, Romano), rancid is one of the most objectionable off-flavours to occur in milk. With oils and nuts, 'rancid' often refers to oxidative rancidity, but in the context of milk, the term rancid refers to hydrolytic rancidity, resulting from the cleavage of fatty acids from the glycerol backbone of triacylglycerols (triglycerides) by the enzyme lipoprotein lipase. Short-chain volatile free fatty acids, including butyric (C4), caproic (C6), caprylic (C8), capric (C10) and lauric acid (C12), contribute to the baby vomit-like smell. The enzyme lipase is a natural component of milk, but is also a component of many bacteria. Generally, lipase is separated from its substrate, milk fat, by the protective milk fat globule membrane. The rancid off-flavour is induced when the milk fat globule membrane is compromised and lipase has access to fatty acids to catalyse hydrolysis. Causes of rancidity include (1) overly rapid cooling or freezing, (2) foaming, (3) homogenizing raw milk and (4) adding raw milk to pasteurized–homogenized milk. Lipase is destroyed by the heat of pasteurization, but lipolytic damage done to milk before pasteurization will not be improved by pasteurization. Additionally, if milk is contaminated with psychrotrophic bacteria that produce lipase, the off-flavour is likely to manifest during storage. Although the acid degree value (ADV) has been used as a measure of free fatty acids, and an indirect measure of milk hydrolytic rancidity, Duncan and Christen (1991) demonstrated that ADV does not measure the short-chain fatty acids that contribute to the threshold sensory detection of rancid off-flavour.

3.4 Chemical

The term cooked can be described as 'eggy', 'sulphur', 'custard', 'steamed' or 'barista milk'. There is a wide range of consumer tolerance or acceptability for the off-flavour 'cooked'. Most fluid milk in the United States is high-temperature short-time (HTST) pasteurized (at or above 72°C, held thereat for at least 15 sec) and some is low-temperature long-time (LTLT) pasteurized (at or above 63°C, held thereat for at least 30 min). At those temperatures, milk is only mildly cooked. Milk naturally contains sulphur-containing amino acids. In the raw form, proteins in their native structure do not present aromatic sulphur compounds. However, when cooked, proteins unravel, enabling exposure and volatilization of the sulphur compounds. UHT milk, in particular, contains significantly higher concentrations of hydrogen sulphide, methanethiol carbon disulphide, dimethyl trisulphide and dimethyl sulfoxide than raw or pasteurized milk (Vazquez-Landaverde et al. 2006). In a study specifically designed to address the concern about cooked flavour, Gandy and others (2008) demonstrated that consumers (n = 298) preferred milk pasteurized at 79°C to milk treated at 77, 82 or 85°C on 'day zero', but differences were not as notable beyond six days of storage. Some clusters of consumers liked the cooked flavour of milk treated at 82 or 85°C, but others did not (Gandy et al. 2008). For many, cooked is a pleasing aroma/flavour, and it does not become objectionable until milk is ultra-pasteurized (UP) or UHT pasteurized. UP and UHT pasteurized milk must be thermally processed at or above 138°C for at least 2 s, either before or after packaging (USDHHS et al. 2011). The main difference between UP and UHT is packaging; UHT milk is either thermally processed and then packaged aseptically or thermally processed in the package (commercially sterile milk). UHT milk has an unrefrigerated, unopened shelf-life of approximately 6 months. UHT milk is not common in the United States, in part because of the strong cooked flavour, but outside the United States, UHT milk is common and popular (Chavan et al. 2011). Further, minimal processing, a current trend in the United States limits the appeal of UHT. Modern technologies, including microfiltration, high hydrostatic pressure, pulsed electric fields and ultrasound, have been considered as adjuncts to pasteurization, but have not found industrial application to date.

The most common off-flavour in milk in the United States is light oxidized, because so much of the milk is packaged in translucent HDPE plastic gallon and half-gallon containers. Unfortunately, milk in HDPE packaging, as well as milk packaged in glass, is prone to light oxidation. Light-oxidized off-flavour has been described as 'wet cardboard', 'burnt feathers', 'burnt hair', 'mushroom', 'old vegetable oil' and even 'plastic-like'. Light-induced oxidation (or 'sunlight flavour') results from chemical reactions initiated by light and facilitated by riboflavin and possibly another photosensitizer, upon amino acids and/or unsaturated fatty acids (Chapman et al. 2002; Webster et al. 2009). Havemose et al. (2004) demonstrated that protein oxidation occurs independently of lipid oxidation, and milk from cows fed grass silage is more vulnerable to lipid oxidation, while milk from cows fed corn silage is more vulnerable to protein oxidation. Chemically, accumulation of dityrosine can serve as a marker for protein oxidation, and accumulation of lipid hydroperoxides, pentanal, hexanal and heptanal can serve as markers for lipid oxidation (Havemose et al. 2004, 2006; Hedegaard et al. 2006). Chapman et al. (2002) demonstrated that trained panelists could note oxidized off-flavour in milk after exposure to light for 15 to 30 minutes, while untrained consumers notice the oxidized off-flavour after 54 minutes to 2 h of light exposure. It was also reported that approximately 50% of the plastic milk containers remain in lighted dairy cases for at least 8 h (Chapman et al. 2002). Webster et al. (2009) demonstrated varied effectiveness of overwraps against oxidized off-flavour when milk was subjected to fluorescent light. For instance, iridescent film material that blocked certain wavelengths of light was not very effective at preventing the off-flavour. Only aluminium foil was effective at blocking all UV and visible wavelengths of light and at preventing the off-flavour. Hexanal, pentanal and heptanal concentrations in milk increased, and riboflavin concentration decreased, in all packaging scenarios except when glass bottles were overwrapped with aluminium foil (Webster et al. 2009). In recent years, pigments (including but not limited to titanium dioxide) have been added to HDPE milk packaging to slow the oxidation process. Similar to what Webster et al. (2009) showed, the tinted cartons only slow milk oxidation, rather than completely preventing it. Paperboard cartons do not permit light penetration of the cartons.

The metal-oxidized off-flavour is similar to light oxidized, with an added 'tingling' or 'prickly' sensation on the tongue. Some have described the aftertaste as similar to blood (after losing a tooth). One may recognize the flavour by rubbing a bunch of copper pennies in the hands, then sniffing. Reactive metals, including copper, iron and manganese, catalyse autoxidation by way of free radial formation, autoxidation of unsaturated fatty acids in milk and production of aldehydes, ketones and other offensive off-flavours. Metal-induced oxidation is not as common today as it was in previous decades because most copper has been replaced with stainless steel. However, the presence of reactive metals in hot water used for diluting cleansers and sanitizers for dairy equipment can cause the off-flavour. Metal-induced oxidation can also result from an imbalance of minerals in the cows' diet. This happened to an Idaho farmer, in reality, in 2005. Processed milk, as well as raw milk directly from animals, tasted extremely metal-oxidized. Upon copper analysis, raw hand-milked milk measured 2.64 mg/L copper, though raw milk copper is supposed to be approximately 0.02 mg/L (Lopez et al. 1985). Although no feed or mineral ration changes were made by the farmer, calculation errors were made by the nutritional supplement supplier such that the nutritional supplement had 702 ppm copper and 2468 ppm iron; they should have been 285 ppm copper and 190 ppm iron. After the farmer obtained a properly balanced mineral ration from a new supplier, the cows produced more milk, milk flavor improved, and reproductive problems were rectified.

Spontaneous oxidation is a general term used to describe an oxidized off-flavour that tastes like a cross between light-induced and metal-induced oxidation, resulting from unknown source(s). Milk neither exposed to reactive metals nor light, which tastes oxidized within a few days of production, is called spontaneous oxidized. Cow diet, seasonal changes and/or lack of balance between pro- and anti-oxidants in milk may be involved in spontaneous oxidation but no single source has been conclusively implicated (Granelli et al. 1998; Havemose et al. 2006; Juhlin et al. 2010; Testroet et al. 2015).

The salty taste is noted typically, though not exclusively, on the tip and sides of the tongue and can provide a warming sensation. Although milk naturally contains a number of salts (Table 1), milk is not perceived as 'salty' unless there is a breakdown of

Table 4 Off-flavours in milk, along with descriptions, causes and remedies

	Off-flavour	Description	Cause	Remedy
Absorbed	Barny	Faecal aroma, unclean flavour.	Poorly maintained barn aromas transmitted to milk by cows inhaling air laden with volatile compounds characteristic of a barnyard.	Ensure appropriate manure management and proper ventilation in cow housing.
	Cowy	An unpleasant medicinal odour and/or chemical aftertaste.	Accumulation of ketone bodies in milk; a sign of physiological malfunction in cows (acetonaemia/ketosis).	Balance cow diets properly to maintain health.
	Feed	Aroma and/or flavour and/or aftertaste of alfalfa, grass, corn silage, molasses or herbs.	Aromatic compounds in feed are transmitted to milk.	Time between feeding and milking should be greater than 30 min. Minimize cow exposure to flavourful weeds if on pasture.
	Garlic/onion	Aroma and/or flavour and/or aftertaste of garlic or onions.	Aromatic compounds absorbed into milk after cows consume wild garlic or onions. Storage of milk next to cut/chopped onions or garlic in refrigerator.	Control wild onions in cow pastures. Wrap cut/chopped onions and garlic well if stored in refrigerator.
	Lacks freshness	Milk tastes 'stale'.	Milk may have absorbed flavours from foods (e.g., green peppers) in refrigerator. A bacterial cause may also be at work (i.e., psychrotrophic bacteria).	Regularly clean refrigerator and wrap cut/chopped highly-flavoured foods.
Bacterial	Acid/sour	Acid/sour is a basic taste and may be experienced as a tingling sensation on the sides of the tongue.	LAB convert lactose to lactic acid at warm temperatures (>10°C/50°F). A buttery aromatic component is notable if diacetyl is also produced by the bacteria. Post-pasteurization contamination is the primary cause of sour/acid milk today.	Ensure proper cleaning and sanitizing. Cool milk quickly after collection (to 4°C within 2 h) and process/package within 48 h. Prevent post-pasteurization contamination.
	Fruity/fermented	The fruity/fermented aroma may be like apples, sauerkraut, pineapples, strawberries, etc.	Certain psychrotrophic bacteria (e.g. *Pseudomonas fragi*), produce volatile compounds, including ethanol, coupled with enzymatic action upon lipids during extended storage.	Ensure proper cleaning and sanitizing to minimize the number of psychrotrophic microorganisms.
	Malty	The malty aroma resembles Grape Nuts® or other malted grain-based cereal.	Contact with improperly sanitized equipment contaminated with *Lactococcus lactis* ssp. *maltigines* followed by temperature abuse (18°C).	Ensure proper cleaning and sanitizing. Cool milk quickly after collection (to 4°C within 2 h) and process/package within 48 h.
	Unclean	A lingering unpleasant aftertaste that may involve multiple off-flavours/aromas.	High numbers of heat-tolerant psychrotrophic microorganisms are transferred into milk from unclean equipment/utensils, multiply and produce off-flavours with storage.	Clean and sanitize immediately after milk collection and processing. Sanitize immediately before milk collection and processing.

Bacterial/chemical	Bitter	Bitter is a basic taste and yields a somewhat numbing sensation on the back of the tongue.	Protein degradation (proteolysis) by enzymes (proteases) generates bitter peptides and amino acids. The natural heat-stable plasmin system in milk generates bitter peptides and amino acids with extended storage. Spore-forming psychrotrophic bacteria (grow at 7°C) endure pasteurization and their enzymes cause bitterness, especially with extended storage. Weeds in cows' diet may also impart bitterness.	Ensure proper cleaning and sanitizing to minimize the number of proteolytic microorganisms.
	Rancid (hydrolytic)	May smell like baby vomit, a male goat, or Romano or feta cheese. The flavour may seem unclean, and lingers.	Hydrolysis of aromatic fatty acids (butyric, caproic, caprylic, capric lauric acids) by the enzyme lipase, which is naturally present in milk and can be microbially derived. Lipase is inactivated by pasteurization. However, overly rapid cooling, excessive agitation or homogenization before pasteurization disrupt the protective milk fat globule membrane, increasing susceptibility to rancidity.	Handle milk carefully, avoiding over-agitating and homogenization of raw milk. Do not mix raw milk into homogenized milk.
Chemical	Cooked	A somewhat nutty, custard-like (cooked eggs) aroma and mildly sweet taste. Excessive heating can lead to burnt hair or burnt feathers aroma.	Heating milk above standard pasteurization temperatures. Sulphide compounds may arise with excessive heating (UHT pasteurization may yield scorched off-flavour).	Process milk at the recommended time/temperature combinations to ensure safety and desired shelf-life.
	Oxidized, light-induced	Milk oxidized by light may smell/taste like wet cardboard, wet paper or even cooked cabbage.	Sunlight/UV light exposure leads to breakdown of sulphur-containing proteins or lipid oxidation (autoxidation). Aldehydes and/or ketones are formed.	Package milk in containers that block light (paperboard). Tinted plastic packaging only partially blocks light, to slow oxidation.
	Oxidized, metal-induced	Milk oxidized by metal may smell/taste like a copper penny.	Oxidation of unsaturated fatty acids occurs after contact with copper or other metals.	Prevent milk from contacting copper and/or other unprotected metals.
	Oxidized, spontaneous	Tastes like a combination of light- and metal-induced oxidation, but of different origin.	The source is not completely understood, but it appears to be diet related. A combination of high polyunsaturated fatty acids and low vitamins A and E have been implicated.	Avoid drastic changes to cow diet, milk foaming or extremes in temperature.
	Salty	Salty is a basic taste experienced on the front and sides of tongue.	May be present in the milk of cows with mastitis or in late lactation because of the increase of milk secretion cell (alveoli) membrane permeability.	Properly treat cows exhibiting mastitis and follow withholding times before introducing their milk to the bulk tank.
Delinquency	Flat	Flat milk lacks creamy aroma or full-bodied flavour.	Adulteration of milk with water will dilute the full-bodied flavour of milk.	Prevent adulteration of milk with water. Run a cryoscope (freezing point) analysis regularly.
	Foreign	The foreign term is used when milk had an odour or flavour not associated with pure milk.	The defect may result from contamination of milk with chemical sanitizers, detergents, medications, etc. Using the wrong packaging (2% milk) on a flavoured milk line (vanilla milk) would be another example. Use of poor-quality vitamin solution can yield a foreign off-flavour.	Properly drain holding tanks of water, cleanser and sanitizer before use. Be alert to line changes to ensure proper packaging.

the semipermeable membrane that maintains a salt balance between blood and milk. If a cow has mastitis, the semipermeable membrane becomes more permeable to salt, and the composition of salt ions in milk increases. Salty is a rare defect in commercial milk because cows with mastitis are generally caught and treated, and their milk is not commingled in the bulk tank.

3.5 Delinquency

Flat is essentially a lack of aroma/flavour. Upon smelling, the milk may lack the expected fatty aroma; upon tasting, the milk may seem 'watered down'. Flat results from two sources: added water or reduced fat. Because fat provides good flavour to milk, 1% and non-fat milk taste 'flat' compared with 2% and whole milk – that is expected. Flat is only a concern if it results from added water. Leaving rinse water in a bulk tank or a line will yield flat milk. Some processing plants test milk for added water, using a cryoscope, before off-loading milk from tankers.

Similar to flat, the foreign off-flavour is a result of delinquency by someone at the farm or in the processing plant. Foreign is a catch-all term to describe any off-flavour that can be explained by any foreign object or chemical that is not supposed to be in milk. An exception would be vitamins, as vitamins A and D are permissible for addition to whole milk, and required for reduced fat, low-fat and fat-free milk. The foreign 'vitamins' flavour can be described as tasting like carrots. A 'fishy' off-flavour can result from incorporation of omega-three fatty acids into milk. A chlorine or iodine flavour can result from the misuse of sanitizers (improper drainage). A plastic off-flavour could result from spent gaskets running through a line. It is even possible that white milk may have a vanilla or coffee or other flavour if someone forgot to switch out the packaging on the line.

3.6 Summary of off-flavours, their causes and their remedies

Table 4 provides a detailed summary of the defects associated with the different categories of off-flavours, along with how to diagnose and prevent the occurrence of such off-flavours.

4 Sensory shelf-life testing

The code dates printed on the top of milk cartons are not arbitrary; they are selected by individual companies within a confidence interval of assurance that the milk will taste good through that date. Unfortunately, many consumers interpret 'best by', 'best if used by' and 'sell by' dates as 'expiration' dates, and assume, wrongly, that milk will taste bad, or harm them on that date. By definition, the code date signifies the last date at which the product can be sold for full price. In practice, many stores remove products from shelves within a week of that date, especially for refrigerated products. Sadly, most of those products enter the waste stream.

The shelf-life of any milk is dictated partially by microbial quality and partially by enzyme activity, which are influenced by cow and season, raw milk quality, processing conditions, packaging materials, temperature abuse and exposure to light. While milk with low microbial counts may have limited shelf-life if enzyme activity is high, milk with poor initial microbial quality will only get worse with time.

For decades, a number of raw milk tests, including the somatic cell count, standard plate count, ropy milk test, coliform count, Gram-negative non-coliform count, laboratory pasteurization count, preliminary incubation count, coagulase-negative staphylococcal count, environmental streptococcal count, direct microscopic count, psychrotrophic/psychrophilic or spore pasteurization count, bioluminescence, impedance microbiology, limulus amoebocyte lysate, direct reflectance colorimetry, Virginia Tech shelf-life procedure and the Moseley keeping quality test, have been used to help predict milk shelf-life (White 1993). Jayarao et al. (2004), who conducted somatic cell count and 10 different bacterial counts, showed that such tests could serve as indicators of, and facilitate monitoring of, herd udder health and milk quality. They showed that dairy herds that used automatic milking detachers, sand as bedding material, dip cups for teat dipping instead of spraying and practiced pre- and post-dipping had significantly lower bulk tank and/or bacterial counts (Jayarao et al. 2004). Later, Miller and colleagues (2015) sampled bulk tank milk from 33 farms in New York every month for one year. They reported that low mesophilic spore levels were associated with large herd size, use of sawdust or sand bedding, and not fore-stripping during the pre-milking routine; low thermophilic spore levels were associated with large herd size, use of straw bedding, spray-based application of the post-milking disinfectant and dry massaging the udder during the pre-milking routine (Miller et al. 2015). However, the reader is cautioned that raw milk tests do not always effectively predict the sensory or microbiological shelf-life performance of commercially pasteurized fluid milk (Martin et al. 2011), and should be accompanied with sensory quality evaluation.

Pasteurized milk microbial testing is routinely conducted. Not only is testing required for Grade 'A' milk, with clear bacterial limits (<20000 total aerobic bacteria per mL and <10 coliform bacteria per mL), but microbial counts can be indicative of contamination at various stages in the milk collection and processing stream. In a study to characterize spoilage bacteria in pasteurized milk collected from three commercial dairy plants, Fromm and Boor (2004) identified the predominant spoilage microorganisms as Gram-positive, heat-resistant, psychrotrophic rods including *Paenibacillus*, *Bacillus* and *Microbacterium* species, which caused undesirable flavours in milk. Ranieri and Boor (2009) obtained pasteurized 2% milk from 18 dairy plants from 5 geographical regions of the United States and characterized 589 bacterial isolates with DNA-sequence-based subtyping methods. More than 58% of the isolates were Gram-positive spore-forming isolates; 84% of those characterized within 10 days of production were of the genus *Bacillus*; and more than 92% of isolates characterized at 17 days of shelf-life were of the genus *Paenibacillus*. Similarly, Martin et al. (2011) characterized isolates from pasteurized milk of four New York State processing plants over a 1-year period into three predominant genera: *Pseudomonas*, *Bacillus* and *Paenibacillus*. Most recently (Ivy et al. 2012), out of 1288 isolates obtained from raw and pasteurized milk and from dairy farm environments, two major clusters predominated: the genus *Paenibacillus* (737 isolates, including the species *Paenibacillus odorifer*, *Paenibacillus graminis* and *Paenibacillus amylolyticus*) and *Bacillus* (467 isolates, including *Bacillus licheniformis*, *Bacillus pumilus* and *Bacillus weihenstephanensis*). These microorganisms, which survive pasteurization and are psychrotolerant, are associated with a variety of bacterial off-flavours, including bitter, fruity/fermented, rancid and unclean. Control strategies to reduce the introduction of these microorganisms will be essential to increase the shelf-life of milk.

To complicate matters, contrary to what might be expected, reduction of pasteurization temperature leads to lower bacterial outgrowth in pasteurized fluid milk during refrigerated storage. Martin et al. (2012) enumerated bacteria in HTST commercially pasteurized fluid milk over refrigerated shelf-life. Milk was pasteurized at 79.4°C for 15 months, followed by 15 months at 76.1°C. Mean total bacterial counts were significantly lower immediately after processing and after 21 days of storage in samples pasteurized at 76.1°C than those pasteurized at 79.4°C. Additionally, bacterial growth was lower throughout shelf-life for products pasteurized at the lower temperature. The authors hypothesized that at the higher temperature, lactoperoxidase (a naturally occurring enzyme in milk that is bacteriostatic against Gram-positive and Gram-negative bacteria) was inactivated at the higher temperature (Martin et al. 2012). No sensory evaluation of products was conducted to evaluate sensory shelf-life, but authors encouraged pasteurization at temperatures near the minimum specified by law.

Because raw and pasteurized milk microbiological testing are only indirect measures of milk quality, it is more appropriate to estimate and validate sensory shelf-life using actual sensory evaluation.

The sensory shelf-life of raw milk in the United States is typically close to seven days. Pasteurization kills all pathogenic bacteria and many spoilage bacteria, extending the shelf-life, but thermoduric spore-forming bacteria survive, and the plasmin enzyme system remains functional. The shelf-life of packaged LTLT or HTST pasteurized milk or pasteurized–homogenized milk in the United States is expected to be approximately 14–21 days. UP milk has a shelf-life of approximately 35–45 days, and UHT milk can last for approximately six to nine months on the unrefrigerated shelf before enzymatic breakdown (by the plasmin enzyme system) reduces the quality significantly.

While some companies select a code date based on these ranges, the more prudent behaviour is to conduct shelf-life studies in-house to confirm product shelf-life before product release. To properly set code date, companies must conduct at least three shelf-life studies on different lots of milk, on all products and all packaging types, to determine the mean time to failure for each product. Sufficient numbers of unopened packages of products must be stored, at one or more temperatures (e.g., 4°C and/or 6°C and/or 8°C), and evaluated on the expected code date (e.g., 21 or 35 days) and one week beyond the expected code date (e.g., 28 or 42 days) to confirm the quality is acceptable for seven days beyond the printed code. When initially selecting the proper code date, milk may be tasted on several days around the predicted code date (e.g., days 18, 21, 24, 27, 30 and 33), in order to properly select the appropriate code date. This necessitates storage of a lot of milk initially, but helps producers select a proper code date. Two negative consequences necessitate careful selection of code date: if shelf-life is set too short, companies will lose money in unsold product; if shelf-life is set too long, companies will lose customers in rejected product and lost future sales. Consider the two scenarios in boxes.

Staff must be effectively trained to recognize off-flavours, tolerance for severity of off-flavours must be agreed upon, and point of failure must be clearly defined by the company before conducting tests in order to set and confirm the appropriate code date to maximize shelf-life and limit customer complaints. Along with trained staff, untrained consumers are sometimes recruited to help determine product shelf-life. Using a combination of trained panelists and consumers, Richards et al. (2014) utilized multivariate accelerated shelf-life test to show how high temperature storage negatively affected shelf-life of low-fat UHT milk. While the shelf-life of UHT milk was 211 days at

Milk shelf-life scenario 1

Description:

To be conservative, and to ensure the highest quality of milk for all consumers, Amy's Dairy sets the code date of 2% milk at 14 days. As per order, 100 cartons of 2% milk are delivered to Hal's Grocery, two days after production. Hal's Grocery staff are directed to pull all unsold product off the shelf within four days of the code. On that day, 20 are shipped back to Amy's Dairy, and discarded.

Interpretation/consequence:

Customers of Hal's Grocery only have a window of eight days to buy Amy's Dairy milk. Amy's Dairy loses money in discarded product. Staff at Hal's Grocery may decide they made a mistake in purchasing 100 cartons and order 80 for subsequent shipments. Thus, that is loss in product (waste) and loss in potential future revenue for Amy's Dairy.

Lesson:

If the staff at Amy's Dairy conducted a sensory shelf-life study and discovered product quality was high for 28 days, a 21-day code would be appropriate. With 15 days in which to buy the Amy's Dairy milk, perhaps 20 more cartons could sell, which would align the order by Hal's Grocery with the sales, reduce waste and contribute positively to Amy's Dairy' revenue.

Milk shelf-life scenario 2

Description:

A new processor, DariFresh sets the shelf life of their ultra-filtered pasteurized milk at 40 days. On day 35, customer complaints start coming into the grocery store and the company hotline.

Interpretation/consequence:

The shelf life of the milk is closer to 30 days than 40 days. Customers have many products to select from and may turn away from DariFresh for a short time, or worse, permanently, because of their bad sensory experience.

Lesson:

DariFresh should have conducted in-house sensory shelf-life tests to select the appropriate code date that is neither too long nor too short.

25°C, it was shortened to 73 (+/−3) and 27 (+/−3) days when stored at 35°C and 45°C, respectively (Richards et al. 2014).

In addition to conducting an in-house study before setting code dates on packages, prudent companies also track product quality throughout shelf-life. Whether setting or confirming shelf-life to conduct a sensory shelf-life study, unopened cartons are pulled from the production line(s) and are held in one or more refrigerated cases (e.g., 4°C and/ or 6°C and/or 8°C and/or 10°C). While 4°C is an appropriate milk storage temperature to maximize shelf-life, it is not very reflective of the average temperature that milk experiences during its shelf-life. A temperature between 6°C and 8°C is more reflective of home refrigerator temperatures or the average temperature milk experiences during shelf-life. A temperature of 10°C is abusive, and hopefully not reflective of home refrigerators, but may reflect the kind of abuse milk undergoes short-term at points during shelf-life; 10°C storage in-house would cause accelerated spoilage of milk. By holding milk at an elevated temperature, the dairy plant, with scheduled sensory evaluation, discovery that a product may not taste good through the code date can be caught before customer complaints occur. Such a discovery may enable a company to proactively pull a product, or put a product 'on sale' to flush that product out of stores more rapidly than if at full price.

In the scenario summarized in Table 5 (and used for sensory shelf-life case studies 1 and 2), a fictional company makes three kinds of products: whole, 2% and skim milk, processed daily in that order. Incoming raw milk is separated, standardized and simultaneously pumped to two different processing and filler lines: one for HTST gallon HDPE containers and one for UP quart-sized paperboard cartons. The gallon size of each product is most popular; they sell faster, so HTST pasteurization is appropriate. The quarts sell more slowly; a longer shelf-life is desired, so UP is used. In this scenario, the company

Sensory shelf-life case study 1

Description:

A trained QA/QC staff member evaluated milk samples stored according to the scheme in Table 5. All samples were criticism-free on day 14. However, by day 21, when HTST products were evaluated, all tasted slightly fruity/fermented.

Interpretation:

The off-flavour should be verified by at least one other trained staff member. The off-flavour is indicative of the presence of high numbers of psychrotrophic bacteria, which suggests cleaning/sanitation issues either at the farm and/or processing facility. The type of microorganisms responsible for the off-flavour can be verified with plating of milk samples. Improving sanitation practices at one or both places may be necessary.

Response:

Plant personnel should be notified that a milk quality problem has been encountered and measures should be taken to immediately remedy the situation so milk spoilage is not propagated in subsequent batches. Since the milk is already in stores, a decision has to be made on how to handle that milk. Because no off-flavour was noted in the same milk stored for 14 days, and the off-flavour was only slight in milk stored at 8°C for 21 days, putting the milk on sale is an option. Removing remaining milk from stores may be an appropriate response, to avoid losing repeat customers. Since the same milk was used to make the UP products (though the processing temperature was higher for UP), UP products should also be placed on sale or removed from stores.

Sensory shelf-life case study 2

Description:

A trained QA/QC staff member evaluated milk samples stored according to the scheme in Table 5. All milks, evaluated through day 14, were criticism-free. On day 21, the milk in gallon containers received no criticism. However, when the UP paperboard quart containers were evaluated on day 21, all tasted slightly acid/sour. A second trained staff member confirmed the defect.

Interpretation:

Because LAB are responsible for the acid/sour defect, and LAB are killed by pasteurization, the off-flavour is indicative of post-pasteurization contamination. Since the defect was only noted in milk packaged in paperboard cartons, it suggests cleaning/sanitation issues at the filler. The source of the problem can be verified with equipment swabbing and microbiological plating.

Response:

Measures should be taken to immediately remedy the situation. Both production lines should be swabbed to confirm if LAB contamination is isolated to one line. Because no off-flavour was noted in the same milk stored for 21 days, and the off-flavour was only slight in milk stored for 21 days, putting milk on sale is a good option. Removing remaining milk from stores may be an appropriate response, to avoid losing repeat customers.

Table 5 Example sensory shelf-life evaluation scheme for a fictional company that produced whole milk, 2% milk and skim milk processed with high temperature short time (HTST) or ultra-pasteurization (UP)

Product	Predicted shelf-life	Evaluation days for samples stored at 8°C
Whole milk HDPE gallon (HTST)	21	14, 21
Whole milk paperboard quart (UP)	35	21, 28, 35
2% milk HDPE gallon (HTST)	21	14, 21
2% milk paperboard quart (UP)	35	21, 28, 35
Skim milk HDPE gallon (HTST)	21	14, 21
Skim milk paperboard quart (UP)	35	21, 28, 35

predicts a 21-day shelf-life for the HTST milk. They hold three unopened containers, from the end of each production run, in storage at 8°C. For the UP milk, the company predicts a 35-day shelf-life. Similarly, four containers are stored at 8°C. For the container types, the evaluation schedule differs, such that HTST products are evaluated on days 14 and 21 and the UP milk samples are evaluated on days 21, 28 and 35.

Obviously, the scheme outlined in Table 5 is not feasible for all milk-processing facilities. In this scenario, with only six products produced at a facility, quality control/quality assurance staff would open and evaluate three cartons on days 14, 28 and 35, and six on day 21, not including the other lots of milk that would also need tasting on the same days. The scheme can be modified to meet the needs of any given facility. For instance, milk can be evaluated less frequently. The important point is that every facility must routinely evaluate products to ensure consistent, high quality to consumers.

5 Conclusion

Providing safe, wholesome, high-quality milk to consumers is an immense responsibility. One of nature's most perfect foods, milk is a source of nine essential nutrients to humans. It also serves as a great growth medium for bacteria and is susceptible to breakdown by native enzymes. Any miss-step in milk handling, from cow to consumer, contributes to reducing the quality of milk. Although microbiological and chemical tests have been and will continue to be utilized to help predict and track milk quality, ultimately, sensory evaluation is key – consumers will be the ultimate judges and they will adjudicate with their money. As milk producers and processors, it is our job to practice care at every step and understand how to guarantee the best tasting products in order to ensure milk's consumption in future generations.

6 Where to look for further information

Barbano, D. M., Y. Ma and M. V. Santos (2006), Influence of raw milk quality on fluid milk shelf life. *J. Dairy Sci.* 89 (Suppl): E15–E19.

Clark, S., M. Costello, M. Drake and F. Bodyfelt (Eds) (2009), *The Sensory Evaluation of Dairy Products*. New York, NY: Springer.

Dairy Practices Council (2004), DPC 10, Maintaining and Testing Fluid Milk Shelf Life. Available online at: http://www.dairypc.org/catalog/maintaining-and-testing-fluid-milk-shelf-life

Duyvesteyn, W. S., E. Simoni and T. P. Labuza (2001), Determination of the end of shelf-life for milk using the Weibull Hazard Method. *LWT-Food Science and Technology*. 34(3): 143–8.

Kilcast, D. and P. Subramaniam (2011), *Food and Beverage Stability and Shelf Life*. Cambridge, UK: Woodhead Publishing Limited.

Martin, N., N. Carey, S. Murphy, D. Kent, J. Bang, T. Stubbs, M. Wiedmann and R. Dando (2016), Exposure of fluid milk to LED light negatively affects consumer perception and alters underlying sensory properties. *J. Dairy Sci.* 99: 4309–24.

Murphy, S. C. (2009), Shelf-life of fluid milk products-Microbial spoilage-The evaluation of shelf-life. *Dairy Foods Science Notes*. Available online at: https://foodsafety.foodscience.cornell.edu/sites/foodsafety.foodscience.cornell.edu/files/shared/documents/CU-DFScience-Notes-Bacteria-Milk-Shelf-Life-Evaluaton-06-09.pdf.

Wiedmann, M. (2011), Raw milk quality tests-do they predict fluid milk shelf-life or is it time for new tests? Available online at: http://www.dairypc.org/assets/2011-Speaker-Presentations/Wiedmann-DPC-11-2011.pdf.

7 References

Black, R. E., S. M. Williams, I. E. Jones and A. Goulding (2002), Children who avoid drinking cow milk have low dietary calcium intakes and poor health. *Am J Clin Nutr.* 76: 675–80.

Chapman, K. W., L. J. Whited and K. J. Boor (2002), Sensory threshold of light-oxidized flavor defects in milk. *J. Food Science.* 67(7): 2770–3.

Chavan, R. S., S. R. Chavan, C. D. Khedkar and A. H. Jana (2011), UHT milk processing and effect of plasmin activity on shelf life: A review. *Comp. Rev. Food Sci. Food Safe.* 10: 251–68.

Costello, M. and S. Clark (2009), Appendix F: Preparation of samples for instructing students and staf in dairy products evaluation, in S. Clark, M. Costello, M. Drake and F. Bodyfelt (Eds), *The Sensory Evaluation of Dairy Products: Second Edition.*, pp. 551–60. New York: Springer.

Davoodi, H., S. Esmaeili and A. M. Mortazavian (2013), Effects of milk and milk products consumption on cancer: A review. *Comp. Rev. Food Sci. Food Safe.* 12: 249–64.

Duncan, S. E. and G. L. Christen (1991), Sensory detection and recovery by acid degree value of added to milk. *J. Dairy Sci.* 74: 2855–9.

Elwood, P. C., J. E. Pickering and A. M. Fehily (2007), Milk and dairy consumption, diabetes and the metabolic syndrome: The Caerphilly prospective study. *J. Epidemiol. Community Health.* 61: 695–8.

Elwood, P. C., D. I. Givens, A. D. Beswick, A. M. Fehily, J. E. Pickering and J. Gallacher (2008), The survival advantage of milk and dairy consumption: An overview of evidence from cohort studies of vascular diseases, diabetes and cancer. *J. Am. Coll. Nutr.* 27(6): 723S–734S.

Elwood, P. C., J. E. Pickering, D. I. Givens and J. E. Gallacher (2010), The consumption of milk and dairy foods and the incidence of vascular disease and diabetes: An overview of the evidence. *Lipids.* 45(10): 925–39.

Elwood, P. C., J. J. Strain, P. J. Robson, A. M. Fehily, J. Hughes, J. Pickering and A. Ness (2005), Milk consumption, stroke, and heart attack risk: Evidence from the Caerphilly cohort of older men. *J. Epidemiol. Community Health.* 59: 502–5.

Fromm, H. I. and K. J. Boor. (2004), Characterization of pasteurized fluid milk shelf-life attributes. *J. Food Sci.* 69(8): M207–14.

Gandy, A. L., M. W. Schilling, P. C. Coggins, C. H. White, Y. Yoon and V. V. Kamadia (2008), The effect of pasteurization temperature on consumer acceptability, sensory characteristics, volatile compound composition, and shelf-life of fluid milk. *J. Dairy Sci.* 91: 1769–77.

Go, A. S., D. Mozaffarian, V. L. Roger, E. J. Benjamin, J. D. Berry, W. B. Borden, D. M. Bravata, S. Dai, E. S. Ford, C. S. Fox, S. Franco, H. J. Fullerton, C. Gillespie, S. M. Hailpern, J. A. Heit, V. J. Howard, M. D. Huffman, B. M. Kissela, S. J. Kittner, D. T. Lackland, J. H. Lichtman, L. D. Lisabeth, D. Magid, G. M. Marcus, A. Marelli, D. B. Matchar, D. K. McGuire, E. R. Mohler, C. S. Moy, M. E. Mussolino, G. Nichol, N. P. Paynter, P. J. Schreiner, P. D. Sorlie, J. Stein, T. N. Turan, S. S. Virani, N. D. Wong, D. Woo, M. B. Turner and on behalf of the American Heart Association Statistics Committee and Stroke Statistics Subcommittee. (2013), Heart disease and stroke statistics – 2013 update: a report from the American Heart Association. *Circulation.* 127: e6–e245.

Granelli, K., P. Barrefors, L. Bjorck and L.-A. Appelqvist (1998), Further studies on lipid composition of bovine milk in relation to spontaneous oxidized flavour. *J. Sci. Food Agric.* 77: 161–71.

Havemose, M. S., M. R. Weisbjerg, W. L. P. Bredie and J. H. Nielsen (2004), Influence of feeding different types of roughage on the oxidative stability of milk. *Int. Dairy J.* 14: 563–70.

Havemose, M. S., M. R. Weisbjerg, W. L. P. Bredie, H. D. Poulsen and J. H. Nielsen (2006), Oxidative stability of milk induced by fatty acids, antioxidants, and copper derived from feed. *J. Dairy Sci.* 89: 1970–80.

Hedegaard, R. V., D. Kristensen, J. H. Nielsen, M. B. Frost, H. Ostdal, J. E. Hermansen, M. Kroger-Ohlsen and L. H. Skibsted (2006), Comparison of descriptive sensory analysis and chemical analysis for oxidative changes in milk. *J. Dairy Sci.* 89: 495–504.

Huncharek, M., J. Muscat and B. Kupelnick (2008), Impact of dairy products and dietary calcium on bone-mineral content in children: Results of a meta-analysis. *Bone.* 43: 312–21.

Ismail, B. and S. S. Nielsen (2010), Invited Review: Plasmin protease in milk: Current knowledge and relevance to dairy industry. *J. Dairy Sci.* 93: 4999–5009.

Ivy, R. A., M. L. Ranieri, N. H. Martin, H. C. den Bakker, B. M. Xavier, M. Wiedmann and K. J. Boor (2012), Identification and characterization of psychrotolerant sporeformers associated with fluid milk production and processing. *Appl. Environ. Microbiol.* 78(6): 1853–64.

Jayarao, B. M., S. R. Pillal, A. A. Sawant, D. R. Wolfgang and N. V. Hegde (2004), Guidelines for monitoring bulk tank milk somatic cell and bacterial counts. *J. Dairy Sci.* 87: 3561–73.

Juhlin, J., W. F. Fikse, I.-L. Orde-Ostrom, P. Barrefors and A. Lunden (2010), Factors relating to incidence of spontaneous oxidized flavor and copper in cow's milk. *Acta Agr. Scand. Section A.* 60: 94–103.

Kliem, K. E. and D. I. Givens (2011), Dairy products in the food chain: Their impact on health. *Annu. Rev. Food Sci. Technol.* 2: 21–36.

Lopez, A., W. F. Collins and H. L. Williams (1985), Essential elements, cadmium, and lead in raw and pasteurized cow and goat milk. *J. Dairy Sci.* 68: 1878–86.

Martin, N. H., M. L. Ranieri, S. C. Murphy, R. D. Ralyea, M. Wiedmann and K. J. Boor (2011), Results from raw milk microbiological tests do not predict the shelf-life performance of commercially pasteurized fluid milk. *J. Dairy Sci.* 94: 1211–22.

Martin, N. H., M. L. Ranieri, M. Wiedmann and K. J. Boor (2012), Reduction of pasteurization temperature leads to lower bacterial outgrowth in pasteurized fluid milk during refrigerated storage: A case study. *J. Dairy Sci.* 95: 471–5.

Metwalli, A. A. M., H. H. J. de Jongh and M. A. J. S. van Boekel (1998), Heat inactivation of bovine plamin. *Int. Dairy J.* 8: 47–56.

Miller, R. A., D. J. Kent, K. J. Boor, N. H. Martin and M. Wiedmann (2015), Different management practices are associated with mesophilic and thermophilic spore levels in bulk tank raw milk. *J. Dairy Sci.* 98: 4338–51.

Moschonis, G., I. Katsaroli, G. P. Lyritis and Y. Manios (2010), The effect of a 30-month dietary intervention on bond mineral density: The Postmenopausal health study. *Brit. J. Nutr.* 104: 100–7.

Mutukumira, A. N., J. A. Narvhus, K. Aaby, S. B. Feresu and R. K. Abrahamsen (2009), Characterization of a malty-compound producing Lacococcus lactis subsp lactis biovar. diacetylactis C1 strain isolated from naturally fermented milk. *Milchwissenschaft-Milk Sci. Int.* 64: 26–9.

Ranieri, M. L. and K. J. Boor (2009), Short communication: Bacterial ecology of high-temperature, short-time pasteurized milk processed in the United States. *J. Dairy Sci.* 92: 4833–40.

Richards, M., H. L. De Kock and E. M. Buys (2014), Multivariate accelerated shelf-life test of low fat UHT milk. *Int. Dairy J.* 36: 38–45.

Rockell, J. E., S. M. Williams, R. W. Taylor, A. M. Grant, I. E. Jones and A. Goulding (2005), Two-year changes in bone and body composition in young children with a history of prolonged milk avoidance. *Osteoporos. Int.* 16: 1016–23.

Siefarth, C. and A. Buettner (2014), The aroma of goat milk: seasonal effects and changes through heat treatment. *J. Ag. Food Chem.* 62: 11805–17.

Testroet, E. D., G. Li, D. C. Beitz and S. Clark (2015), Feeding dried distillers grains with solubles affects composition but not oxidative stability of milk. *J. Dairy Sci.* 98: 2908–19.

United States Department of Agriculture Agricultural Research Service (USDA) (2015), National Nutrient Database for Standard Reference Release 28. Available at: http://ndb.nal.usda.gov/ndb/foods/show/180?fgcd=&manu=&lfacet=&format=&count=&max=35&offset=&sort=&qlookup=milk%2C+whole (accessed on 31 December 2015).

U.S. Department of Health and Human Services, Public Health Service and Food and Drug Administration (USDHHS et al.) (2011), Grade 'A' Pasteurized Milk Ordinance. Available at: http://www.fda.gov/downloads/Food/GuidanceRegulation/UCM291757.pdf date (accessed on 15 February 2016).

Vazquez-Landaverde, P. A., J. A. Torres and M. C. Qian (2006), Quantification of trace volatile sulfur compounds in milk by solid-phase microextraction and gas chromatography-pulsed flame photometric detection. *J. Dairy Sci.* 89: 2919–27.

Webster, J. B., S. E. Duncan, J. E. Marcy and S. F. O'Keefe (2009), Controlling light oxidation flavor in milk by blocking riboflavin excitation wavelengths by interference. *J. Food Science*. 74(9): S390–8.

Wehr, M. H. and J. F. Frank (Eds) (2004), Standard Methods for the Examination of Dairy Products, 17th Edition. Washington, DC.: American Public Health Association, 570 p.

White, C. H. (1993), Rapid methods for estimation and predition of shelf-life of milk and dairy products. *J. Dairy Sci.* 76: 3126–32.

Yue, J., Y. Zheng, Z. M. Liu, Y. Deng, Y. F. Jing, Y. L. Luo, W. J. Yu and Y. Y. Zhao (2015), Characterization of volatile compounds in microfiltered pasteurized milk using solid-phase microextraction and GCxGC-TOFMS. *Int. J. Food Properties*. 18(10): 2193–212.

Part 2

Genetics, breeding and other factors affecting quality and sustainability

Chapter 6

Using genetic selection in the breeding of dairy cattle

Julius van der Werf, University of New England, Australia and Jennie Pryce, Department of Economic Development, Jobs, Transport and Resources (Government of Victoria) and La Trobe University, Australia

1 Introduction

Genetic change in dairy populations has been dramatic over the last five decades. Over the long term, selective breeding had been one of the most powerful tools to change the constitution and productivity of the dairy herd. Two main technologies have shaped dairy breeding programmes during the last five decades. Artificial insemination (AI) has been applied since the 1950s and had an important impact on the selection intensity of males and the dissemination of the best genetic material across the population. Genomic selection was introduced in the last decade and has been a second wave of influential technology that has affected the rates of gain in breeding programmes. Herd recording of an increasing number of dairy characteristics and genetic evaluation methods, notably best linear unbiased prediction (BLUP) (Henderson, 1973), have developed alongside these breeding and selection technologies and as such have become very powerful instruments in the genetic selection of dairy populations. AI, and

http://dx.doi.org/10.19103/AS.2016.0005.15

to a smaller extent the female reproductive technologies, have made dairy breeding a very dynamic activity where genetic material can easily be acquired or exchanged over long distances. This has made possible the large-scale introduction of Holstein Friesian genes into many dairy populations throughout the world a few decades ago, while there is currently still a large open world market for genetic material, which is supported by international genetic evaluation systems (Banos, 2010). With it, of course, has developed the threat that the genetic diversity of the world's dairy population has been narrowing to suboptimal levels.

The genetic change of dairy populations has largely been driven by the increased milk productivity per cow. The milk production per cow has more than doubled since the introduction of AI in the 1950s, with an annual increase of about 2%. In the United Kingdom, the average milk production per cow per year was 5151 kg in 1990, whereas it was 7899 kg in 2014 (AHDB Dairy), which is a 53% increase. In the United States, the production per cow per lactation increased by 72% between the years 1985 and 2015 (USDA, Fig. 1). The genetic trend between 1990 and 2000 was 1092 kg (VanRaden, 2004), implying that about 70% of the phenotypic increase was due to genetic selection. The emphasis on productivity increases have led to associated changes in other traits, notably a decrease of reproductive performance. Selection indexes have been augmented since the 1990s to also include more objective measurements of reproductive performance, health traits and efficiency and longevity.

This chapter will emphasize two main aspects of breeding programmes in dairy. First, we will discuss the development of structures of breeding programmes. Then we will discuss the main factors that drive genetic change, in particular, the dominating progeny testing schemes, and how these factors have changed over time with the introduction of new technologies. The second part of this chapter will focus on the evolution of breeding objectives and to what extent this has affected selection response for the various traits in dairy production. This section will include some principles about selection response, particularly how multiple traits may change depending on the breeding objectives used and the recorded information that is available about the various traits. This section is important because there is often a lack of understanding of how a change in selection emphasis can be quite different from a change in selection response. Many breeding programmes lack good phenotypic information about non-production traits and as a result the genetic change is still dominated by an increase in yield, in spite of an increased selection emphasis on other traits.

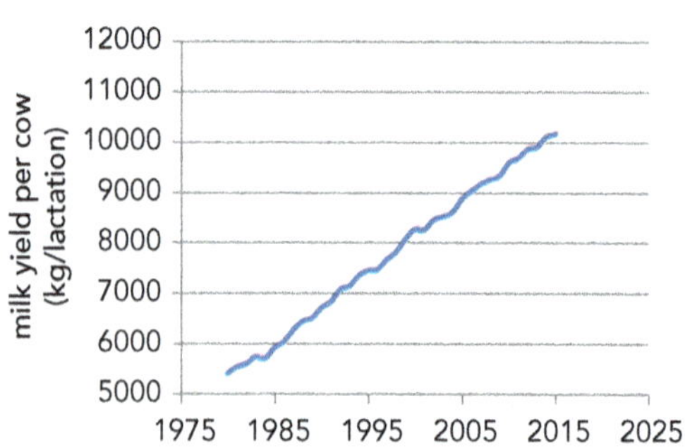

Figure 1 Milk production per cow in the United States (www.ers.usda.gov/datafiles).

2 Breeding programmes: AI, progeny testing, embryo transfer and in vitro fertilization

2.1 AI and progeny testing

An important starting point in a discussion of dairy breeding programmes is the paper by Rendel and Robertson (1950). They discuss the potential to achieve genetic change in breeding programmes, and question the progress that has been made, including the issues of separating progress due to selection and progress due to improved management. This paper also introduced the concept of selection pathways, arguing that genetic improvement could be predicted by adding the improvements from cow and bull selection and by distinguishing four different pathways of selection. On the basis of more intense selection, the very best bulls would be mated to the very best cows and from the offspring the males would be chosen to sire the herds. Cows could be selected on the basis of their milk performance, while bulls could be selected on the basis of the performance of their dam or progeny if they had some. Less intense selection could be practised to choose the young bulls and cows that produce the female replacements for the herd. This will lead to the well-known Rendel and Robertson formula, which states that the total genetic change is the sum of the selection differentials over the various pathways, divided by the sum of the generation intervals in each pathway. The selection differential depends on selection intensity, selection accuracy and the amount of genetic variation that is available. So the gain per year is

$$\delta G_{\text{year}} = \frac{\sum_{i=1}^{\text{nr\_of\_paths}} \text{intensity*accuracy*}\sigma_A}{\sum_{i=1}^{\text{nr\_of\_paths}} \text{generation\_interval}}$$

where σ_A is the additive genetic standard deviation of the trait (or aggregate of traits) under selection. Rendel and Robertson predicted a maximum genetic gain of 1% per year due to selection. A second paper from Robertson and Rendel (1950) focused on the progeny testing scheme. Progeny testing was not obviously a better scheme, as earlier pointed out by Dickerson and Hazel (1944), because relatively large coordinated breeding programmes were required to allow enough test matings of young bulls. With few test matings and larger generation intervals, progeny testing was unlikely to be competitive. However, Robertson and Rendel (1950) were inspired by the upcoming AI technology and considered larger breeding units with 2000 dairy cows, for example, from a number of farms.

2.2 Example: rate of genetic gain in the 4-pathway structure of a progeny testing scheme

This section discusses an example of a national breeding programme that can achieve a rate of genetic improvement of 1.25%.

Assumptions

- A commercial dairy cow population of 1 million animals
- Fifty breeding bulls needed per year
- Five hundred young bulls tested
- Two thousand elite cows mated, selected out of 300 000 herd-recorded cows (30% is assumed to be recorded and suitable as a bull dam)
- Five sires selected for elite matings
- Seventy per cent of female calves kept as herd replacements

Selection accuracies

Let 20% of the population be used for test matings, that is, 200 000 cows, giving 400 test matings per young bull, giving 100 daughters per tested sire completing a first lactation. Selection accuracy is 0.87 for males (based on 100 progeny, heritability = 0.25) and 0.50 for females (based on one's own performance).

Generation intervals

The average age of the parents when their progeny are born: cows 4.5 years, bulls 6.5 years.

Table 1 summarizes the key parameters needed to predict the gain per year.

The Rendel and Robertson (1950) formula for genetic gain in a 4-pathway breeding structure is

$$\delta G_{year} = \frac{\sum_{i=1}^{nr\_of\_paths} \text{intensity} * \text{accuracy} * \sigma_A}{\sum_{i=1}^{nr\_of\_paths} \text{generation\_interval}} = \frac{(i_{SS}.r_{SS} + i_{DS}.r_{DS} + i_{SD}.r_{SD} + i_{DD}.r_{DD}) * \sigma_A}{L_{SS} + L_{DS} + L_{SD} + L_{DD}}$$

$$= \frac{(2.65 * 0.87 + 2.79 * 0.5 + 1.76 * 0.87 + 0.47 * 0.5) * \sigma_A}{6.5 + 4.5 + 6.5 + 4.5} = \frac{5.47}{22}\sigma_{BO} = 0.25\sigma_A$$

Table 1 Genetic contribution for each of the four selection paths in a dairy cattle breeding programme

Selection path	Selection proportion	Selection intensity (i)	Selection accuracy (r)	Generation interval (L)	Selection differential (in σ_A)	% contribution to genetic gain
Sires for Sires	5/500	2.65	0.87	6.5	2.31	45
Dams for Sires	2000/300,000	2.79	0.50	4.5	1.40	26
Sires for Dams	50/500	1.76	0.87	6.5	1.53	28
Dams for Dams	70%	0.47	0.50	4.5	0.24	4
Total				22	5.47	

Hence, we may expect an annual genetic improvement equal to be one-quarter of a genetic standard deviation (σ_A). This is equal to about 1.25% of the mean (given that $\sigma_A = h^*\sigma_P$ and assuming a heritability (h^2) = 0.25 and σ_P (phenotypic standard deviation) is 10% of the mean.

The paper by Rendel and Robertson (1950) laid the basis for progeny testing schemes where they pointed out that the number of cows used for test matings and the number of young bulls to be progeny tested could be optimized. They predicted that a rate of genetic gain of 1.5% was possible. They also proposed to mate young bulls to 20 of their daughters to test the bull for carrying deleterious recessives. With the strong growth of AI in the dairy industry and consequently the larger breeding units, the progeny test programme became a dominating feature of dairy selection in the next 60 years.

2.3 Breeding programmes using embryo transfer and in vitro fertilization

A second revolutionary insight in dairy breeding programme design was introduced by the classic paper of Nicholas and Smith (1983), and this time it was motivated by the introduction of female reproductive technologies. On the basis of an earlier idea for beef cattle schemes (Land and Hill, 1975), they proposed a dairy breeding scheme that differed radically from the present classical progeny testing scheme. During the early 1980s, it became possible that females could produce a larger number of offspring via multiple ovulation and embryo transfer (MOET), and therefore, fewer females needed to be selected for breeding. It even became possible to harvest oocytes from juvenile females and create embryos in vitro, after which the embryos were implanted in recipient cows: juvenile in vitro fertilization and embryo transfer (JIVET). Initially, this was thought to be more suitable for beef cattle breeding where relevant traits can be measured in both sexes in an early stage of their life. MOET could be used for cows used in elite matings, but selection intensities in dairy cows were already high. For example, if 400 rather than 2000 cows were used in the DS path, the selection intensity would be 3.0 instead of 2.79 and the overall gain would only be 2% higher, which seemed not enough for a costly technology.

Nicholas and Smith (1983), however, sought to improve upon the classical progeny test scheme, and pointed out that MOET schemes produce full sib families and information from female sibs can be used to select bulls at a much younger age. As the genetic gain is a balance between selection accuracy and generation interval, such a scheme could give more gain per year, even though the accuracy of selection is much lower than in progeny testing schemes. In breeding schemes using JIVET, selection could be on the average breeding value of the parents and generation intervals could be as low as 15 months. Initial estimates of additional benefits were high with a predicted 50% increase in genetic gain per year. However, these predictions were too optimistic. First, it became clear that a selection strategy that relies heavily on family information was strongly affected by the reduction of genetic variance as a result of that selection. This is referred to as the Bulmer effect (Bulmer, 1971), which shows that in a typical breeding scheme the genetic variation would be reduced by about 25% as a result of selection. The variance among selected males is drastically reduced and the accuracy of selection based on information on paternal half siblings is easily halved when compared to the accuracy that does not take selection into account (Van Arendonk and Bijma, 2003). Second, as already pointed out by Nicholas and Smith (1983), the MOET and JIVET schemes were likely going to lead to more inbreeding and additional benefits of these schemes might be lower if inbreeding rates had to be constrained.

Many scientific studies appeared in the 1980s and 1990s on optimizing MOET breeding schemes, showing smaller gains, roughly between 10% and 25% (Nicholas, 1997). However, the higher inbreeding was still a problem, and when methods to constrain inbreeding in breeding programmes became available (Meuwissen, 1997), much of the predicted gains disappeared (Van Arendonk and Bijma, 2003). Nevertheless, the in vitro embryo production became widespread in the 1990s; in the year 2000, more than 100 000 embryos were transferred in dairy cattle in Europe (figure cited by Van Arendonk and Bijma, 2003). Many breeding companies started to use MOET to more efficiently use elite cows to generate young bulls for progeny testing. At that time, MOET breeding programmes did not replace progeny testing schemes, probably due to factors such as the variation in embryo yield (Nicholas, 1997), cost and the love affair of the dairy industry with progeny tested bulls. It was also clear that the AI companies were reluctant to market semen from bulls that had no progeny test information. A few companies used schemes where initial selection was based on the performance of siblings, although the key to widespread use was still large progeny groups (e.g. the MOET scheme operated by Genus in the United Kingdom (McGuirk, 1990)).

It could be said that these centralized nucleus programmes arrived too soon, as one aspect that might have made them more popular was the increased need for a more balanced breeding programme, where selection based on production traits was gradually replaced by a wider breeding objective where functional traits such as health and fertility became more important. Selection for functional traits requires a large amount of well-recorded phenotypic data. In fact, there has been limited success in many countries in developing breeding values, for health traits in particular, because of scarcity of data in progeny test herds. Often these data are only available in a subset of herds that have a particular interest in data recording. It was not until around 2008 that there was a renewed interest due to the potential created by genomic selection that can make use of these data-rich herds.

3 The structure of dairy breeding programmes

At this stage, it is useful to consider the structure of dairy breeding programmes, as this is affected by reproductive technologies. The structure of breeding programmes is usually described as a pyramid, with a breeding nucleus at the top (Fig. 2). In the nucleus, selection is made on the basis of investment in the measurement of phenotype, pedigree and now genomic testing. The genetic mean of the nucleus improves continuously due to this selection process. Animals born in the nucleus, but not used as parents, can be transferred to lower tiers to ultimately disseminate the genetic improvement to commercial producers. In pig and poultry breeding schemes, the breeding programmes have a distinct tier structure, with a closed centralized nucleus, and often one or two multiplier tiers. The number of animals in the nucleus is low relative to the number of commercial animals they ultimately affect. This is due to the higher fecundity of these species. As a result, nucleus breeding programmes for pigs and poultry are run by a few large breeding companies that dominate the world market. Furthermore, pig and poultry breeding programmes use several breeding lines and ultimately sell a crossbred animal. By contrast, breeding programmes for cattle and sheep have a much less distinct structure.

The dairy breeding programme is characterized by having an open dispersed nucleus, consisting of elite cows that are mated to elite bulls. These elite matings result in young bulls and ultimately tested bulls that are sold to commercial farmers. The nucleus dispersed as elite cows could be located at many different dairy herds, often mixed with commercial dairy cows. The female offspring of elite matings often are candidates for elite matings themselves, but the elite commercial cows are also candidates, hence the term 'open' nucleus. The bulls are usually owned by the AI companies. At the time when embryo technologies became commercialized, a number of breeding companies also started to own females, and in essence created a centralized nucleus. There were two main reasons for this development: the first is that the logistics of running an efficient MOET programme was easily implemented when the donor cows were physically together in one place. The other reason was that a centralized nucleus allowed centralized testing of females. This was an advantage over selection of the elite cows from the herd recording schemes, as there was a perception that the breeding value of these elite cows was usually overpredicted. This was evident when the parent average estimated breeding value (EBV) of young bulls usually dropped once a progeny EBV based on their progeny test became available. In addition, it was easier to use a centralized nucleus to record traits that were hard to measure in commercial herds, for example, feed efficiency.

As with female reproductive technologies, the number of females needed in a nucleus is much smaller and therefore the technology had the potential to drive dairy breeding programmes more towards the centralized and potentially closed nucleus programmes such as in poultry and pigs. In spite of these arguments, establishing commercial dairy nucleus herds does not seem to be sustainable; some were only temporarily profitable, as they were funded out of additional embryo sales to commercial farmers or large semen exports outside a targeted commercial population. In essence, a specialized dairy breeding nucleus is likely to be too expensive, although we are not aware of any published cost–benefit studies. Important factors are likely to include the low reproductive rate of females and the cost of increasing this with technology. The main way of disseminating the genetic superiority created is, therefore, through the sale of semen. However, the margins in the semen market are small, whereas the price elasticity may be high. In other

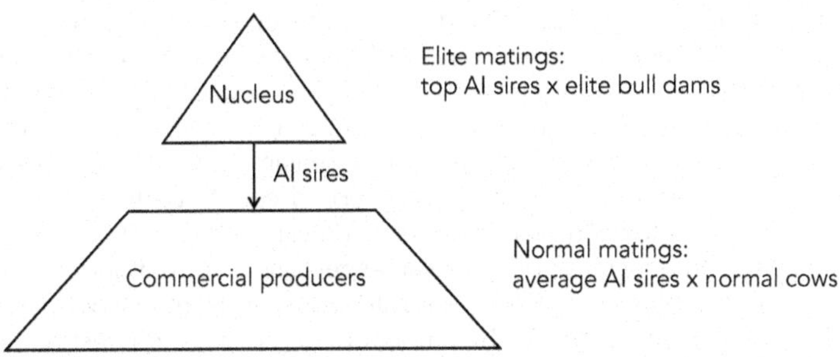

Figure 2 The two-tier breeding structure.

words, running a complete centralized dairy nucleus breeding programme may simply be too expensive.

Nicholas (1997) pointed out that there could be a role for MOET nucleus herds, even though the more developed dairy breeding programmes seemed to hang on to their progeny testing paradigm. The higher reproductive rate allows managing a smaller size nucleus, which is easier to maintain from a commercial point of view. If there were many of such breeding units, he thought that could circumvent the inbreeding problem, and he suggested that commercial animals could be produced by crossing animals from different breeding schemes. Nicholas (1997) also referred to the finding by Smith (1988), who pointed out that such nucleus herds might be very suitable for breeding programmes in developing countries as it could be a way to focus on the breeding programme investments more efficiently because the nucleus is relatively small and centralized.

Bichard (2002) presented an 'outsider view' on dairy breeding programme structures and suggested that there might be an overemphasis on the role of progeny testing in dairy breeding programmes. He argued that as a result of a push for very high sire proofs, it has become more difficult to test young bulls, as farmers are used to highly accurate figures for EBV. The large number of progeny that needed to be recorded per sire also made it harder to have accurate sire proofs for other traits than milk production as these other traits are usually less easy to record. Finally, he observed that progeny testing schemes have resulted in large data-recording operations requiring complex statistical analysis to provide accurate data and that much of the intellectual power devoted to dairy genetics has been used to further develop such genetic evaluation systems, rather than developing a broader view on breeding programme alternatives, for example, where more traits are measured on fewer animals. Milk production data are used primarily for reasons other than genetic improvement, for example, for making good management decisions, and recorded in much larger quantities than what would be needed for a genetic improvement programme to select superior bulls. Whereas in most countries there has always been a lack of good recorded data for fertility and health traits, it is possible that in the current era of genomic selection, there is an opportunity to redress this imbalance, with most traits of economic value recorded in specialized resource herds.

Another intriguing contrast between genetic improvement programmes in dairy and other livestock species is the apparent lack of crossbreeding in dairy breeding schemes. One exception is New Zealand, where over half of all dairy replacements are crossbred. The reason for the popularity of crossbreeding in New Zealand is likely to be the importance given to reproductive performance in pasture-based systems in order to match pasture availability to lactation requirements. The crisis in poor reproductive performance in dairy cows (especially on pasture) has generated a lot of interest in crossbreeding, especially in the United States, Ireland and New Zealand. Crossbreds have consistently higher reproductive performance than their purebred counterparts and more profitable performance because of lower replacement rates (Buckley et al., 2014). Crossbreeding on dairy farms may become increasingly popular, especially in current and likely future dairy markets where the global price of milk is, and will be, low and farmers look at opportunities to reduce the cost of production. As the amount of heterosis is greatest in the first cross, most of the benefits of crossbreeding are realized in the first cross. However, rotational crosses (of two or three breeds) are also becoming popular, with evidence to suggest that these are more profitable than straightbred herds in New Zealand (Lopez-Villalobos et al., 2000).

4 The exchange and selection of genetic material

4.1 Exchange of genetic material

Genetic material (bull semen, embryos and even animals) have been traded internationally for around 60 years. Breeding companies are also multinational and their focus is to provide farmers worldwide with the best bulls from around the world. National genetic evaluations have been around for almost as long as genetic material has been exchanged nationally and internationally. To help breeders and farmers compare bulls from different countries, Interbull was established in 1983 to support international genetic evaluations (Banos, 2010).

4.2 Selection on merit versus genetic diversity

On account of the internationalization of the bull semen market and across-country genetic evaluation, it became more common that worldwide only sons of the very best bulls were used. Therefore, the genetic basis of dairy breeding populations became rapidly smaller, especially in the Holstein Friesian breed, with estimates of the effective population size of the Holstein breed being less than 100 (McParland et al., 2007). Another way of looking at the diversity problem was that more than 80% of the young bulls tested in 2000 were grandsons of only five influential sires. Fortunately, selection theory was developed in the 1990s allowing 'optimal contribution selection' (Wray and Goddard, 1994; Meuwissen, 1997), where selection was aimed at not only maximizing genetic merit, but also allowing for a sufficient level of genetic diversity. This strategy allows controlling diversity by constraining the average co-ancestry of selected parents, which is a prediction of half the rate of inbreeding. Meuwissen showed that about 60% more genetic gain could be achieved at the same rate of inbreeding with optimal contribution selection, compared to truncation selection on merit and imposing 'ad hoc' selection rules to control inbreeding. Although the theory for optimal selection exists, selection is rarely optimized at the population level, as most selection decisions are made by individual breeding organizations that compete with each other in the dairy bull semen market. However, for individual breeding programmes, there is an incentive to consider diversity, as it is synonymous with risk.

5 Genomic selection

In the 1960s, it was noted that selection could possibly be based on 'genetic markers' in the form of blood groups (Neimann-Sorensen and Robertson, 1961). When the number of DNA markers increased rapidly due to the development of molecular biology, the prospect of marker-assisted selection seemed to become a realistic addition to the breeding programme, and early papers showed how genetic markers could be incorporated in genetic evaluation (Fernando and Grossman, 1989) and into breeding programmes (Lande and Thompson, 1990). Selection would be based on information from genetic markers that are in linkage disequilibrium with genomic regions with a large effect on quantitative trait, the so-called quantitative trait loci (QTL) (Meuwissen and Goddard, 1996). Although initial estimates of QTL effects seemed promising (e.g. Georges et al., 1995), it became clear that QTL effects were often overestimated, and the number of QTL with a large effect that could be useful in marker-assisted selection appeared to be disappointingly low. The largest QTL effect found in dairy cattle is the DGAT1 mutation (Grisart et al., 2002) with a major effect on milk fat and other milk constituents. For example, the DGAT1 allele that encodes lysine at position 232 is associated more with fat, implying that selection on this gene can alter fat composition of cows' milk (Schennink et al., 2007). However, DGAT1 is usually not a direct selection target because it is more effective to select for breeding values for traits predicted from multiple genetic markers simultaneously, that is, selecting for breeding values for fat and protein yield as target traits.

Since the turn of the century, many studies based on increasingly dense marker panels have revealed that most of the observed genetic variation on most quantitative traits is due to a large number of genes, each with a small effect. This probably explains why the impact of identified QTL in marker-assisted breeding programmes has been small to negligible. Instead, another approach to use marker information has started to revolutionize breeding programmes. Meuwissen et al. (2001) proposed to use all marker information across the whole genome in a single analysis to predict breeding values. They showed in a simulation study that the reliability of a breeding value could be as high as 64% when using a large training population and dense genetic markers. It was not until about five years later that the first single nucleotide polymorphism (SNP)-chip was released by Affymetrix as a tool to genotype individuals for about 10000 genetic markers. Soon followed the first Illumina SNP-chip that contained ~50k markers, and by 2009, Illumina released the high-density cattle chip with ~800k markers. A growing area is the use of low-density SNP panels (<10k), which are especially popular for screening large numbers of young bulls and dairy cows (Fig. 3A). It is noteworthy that genotyping of cows is becoming particularly popular (Fig. 3A), as farmers begin to use genomic breeding values for management decisions, such as which animals to select as herd replacements. Generally, the low-density genotypes are imputed to 50k to calculate genomic breeding values. However, customized chips that do not require imputation are also being used on a large scale. For example, genotyping is mostly complete for around a million cattle in Ireland, with completion anticipated by the end of this year through the use of a customized SNP panel of the 40k that is currently in vogue in dairy genomic selection, and this technique obviates the need for imputation (Berry, 2016 personal communication).

Many dairy bulls have been genotyped, as illustrated in Fig. 3A, and their EBVs served as the phenotypic information that is required for genomic predictions. By 2008, it had become clear that the breeding value of young bulls could be predicted with a reliability of at least 50% for most milk production characteristics. This was a big improvement over the reliability that can be achieved from an EBV based on the mean of the parents (about 25%, or less due to the effect of selection), although it was still short of what is typically achieved in a first-proof of around 50 daughters (80%). But because the genomic prediction could be made available at an early age, and selection of bulls at an early age could now be based on an EBV with reasonable prediction accuracy, this could have a large impact on dairy breeding programmes. Early modelling studies showed that this had the potential to double rates of genetic gain (Schaeffer, 2006). Schaeffer (2006) also pointed out that genomic selection would be attractive for the AI companies simply from a cost savings perspective, and he showed they could save about 95% of the cost of a progeny test scheme. Hence, there were two strong arguments to quickly adopt the technology. It seems that after 60 years, the classic progeny testing schemes proposed by Rendel and Robertson have finally been overtaken by the paradigm of genomic selection.

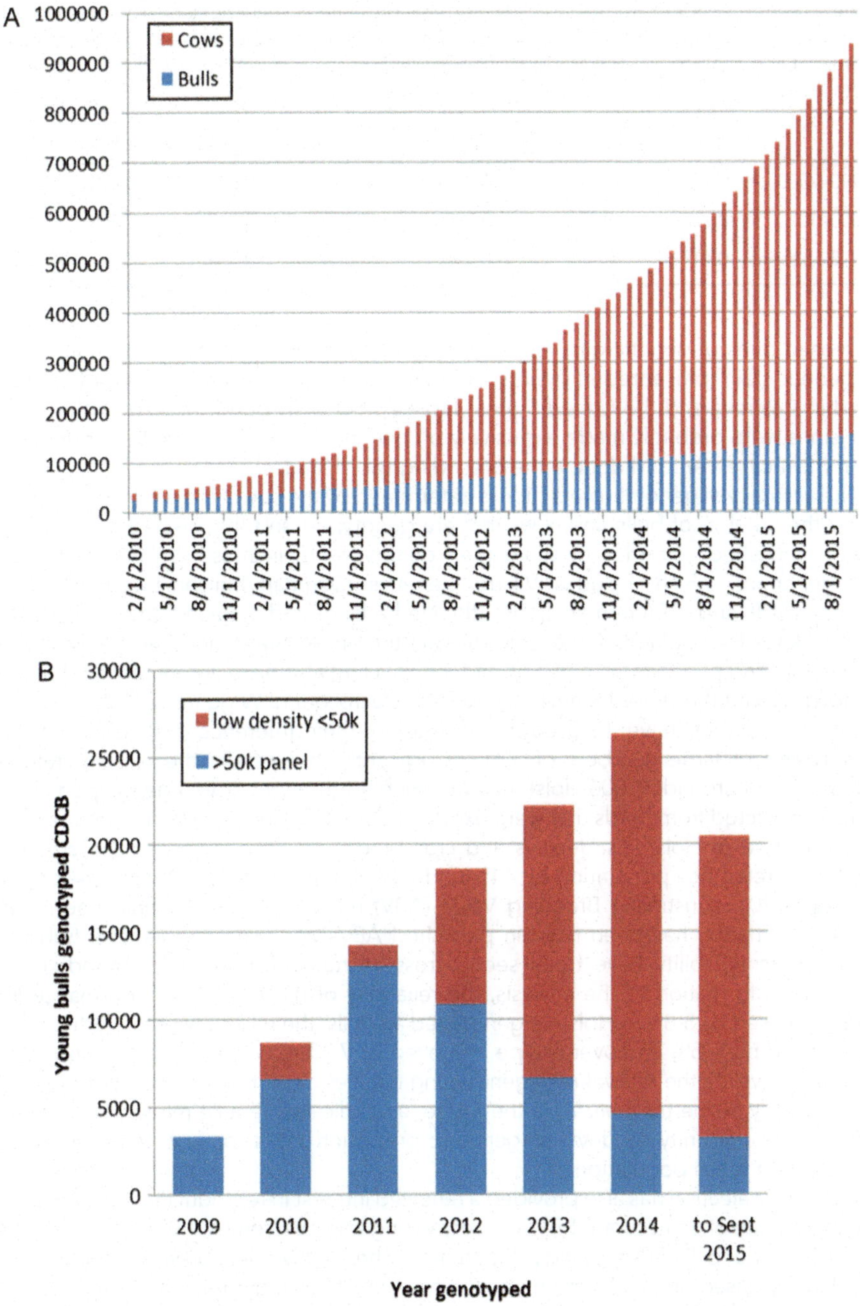

Figure 3 Genotypes included in Council On Dairy Cattle Breeding (USA) Holstein evaluations by (A) sex and evaluation date and (B) SNP panels used for genotyping of young Holstein bulls (unproven) (accessed November 2015; source: https://www.cdcb.us/Genotype/cur_density.html).

Table 2 Genetic trends of Holstein bulls by birth year presented in genetic standard deviations for the following time intervals: 1990–2000, 2000–10, 2010–present; for all available bulls with progeny in the country of origin (Australia – BPI, HWI, TWI (n = 7412); Canada – USA – TPI (n = 912), NM (n = 151, 246); LPI (n = 7912) (accessed June 2015)

	BPI (AUS)	HWI (AUS)	TWI (AUS)	TPI (USA)	NM (USA)	LPI (CAN)
1990–2000	0.198	0.172	0.204	0.290	0.252	0.223
2000–10	0.223	0.220	0.241	0.300	0.319	0.320
2010–present	0.418	0.435	0.477	0.678	0.404	Data only till 2010

Increased rates of genetic gain are beginning to be realized, as can be seen in the rates of genetic gain (presented in genetic standard deviations) for national selection indices used in Australia and the United States (Table 2). For bulls born since 2010, that is, after decisions were probably made exclusively on genomic selection, the rate of genetic gain is, in most cases, around double the preceding two decades.

The accuracy of genomic prediction is determined by the size of the reference population or by the number of bulls and cows that are genotyped and that also have phenotypes. This has pushed countries to share their genotypes to increase the numbers. One is Eurogenomics, which includes Viking Genetics (Denmark/Finland/Sweden), UNCEIA (France), DHF and VIT (Germany), CRV (the Netherlands and Belgium), Conafe (Spain) and Genomika Polska (Poland). Other countries are the United States and Canada, which share genotypes and phenotypes of males and females and also have agreement to share male genotypes with the United Kingdom and Italy (Ducrocq and Wiggans, 2014).

Other countries have focused on increasing their reference populations through genotyping of large numbers of females. Recently in Australia the existing reference population of around 15000 Holsteins was doubled through adding genotyped Holstein females selected from herds that were excellent data recorders, known as Ginfo. The cows are a mixture of Holsteins, Jerseys and crossbreds and have increased the size of the Australian reference population by 44% and 38% for Holsteins and Jerseys, respectively. The April 2016 Australian Breeding Value (ABV) release will be the first time in which Ginfo information has been used in published ABVs and breeding indices. Substantial increases in reliability have been seen across all traits. For example, by adding only genotype information to the analysis, the reliability of the balanced performance index (BPI) increased by 5.8% in Holstein genotyped animals; the reliability of the fertility ABV(g)s improved by 4–5% and overall type improved by 7.1% (Pryce et al., 2016; unpublished results). However, the real value of genotyping females as part of the reference population will probably be better realized in the future, as dedicated female reference populations offer the opportunity to develop genomic prediction tools for new traits that can be measured in these populations.

Genomic selection has also provided a renewed interest in reproductive technologies, as the availability of an accurate EBV at a very young age overcomes the potential weakness of the MOET and JIVET breeding programmes, where early selection was not accurate and was largely based on family information. It has been shown by simulation and in real data that genomic selection leads to lower levels of inbreeding (Daetwyler et al., 2007; Clark et al., 2013). Granleese et al. (2015) showed that in an optimized dairy breeding nucleus applying MOET and JIVET with optimal contribution selection the genetic gain could be

81% higher under a constrained inbreeding rate of 1% per generation, compared to a genomic selection programme without female reproductive technologies. The allocation of matings to MOET and JIVET was about 28% and 34%, respectively, with the rest being normal AI matings. With the increased use of embryo technologies in nucleus breeding programmes, other technologies such as cloning of cell lines and gene editing can more easily be implemented, and if accepted by regulators and the society at large, these technologies could soon become a commercial reality in dairy breeding programmes.

6 Multi-trait selection

6.1 Choice of selection index

One of the most disputed topics in the science of dairy breeding programmes is the choice of the selection index. Since the genetic change in the population is driven by the choice of traits considered important in the selection of animals, there is no doubt that the correct set of traits, along with their relative value, is of critical importance in breeding programmes. Formally, the derivation of a selection index starts with the definition of the breeding objective. The breeding objective, or the breeding goal, is represented as the sum of the breeding value for each trait (g_i) multiplied by a relative economic value v_i

Breeding objective: $H = v_1g_1 + v_2g_2 + \ldots\ldots + v_mg_m$

This definition assumes already that the objective is defined in economic terms. The economic value is derived from regressing profit on the breeding value of each trait, or, similarly, as the partial derivative of the profit function towards the breeding value of a trait. The extensive literature about the methodology and issues related to the derivation of economic values is well summarized by Goddard (1998). An important principle is that all traits that have economic value should be included in the objective, even if they are not measured. An example of a trait with a (negative) economic value is feed intake. If feed intake was ignored, we would overestimate the economic value of improving milk yield. If feed intake was never measured, then instead of including it in the breeding objective, it could also be accounted for by adjusting the economic value for other traits with a genetic regression on feed intake. In other words, the economic value for milk production would be the extra revenue for the extra milk produced minus the extra feed that will be required to produce that extra milk.

The purpose of the selection index method is to combine information from different sources such that an optimal selection criterion is achieved. Information sources can be different measurements on different traits on an animal or measurements on related animals. Selection indexes are constructed on the basis of the set of traits defined in the breeding objective. Index traits are those for which trait recordings exist, that is, for which EBVs can be predicted. These are selection criteria traits. There is usually a huge overlap between selection criteria traits and breeding objective traits, as after all, if a trait has to be improved, it is best to measure it. However, some objective traits may be hard to measure, for example, feed intake or resistance to mastitis. Selection criteria could be added if they are easier to measure, and if they are correlated to objective traits. For example, a somatic cell count in milk predicts mastitis resistance and body weight could predict feed intake.

Combining measurement information on related animals on a single trait is exactly what is done by BLUP, which is the standard method to predict breeding values. A selection index for m traits can be written as

$$\text{Index} = b_1\text{EBV}_1 + b_2\text{EBV}_2 + b_3\text{EBV}_3 + \ldots\ldots + b_m\text{EBV}_m$$

where the b-values are the index weights. These weights can be derived as optimal weights, where 'optimal' is defined as 'maximizing the correlation between index and aggregate true genotype'. Selecting on this index gives the highest selection overall response in true genetic merit. Since the breeding objective is an aggregate genotype, one can interpret this as ranking for the highest profit.

A proper derivation of index weights can be achieved on the basis of selection index theory, where the weights are calculated as $b = P^{21}Gv$, where P is a covariance matrix among the EBVs in the index and G is a matrix with covariances between these EBVs and the trait in the aggregate genotypes. If EBVs are the results of multiple trait BLUP evaluation, then P and G are identical; hence, b = v and the index weights would be the same as the economic values. We could then include EBVs of traits that are never recorded, because the multiple trait BLUP evaluation would predict such EBVs from the correlations with recorded traits. Traits that are only selection criteria and have no economic value would have a weight of zero and could be left out of the index. However, because of the size of the data sets analysed in dairy, it is rare to have a multiple trait analysis of all relevant traits.

When using single-trait EBVs in an index and using economic values as index weights, the value of responses in correlated traits are not fully accounted for. A selection index approach would derive index weights for a single-trait EBV that combines the economic value of the trait with the value of correlated responses in other traits, if those traits have incomplete information about their true merit (i.e. EBV reliability less than 1). For example, if the accuracy of the fertility ABV is only 40%, then an index weight for production traits would be reduced somewhat to account for a negative response in fertility when selecting for production, because the information on fertility is insufficient to account for that. The adjustment would be smaller if the reliability of the fertility EBV would be higher. This becomes clear in an extreme case where an objective trait has a zero accuracy from single-trait evaluation because it was not measured. The index weight for the EBV of other traits is then adjusted by adding the value of the correlated response in feed intake, calculated as the genetic regression of feed intake on the other traits multiplied by the economic value of feed intake.

For some traits, index weights could deviate substantially from their economic value and this difference is not always easy to explain as these weights depend on a complex multivariate correlation structure. Furthermore, optimal weights can vary between bulls as they depend on the EBV reliabilities. Therefore, it is not straightforward how to implement correct weights in a selection index based on single-trait EBVs. For older bulls with higher reliability for most traits, the optimal weights for the index traits would be closer to their economic values.

6.2 The selection emphasis in dairy indexes

Dairy breeding programmes are mainly driven by a huge emphasis on 'bulls proofs', that is, the EBVs of the bulls that are marketed by the AI companies. Until about 1990, the selection emphasis was almost exclusively on milk production traits: (i) milk yield, (ii) milk fat and (iii) milk protein (summarized by VanRaden, 2004). There were debates about whether to include yield or percentage traits for the milk components, and how to adjust indexes for the quota systems that were introduced in Europe in 1984. However, very few non-production traits were included in a selection index. At the World Congress on Genetics Applied to Livestock Production (WCGALP) in Edinburgh (1990), a workshop was held on dairy breeding programmes, and in this workshop only one paper was devoted to fertility traits and one to feed efficiency. By contrast, the WCGALP in 2014 had two complete sessions devoted to reproduction and two sessions to feed conversion efficiency. In the decade that followed the 1990 congress, research on widening the breeding goals flourished, and not only fertility, but also udder health, other health traits and longevity were the topics of many studies. There was also an attempt to include the type traits in a more rational breeding goal by introducing the term 'functional type traits'. This refers to those type traits that could be predictors of functional breeding goal traits such as health and longevity (e.g. Rogers and McDaniel, 1989; Short et al., 1992; Dekkers et al., 1994). Dekkers and Gibson (1998) gave a comprehensive overview of the challenges associated with constructing selection indexes for dairy cattle, and how to communicate those to breeders that use them. They argued it was more important to focus on individual trait responses when selecting on the index, rather than on index weights for individual traits.

Breeding objectives are now becoming more complex in order to meet challenges set by consumers and society (Boichard and Brochard, 2012; Martin-Collado et al., 2015). For example, the growing human population places more pressure on limited resources and global changes may mean hotter, drier conditions to manage livestock; there is also increased consumer awareness of animal welfare and farming conditions. So, future agriculture needs to be more sustainable so that economic, societal and environmental requirements underpin future breeding goals (Boichard and Brochard, 2012). However, over the last couple of decades, we have already seen a rapid evolution in the number of traits that are available for farmers to select on. Almost without exception these breeding values rely on large amounts of field data that are freely available through current recording systems. However, not all important traits are well recorded and we are likely to see increased use of data (to estimate breeding values) that originate from research herds or commercial herds with much more in-depth phenotyping than has been possible before. In the next section, we discuss current and potential breeding objectives in more detail.

7 Breeding objectives

7.1 Current breeding objectives

7.1.1 Overview of current objectives

Common current breeding objectives in dairy cattle include: milk production traits, conformation, mastitis resistance through somatic cell count, longevity, calving ease, female fertility and workability traits. In fact, Interbull calculates breeding values for these traits using multi-trait across country evaluations (MACE). Currently, 30 countries participate in MACE for milk production traits in Holsteins (Table 3), with 14 for Red Dairy Cattle (RDC), 11 for Jersey and Brown Swiss and 6 for Guernsey. There are almost as many countries participating in MACE for somatic cell count and some of these countries

Table 3 Number of countries participating in Interbull evaluations (December 2015; http://www.interbull.org/ib/maceev_archive)

Trait categories	Brown Swiss	Guernsey	Holstein	Jersey	Red Dairy Cattle
Milk production	11	6	30	11	14
Conformation‡	9	4	24	9	9
SCC/Mastitis‡	10	6	29	8	13
Longevity‡	10	6	20	9	10
Direct calving ease*	6		15		7
Female fertility‡	9	6	19	9	11
Milking speed	7		10	5	6
Temperament			9		6

*Direct and maternal; ‡The trait with the largest number of countries included in evaluations is presented.

also provide breeding values for clinical mastitis. Traits such as female fertility and the workability traits are reasonably recent additions to Interbull evaluations, being included in MACE from 2007 to 2009, respectively. The number of traits that are actually recorded and evaluated is even larger (30–40) (Banos, 2010); for example, there are 21 conformation traits that are evaluated by Interbull.

7.1.2 Fertility

Fertility is a trait of great importance in dairy cattle breeding as lactation is dependent on parturition (i.e. reproduction). It is also a good example of the unintended consequences of selection, as dramatic reductions in phenotypic and genetic fertility performance have been universally documented (Pryce et al., 2014).

Heritability estimates of traditional fertility traits are generally low (<0.1); yet selection for fertility can lead to worthwhile changes, because the trait is highly variable (Pryce and Veerkamp, 2001; Berry et al., 2014). However, most genetic correlations between milk production and fertility traits are antagonistic (Berry et al., 2014). Therefore, an inevitable consequence of selection for production (while ignoring all other traits) has been the unfavourable genetic trend in fertility through selection for milk production traits (as illustrated in Table 4).

Although selection for fertility, health, longevity, etc. has been practised for over 30 years in Nordic countries (Heringstad and Østerås, 2013), for most countries, it was only from the 1990s that serious attention was paid to selection to improve fertility (Weigel, 2006). Through the research conducted during this time, it became clear that the observed deterioration in fertility was partly genetic, therefore requiring a genetic solution. Consequently, most breeding programmes have gradually expanded to include fertility traits, with fertility being evaluated in many countries (Table 3).

The most popular measures of fertility is the non-return rate (the proportion of cows not coming back into oestrus within a specified period after breeding, and are thus considered to be pregnant), for example, the 56-day non-return rate, although some, such as the Walloon region of Belgium, use a 90-day non-return rate. A number of countries have developed methodologies to improve the accuracy of selection for fertility by including various predictors of fertility. For example, the UK model includes days to first service, the interval from calving to conception and two measures of the cow's ability to conceive (Wall et al., 2003). In Germany, five fertility traits are used, being (i) interval from first to successful insemination of heifers, (ii) non-return rate to 56 days of heifers, (iii) interval from calving to first insemination of cows, (iv) non-return rate to 56 days of cows and (v) interval first to successful insemination of cows (Liu et al., 2008).

There is variation in the relative importance of fertility in national breeding objectives. In some countries, it is economically more valuable to produce extra milk rather than improve fertility, while in others there is a strong link between pasture utilization and lactation profiles (Pryce et al., 2014b). The weight applied to fertility in a selection index will also affect the response to fertility achieved through selection. Despite a considerable amount of effort invested into improving the accuracy of genetic evaluations for fertility and the fact that the weightings assigned to fertility in breeding objectives have increased in many countries, progress using conventional selection criteria is often still limited due to insufficient fertility data, especially mating and pregnancy records (Sun and Su, 2010). Furthermore, there is evidence that genotype × environment (GxE) interactions exist, which may limit the utility of fertility proofs derived in other countries, especially where the definitions of fertility between countries are very different (Pryce et al., 2014b).

The importance of data can be shown in the following simple selection index example. Consider selection of sires based on a progeny test for milk yield (kg/lact) and fertility (%NR) with phenotypic standard deviations being 800 kg and 46%, respectively, and heritabilities 0.30 and 0.03. Phenotypic and genetic correlations between the traits are −0.1 and −0.25, respectively. Table 4 shows predicted selection responses with varying economic values and varying number of progeny per sire for fertility. The example shows that simply increasing the economic weight for fertility does not necessarily alter the response between milk yield and fertility. In fact, there is no change in response if fertility is not recorded. Unless an increased economic weight is combined with enough data on

Table 4 Selection response for milk yield and fertility (non-return rate) in a simple two-trait selection model, with varying economic weights and number of recorded progeny per sire for each trait. Phenotypic standard deviations are 800 kg for milk yield (kg/lact) and 46% for fertility and heritabilities 0.30 and 0.03, respectively. Phenotypic and genetic correlations between the traits are assumed to be −0.1 and −0.25, respectively

Economic weights		Number of progeny measured (per sire)		Selection response (4 years)	
Milk ($/kg)	Fertility ($/%)	Milk	Fertility	Milk (kg/lact)	Fertility (%NR)
0.2	0	50	–	392	−1.78
0.2	0	50	50	392	−1.75
0.2	3	50	–	392	−1.78
0.2	3	50	50	387	−1.09
0.2	8	50	–	392	−1.78
0.2	8	50	10	381	−1.25
0.2	8	50	50	352	0.17

fertility, the response for fertility is expected to decline. In the era of genomic selection, it has become easier to target resource herds to obtain information on non-production traits that are harder to obtain in traditional milk recording schemes. A more detailed discussion on balancing selection for production and fertility traits in breeding programmes that use genomic selection can be found in the study by Berry et al. (2014).

To conclude this section, fertility is one of the most important traits in dairy cattle breeding and the remarkable decline in fertility observed through the 1990s and part of the 2000s is partly attributable to narrow selection goals that focused on selection for milk production traits. VanRaden (2004) reported that 40% of the phenotypic decline observed in the United States was due to genetics. Comparing the genetic and phenotypic trend in Ireland between 1980 and 2010 reported by Berry et al. (2014), 64% of the phenotypic decline in calving interval was due to genetics. The introduction of fertility breeding values is leading to improvements in fertility. Although there are signs that selection to improve this trait has worked, phenotypic trends in fertility gathered over the next decade will be very valuable in assessing the success of selecting for fertility as part of modern breeding goals. However, as a final note to this section, valuable lessons have been learnt by dairy geneticists and sire analysts on the dangers of narrow breeding goals. In fact, farmers are increasingly demanding selection indexes where additional weight is placed on functional traits (Martin-Collado et al., 2015). In addition to sustaining selection on fertility, welfare and disease resistance traits in particular are becoming key areas where breeding values are being developed for future breeding goals.

7.2 New traits

Egger-Danner et al. (2015) recently reviewed opportunities to genetically improve traits associated with functionality of dairy cows. They identified health traits, traits associated with efficiency and resilience to changing environments as being key areas to focus future selection. They emphasized that the success of future breeding strategies relies on balancing the effort required to record data with the realized benefits. Another review by Chesnais et al. (2015) explored opportunities to improve udder health, hoof health, other health traits, feed efficiency and methane emissions in North America and focused on options to improve rates of genetic gain through different genomic selection strategies. As part of the last National Breeding Objective review in Australia, farmer preferences for improvements in key traits were assessed (Martin-Collado et al., 2015) and the ranking was as follows: (1) mastitis; (2) longevity; (3) fertility; (4) mammary system; (5) lameness; (6) protein yield; (7) type; (8) feed efficiency; (9) calving ease; (10) temperament; (11) lactation persistence and (12) live weight. Clearly, functional traits, such as mastitis resistance, longevity, fertility and lameness, are traits that farmers would like to select. Results of this survey showed that improving production is an area that can be managed easily through improved feeding and other practices, while it is more difficult to improve health and fertility. Furthermore, there is a level of satisfaction with levels of milk production, while farmers have been especially disappointed with fertility and health in their herds. Currently, milk prices are very low, which may lead to decisions being made on the basis of semen price instead of future benefits from multi-trait selection on economic indexes. While these decisions may make sense in the short term, genetics should be viewed as a long-term investment, with the benefits of good selection decisions being accrued in the future.

7.2.1 Health traits

In 1988, the first major review of data recording opportunities and consequently breeding strategies to improve production diseases was published (Emanuelson, 1988). However, quite a lot has changed since then. Notably, computerized farm recording has led to a large increase in data available on these traits and consequently studies on genetic parameter and breeding value estimation. Consequently, several countries around the world have implemented routine genetic evaluations of health traits using (predominantly) farm-recorded data (Egger-Danner et al., 2015).

Mastitis

One of the most important diseases in dairy production is mastitis. Selection to improve this trait is already being practised in many countries (Table 2) through selection for reduced somatic cell count (SCC). Cell count can be quantified from routinely assessed milk samples and has a genetic correlation of around 0.7 with mastitis (Mrode and Swanson, 1996). However, several studies have shown that direct selection for mastitis is more effective than reliance on predictor traits (such as SCC) (Heringstad et al., 2007; Egger-Danner et al., 2012; Parker Gaddis et al., 2014). The Nordic countries have a long history of recording health traits; for example, there has been a stipulation in Norway that from the year 1975, veterinary treatments have to be registered on an individual basis (Heringstad and Osteras, 2012), with a similar type of recording being established in Denmark, Finland and Sweden during the 1980s. In addition to the Nordic countries, routine genetic evaluations of mastitis have been in place in Austria and Germany since 2010, in France from 2012 and in Canada from 2012 (Egger-Danner et al., 2015).

In addition to SCC, other predictors of clinical mastitis could be used to improve the accuracy of breeding values. Examples include electrical conductivity from automated milking systems (Norberg, 2005) and lactate dehydrogenase, which is a potential biomarker for mastitis (Friggens et al., 2007). Although Norberg (2005) noted that there were practical issues around collating sufficient electrical conductivity data for genetic evaluation purposes, if this can be overcome, she illustrated there are opportunities to use these data in breeding programmes. Electrical conductivity requires specialist machinery to evaluate, while SCC and lactate dehydrogenase usually have to be analysed by laboratories. For example, portable devices for evaluating SCC are becoming available (Persson and Olofsson, 2011).

Lameness

Feet and leg issues are common reasons for culling in dairy cattle (Egger-Danner et al., 2015). Selection to reduce lameness has historically focused on feet and leg conformation traits, as this is routinely recorded by breed societies, and many countries are already into the process of calculating breeding values for these traits (Table 2). However, Koenig et al. (2005) showed that the accuracy of breeding values for claw health or resistance to lameness increased when claw health data were included.

Metabolic disease

Metabolic disorders, such as ketosis, displaced abomasum, milk fever and tetany, are disturbances to one or more of the metabolic processes in dairy cattle. Under-recording and difficulty in diagnosing subclinical cases are amongst the reasons why there is growing interest in using easily measurable predictors of metabolic diseases, either recorded 'on-farm' by using sensors and milk tests or 'off-farm' using data collected from routine milk recording (Pryce et al., 2016). Some countries have already initiated genetic evaluations of metabolic disease traits and most of these currently use clinical observations of disease.

7.2.2 Heat tolerance

Increases in ambient temperature, humidity, air flow and radiation exceeding the comfort zone are known to lead to heat stress in dairy cattle. This can lead to reduced appetite, production and compromised health and fertility. The issue obstructing the development of breeding values for heat tolerance is defining a trait that can be measured on a large scale. Direct measurements such as core body temperatures and respiration rate are feasible only in small-scale populations due to difficulty in measurement and cost. However, it may be possible to evaluate heat tolerance by quantifying reductions in milk yield that arise at times of heat stress if milk records are matched to weather station information. This is exactly the approach recently taken by Nguyen et al. (2016), who calculated that the accuracy of genomic prediction of heat tolerance is around 0.5 in Australian dairy cattle.

7.2.3 Feed efficiency and methane emissions

Feed is a major component of variable costs associated with dairy systems; therefore, feed efficiency is a trait of great interest in dairy breeding where the objective is to maximize profitability. Currently, very few countries include feed efficiency as a selection criterion because it is expensive to accurately measure on large numbers of cows. However, this makes feed efficiency an ideal candidate for genomic selection, as long as dry matter intake (DMI) is measured together with other traits such as milk production and live weight in a subset of genotyped animals that are representative of the commercial population (Pryce et al., 2014a). The genomic prediction equation that is derived can then be applied to other animals whose genotypes but not phenotypes are known (i.e. no individual measurements for DMI). Feed-saved breeding values have recently been released in Australia (Pryce et al., 2015). The breeding value includes a genomic breeding value for residual feed intake, which is available for Holsteins only, combined with either genomic or pedigree EBVs for maintenance requirements predicted using type traits.

Selection for reduced methane emissions is a growing area of interest. However, building a sufficiently large data set for genetic parameter estimation has been challenging, as phenotype data are scarce. Another complicating factor is the influence of methanogens in the rumen microbiome on methane emissions (Leahy et al., 2013; Ross et al., 2013). New methods of detecting methane emissions are being developed, including gas sensors and radioactive tracers (SF6), which will enable enough phenotypes to be collected to estimate genetic parameters. For example, using a portable air sampler and analyser unit to measure methane emissions on 3121 cows from 20 herds, Lassen et al. (2012) estimated that the heritability of methane emissions varied between 0.16 (s.e. 0.04) and 0.21 (s.e. 0.06) for various methane emission traits. The genetic correlation with fat- and protein-corrected milk was high, indicating that selection for production will lead to an improvement (reduction) in methane emissions. However, including methane emissions in the selection objective may further reduce greenhouse gas emissions at a small economic cost. Selecting on traits that improve the efficiency of farm systems, for example, residual feed intake and longevity, will also have a favourable effect on overall emissions (Wall et al., 2010). Manipulating diet (Ross et al., 2013) and understanding the relationship that methanogens have with other rumen microbes also play an important part in strategies to mitigate emissions (Leahy et al., 2013).

8 Genomic selection for functional traits

While genomic selection has been implemented for many traits (production, fertility, cell count, longevity, etc.), there are still obstacles in applying it to several new traits, associated with the heritability of the trait, the number of animals with records, etc. For cheap and easy-to-measure phenotypes, reasonable reliabilities can be achieved using progeny testing. In case of genomic selection, reference populations comprising genotyped bulls with progeny groups can be used. For traits that are expensive to measure, or where data are sparse, the best option is to obtain phenotypes on nucleus populations of genotyped cows (Chesnais et al., 2015).

Genomic selection opens powerful, new opportunities to select for traits that are difficult and/or expensive to measure, such as phenotypes associated with feed efficiency, methane production and health traits that are difficult or expensive to measure. Accuracy of trait prediction is limited by the size of the reference population. One possible solution is to collaborate with other researchers to establish an even larger reference population. In fact, de Haas et al. (2012) combined dry matter intake phenotypes from Dutch and UK cows with Australian heifer phenotypes and found that the accuracy of genomic prediction was 5.5% higher when a multicountry reference population was used, compared to single-trait models. Since then, there has been further international collaboration through the global dry matter initiative to build an even larger reference population (Berry et al., 2014) with promising results. There are also good prospects to develop genomic breeding values for traits such as tick, parasite and general disease resistance. One option is to customize future breeding goals to include appropriate traits for the prevailing management and environmental conditions. For example, selection for heat tolerance is of great importance in emerging (and many existing) dairy regions.

9 Conclusion

Dairy breeding programmes have been extremely successful in improving milk productivity of dairy cows, with genetic trends in production increasing between 1% and 2% per year over the last few decades. For many years, these breeding systems were based on progeny testing, but only in the last five years is there a major change in the application of genomic selection. Breeding objectives have, in the last two decades, gradually shifted from an emphasis on production traits to functional traits and traits related to efficiency increases. Poor fertility was a major issue caused, in part, by selection for production. However, the availability of breeding values for fertility has resulted in many countries observing positive genetic trends for fertility in recent years. The next challenge is to focus on other functional traits, in particular traits associated with health, efficiency and environmental impact. A balanced genetic change requires recording of traits. With genomic selection there is now more opportunity to use resource herds specifically designed to record hard-to-measure traits.

10 Where to look for further information

New journal articles on selection in dairy cattle appear mostly in the *Journal of Dairy Science* (JDS) and in *Genetics, Selection and Evolution* (GSE). Also Interbull has annual meetings with the latest finding and discussions: http://www.interbull.org. There's a JDS collection on balanced breeding, http://www.journalofdairyscience.org/content/balancedbreeding

11 Acknowledgements

JvdW likes to acknowledge Robert Banks and John Gibson for helpful discussions on breeding programme structures.

12 References

Banos, G. (2010). Past, present and future of international genetic evaluations of dairy bulls. *Proceedings of the 9th World Congress of Genetics Applied to Livestock Production*, Paper.

Berry, D., Wall, E. and Pryce, J. (2014). Genetics and genomics of reproductive performance in dairy and beef cattle. *Animal* 8, 105–21.

Bichard, M. (2002). Genetic improvement in dairy cattle – an outsider's perspective. *Livestock Production Science*. Elsevier.

Boichard, D. and Brochard, M. (2012). New phenotypes for new breeding goals in dairy cattle. *Animal* 6, 544–50.

Buckley, F., Lopez-Villalobos, N. and Heins, B. J. (2014). Crossbreeding: implications for dairy cow fertility and survival. *Animal* 8(Suppl. 1), 122–33.

Bulmer, M. G. (1971). The effect of selection on genetic variability. *Am. Nat.* 105, 201–11.

Chesnais, J. P., Cooper, T. A., Wiggans, G. R., Sargolzaei, M., Pryce, J. E. and Miglior, F. (2015). Using genomics to enhance selection of novel traits in North American dairy cattle. *J. Dairy Sci.* (in press).

Clark, S. A., Kinghorn, B. P., Hickey, J. M. and van der Werf, J. H. J. (2013). The effect of genomic information on optimal contribution selection in livestock breeding programs. *Genetics Selection Evolution* 45, 44.

Daetwyler, H. D., Villanueva, B., Bijma, P. and Woolliams, J. A. (2007). Inbreeding in genome-wide selection. *J. Anim. Breed Genet.* 124, 369–76.

Dekkers, J. C. M. and Gibson, J. P. (1998). Applying breeding objectives to dairy cattle improvement. *J. Dairy Sci.* 81(Suppl 2), 19–35.

Dekkers, J. C. M., Jairath, L. K. and Lawrence, B. H. (1994). Relationships between sire genetic evaluations for conformation and functional herd life of daughters. *J. Dairy Sci.* 77, 844–54.

Dickerson, G. E. and Hazel, L. N. (1944). Effectiveness of selection on progeny perfrmances as a supplment to earlier culling of livestock. *J. Agric. Res.* 69, 459.

Ducrocq, V. and Wiggans, G. (2014). Genetic improvement of dairy cattle. In D. J. Garrick and A. Ruvinsky (Eds), *The Genetics of Cattle* (2nd Edition). C.A.B. International.

Egger-Danner, C., Cole, J., Pryce, J., Gengler, N., Heringstad, B., Bradley, A. and Stock, K. (2015). Invited review: overview of new traits and phenotyping strategies in dairy cattle with a focus on functional traits. *Animal* 9, 191–207.

Fernando, R. L. and Grossman, M. (1989). Marker assisted selection using best linear unbiased prediction. *Genet. Sel. Evol.* 21, 467–77.

Friggens, N., Chagunda, M., Bjerring, M., Ridder, C., Hojsgaard, S. and Larsen, T. (2007). Estimating degree of mastitis from time-series measurements in milk: a test of a model based on lactate dehydrogenase measurements. *J. Dairy Sci.* 90, 5415–27.

Georges, M., Nielsen, D., Mackinnon, M., Mishra, A., Okimoto, R., Pasquino, A., Sargeant, L., Sorensen, A., Steele, M. R., Zhhao, X., Womack, J. E. and Hoeschele, I. (1995). Mapping quantitative trait loci controlling milk production in dairy cattle by exploiting progeny testing. *Genetics* 139, 907–20.

Goddard, M. E. (1998). Consensus and debate in the definition of breeding objectives. *J. Dairy Sci.* 81, 6–18.

Granleese, T., Clark, S. A., Swan, A. A. and van der Werf, J. H. J. (2015). Increased genetic gains in sheep, beef and dairy breeding programs from using female reproductive technologies combined with optimal contribution selection and genomic breeding values. *Genetics Selection Evolution* 47, 70.

Grisart, B., Coppieters, W., Farnir, F., Karim, L., Ford, C., Berzi, P., Cambisano, N., Mni, M., Reid, S., Simon, P., Spelman, R., Georges, M. and Snell, R. (2002). Positional candidate cloning of a QTL in dairy cattle: identification of a missense mutation in the Bovine DGAT1 gene with major effect on milk yield and Composition. *Genome Res.* 12, 222–31.

Henderson, C. R. (1973). Sire evaluation and genetic trends. *J. Animal Science*. Symposium, pp. 10–41.

Heringstad, B., Klemetsdal, G. and Steine, T. (2007). Selection responses for disease resistance in two selection experiments with Norwegian Red cows. *J. Dairy Sci.* 90, 2419–26.

Heringstad, B. and Østerås, O. (2013). More than 30 years of health recording in Norway. In ICAR Technical Series no. 17, pp. 39–46.

Koenig, S., Scharifi, A. R., Wentrot, H., Landmann, D., Eise, M. and Simianer, H. (2005). Genetic parameters of claw and foot disorders estimated with logistic models. *J. Dairy Sci.* 88, 3316–25.

Land, R. B. and Hill, W. G. (1975). The possible use of superovulation and embryo transfer in cattle to increase response to selection. *Anim Sci.* 21, 1–12.

Lande, R. B. and Thompon, R. (1990). Efficiency of marker-assisted selection in the improvement of quantitative traits. *Genetics* 124, 743–56.

Lassen, J., Løvendahl, P. and Madsen, J. (2012). Accuracy of noninvasive breath methane measurements using Fourier transform infrared methods on individual cows. *Journal of Dairy Science* 95(2), 890–8.

Leahy, S., Kelly, W., Ronimus, R., Wedlock, N., Altermann, E. and Attwood, G. (2013). Genome sequencing of rumen bacteria and archaea and its application to methane mitigation strategies. *Animal* 7, 235–43.

Liu, Z., Jaitner, J., Reinhardt, F., Pasman, E., Rensing, S. and Reents, R. (2008). Genetic evaluation of fertility traits of dairy cattle using a multiple-trait animal model. *J. Dairy Sci.* 91, 4333–43.

Lopez-Villalobos, N., Garrick, D. J., Holmes, C. W., Blair, H. T. and Spelman, R. J. (2001). Profitabilities of some mating systems for dairy herds in New Zealand. *J. Dairy Sci.* 83, 144–53.

Martin-Collado, D., Byrne, T., Amer, P., Santos, B., Axford, M. and Pryce, J. (2015). Analyzing the heterogeneity of farmers' preferences for improvements in dairy cow traits using farmer typologies. *J. Dairy Sci.* 98, 4148–61.

McGuirk, B. (1990). Operational aspects of a MOET nucleus dairy breeding scheme. *Proceedings of the 4th World Congress on Genetics Applied to Livestock Production*, Edinburgh, 23–27 July 1990. XIV. Dairy Cattle Genetics and Breeding, Adaptation, Conservation.

McParland, S., Kearney, J. F., Rath, M. and Berry, D. P. (2007). Inbreeding trends and pedigree analysis of Irish dairy and beef cattle populations. *J. Anim. Sci.* 85, 322–31.

Meuwissen, T. H. E. (1997). Maximizing the response of selection with a predefined rate of inbreeding. *J. Anim. Sci.* 75, 934–40.

Meuwissen, T. H. E. and Goddard, M. E. (1996). The use of marker haplotypes in animal breeding schemes. *Genetics Selection Evolution* 28, 161–76.

Meuwissen, T. H. E., Hayes, B. J. and Goddard, M. E. (2001). Prediction of total genetic value using genome-wide dense marker maps. *Genetics* 157, 1819–29.

Mrode, R. A. and Swanson, G. J. T. (1996). Genetic and statistical properties of somatic cell count and its suitability as an indirect means of reducing the incidence of mastitis in dairy cattle. *Animal Breeding*. Abstracts, UK.

Neiman-Sorensen, A. and Robertson, A. (1961). The association between blood groups and several production characters in three Danish cattle breeds. *Acta Agric. Scand.* 11, 163–96.

Nguyen, T. T. T., Bowman, P. J., Haile-Mariam, M., Oryce, J. E. and Hayes, B. J. (2016). Genomic selection for tolerance to heat stress in Australian dairy cattle. *J. Dairy Sci.* 99, 2849–62.

Nicholas, F. W. and Smith, C. (1983). Increased rates of genetic change in dairy cattle by embryo transfer and splitting. *Anim. Sci.* 36, 341–53.

Norberg, E. (2005). Electrical conductivity of milk as a phenotypic and genetic indicator of bovine mastitis: a review. *Livestock Production Science* 96, 129–39.

Parker Gaddis, K. L., Cole, J. B., Clay, J. S. and Maltecca, C. (2014). Genomic selection for producer-recorded health event data in US dairy cattle. *J. Dairy Sci.* 97, 3190–99.

Persson, Y. and Olofsson, I. (2011). Direct and indirect measurement of somatic cell count as indicator of intramammary infection in dairy goats. *Acta Veterinaria Scandinavica* 53, 1.

Pryce, J. E., Gonzalez-Recio, O., Nieuwhof, G., Wales, W., Coffey, M., Hayes, B. and Goddard, M. (2015). Hot topic: Definition and implementation of a breeding value for feed efficiency in dairy cows. *J. Dairy Sci.* 98, 7340–50.

Pryce, J. E. and Veerkamp, R. F. (2001). The incorporation of fertility indices in genetic improvement programmes. BSAS Occ. Publ. *Fertility in the High Producing Dairy Cow* 26, 237–50.

Pryce, J. E., Wales, W., De Haas, Y., Veerkamp, R. and Hayes, B. (2014a). Genomic selection for feed efficiency in dairy cattle. *Animal* 8, 1–10.

Pryce, J. E., Woolaston, R., Berry, D. P., Wall, E., Winters, M., Butler, R. and Shaffer, M. (2014b). World trends in dairy cow fertility. *Proceedings 10th World Congress of Genetics Applied to Livestock Production.* Paper No.154. Vancouver, Canada.

Rendel, J. M. and Robertson, A. (1950). Estimation of genetic gain in milk yield by selection in a closed herd of dairy cattle. *J. Genet.* 50, 1–8.

Robertson, A. and Rendel, J. M. (1950). The use of progeny testing schemes to improve rates of genetic progress and decrease rates with artificial insemination in dairy cattle. *J. Genet.* 50, 21–31.

Rogers, G. W. and McDaniel, B. T. (1989). The usefulness of selection for yield and functional type traits. *J. Dairy Sci.* 72, 523–7.

Ross, E., Moate, P., Marett, L., Cocks, B. and Hayes, B. (2013). Investigating the effect of two methane-mitigating diets on the rumen microbiome using massively parallel sequencing. *J. Dairy Sci.* 96, 6030–46.

Schennink, A., Stoop, W. M., Visker, M. H. P. W., Heck, J. M. L., Bovenhuis, H., Van Der Poel, J. J., Van Valenberg, H. J. F. and Van Arendonk, J. A. M. (2007). DGAT1 underlies large genetic variation in milk-fat composition of dairy cows. *Anim. Genet.* 38, 467–73.

Short, T. H. and Lawlor, T. J. (1992). Genetic parameters of conformation traits, milk yield, and herd life in Holsteins. *J. Dairy Sci.* 75, 1987–98.

Smith, C. (1988). Genetic improvement of livestock in developing countries using nucleus breeding units. *World Anim. Rev.* 65, 2–10.

Sun, C. and Su, G. (2010). Comparison on models for genetic evaluation of non-return rate and success in first insemination of the Danish Holstein cows. *Livest. Sci.* 127, 205–10.

van Arendonk, J. A. M. and Bijma, P. (2003). Factors affecting commercial application of embryo technologies in dairy cattle in Europe – a modelling approach. *Theriogenology* 59, 635–49.

VanRaden (2004). Invited review: Selection on net merit to improve lifetime profit. *J. Dairy Sci.* 87, 3125–31.

Wall, E., Brotherstone, S., Woolliams, J., Banos, G. and Coffey, M. (2003). Genetic evaluation of fertility using direct and correlated traits. *J. Dairy Sci.* 86, 4093–102.

Wall, E., Simm, G. and Moran, D. (2010). Developing breeding schemes to assist mitigation of greenhouse gas emissions. *Animal* 4, 366–76.

Wray, N. R. and Goddard, M. E. (1994). Increasing long-term response to selection. *Genet Sel. Evol.* 26, 431–51.

Chapter 7

Genetic factors affecting fertility, health, growth and longevity in dairy cattle

Joel Ira Weller, Agricultural Research Organization, The Volcani Center, Israel

1 Introduction

Traditionally most of the emphasis of breeding in dairy cattle has been for milk and fat content. However, during the early 1980s, the emphasis shifted towards milk protein production, following the introduction of machines that could inexpensively assay protein content on thousands of samples. In addition to production, there was generally some emphasis on conformation or 'type' traits. (The economic value of breeding for these traits will be considered in Section 7). As shown in Fig. 1, in 1994, only two indices used in advanced countries emphasized more than 10% on selection of traits other than production and type: the US total merit index (TMI) and the Danish S-index (Leitch, 1994). Inclusion of secondary or 'functional' traits in breeding objectives began in earnest only

http://dx.doi.org/10.19103/AS.2016.0005.13

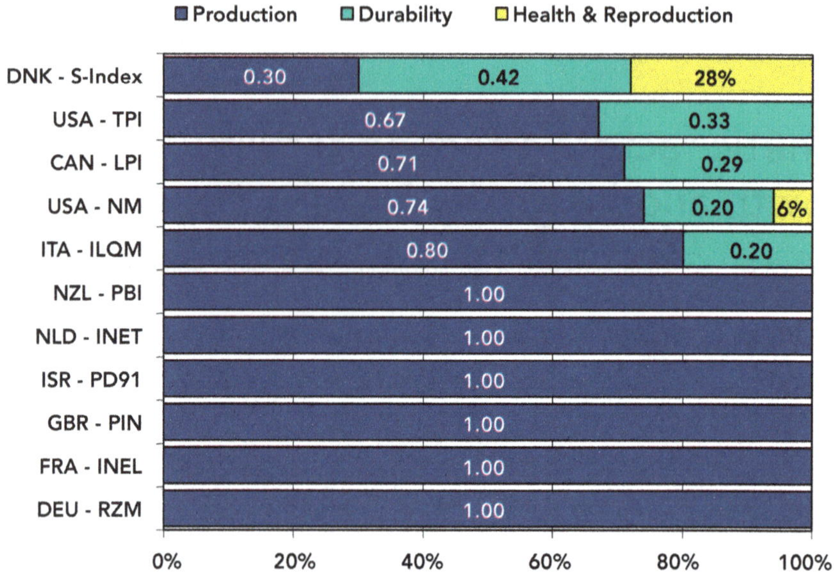

Figure 1 Selection indices used by the major commercial dairy cattle populations divided into production, durability and health and production traits in 1994, according to Leitch (1994).

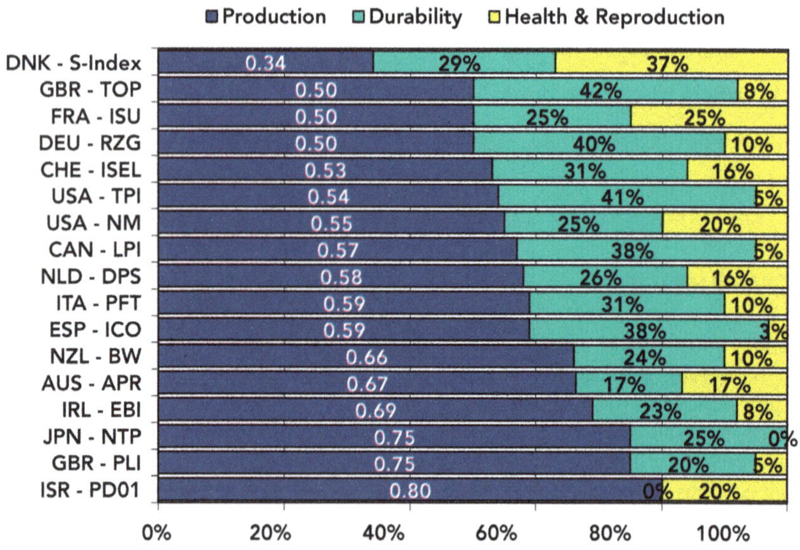

Figure 2 Selection indices used by the major commercial dairy cattle populations divided into production, durability and health and production traits in 2003, according to Miglior et al. (2005).

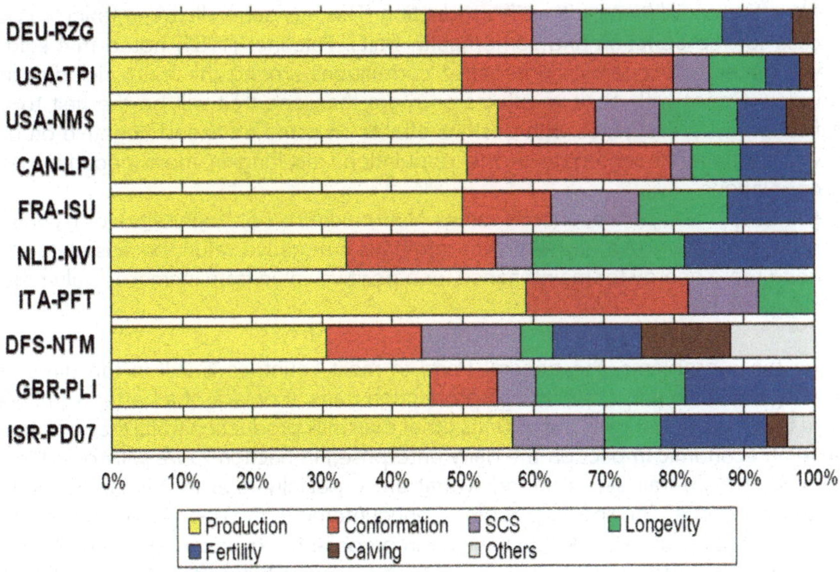

Figure 3 Selection indices used by the major commercial dairy cattle populations divided into production, conformation, somatic cell score (SCS), longevity, fertility, calving traits and other traits in 2009, according to Stefan Rensing, Vereinigte Informationssysteme Tierhaltung (VIT) w.V., Verden (Germany).

in the late 1990. By 2003, production traits accounted for less than 60% of the indices in most advanced countries (Fig. 2, adapted from Miglior et al., 2005). By 2009, production traits accounted for less than 60% in all countries and averaged less than 50% (Fig. 3, adapted from Rensing, VIT).

Although there is nearly complete consensus of the economic importance on fertility, health traits and longevity, genetic evaluation and inclusion of these traits in selection indices was hindered by several factors:

1. *Difficulty of measurement*: Health and fertility traits were generally recorded by the farmer, and their validity varied widely. Herd-life (HL) can only be actually recorded after the cow's removal from the herd, and by this time breeding decisions have already been made.
2. *Non-normal distributions*: Nearly all health traits have discrete or even dichotomous distributions (e.g. sick vs. well). Somatic cell concentration (SCC) has a continuous distribution, but with very strong negative skewness. As will be seen in Section 3, a trait with a continuous non-normal distribution can generally be handled by a transformation of the data.
3. *Low heritability*: Nearly all of the traits that are considered in this chapter have low to negligible heritability.
4. *Lack of consensus on trait definition*: This is especially the case with respect to fertility, for which numerous definitions have been proposed in the literature, but is also the case for HL, which has been scored as total HL, HL corrected for milk production or survival until a particular age or event.

5. *Negative correlations with milk production*: This has generally been found to be the case for reproduction and some health traits. Falconer (1981) noted that selection will generate economically negative correlations among the traits included in the index. Genes with positive effects on more than one trait will be the first to reach fixation, whereas genes with positive effects on some traits and negative on others will continue to segregate in the population, resulting in more negative genetic correlations.
6. *Difficulty to compute economic values*: How much is a 'unit' of fertility worth compared to a kg of milk? Also, does growth rate have a negative value, because larger cows need more energy for maintenance, or a positive value from increased value of male calves for beef?

Despite these problems, the general trend in most countries is still in the direction of increasing selection emphasis on non-production traits. (For lack of a better term, we will use the term 'functional traits' for all traits other than milk production traits that are assumed to have an economic impact on the dairy enterprise.) In Section 2, we will consider some principles of multi-trait selection index that are especially relevant to current breeding goals in dairy cattle. The specific problems encountered in the genetic evaluation of non-production traits and methods that have been proposed to overcome these problems will be considered in Section 3. In Sections 4, 5 and 6, we will consider the main functional traits that are currently analysed in most commercial dairy cattle populations. In Section 7, we will consider alternative methods to improve functional traits, including selection on conformation traits and cross-breeding, and in Section 8, we will consider methods to identify the individual genes responsible for the observed variation, and the results that have been obtained to date.

2 Important principles of multi-trait selection index

2.1 Derivation of the optimum linear selection index

We will use the following conventions throughout the chapter: vectors will be denoted in lower case **bold type**, matrices in upper case **BOLD TYPE** and a transpose of a vector or matrix will be denoted by an apostrophe. Hazel (1943) formulated the principles of selection index, based on the assumption that for each individual there is a vector **y** of length *m* consisting of the individuals breeding values for the traits with economic values. In addition, he assumed that there is a vector **a** also of length *m* consisting of the economic values for the traits in **y**. The aggregate economic value, *H*, can then be computed as **a'y**. Although the values included in **y** are not known, estimated breeding values can be computed. The optimum linear selection index, I_o can then be computed as **b'x**, where **x** is the vector of trait values (either phenotypic or estimated breeding values) and **b** is a vector of coefficients that maximizes the correlation with *H*. Henderson (1973) proved that if the estimated breeding values included in **x** have best linear unbiased prediction (BLUP) properties, then the optimum linear selection index will be **a'x**, the vector of economic values multiplied by the vector of BLUP genetic evaluations.

Table 1 Relative weights of traits included in the current Israeli breeding index, PD11

Trait	Index coefficient	Genetic SD	Contribution to the index	% of index
Milk (kg)	0	910	0	0
Fat (kg)	7.9	32.1	253.6	19.2
Protein (kg)	23.7	22.6	535.6	40.5
Somatic cell score*	−300.0	0.47	141.0	10.7
Female fertility (%)	26.0	7.1	184.6	14.0
Herd life (days)	0.6	205	123.0	9.3
Persistency (%)	10.0	5.44	54.4	4.1
Dystocia (%)*	−3.0	5.45	16.4	1.2
Stillbirth (%)*	−6.0	2.21	13.3	1.0
Total			1321.8	100.0

*Negative values are economically favourable.

In Section 1, we introduced the following concept: the fraction of the index explained by the individual traits. This will now be described with the aid of Table 1, which presents the vector of index coefficients, **a**, for the current Israeli breeding index, PD11. PD11 includes 9 traits. The economic values (**a** vector) are given in the second column, and the genetic standard deviations are given in the third column. 'Contributions to the index', computed as the absolute value of the product of columns 2 and 3, are given in column 4. Note that for three traits, somatic cell score (SCS), dystocia (DC) and stillbirth (SB), the economic values are negative; that is, low values are economically favourable, while all values in the column are positive. 'Percent of the index' given in the last column is computed as the contribution to the index of each trait divided by the sum of the contributions to the index of all 9 traits multiplied by 100. As can be seen, protein and fat production contribute 60% to the index, while female fertility has the next largest contribution, 14%.

Optimally, **a** should be computed as the derivative of the economic objective with respect to a unit change in the trait value (Weller, 1994). For the individual farmer, the 'economic objective' will generally be his profit. That is the elements of **a** are computed as the farmer's profit change due to a unit change in a specific trait. Direct computation of economic values for non-production traits is generally not a realistic option, first because it is difficult to compute an accurate profit function. Also profit will not be a linear function of the trait value, and economic values will usually differ across farms and even for different cows. Therefore, alternative methods have been proposed, including maximization of biological or economic efficiency, and restricted selection indices (see Weller, 1994, for a detailed discussion of these methodologies).

Until now, we have considered only linear selection indices of the form **b'x**, where both **b** and **x** are vectors. Various studies have proposed application of non-linear indices, based on the premise that economic values are generally not constant, but are functions of the trait values. This is of course true if there is an economically optimal trait value. For example, intermediate size may be economically optimal, though it should be noted that long-term genetic gain for any economic objective will always be optimized by a linear selection index (Goddard, 1983).

2.2 Computation of expected changes due to selection and restricted selection indices

The vector of expected genetic changes due to selection on an index, Φ, can be computed from the following equation:

$$\Phi = i\mathbf{Cb}/\sigma_{is} \tag{13.1}$$

where *i* is the selection intensity, **C** is the genetic variance matrix among the traits included in the selection index and σ_{is} is the standard deviation of the index, computed as $(\mathbf{b'Pb})^{0.5}$, where **P** is the phenotypic variance matrix. In a variance matrix, the trait variances are on the diagonal, and the covariances between each pair of traits are on the off-diagonals. The selection intensity is a function of the fraction of individuals selected as parents for the next generation.

We will now illustrate the application of this equation with the data from the Israeli dairy industry given in Table 1. In dairy cattle breeding programmes, the selection intensities will be different over the four paths of inheritance: sire-to-sire, sire-to-dam, dam-to-sire and dam-to-dam. In an advanced breeding programme, the overall value of *i* for ten years, which is approximately equal to 2 generations, will be approximately 4. Using this value, the expected genetic gains for ten years based on the Israeli breeding index, PD11, are given in Table 2. The expected gain for PD11 was computed by summing the absolute values of the individual trait gains. The realized genetic gains for the last ten years,

Table 2 The expected and realized genetic gains during the last ten years with selection based on the current Israeli breeding index, PD11

	Ten-year genetic trends		
Trait	Expected	Realized	Realized/expected
Milk	1035	775	0.75
Fat	41	24.5	0.60
Protein	32	31.6	0.99
Somatic cell score*	−0.12	−0.12	1.00
Female fertility	0.20	2.25	11.25
Herd life	140	164	1.17
Persistency	3.40	1.42	0.42
Dystocia*	−1.10	0.70	−0.64
Calf mortality*	−0.41	−0.10	0.24
PD11	**1247**	**1150**	**0.92**

*Negative values are economically favourable.

computed as the regression of the cows' estimated breeding values on their birthdates, for the entire milk-recorded population and the ratio of realized to expected gains are also given. First, note that the realized genetic gain PD11 given in the bottom row was 92% of the expected gain. Thus, it can be assumed that selection was reasonably efficient in this population. Realized genetic gains for protein SCS and HL were very close to the expected values. Realized gains for fat, persistency of production and the calving traits were less than the expected values, but the latter 3 traits were only added to the index in 2007. The realized gain for fertility was more than ten times the expected gain, but both values are low relative to the genetic standard deviation for this trait. The expected gain in milk production is 1035 kg, even though the coefficient for this trait in PD11 is zero. The large genetic gain for milk is due to the high genetic correlation between milk and protein production.

As noted previously, one of the alternatives to the optimum selection index, computed from the economic values of all traits, is a restricted selection index. A 'restricted index' is one in which the genetic change of some traits is fixed to a specific value. The simplest type of restriction proposed is to hold genetic change for a specific trait to zero. For example, many studies have found negative genetic correlations between fertility and milk production traits. Thus, if fertility is not included in the index, then it should decline, similar to the situation for milk production in Table 2, which increased dramatically, even though the selection index coefficient for milk is zero. On an intuitive level, construction of a restricted index consists of using eq. (13.1) to solve for **b** with the elements of ϕ fixed at the desired values. For example, the expected change for fertility could be set to zero. This is in fact nearly the case for PD11, as the expected gain for fertility is only 3% of the genetic standard deviation. Another type of restriction that has been proposed in relation to production traits has been the construction of an index that results in genetic gains for production traits in proportion to market demands. That is, if the market demand for protein is 80% of demand for fat, then construct the index that results in gains of the desired ratio.

A final point that should be noted with respect to selection indices is that expected gains are very robust with respect to the index coefficients (Weller, 1994). That is, very large changes in the relative values of the coefficients are required to generate economically important changes in the expected gains. It then follows that even a very approximate estimate of the relative trait values is probably sufficient.

3 Statistical methods for the genetic analysis of non-production traits

Since the 1970s, BLUP has been the method of choice for genetic evaluation of field data. The properties of BLUP were summarized by Henderson (1973). BLUP or the 'mixed model', can deal with a situation in which some of the independent variables are fixed and some are random. It is generally assumed in BLUP analyses that the variance due to random variables is known without error, and that variances are equal for all levels of the random effects. In single-trait analyses, it is generally assumed that residual variances are equal for all records and are uncorrelated, although BLUP is sufficiently flexible to deal with situation in which this is not the case, for example, multiple trait analyses. In this case, residual variances will be different for each trait, and residuals among traits recorded on the same individual can be correlated. However, even in multi-trait analyses, BLUP methodology generally assumes that the residuals for the dependent variable have a normal distribution.

Since the 1980s, the standard model for genetic analysis of dairy cattle has been the 'individual animal model' (Henderson, 1973; Westell et al., 1988). Each cow is assumed to produce a single record for each parity. For milk production traits, the records are total milk, fat and protein production per lactation. In the 'repeatability' model, a genetic effect and a permanent environmental effect are estimated for each cow with records. Both effects are assumed to be random, but can be distinguished, due to the inclusion of the inverse of numerator relationship matrix in the model. That is, the genetic effect accounts for genetic relationships among cows, while the permanent environmental effect is not affected by relationships among animals. Genetic effects for animals that did not produce records, including males, are computed via genetic relationships between animals with and without records. An alternative to the animal repeatability model is the 'multi-trait animal model'. In this model, records from different parities are considered to be different traits. These different parities are assumed to have both 'environmental' and genetic correlations. The 'environmental' correlations are the correlations among the records of a specific cow on different parities and are not those related to genetic effects common to different parities. Genetic correlations can be estimated based on sire effects in the different parities. In this model, genetic evaluations are computed for each parity. To obtain multi-parity evaluations the individual parity evaluations are combined into a selection index.

Several important functional traits violate the basic assumptions of standard BLUP methodology. In Section 1, we mentioned the problem that many functional traits have non-normal distributions. This problem can be divided into traits that have continuous non-normal distributions, such as SCC and HL, and traits that have discrete distributions, including fertility, calving and health traits. For continuous traits, this problem can generally be solved by a transformation of the trait values to an approximately normal scale, as will be described in detail with respect to SCS. For traits with discrete distributions, three alternatives have been proposed in the literature:

1. Apply standard BLUP methodology on the trait scores. For example, for health traits, score healthy animals as 0 and diseased animals as 1. Although the residuals will not have a normal distribution, many studies have shown that BLUP is relatively robust to violation of this assumption. The problem of non-normal residuals is most severe for dichotomous traits in which one category has a much higher frequency than the other. In this case, the genetic evaluations will also have a skewed distribution, which is an artefact of the analysis model.
2. *Log-linear models*: In standard additive models, the value for the dependent variable is assumed to be the sum of the effects for the independent variables. In log-linear models, an underlying multiplicative model is assumed, in which the probability of a given state for the dependent variable (e.g. 'healthy') is assumed to be the product of the probabilities of the values of independent variables. Computation is facilitated if the model is log transformed, so that the products of the independent variables become sums. The main problem with these models for analysis of field data is how to deal with combinations of effects with zero frequencies. For example, assume that the dependent variable is the cow's phenotype for a disease, and the independent variables are the cows' sire and herd. Most sires will have records in only a few

herds, so that for most sire-by-herd combinations, the probabilities will be zero. In most statistical packages, this problem is dealt with by assuming a very low positive probability.

3. *Threshold models*: In these models, the discrete dependent variable is assumed to be the results of a non-observable continuous variable, generally assumed to have a normal distribution. In the case of a dichotomous trait, the distribution will have a single threshold. All individuals with values below the threshold will display one value for the observed discrete traits, while all individuals with values above the threshold will display the other threshold. The independent variables affect the mean of the continuous underlying variable, which affects the probability to obtain a specific state for the dependent variable. This is illustrated in Fig. 4 with an example of the effects of two sires on their daughter values for the underlying and discrete variables. The underlying normal variables for the two sires have means of μ_1 and μ_2. The threshold is at point *T*. All individuals with values <*T* will have value '1' for the discrete variable and all individuals with values >*T* will have value '2' for the discrete variable. Since the two sires differ in their means, the probabilities for the two values of the discrete variable in their daughters will differ. The objective of analysis of the threshold model is then to estimate the difference between the means of the two sire distributions, based on the frequencies of the discrete scores for their daughters. If the model includes both fixed and random variables, the algebra becomes rather complicated, but has been solved, based on equations that are remarkably similar to the mixed model equations for linear models (Gianola and Foully, 1983).

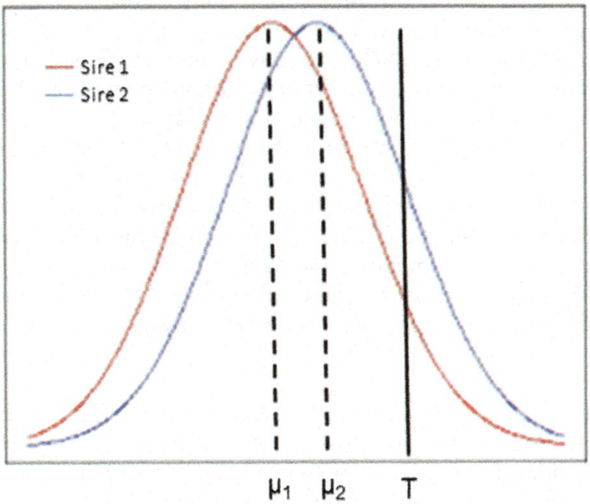

Figure 4 The threshold model is illustrated with the distributions on the underlying variable for the effects of two sires with means of μ_1 and μ_2. The threshold is at point *T*. All individuals with values <*T* will have value '1' for the discrete variable and all individuals with values >*T* will have value '2' for the discrete variable. Since the two sires differ in their means, the probabilities for the two values of the discrete variable in their daughters will differ.

4 Non-production traits and selection strategies: fertility

Discussion of the genetics of fertility should be divided into three groups of traits, which have nearly independent genetic control:

1. *The cow's ability to conceive*: This is generally denoted 'female fertility'.
2. *The bull's ability to impregnate*: This is generally denoted 'male fertility'.
3. *The cow's ability to produce a live, healthy calf once it is pregnant*: The rate of SB is highly correlated genetically with DC, and therefore both traits will be discussed jointly.

With respect to female fertility, conception rates of virgin heifers and milking cows are nearly always analysed separately. Conception rates of heifers are generally 20% higher than milking cows, genetic and environmental correlations between heifers and cows are low and environmental effects such as season or herd are different.

4.1 Female fertility

All female fertility traits have low to negligible heritability. Many studies have reported on negative genetic correlations between female fertility and milk production traits (reviewed by Weller, 1989). Reproductive efficiency has declined for dairy cows worldwide, and this has been the trend for dairy cow fertility since 1957 (Lucy, 2001, 2007). The first genetic analyses of female fertility were performed by Janson (1980) and Janson and Andreasson (1981) on Swedish dairy cattle. Until the 1990s, the most common fertility traits were interval between calvings, days open (DO), non-return rate and number of inseminations per lactation. 'Days-open' is defined as the number of days from freshening until the insemination that resulted in conception. 'Non-return rate' is the probability that an inseminated cow is not re-inseminated by a given number of days after the first breeding. As noted by Weller (1989), all four traits are problematic, especially with respect to their relationship to production traits. Cows that are culled due to non-conception do not generate a calving interval or a DO record. Thus, the cows with the most problematic fertility are removed from the analysis. Furthermore, if the cow is inferior for production, the farmer is more likely to cull, rather than re-inseminate. Thus, with respect to non-return rate and the number of inseminations, a culled cow is treated in the same way as a cow that was not re-inseminated because of conception. This results in an artificially high negative correlation between fertility and production.

Since the early 1970s, fertility data in Israel were unique, in that all cows in communal herds that were inseminated and were not culled were checked for pregnancy approximately 45 days after the last insemination. Thus, conception status (CS) was known for most inseminations. Ron et al. (1984) performed a fixed model analysis, in which the dependent variable was conception per insemination and the independent variables were the effects of sire of cow, inseminating sire and technician that inseminated the cow. Variance components were not estimated.

Weller (1989) analysed a lactation measure of fertility termed 'conception status' (CS) defined as 100/(number of inseminations to conception). Cows that were inseminated,

Table 3 Heritabilities and genetic and environmental correlations for female fertility in parities 1–5 in the Israeli Holstein population (Weller and Ezra, 2004)*

Parity	Parity				
	1	2	3	4	5
1	**0.02**	0.74	0.86	0.66	0.54
2	0.05	**0.03**	0.82	0.93	0.79
3	0.05	0.05	**0.02**	0.88	0.75
4	0.04	0.05	0.06	**0.03**	0.92
5	0.05	0.06	0.08	0.09	**0.02**

*Heritabilities are on the diagonal, genetic correlations are above the diagonal and environmental correlations are below the diagonal.

but did not conceive, as determined by veterinary examination, were scored as 0, and cows with >5 inseminations were given the same score as cows with 5 insemination. Weller and Ezra (1997) modified CS, so that for cows that were inseminated, but did not conceive, the number of inseminations to conception was replaced with its expectation, based on the number of recorded inseminations and parity. Correlations with milk and fat production were negative, but <0.05. First parity heritability of CS was still only 0.02 and similar for the other parities. Weller and Ezra (2004) analysed CS by the multi-trait animal model described previously, with each parity being considered a separate trait. The heritabilities by parity and the genetic and environmental correlations among parities based on Weller and Ezra (2004) are given in Table 3. Despite the low heritability, genetic correlations among parities 1–5 by using the multi-trait animal model were all >0.7, while the environmental correlations were all <0.1. Thus, analysis by using the multi-trait animal model appears to be most appropriate.

Phenotype and estimated breeding value (EBV) means of CS in the Israeli Holstein population by the cows' birth year are given in Fig. 5. Consistent with most other studies there was a reduction of approximately 8.5% from 1983 through 2008, although since then, there has been an increase. CS was added to the Israeli index in 2000. Since 2001, there has been a genetic increase of approximately 1.9% in the mean of the cows' EBV.

Currently, the United States analyses three female fertility traits: heifer conception rate (HCR), cow conception rate (CCR) and daughter pregnancy rate (DPR) (Kuhn et al., 2006; VanRaden et al., 2004). HCR is defined as percentage of inseminated heifers that become pregnant at each service. Similarly, CCR is defined as percentage of lactating inseminated cows that become pregnant at each service up to 7. The linear model estimate of heritability for HCR on first service was 0.5% and the threshold model estimate was 1.0% (Kuhn et al., 2006). Heritability of CCR was estimated at 1.6% (http://aipl.arsusda.gov/reference/Form_GE_FFertility_1209.pdf). For DPR, DO records are transformed to pregnancy rate using the simple linear function:

$$DPR = 0.25 \times (233 - DO) \tag{13.2}$$

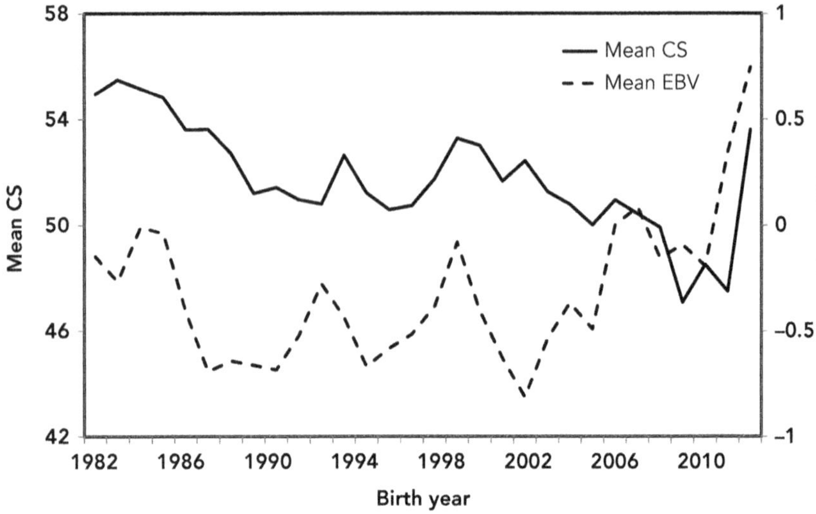

Figure 5 First parity mean CS and mean EBV of CS in the Israeli Holstein population by the cows' birth year.

A DPR of '1' implies that daughters from this bull are 1% more likely to become pregnant during that oestrus cycle than those from a bull with an evaluation of zero. Each increase of 1% in PTA DPR equals a decrease of four days in PTA DO. The heritability of DPR is 4%.

4.2 Male fertility

There has been much less research on the genetics of male fertility. The consensus of most studies is that the heritability of male fertility is virtually zero (Kuhn and Hutchison, 2008). Weller and Ron (1992) found that the service sire accounted for 0.15 and 0.29% of the variance per insemination for heifers by linear and threshold models, and 0.23 and 0.37% of the variance for cows. Cows or heifers that were inseminated and were not culled were checked for pregnancy approximately 45 days after the last insemination. Only records for which CS could be determined were included in the analyses.

4.3 Calving traits

The genetic literature on the incidence of SB and DC is extensive. Both traits affect the profitability of dairy production. DC reduces the productive life of the cow and decreases the probability of survival of the calf. The DC on first parity calvings reduced HL of Israeli Holsteins by approximately 180 d (Weller and Ezra, 2015). Despite the economic value of both traits, genetic evaluation is problematic, because both traits are categorical, rather than continuous; both have low heritability. In addition, incidence of both DC and SB are considerably higher for first parity calvings, and genetic correlations among parities are incomplete (e.g. Liu et al., 2012). Finally, for both traits the phenotype is determined by both the cow calving and the sire of the calf (SC). Correlations between these two effects are at best minimal, and may even be antagonistic (Weller et al., 1988, Weller and Gianola,

1989). The effect of the cow calving can only be controlled genetically by inclusion of these traits in a selection index, while the SC effect can be controlled by a selective mating programme. A selective mating plan for virgin heifers to reduce incidence of these traits was introduced in Israel in the 1960s (Bar-Anan et al., 1976).

Because of the discrete nature of the records, various studies have proposed analysis by threshold models (Berger, 1994; Djemali et al., 1987; Weller et al., 1988). Genetic evaluations for DC were first computed in the United States in 1977 (Van Tassell et al., 2003). A sire-maternal grandsire (MGS) threshold model in which calvings for all parities were considered to be the same trait was introduced for DC in 2003 (Van Tassell et al., 2003; Wiggans et al., 2003) and for SB in 2006 (Cole et al., 2007a, b). Cue and Hayes (1985) analysed DC and SB jointly in a two-trait sire model, but computed separate analyses for first and later parities, and separate analyses for the direct and maternal effects. Wiggans et al. (2008) proposed a multi-trait sire-MGS model in which first and later parities were analysed as correlated traits. The DC and SB were analysed separately. They assumed that genetic correlations among later parities were equal to unity, although this assumption was not tested. Several studies have shown that DC and SB are correlated both phenotypically and genetically (Cue and Hayes, 1985; Ron et al., 1986; Weller et al., 1988; Weller and Gianola, 1989, Liu et al., 2012). Thus, optimally both traits should be analysed jointly. Boelling et al. (2007) analysed calving traits by a multi-trait sire-MGS model including DC, SB and size of calf in first and later parities, but did not account for relationships. Weller and Ezra (2016) developed a multi-trait individual animal model algorithm for joint analysis of DC and SB in first and second parities. Thus, genetic evaluations for the effect of cow calving were derived for all milk-recorded cows in the population. Relationships matrices were included both for the effects of cow and sire of calf. All heritabilities were quite low, but higher for DC than for SB and higher in first parity.

5 Non-production traits and selection strategies: health

The main reasons for cow culling are low production, non-conception and illness, chiefly mastitis. Because of the major economic importance of mastitis relative to other health traits, we will first discuss this trait, and then survey the genetics of other health traits.

5.1 Mastitis and somatic cell concentration

Estimates of the economic loss from mastitis in the 1980s ranged from \$100 to \$200 per cow per lactation (Young, 1992). In 1987, the additive genetic standard deviations for the cost of clinical mastitis in first and later parities were estimated as \$13.49 and \$56.20, compared to with \$82 for mature equivalent milk production (Heuven, 1987). Like nearly all disease traits, mastitis is scored dichotomously; that is, the cow is scored as either 'healthy' or 'diseased'. Heritabilities for all direct measures of udder infection are mostly in the range of 1 to 10% (Schutz, 1994). Numerous studies have indicated a positive relationship between milk SCC and udder disease. The mean of the estimates of the genetic correlation between SCC and clinical mastitis is about 0.60 (Shook and Schutz, 1994). Weller et al. (1992) found that the genetic correlation between SCC and bacterial infection status is nearly complete. Based on these correlations and the fact that SCC data can be collected automatically together with fat, protein and lactose concentration, Shook and Schutz (1994) proposed genetic evaluation and selection for SCC as an alternative to direct selection on incidence of mastitis. In addition, SCC has an intrinsic economic value. High SCC has an adverse effect of cheese production. An increase of SCC from 100 000 to 500 000/ml was associated with a decrease of approximately 5.0 and 11.0% in adjusted yield and efficiency of yield, respectively (Politis and Ng-Kwai-Hang, 1988).

A difficulty with direct analysis of SCC is that this trait has a highly skewed distribution. Wiggans and Shook (1987) proposed converting SCC to SCS by the following formula:

$$SCS = \log 2\,(SCC/100) + 3 \quad (13.3)$$

where SCC are in units of cells per microliter. The relationship between SCC and SCS is given in Table 4. Frequency distributions of SCC and SCS in the Israeli Holstein population, 1986–96, are shown in Fig. 6. As can be seen, the distribution of SCC is highly skewed, while the distribution of SCS is approximately normal.

Similar to milk production traits, SCS is affected by both the cow's parity and days in milk. However, unlike the milk production traits, which peak between one and two months in milk, SCS declines from the beginning of the lactation, reaches a minimum, and then continues to increase with increase in days in milk. Wiggans and Shook (1987) proposed that a lactation measure of SCS could be computed by summing the individual monthly SCS scores corrected for the effect of days in milk and parity. First parity heritability of lactation SCS is in the range of 0.2–0.25, which is considerably higher than for incidence of mastitis. Similar to milk production traits, genetic correlations among parities are > 0.8, while environmental correlations are <0.5. Heritabilities and genetic and environmental correlations for SCS in parities 1–5 in the Israeli Holstein population based on Weller and Ezra (2004) are given in Table 5. First parity mean SCS and mean EBV of SCS in the Israeli Holstein population by the cows' birth year are shown in Fig. 7. Both the phenotypic mean and the mean EBV peaked in 1994. SCS was added to the Israeli breeding index in 1996. As shown in Table 2, the realized decline in SCS was equal to the predicted decline over the last ten years, due to selection on PD11. Although too great a decline in SCS may have

Table 4 The relationship between somatic cell concentration and SCS

Somatic cell concentration (cells/ml)	SCS
25 000	1
50 000	2
100 000	3
200 000	4
400 000	5
800 000	6
1 600 000	7
3 200 000	8
6 400 000	9
12 800 000	10

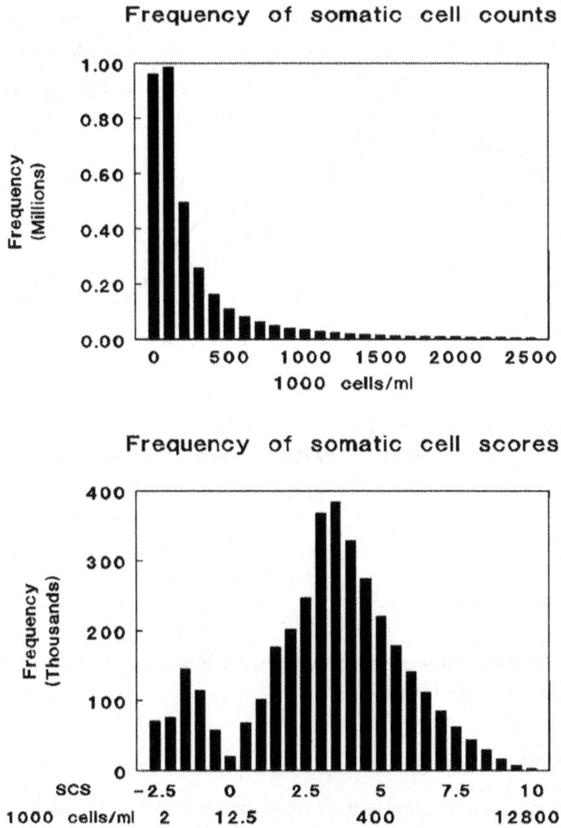

Figure 6 Frequency distributions of somatic cell counts and SCS in the Israeli Holstein population, 1986–96.

Table 5 Heritabilities and genetic and environmental correlations for SCS in parities 1–5 in the Israeli Holstein population (Weller and Ezra, 2004)*

Parity	Parity				
	1	2	3	4	5
1	**0.25**	0.85	0.84	0.82	0.81
2	0.31	**0.22**	0.98	0.95	0.93
3	0.26	0.44	**0.23**	0.99	0.97
4	0.24	0.37	0.48	**0.23**	0.99
5	0.23	0.33	0.40	0.49	**0.18**

*Heritabilities are on the diagonal, genetic correlations are above the diagonal and environmental correlations are below the diagonal.

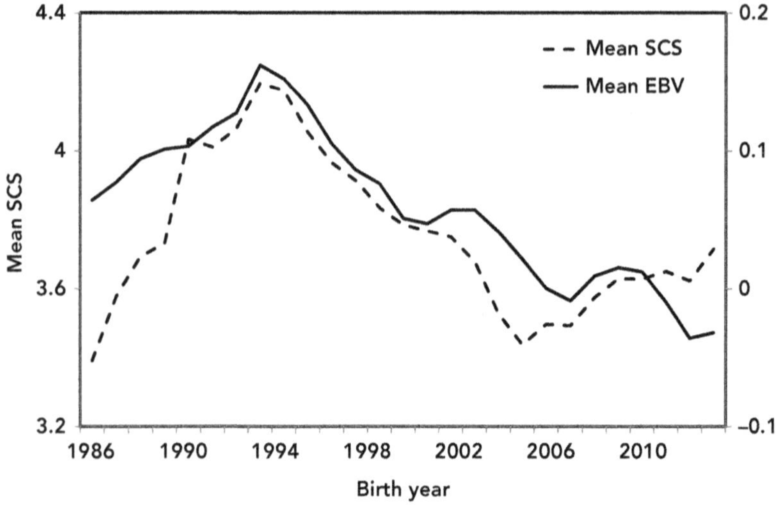

Figure 7 First parity mean SCS and mean EBV of SCS in the Israeli Holstein population by the cows' birth year.

a negative effect on the ability of the cow to resist infection, this does not appear to be a problem for the foreseeable future.

5.2 Other disease traits

Hoof disorders are also associated with considerable costs and have been identified as the third most costly disease in dairy farming after mastitis and infertility (Haggman and Juga, 2013). The model for hoof health included in the Nordic TMI is a linear multiple-trait animal model, which includes seven hoof disorders or groups of hoof disorders from the three first lactations. Heritability estimates from univariate logistic models for 8 hoof disorders ranged from 0.02 for sole haemorrhage, chronic laminitis and heel horn erosion to 0.13 for digital dermatitis. Haggman and Juga (2013) concluded that indirect selection for better hoof health using feet and leg conformation traits as indicator traits is not efficient in the Finnish Holstein breed.

In addition to mastitis and hoof disorders, the main disease traits that have been analysed genetically are udder oedema, milk fever, retained placenta, metritis, ketosis, lameness, cystic ovaries and displaced abomasum (Berry et al., 2011; Van Dorp et al., 1998). Heritability estimates for these nine disease traits from four different studies are given in Table 6. Considering that studies differ in the populations analysed, the analysis models, whether the data was collected by the farmers or veterinarians, it is not surprising that there are major differences for heritability estimates for the same traits. All estimates were <0.2, except for metritis in the study of Van Dorp et al. (1998) and for displaced abomasum in the study of Neuenschwander et al. (2012). The relatively low heritabilities and the difficulty to obtain accurate recording are the main reasons that these traits are generally not included in national selection indices. However, because of differences in exposure rates, as well as the lack of complete sensitivity and specificity of tests, heritability

Table 6 Heritability estimates for nine disease traits

Trait	Heritability	Model*	Parities	Reference
Udder oedema	0.05	AM	1	Van Dorp et al. (1998)
Milk fever	0.04	AM	1	Van Dorp et al. (1998)
	0.10	TAM	1	Hossein-Zadeh and Ardalan (2011b)
	0.18	STM	All	Neuenschwander et al. (2012)
Ketosis	0.39	AM	1	Van Dorp et al. (1998)
	0.06	STM	All	Zwald et al. (2004)
	0.09	STM	All	Neuenschwander et al. (2012)
Displaced abomasum	0.00	AM	1	Van Dorp et al. (1998)
	0.15	STM	All	Zwald et al. (2004)
	0.05	TAM	1	Hossein-Zadeh and Ardalan (2011a)
	0.21	STM	All	Neuenschwander et al. (2012)
Retained placenta	0.01	AM	1	Van Dorp et al. (1998)
	0.07	TAM	1	Hossein-Zadeh and Ardalan (2011b)
	0.08	STM	All	Neuenschwander et al. (2012)
Metritis	0.02	AM	1	Van Dorp et al. (1998)
	0.07	STM	All	Zwald et al. (2004)
	0.09	TAM	1	Hossein-Zadeh and Ardalan (2011b)
	0.02	STM	All	Neuenschwander et al. (2012)
Cystic ovaries	0.02	AM	1	Van Dorp et al. (1998)
	0.05	STM	All	Zwald et al. (2004)
	0.09	TAM	1	Hossein-Zadeh and Ardalan (2011a)
	0.05	STM	All	Neuenschwander et al. (2012)
Mastitis	0.04	AM	1	Van Dorp et al. (1998)
	0.09	STM	All	Zwald et al. (2004)
	0.06	TAM	1	Hossein-Zadeh and Ardalan (2011b)
	0.05	STM	All	Neuenschwander et al. (2012)
Lameness	0.16	AM	1	Van Dorp et al. (1998)
	0.06	STM	All	Zwald et al. (2004)
	0.15	TAM	1	Hossein-Zadeh and Ardalan (2011a)
	0.05	STM	All	Neuenschwander et al. (2012)

*AM, linear animal model; TAM, threshold animal model; STM, sire threshold model.

estimates for health and disease resistance traits are likely to be underestimates of the true heritability (Bishop and Woolliams, 2010).

In addition to estimating heritability of these traits, several studies have estimated genetic correlation among disease traits, genetic correlations between disease and production traits, and genetic correlations between disease and conformation traits. Although natural selection should work to reduce incidence of illness, this may not be the case if incidence of disease is positively correlated with production (e. g. Hagiya et al., 2014). Based on genetic correlations computed over several studies, selection for increased milk production is expected to increase the incidence of mastitis, lameness, cystic ovaries, ketosis and metritis (Berry et al., 2011).

One of the main justifications for inclusion of conformation traits in selection indices is the assumption that these traits, which are relatively easy to score, are genetically correlated with disease traits. As note previously for hoof disorders, genetic correlations between conformation and disease traits are low. However, if many conformation and disease traits are included in the analysis, some of the correlations estimates will be nominally 'significant', by chance, unless the analysis is corrected for multiple comparisons. The problem of determining meaningful significance values with multiple comparisons will be considered in detail in Section 8. Van Dorp et al. (1998) found genetic correlations of 0 to 0.37 between udder conformation traits and mastitis, from –0.38 to 0.09 between leg conformation traits and lameness, and from –0.11 to 0.38 between rump conformation and retained placenta. Thus, selection on conformation traits may slightly reduce the incidence of disease.

6 Non-production traits and selection strategies: growth rate and longevity

6.1 Growth rate

Numerous studies have considered the economic consequences of animal size for dairy cows (reviewed by Brothersone et al., 2007). Most studies have concluded that increased cow size has a negative effect on profitability, and several countries have included negative weights for various measures of cow size in selection indices (Miglior et al., 2005). However, the economic value for growth rate may also be positive for countries in which meat production of surplus calves from the dairy herd is economically important.

Although a relatively large number of studies have analysed genetic parameters of growth rate for beef cattle strains, and several of these studies are based on large data sets (e. g. Crews et al., 2004; de Mattos et al., 2000; Giannotti et al., 2005), only a few studies have analysed the growth rate of dairy breeds, and these studies have been based on small samples from experimental herds (e.g., Brothersone et al., 2007).

Weight of male and female Israeli Holstein calves and yearling gain of 285 800 records from 105 935 animals from 458 herds recorded between 1994 and 2007 were analysed by Weller and Ezra (2008). Yearling gain of male and female calves were highly correlated genetically, thus records from both sexes were combined into a joint genetic analysis. Heritability and repeatability were 0.33 and 0.73. The genetic trend for yearling gain was 0.16 kg/yr, while phenotypic trends for first and last weighing were both negative.

Genetic and environmental correlations between growth rate and other economic traits were computed in various studies, and these generally tend to be economically negative or negligible (Brothersone et al., 2007). Both positive and negative correlations have been obtained between cow size and milk production (Brotherstone et al., 2007). The trait with the genetic correlation with the highest absolute value was female fertility, −0.4. Brotherstone et al. (2007) found a genetic correlation between growth during weaning and mastitis, but not with fertility disorders. Also, comparing two cow lines selected for large and small size, Hansen et al. (1999) found higher fertility in the cow line selected for small size as compared to that of large size. Weller and Ezra (2008) found that yearling gain was positively correlated with milk, fat and protein production; with SCS and with both the SC and cow calving effects on DC, but not with calf mortality. Yearling gain was negatively correlated with fertility and cow survival.

6.2 Longevity

Cows that live longer are more profitable for the following reasons:

1. Fewer replacement heifers are required, thus reducing replacement costs (allowing surplus heifers to be sold or enabling a larger proportion of cows to conceive to a beef bull).
2. More intensive selection along the dam-to-daughter path is possible. This possibility has become more important within the last decade, due to low cost genomic evaluation of young animals (Calus et al., 2015).
3. A greater proportion of potential lactation milk yield is captured from mature cows.

Literature estimates of the economic value of HL vary among studies, but centred on a relative value of 2.5:l for production: HL in units of the genetic standard deviations of both traits (VanRaden and Wiggans, 1995). The economic value of HL relative to production is greater than this value for three of the indices presented in Fig. 3. Using the values from Table 1 for milk production and HL, the value of a genetic standard deviation for HL (205 d) should be approximately equal to 364 kg milk. In PD11, HL accounts for 9.3% of the index versus 60% for fat and protein production.

Direct selection for HL is problematic; first, because actual measurement of HL is only possible after the cow has been culled, and most selection decisions, especially with respect to bulls, are made long before the last of their daughters are culled. Second, although there is disagreement as to the optimal measure of cow survival, all measures have relatively low heritability (e.g. Vollema and Groen 1996). Unlike milk yield traits, SCS and fertility, in which a record is generated for each parity, there is only one HL record per cow. Two approaches have been suggested to include HL in a breeding programme: inclusion of a direct measure of HL in the selection index (VanRaden and Wiggans 1995), or selection for other traits, which are correlated with HL, but have higher heritability and are expressed earlier (Vollema and Groen, 1997).

Four basic strategies have been suggested to evaluate HL for live cows. First, cow survival to a specific age can be analysed as a binary trait by either linear or threshold models (Harris et al., 1992; Jairath et al., 1998; Vollema and Groen 1998). Second, VanRaden and Klaaskate (1993) proposed estimating life expectancy of live cows and including these records in a linear model analysis. Estimates based on incomplete data are regressed towards the mean, and therefore have lower heritability and variance than do complete records (Settar and Weller, 1999; VanRaden et al., 1991). Therefore, incomplete records are multiplied by a factor to bring all records to an equal genetic variance. Because the incomplete records have lower heritability, this strategy results in greater residual variance for the incomplete records. The records are then weighted in the mixed model equations according to their residual variances. The third method is survival analysis or consideration of cows still alive as 'censored' records (Vollema and Groen, 1998) and the fourth method is random regression with multi-trait models of different orders (van Pelt et al., 2015).

There have been numerous suggestions for a definition of the longevity trait, based chiefly either on the number of parities or the actual HL (Vollema and Groen, 1996). VanRaden and Klaaskate (1993) evaluated months in milk up to a maximum of 84 months, but counted only a maximum of ten months per lactation, because records >305 d are not stored. Because genetic differences for milk yield have a major effect on direct measures of survival, many studies have proposed analysing 'functional' HL, which is generally computed as HL adjusted for milk yield (Dekkers, 1993; Jairath et al., 1998; Vollema and Groen, 1996, 1998). This trait accounts only for culling that is due to causes other than milk yield. The problem of double counting of production in a selection index that includes both milk yield and uncorrected HL correlated with yield can also be handled by computing appropriate economic values for these traits (VanRaden and Wiggans, 1995). Analysing HL adjusted for yield is problematic because, as noted in Section 1, selection goals change over time. In addition, since milk production has significantly higher heritability than all measures of HL, functional HL will have significantly lower heritability than unadjusted HL (e.g. Zavadilová and Štípková, 2012).

Settar and Weller (1999) studied the effect of incorporating data on pregnancy, DO, and protein yields in the computation of expected HL. They derived adjustment factors from variance components for complete and incomplete HL records estimated by a multi-trait animal model analysis with records truncated at different lengths considered as correlated traits. Based on this model, Weller and Ezra (2015) analysed HL of Israeli Holsteins born between 1983 and 2003. Estimated the phenotypic and genetic trends were 5.7 and 16.8 d/yr. The genetic trend was nearly linear, while the phenotypic trend showed 4 peaks and 3 valleys. DC and twinning on first parity calving reduced HL by approximately 180 and 120 d, but the interaction effect increased HL by 140 d. Heritability for HL was 0.14, similar to the value of 0.15 found by Van Pelt et al. (2015) by a random regression analysis, but more than double the value of 0.06 found by Pritchard et al. (2013).

7 Alternative methods to genetically improve functional traits

According to selection index theory, the optimum method for breeding within a population is to compute genetic evaluations for all traits and inclusion of these traits in a selection index. This is generally the case for milk production traits in all advance commercial populations. In Section 1, we considered in detail the problems with respect to direct breeding for non-production traits. Because of these problems, two major alternatives have been proposed to increase the economic value of cows for secondary traits, or at least minimize regression, due to economically negative correlations between non-production and production traits:

1. Evaluation and selection on traits that are genetically correlated with 'functional' traits, but are more amenable to genetic evaluation.

2. Crossbreeding breeds that are superior for production to breeds with economically higher genetic levels for secondary traits.

Each of these alternatives will now be considered in detail.

7.1 Breeding for secondary traits via selection on correlated traits

In eq. (13.1), we presented the expected genetic changes due to selection on a linear selection index. It is possible to include traits in **C**, the genetic variance matrix that are not scored, and therefore have index coefficients of zero. Thus, by using this equation, it is possible to estimate the expected change in a trait not under selection due to genetic correlations with other traits. If a trait is very expensive to score, or there are other limitations to recording, then selection on correlated traits may be the only economically viable alternative. Generally, selection on correlated traits is only efficient if the correlated traits have at least moderate heritability, can be scored at low cost, and the genetic correlations with the economic trait are moderate to large. It should also be noted that there is an intrinsic cost to including more traits in the index. The greater the number of traits included in the index, the less the genetic gain that will be achieved for each trait. Although many studies have considered the possibility of selection on correlated traits, very few have actually attempted to estimate the efficiency of this procedure. An exception is Boldman et al. (1992), who estimated the reliability of evaluations for HL computed from an index of 18 type traits.

In Section 5.1, we noted that the main justification for inclusion of SCS in most national selection indices is the relatively high genetic correlation between SCS and mastitis, although SCS does have an intrinsic economic value. SCS meets the additional criteria for selection on a correlated trait, in that its heritability is in the range of 0.2–0.25, and can be scored automatically by the same equipment used to measure fat and protein concentration. As far as I am aware, there is no study that attempted to estimate the genetic change in mastitis due to inclusion of SCS in a selection index.

Many studies have proposed that genetic gain for health traits and HL can be achieved by selection on conformation traits. Most of these studies have not gone beyond estimating genetic correlations among economic and conformation traits. Two problems should be noted with these analyses. First, unless sample sizes are very large standard errors of the genetic correlations (which are difficult to estimate accurately) will also be quite large. Second, the numbers of economic traits and conformation traits that are included in most studies are quite large. Thus, some combinations will be nominally 'significant' purely by chance. Assume, for example, that correlations are computed between 10 health traits and 20 conformation traits. Then 200 different genetic correlations are estimated, and 10 should display 5% 'significance' purely be chance! This problem can be solved by applying one of several methods to correct for multiple comparisons (Weller, 2009). Alternatively, application of the nominal significance values can be justified, if the same combination of traits is significant in several independent studies, or if there is a clear biological connection between the traits, for example, leg conformation and lameness.

Genetic correlations between disease and conformation traits are low to moderate (Van Dorp et al., 1998). Estimates of genetic and phenotypic correlations between type traits and SCS as an indicator of mastitis were mostly low to moderate (Rogers et al., 1995).

Rogers (1993) found a genetic correlation of −0.04 between lameness and foot angle. In the US-registered Holsteins, Van Dorp (1998) found that genetic correlations ranged from 0 to 0.37 between udder conformation traits and mastitis, from −0.38 to 0.09 between leg conformation traits and lameness, and from −0.11 to 0.38 between rump conformation and retained placenta.

With respect to relationships between conformation traits and HL, the results of Zavadilová and Štípková (2012) are typical. They analysed 4 longevity traits, 21 linear descriptive type traits, six composite traits and height at sacrum measured. Most genetic correlations had absolute values <0.2. Negative genetic correlations with all longevity traits were found for height at the sacrum, stature, dairy form, body conformation and capacity. Final score showed weak genetic correlation with all analysed longevity traits. Body condition score and angularity had genetic correlations of 0.3 and −0.31 with functional HL. Rear legs set (side view) had a genetic correlation of −0.22 with functional HL. The udder traits showed inconsistent and rather weak genetic correlations with longevity traits, with the exception of a genetic correlation of −0.22 between rear udder width and functional HL. The teat traits always showed negative genetic correlations with longevity traits. The strongest correlations were found for rear teat position (−0.28).

Boldman et al. (1992) found that fore udder attachment and udder depth had the highest genetic correlations with both true and functional HL of 18 conformation traits analysed. Genetic evaluations of sires for HL derived from an index of the type traits had a maximum reliability of 0.56. In conclusion, it is unlikely that significant genetic progress for either health traits or HL can be achieved via selection for conformation traits.

7.2 Breeding for secondary traits via cross-breeding among strains

Cross-breeding is a common strategy for most agricultural animal species, including poultry, swine, sheep and beef cattle. Generally, 'economic heterosis' is obtained; that is, the progeny of the cross is more profitable than either parental breed (Weller, 1994). For dairy cattle, cross-breeding has generally not been considered an economically viable option, due to the very low fertility of females. Unlike poultry, cross-bred female cows must also be used as dams for the following generation. If the first-generation cross-bred animals are bred back to either parental breed, then the heterosis effect is reduced in the backcross generation. Swan and Kinghorn (1992) attributed the dearth of cross-breeding in dairy cattle to the higher milk production of purebred Holsteins compared to other strains, although Buckley et al. (2014) presented examples of high economic heterosis.

Interest in cross-breeding Holstein cows to other breeds has increased in recent years; mainly to improve secondary traits, especially those related to female fertility (Weigel and Barlass, 2003; Funk, 2006; Buckley et al., 2014). Several studies analysed progeny of crosses between Holsteins and Normande, Montbéliarde, Scandinavian Red, Jersey and Brown Swiss (Dechow et al., 2007; Heins and Hansen, 2012; Heins et al., 2006a, b, c, 2012; Malchiodi et al., 2014; Olson et al., 2011). Generally, these studies have found that purebred Holsteins are superior for milk production, but somewhat inferior for fertility and HL. However, only a few hundred cows of each cross were generally included in the analyses (Dechow et al., 2007; Heins and Hansen, 2012; Heins et al., 2006a, b, c, 2012; Malchiodi et al., 2011, 2014; Olson et al., 2011).

8 Mapping and identification of quantitative trait loci (QTL) affecting functional traits

8.1 Detection of QTL in dairy cattle prior to development of SNP chips

In 1923, Sax demonstrated with beans that the effect of an individual QTL could be isolated though a series of crosses resulting in randomization of the genetic background with respect to all genes not linked to the genetic markers under observation. In dairy cattle, generation of specific crosses to detect QTL is not an option. Thus, all studies have been based on analysis of existing populations. In 1961, Neimann-Soressen and Robertson proposed a half-sib design for QTL detection in commercial dairy cattle populations. It was the first study to use blood groups rather than morphological markers, which attempted to estimate the power to detect QTL, and to consider the problem of multiple comparisons, due to joint analysis of several traits and markers. In 1989, with the discovery of microsatellites, it became possible to scan the genomes of all commercial animal populations for segregating QTL. Weller et al. (1990) demonstrated that reasonable power of detection could be obtained in commercial dairy cattle populations by genotyping several thousand individuals via the 'granddaughter design'. This design has been applied to all major dairy cattle populations, and nearly all traits of any interest have been analysed. The current release of the 'Cattle QTLdb' contains 36 693 QTLs on 492 different traits from 623 publications (http://www.animalgenome.org/cgi-bin/QTLdb/BT/index). However, the vast majority of reports have not been confirmed by independent studies, and significance values were generally not adjusted for multiple comparison. Thus, most reported QTL are likely 'false positives'.

If tens of traits are analysed for thousands of markers, nominal significance levels of 5 or 1% are meaningless. Lander and Kruglyak (1995) developed a formula that accounts for the fact that sites along a chromosome are linked. For a complete genome analysis of the bovine genome with 30 chromosomes, and a map length of 30 Morgans, a 'point-wise' probability value of 5×10^{-5} is required to obtain a 'genome-wide' probability value of 0.05. That is, the overall probability that a 'nominal' probability of 5×10^{-5} ($\log 5 \times 10^{-5} = -4.3$) for a specific genome location will be obtained at least once in the genome is 0.05.

8.2 Detection of segregating QTL as a by-product of genomic evaluation

With the advent of high-density single nucleotide polymorphism (SNP) DNA chips, genomic evaluation of dairy cattle became a reality. The mid-density Illumina BeadChip contains 54 609 SNPs, distributed throughout all 30 bovine chromosomes (http://www.illumina.com/products/bovine_snp50_whole-genome_genotyping_kits.html). Effective genomic evaluation requires genotyping of at least several thousand bulls with genetic evaluations based on a progeny test of 50–100 daughters (VanRaden et al., 2009). A detailed description of methodologies for genomic evaluation is beyond the scope of this chapter. However, all methods are based on estimating the effect of each informative SNP on the trait analysed. (For a review of genomic evaluation methodologies, see Weller, 2016.)

The 'effects' associated with the SNPs are generally due to population-wide linkage disequilibrium between segregating QTL and SNP and not due to the specific polymorphisms genotyped. Although the overwhelming majority of 'effects' estimated by high-density SNP chips are not significantly different from zero, SNPs with relatively major effects were found for nearly all traits. 'Manhattan' plots of the effects of all informative SNPs for the U.S. Holstein populations are presented at: https://www.cdcb.us/Report_Data/Marker_Effects/marker_effects.cfm?Breed=HO. These plots include all traits analysed in the United States, including the 5 milk production traits, net merit, 3 fertility traits, 4 calving traits, SCS, HL and 18 conformation traits. Although these plots indicate the approximate locations of segregating QTL, estimates of the effects, which are due to population-wide linkage disequilibrium, are biased downwards by recombination between the markers and the actual quantitative trait nucleotides (QTN).

VanRaden et al. (2011) proposed that recessive lethal alleles could be detected in genome scans by the lack of homozygotes for a specific haplotype. A requirement of this method is that the frequency of the haplotype in the population is sufficiently high, so that the lack of homozygotes is significantly different from their expected Hardy–Weinberg frequency. In addition, they postulated that a recessive lethal allele resulting in an early abortion would yield an observed reduction in fertility for heterozygotes. The reduction in fertility can be estimated as follows. Assume that according to Hardy–Weinberg equilibrium, frequency of heterozygotes should be $2p(1-p)$, where p is the frequency of haplotype containing the lethal allele. It is expected that a heterozygote will mate to another heterozygote with a probability of $[2p(1-p)]^2$, and the probability of producing a homozygous progeny for the lethal allele from this mating will be 1/4. Assuming that all these progeny will be aborted before pregnancy is recorded, the observed pregnancy rate for the heterozygotes will be reduced by $[p(1-p)]^2$. VanRaden et al. (2011) found five haplotypes in dairy cattle that met both criteria: lack of homozygotes and significant negative effects on conception rate. Of these, three were in Holsteins: HH1, HH2 and HH3 on chromosomes 5, 1 and 8, respectively. Causative mutations have been identified for HH1 and HH3 but not for HH2 (Daetwyler et al., 2014; McClure et al., 2014).

8.3 The 'a posteriori' granddaughter design

Ron and Weller (2007) proposed application of the 'a posteriori granddaughter design' (APGD) to determine QTL genotypes for bulls from large populations of individuals genotyped using mid- or high-density SNP chips. Similar to the original granddaughter design, sires with many progeny-tested sons are analysed. However, rather than genotype the sons specifically for application of a granddaughter design, the data generated by genotyping many bulls for high-density SNP chips are utilized. Thus, the design is considered 'a posteriori'. Like the original granddaughter design, sufficient statistical power to detect segregating QTL can only be obtained for the largest commercial dairy cattle populations.

Thirty-three traits scored in the U.S. Holstein population were analysed by this design (Weller et al., 2013, 2014; Wiggans and Weller, 2005). The most recent analysis of Wiggans and Weller (2015), based on April 2015 U.S. evaluations included analysis of 71 grandsires with 14 246 sons. The number of genotyped sons per grandsire ranged from 791 to 100. The entire bovine genome including 60 671 SNPs was divided into 621 segments of approximately 100 markers each. Haplotypes were determined using the 'findhap.f90' program (VanRaden, 2011). The sons of each bull were divided into two groups based on which paternal haplotype was passed to each son for the chromosomal region with the putative QTL. The statistical model for each haplotype segment was

$$Y_{ijk} = S_i + H_{ij} + e_{ijk}, \tag{13.4}$$

where Y_{ijk} is the genetic evaluation of bull k, which is a son of sire i that received sire haplotype j, S_i is the effect of sire i, H_{ij} is the effect of haplotype j of sire i and e_{ijk} is the random residual associated with each evaluation. An overall haplotype effect indicates that a QTL is segregating within the haplotype segment or in close proximity to it. A specific within-family haplotype effect indicates that the specific sire is segregating for the QTL. Since the haplotypes segment lengths are approximately 5 centi-Morgans, recombination within each segment over the single generation from sires to sons is minimal, and nearly the entire QTL effect is retained. This was confirmed by Weller et al. (2013), who found that the effects estimated by the APGD were generally greater than the effects associated with individual SNPs in the genomic evaluation of the U.S. Holstein population.

Excluding *DGAT1* and *ABCG2* (Grisart et al., 2002; Winter et al., 2002; Cohen-Zinder et al., 2005), for which causative polymorphisms have been identified, 55 trait × chromosome combinations were significant ($P < 10^{-14}$). All traits had at least one significant effect, except for protein yield, daughter SB rate and four conformation traits. APGD significant

Table 7 A posteriori granddaughter design significant effects ($P < 10^{-15}$) by trait, chromosome number and first SNP in the chromosomal segment with the lowest nominal probability for functional traits (Wiggans and Weller, 2015)

Trait	Chromosome	SNP	Nominal probability
Somatic cell score	5	13 244	1.79×10^{-19}
	6	16 674	1.39×10^{-39}
	14	34 278	2.43×10^{-18}
	16	38 739	1.09×10^{-16}
Productive life	5	13 244	5.05×10^{-19}
	6	16 674	2.47×10^{-26}
Daughter pregnancy rate	1	1863	7.69×10^{-17}
	5	13 838	1.59×10^{-26}
	18	42 650	2.87×10^{-27}
Cow conception rate	5	13 937	1.75×10^{-29}
	18	42 748	4.14×10^{-29}
Heifer conception rate	5	13 343	5.88×10^{-23}
	6	16 965	4.87×10^{-22}
Daughter dystocia	18	42 944	2.70×10^{-16}
Service-sire dystocia	5	13 244	1.41×10^{-19}
	18	42 944	2.27×10^{-18}
Service-sire stillbirth rate	5	13 046	5.14×10^{-16}

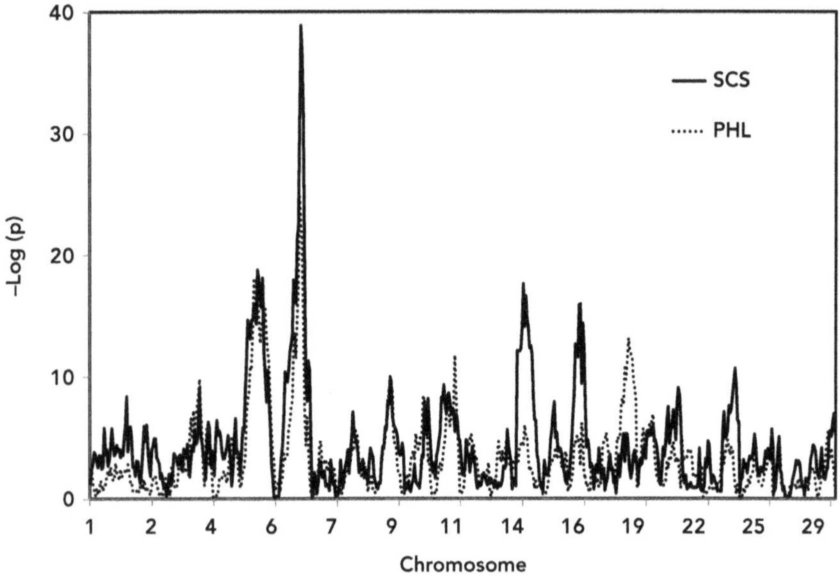

Figure 8 Results of the 'a posteriori' granddaughter design analysis of the U.S. Holstein population for SCS and productive herd-life (PHL). The –log10 of the nominal probability is plotted for each of 605 segments covering all 29 bovine autosomes.

effects (nominal $P < 10^{-15}$) by trait, chromosome number, and first SNP in the chromosomal segment with the lowest nominal probability for functional traits are given in Table 7. With respect to the three recessive lethals found by VanRaden et al. (2011), significant APGD effects were found for CCR and DPR on chromosomes 5 (HH1) and 1 (HH2), but not on chromosome 8. A major QTL on chromosome 18 apparently affects both fertility and DC, both the effect of the sire and the cow calving. The –log10 of the nominal probability for SCS and productive HL are plotted for each of 605 segments, covering all 29 bovine autosomic chromosomes in Fig. 8. Both traits display major peaks on chromosomes 5 and 6. Thus, it is likely that the same QTL are affecting both traits.

8.4 From QTL to QTN

Ron and Weller (2007) presented a schematic strategy for determination if a genetic variant is a QTN in farm animals. As they proposed, the next step to find the QTNs for these effects will be determination of 'concordance' between effects in individual families and specific polymorphisms within the CI. That is to determine that all sires heterozygous by the APGD for a segregating QTL are also heterozygous for a putative QTN, and all sires homozygous for the QTL are also homozygous for the putative QTN. This will require genomic sequencing of the sires. Of the 71 families analysed, 42 sires have been sequenced, and their sequence data are available through the 1000 Bull Genomes Project (Daetwyler et al., 2014).

In addition to concordance, Ron and Weller (2007) have presented additional statistical and biochemical criteria to verify that the QTN has been determined. In cattle, QTN has been determined and verified only for *DGAT1* and *ABCG2* (Grisart et al., 2002; Winter et al., 2002; Cohen-Zinder et al., 2005). *DGAT1* chiefly affects fat concentration, and *ABCG2* chiefly affects protein concentration. Both are missense mutations. That is, the causative polymorphism is a single DNA base pair, which results in a change in the amino acid code. It is not clear at this point if missense mutations are likely to be the rule or the exception for the causative polymorphisms for QTL in cattle. If the QTN are known, then these factors can be included directly in the genomic analysis model, resulting in more accurate genetic evaluations. Thus, determination of the QTN should result in increased rates of genetic gain (Weller and Ron, 2011).

9 Summary

Inclusion of secondary or 'functional' traits in breeding objectives began in earnest only in the late 1990, but by 2003, production traits accounted for less than 60% of the indices in most advanced countries. Although there is nearly complete consensus of the economic importance of fertility, health traits and HL, genetic evaluation and inclusion of these traits in selection indices were hindered by difficulty of measurement and analysis and generally low heritability. Statistical methods have been developed that overcome the problems encountered in the analysis of non-production traits, including discrete distributions, and records that become available only after selection decisions have been made. Current state of the art for genetic analysis of fertility, calving traits, health, growth rate and HL are explained. Genetic parameters and genetic and phenotypic trends for these traits are presented. The advantages and limitations of breeding for disease traits and HL via selection on correlated traits and via cross-breeding are discussed. Finally, we discussed methodologies for detection and analysis of the actual segregating genes that affect functional traits and the current state of the art in the age of genomics.

10 Future trends in research

Prior to 2010, breeding programmes for relatively small breeds, or breeding programmes for Holsteins in the smaller countries were viable. Over the last five years, genomic selection has become the rule for all commercial dairy cattle populations. Since the reliabilities of genomic evaluations are dependent on the number of bulls with both genotypes and progeny tests, consortiums based on several countries have been established and continue to grow. On the other hand, breeding programmes for the smaller breeds are becoming less viable. Thus, small countries are faced with the choice of either joining one of the major consortiums or closing down their breeding programmes.

Genotyping and sequencing costs will continue to decline. In addition to genotyping candidate bulls, genotyping of female calves is also becoming a routine tool to make selection decisions at the individual farm (e.g. Calus et al., 2015). Genotypes for most animals in the population will have other additional valuable by-products, including unequivocal determination of parentage, and the ability to trace food products from 'the farm to the fork'.

The trends for increased emphasis on secondary or functional traits in breeding goals will increase, as breeding for these traits become more effective, due to genomics and better recording systems. Novel traits that affect product quality, cost efficiency and the environmental impact of dairying will probably be included in future selection indices. These traits include hoof health, other health traits, feed efficiency, methane emissions, milking speed, temperament, further dissection of milk composition (e.g. fatty acids and lactoferrin) and cow profitability. All of these traits are under active research in either the United States or Canada (Chesnais et al., 2016).

Emission of methane gas by cows contributes to climate change. Globally, livestock contributes 18% (7.1 billion tons CO_2 equivalent) of global greenhouse gas emissions (http://www.fao.org/agriculture/lead/themes0/climate/en/). Feed accounts for as much as 50% of costs on dairy farms in North America, and feed costs have been increasing recently (Chesnais et al., 2016). Furthermore, genetic increase in feed efficiency is associated with a decrease in methane emission. All methods to directly measure individual feed intake and methane emissions are expensive, and do not allow for data collection on a very large scale (Chesnais et al., 2016). The two options that are likely to provide sufficient selection accuracy, despite the limited number of actual records, are the use of indicator traits and genomic prediction. The global Dry Matter Initiative includes ten research herds from nine countries. The Initiative's goal is to combine feed intake records from different sources to create a large cow reference population for genomic selection of dry matter intake (de Haas et al., 2015). Mid-infrared spectral data is a useful predictor of individual methane emission and feed efficiency of dairy cows (Dehareng et al., 2012; McParland et al., 2014). Yin et al. (2015) developed a strategy that combines deterministic equations and stochastic simulations to predict daily methane emissions based on routinely recorded on-farm data. The heritability of predicted lactation methane emissions is 0.44. Genetic correlations with both milk production and DO are >0.8. Thus, selection for methane emission reduction will improve fertility as measured by DO, but impede selection for milk production.

The fact that breeding has become a worldwide enterprise has of course advantages, but also potential dangers, specifically increase in inbreeding. Despite the nearly 10 million U.S. Holstein cows, the effective population size of the Holstein breed is probably less than 100 (Chesnais et al., 2016). It is a real possibility that within the next decade, most Holstein bulls may be progeny of a single exceptional sire. At present, no one seems to have a practical method to limit inbreeding, and still remain economically viable. The effective population size of the world Holstein population will continue to decline. On the commercial level, it is hard to see that other breeds or cross-breeding programmes can compete with genomic selection programmes within the Holstein breed. Only through international funding will it be possible to maintain land races or even the smaller commercial breeds.

11 Where to look for further information

I know of no books or review articles that specifically deal with genetic analysis of non-production traits in dairy cattle. A good place to look is at the recent *Interbull Bulletins* at: https://journal.interbull.org/. With respect to the principles of quantitative genetics, I recommend Lynch, M. and Walsh, B. (1998), *Genetics and Analysis of Quantitative Traits*, Sinauer Associates, Inc. Sunderland, MA. Lynch and Walsh also have a good explanation of BLUP methodology and methods to estimate variance components. With respect to the specific analysis models in common use for analysis of dairy cattle, see Mrode, R. A. (2014), *Linear Models for the Prediction of Animal Breeding Values*, CABI Publishing, 350 pp. (https://www.bookdepository.com/Linear-Models-for-Prediction-Animal-Breeding-Values-Raphael-Mrode/9781845939816). With respect to principles of selection index and economic aspects of breeding, see Weller, J. I. (1994), *Economic Aspects of Animal Breeding*, Chapman & Hall, 244 pp. For genomic evaluation, see Gondro, C., van der Werf, J., and Hayes, B., Editors (2013), *Genome-Wide Association Studies and Genomic Prediction*, Humana Press, New York, 566 pp. (http://www.springer.com/gp/book/9781627034463), J. E. Womack, Editor (2012), *Bovine Genomics*, Wiley-Blackwell, Ames, Iowa, 284 pp. (http://eu.wiley.com/WileyCDA/WileyTitle/productCd-0813821223.html) and Weller, J. I. (2016), *Genomic selection in Animals*. For analysis of quantitative trait loci and genomic selection, see Weller, J. I. (2009), *Quantitative Trait Loci Analysis in Animals*, second edition. CABI Publishing. London. 272 pp.

12 Acknowledgements

I thank E. Ezra for useful suggestions.

13 References

Bar-Anan, R., Soller, M. and Bowman, J. C. (1976), 'Genetic and environmental factors affecting the incidence of difficult calving and perinatal calf mortality in Israeli Friesian dairy herds', *Anim. Prod.* 22(June): 299–310.

Berger, P. J. (1994), 'Genetic prediction for calving ease in the United States: data, models, and use by the dairy industry', *J. Dairy Sci.* 77(4): 1146–53.

Berry, D. P., Bermingham, M. L., Good, M. and More, S. J. (2011), 'Genetics of animal health and disease in cattle', *Irish Vet. J.* 64: 5.

Bishop, S. C. and Woolliams, J. A. (2010), 'On the genetic interpretation of disease data', *PLoS One* 5(1): e8940.

Boelling, D., Sander Nielsen, U., Pösö, J., Eriksson, J.-Å. and Aamand, G. P. (2007), 'Genetic evaluation of calving traits in Denmark, Finland and Sweden', *Interbull Bull.* 37: 179–84. Dublin, Ireland.

Boldman, K. G., Freeman, A. E., Harris, B. L., and Kuck, A. L. (1992), 'Prediction of sire transmitting abilities for herd life from transmitting abilities for linear type traits', *J. Dairy Sci.* 75(2): 552–63.

Brotherstone, S., Coffey, M. P. and Banos, G. (2007), 'Genetic parameters of growth in dairy cattle and associations between growth and health traits', *J. Dairy Sci.* 90(1): 444–50.

Buckley, F., Lopez-Villalobos, N. and Heins, B. J. (2014), 'Crossbreeding: implications for dairy cow fertility and survival', *Animal* 8(s1): 122–33.

Calus, M. P. L., Bijma, P. and Veerkamp, R. F. (2015), 'Evaluation of genomic selection for replacement strategies using selection index theory', *J. Dairy Sci.* 98(9): 6499–509.

Chesnais, J. P., Cooper T. A., Wiggans, G. R., Sargolzaei, M., Pryce, J. E. and Miglior, F. (2016), 'Using genomics to enhance selection of novel traits in North American dairy cattle', *J. Dairy Sci.* 99 (3): 2413–27.

Cole, J. B., Wiggans, G. R. and VanRaden, P. M. (2007a), 'Genetic evaluation of stillbirth in United States Holsteins using a sire maternal grandsire threshold model', *J. Dairy Sci.* 90(5): 2480–8.

Cole, J. B., Wiggans, G. R., VanRaden, P. M. and Miller, R. H. (2007b), 'Stillbirth (co)variance components for a sire-maternal grandsire threshold model and development of a calving ability index for sire selection', *J. Dairy Sci.* 90(5): 2489–96.

Crews, D. H., Lowerison, M., Caron, N. and Kemp, R. A. (2004), 'Genetic parameters among growth and carcass traits of Canadian Charolais cattle', *Can. J. Anim. Sci.* 8(4): 589–97.

Cue, R. I. and Hayes, J. F. (1985), 'Correlations between calving ease and calf survival', *J. Dairy Sci.* 68(4): 958–62.

Daetwyler, H. D., Capitan, A., Pausch, H., Stothard, P., van Binsbergen, R., Brøndum, R. F., Liao, X., Djari, A., Rodriguez, S. C., Grohs, C., Esquerré, D., Bouchez, O., Rossignol, M.-N., Klopp, C., Rocha, D., Fritz, S., Eggen, A., Bowman, P. J., Coote, D., Chamberlain, A. M., Anderson, C., Van Tassell, C. P., Hulsegge, I., Goddard, M. E., Guldbrandtsen, B., Lund, M. S., Veerkamp, R. F., Boichard, D. A., Fries, R. and Hayes, B. J. (2014), 'Whole-genome sequencing of 234 bulls facilitates mapping of monogenic and complex traits in cattle', *Nat. Genet.* 46(8): 858–65.

Dechow, C. D., Rogers, G. W., Cooper, J. B., Phelps, M. I. and Mosholder, A. L. (2007), 'Milk, fat, protein, somatic cell score, and days open among Holstein, Brown Swiss, and their crosses', *J. Dairy Sci.* 90(7): 3542–9.

de Haas, Y., Prycem, J. E., Calus, M. P. L., Wall, E., Berry, D. P., Løvendahl, P., Krattenmacher, N., Miglior, F., Weigel, K., Spurlock, D., Macdonald, K. A., Hulsegge, B. and Veerkamp, R. F. (2015), 'Genomic prediction of dry matter intake in dairy cattle from an international data set consisting of research herds in Europe, North America, and Australasia', *J. Dairy Sci.* 98(9): 6522–34.

Dehareng, F., Delfosse, C., Froidmont, E., Soyeurt, H., Martin, C., Gengler, N., Vanlierde, A. and Dardenne, P. (2012), 'Potential use of milk mid-infrared spectra to predict individual methane emission of dairy cows', *Animal* 6(10): 1694–701.

Dekkers, J. C. M. (1993), 'Theoretical basis for genetic parameters of herd life and effects on response to selection', *J. Dairy Sci.* 76(5): 1433–43.

de Mattos, D., Misztal, I. and Bertrand, J. K. (2000), 'Variance and covariance components for weaning weight for Herefords in three countries', *J. Anim. Sci.* 78(1): 33–7.

Djemali, M., Berger, P. J. and Freeman, A. E. (1987), 'Ordered categorical sire evaluation for dystocia in Holsteins', *J. Dairy Sci.* 70(11): 2374–84.

Falconer, D. S. (1981), *Introduction to Quantitative Genetics*, 2nd ed. Logman, Inc., New York, NY.

Funk, D. A. (2006), 'Major advances in globalization and consolidation of the artificial insemination industry', *J. Dairy Sci.* 89(4): 1362–8.

Giannotti, J. D. G., Packer, I. U. and Mercadante, M. E. Z. (2005), 'Meta-analysis for heritability of estimates growth traits in beef cattle', *Brazil. J. Anim. Sci.* 34(4): 1173–80.

Gianola, D. and Foulley, J. L. (1983), 'Sire evaluation for ordered categorical data with a threshold model', *Genet. Sel. Evol.* 15(2): 201–23.

Goddard, M. E. (1983), 'Selection indices of non-linear profit functions', *Theor. Appl. Genet.* 64(4): 339–44.

Häggman, J. and Juga, J. (2013), 'Genetic parameters for hoof disorders and feet and leg conformation traits in Finnish Holstein cows', *J. Dairy Sci.* 96(3): 3319–25.

Hagiya, K., Yamazaki, T., Nagamine, Y., Togashi, K., Yamaguchi, S., Gotoh, Y., Kawahara, T., Masuda, Y. and Suzuki, M. (2014), 'Genetic correlations between production and disease traits during first lactation in Holstein cows', *Animal* 8(2): 217–23.

Hansen, L. B., Cole, J. B., Marx, G. D. and Seykora, A. J. (1999), 'Productive life and reasons for disposal of Holstein cows selected for large versus small body size', *J. Dairy Sci.* 82(4): 795–801.

Harris, B. L., Freeman, A. E. and Metzger, E. (1992), 'Analysis of herd life in Guernsey dairy cattle', *J. Dairy Sci.* 75(7): 2008–16.

Hazel, L. N. (1943), 'The genetic basis for constructing selection indexes', *Genetics* 28(6): 476–90.

Heins, B. J. and Hansen, L. B. (2012), '*Short communication:* Fertility, somatic cell score, and production of Normande 3 Holstein, Montbéliarde 3 Holstein, and Scandinavian Red 3 Holstein crossbreds versus pure Holsteins during their first 5 lactations', *J. Dairy Sci.* 95(2): 918–24.

Heins, B. J., Hansen, L. B. and De Vries, A. (2012), 'Survival, lifetime production, and profitability of Normande 3 Holstein, Montbeliarde 3 Holstein, and Scandinavian Red 3 Holstein crossbreds versus pure Holsteins', *J. Dairy Sci.* 95(2): 1011–21.

Heins, B. J., Hansen, L. B. and Seykora, A. J. (2006a), 'Fertility and survival of pure Holsteins versus crossbreds of Holstein with Normande, Montbeliarde, and Scandinavian Red', *J. Dairy Sci.* 89(12): 4944–51.

Heins, B. J., Hansen, L. B. and Seykora, A. J. (2006b), 'Production of pure Holsteins versus crossbreds of Holstein with Normande, Montbeliarde, and Scandinavian Red', *J. Dairy Sci.* 89(7): 2799–804.

Heins, B. J., Hansen, L. B. and Seykora, A. J. (2006c), 'Calving difficulty and stillbirths of pure Holsteins versus crossbreds of Holstein with Normande, Montbeliarde, and Scandinavian Red', *J. Dairy Sci.* 89(7): 2805–10.

Henderson, C. R. (1973), 'Sire evaluation and genetic trends', pp. 10–41 in *Proc. Anim. Breed. Genet. Symp. In Honor of Dr. D. L. Lush.* ASAS and ADSA, Champaign, IL.

Heuven, H. C. M. (1987), 'Diagnostic and genetic analysis of mastitis field data', Ph.D. Diss. Univ. Wisconsin, Madison.

Hossein-Zadeh, N. G. and Ardalan, M. (2011a), 'Bayesian estimates of genetic parameters for cystic ovarian disease, displaced abomasum and foot and leg diseases in Iranian Holsteins via Gibbs sampling', *J. Agri. Sci.* 149(1): 119–24.

Hossein-Zadeh, N. G. and Ardalan, M. (2011b), 'Bayesian estimates of genetic parameters for metritis, retained placenta, milk fever, and clinical mastitis in Holstein dairy cows via Gibbs sampling', *Research Vet. Sci.* 90(1): 146–9.

Jairath, L., Dekkers, J. C. M., Schaeffer, L. R., Liu, Z., Burnside, E. B. and Kolstad, B. (1998), 'Genetic evaluation for herd life in Canada', *J. Dairy Sci.* 81(2): 550–62.

Janson, L. (1980), 'Studies on fertility traits in Swedish dairy cattle. II. Genetic parameters', *Acta Agric. Scand.* 30(4): 427–36.

Janson, L. and Andreasson, B. (1981), 'Studies on fertility traits in Swedish dairy cattle. IV. Genetic and phenotypic correlations between milk yield and fertility', *Acta Agric. Scand.* 31(3): 313–22.

Kuhn, M. T. and Hutchison, J. L. (2008), 'Prediction of dairy bull fertility from field data: use of multiple services and identification and utilization of factors affecting bull fertility', *J. Dairy Sci.* 91(6): 2481–92.

Kuhn, M. T., Hutchison, J. L. and Wiggans, G. R. (2006), 'Characterization of Holstein heifer fertility in the United States', *J. Dairy Sci.* 89(12): 4907–20.

Lander, E. and Kruglyak, L. (1995), 'Genetic dissection of complex traits: guidelines for interpreting and reporting linkage results', *Nature Genet.* 11(3): 241–7.

Leitch, H. W. (1994), 'Comparison of international selection indices for dairy cattle breeding', *Proc. Interbull Ann. Meet.* Ottwa, CA. Bull. 10.

Liu, Z., Pasman, E., Reinhardt F. and Reents, R. (2012), 'A multi-parity animal model for genetic evaluation of calving traits enhanced with genomic information', *Interbull Bull.* 46: 133–40. Cork, Ireland.

Lucy, M. C. (2001), 'Reproductive loss in high-producing dairy cattle: where will it end?', *J. Dairy Sci.* 84(6): 1277–93.

Lucy, M. C. (2007), 'Fertility in high-producing dairy cows: reasons for decline and corrective strategies for sustainable improvement', *Soc. Reprod. Fertil. Suppl.* 64: 237–54.

Malchiodi, F., Cecchinato, A. and Bittante, G. (2014), 'Fertility traits of purebred Holsteins and 2- and 3-breed crossbred heifers and cows obtained from Swedish Red, Montbéliarde, and Brown Swiss sires', *J. Dairy Sci.* 97(12): 7916–26.

Malchiodi, F., Penasa, M., Tiezzi, F., and Bittante, G. (2011), 'Milk yield traits, somatic cell score, milking time and age at calving for pure Holstein versus crossbred cows', *Agriculturae Conspectus Scientificus* 76(3): 259–61.

McClure, M. C., Bickhart, D., Null, D., VanRaden, P., Xu, L. Y., Wiggans, G., Liu, G., Schroeder, S., Glasscock, J., Armstrong, J., John, B. Cole, J. B., Van Tassell, C. P. and Sonstegard, T. S. (2014), 'Bovine Exome Sequence Analysis and Targeted SNP Genotyping of Recessive Fertility Defects BH1, HH2, and HH3 Reveal a Putative Causative Mutation in SMC2 for HH3', *Plos One* 9(3): e92769.

McParland, S., Kennedy, E., Lewis, E., Moore, S. G., McCarthy, B., O'Donovan, M., Butler, S. T., Pryce, J. E. and Berry, D. P. (2014), 'Mid-infrared spectrometry of milk as a predictor of energy intake and efficiency in lactating dairy cows', *J. Dairy Sci.* 97(9): 5863–71.

Miglior, F., Muir, B. L. and Van Doormaal, B. J. (2005), 'Selection indices in Holstein cattle of various countries', *J. Dairy Sci.* 88(3): 1255–63.

Neimann-Sørensen, A. and Robertson, A. (1961), 'The association between blood groups and several production characters in three Danish cattle breeds', *Acta. Agr. Scand.* 11(1): 163–96.

Neuenschwander, T. F.-O., Miglior, F., Jamrozik, J., Berke, O., Kelton, D. F. and Schaeffer, L. R. (2012), 'Genetic parameters for producer-recorded health data in Canadian Holstein cattle', *Animal* 6(4): 571–8.

Olson, K. M., Cassell, B. G., Hanigan, M. D. and Pearson, R. E. (2011), '*Short communication*: Interaction of energy balance, feed efficiency, early lactation health events, and fertility in first-lactation. Holstein, Jersey, and reciprocal F1 crossbred cows', *J. Dairy Sci.* 94(1): 507–11.

Politis, I. and Ng-Kwai-Hang, K. F. (1988), 'Association between somatic cell count of milk and cheese-yielding capacity', *J. Dairy Sci.* 71(7): 1720–7.

Pritchard, T., Coffey, M., Mrode, R. and Wall, E. (2013), 'Understanding the genetics of survival in dairy cows', *J. Dairy Sci.* 96(5): 3296–309.

Rogers, G. W. (1993), 'Index selection using milk yield, somatic cell score, udder depth, teat placement, and foot angle', *J. Dairy Sci.* 76(2): 664–70.

Rogers, G. W., Hargrove, G. L. and Cooper, J. B. (1995), 'Correlations among somatic cell scores of milk within and across lactations and linear type traits of Jerseys', *J. Dairy Sci.* 78(4): 914–20.

Ron, M., Baranan, B. and Wiggans, G. R. (1984), 'Factors affecting conception rate of Israeli Holstein cattle', *J. Dairy Sci.* 67(4): 854–60.

Ron, M. and Weller, J. I. (2007), 'From QTL to QTN identification in livestock – winning by points rather than knock-out: a review', *Anim. Genet.* 38(5): 429–39.

Sax, K. (1923), 'The association of size differences with seed-coat pattern and pigmentation in *Phaseeolus vulgaris*', *Genetics* 8(6): 552–60.

Schutz, M. M. (1994), 'Genetic evaluation of somatic cell scores for United States dairy cattle', *J. Dairy Sci.* 77(7): 2113–29.

Settar, P. and Weller, J. I. (1999), 'Genetic analysis of cow survival in the Israeli dairy cattle population', *J. Dairy Sci.* 82(10): 2170–7.

Shook, G. E. and Schutz, M. M. (1994), 'Selection on somatic cell score to improve resistance to mastitis in the United States', *J. Dairy Sci.* 77(2): 648–58.

Swan, A. A. and Kinghorn, B. P. (1992), 'Evaluation and exploitation of crossbreeding in dairy cattle', *J. Dairy Sci.* 75(2): 624–39.

Van Dorp, T. E., Dekkers, J. C. M., Martin S. W. and Noordhuizen, J. P. T. M. (1998), 'Genetic parameters of health disorders, and relationships with 305-day milk yield and conformation traits of registered Holstein cows', *J. Dairy Sci.* 81(8): 2264–70.

van Pelt, M. L., Meuwissen, T. H. E., de Jong, G. and Veerkamp, R. F. (2015), 'Genetic analysis of longevity in Dutch dairy cattle using random regression', *J. Dairy Sci.* 98(6): 4117–30.

VanRaden, P. M. 2011. findhap.f90. Accessed January 23, 2013. http://aipl.arsusda.gov/software/findhap/.

VanRaden, P. M. and Klaaskate, E. J. H. (1993), 'Genetic evaluation of length of productive life including predicted longevity of live cows', *J. Dairy Sci.* 76(9): 2758–64.

VanRaden, P. M., Olson, K. M., Null, D. J. and Hutchison, J. L. (2011), 'Harmful recessive effects on fertility detected by absence of homozygous haplotypes', *J. Dairy Sci.* 94(12): 6153–61.

VanRaden, P. M., Sanders A., H., Tooker, M. E., Miller, R. H., Norman, H. D., Kuhn, M. T. and Wiggans, G. R. (2004), 'Development of a national genetic evaluation for cow fertility', *J. Dairy Sci.* 87(7): 2285–92.

VanRaden, P. M, Van Tassell, C. P., Wiggans, G. R.,Sonstegard, T. S., Schnabel, R. D., Taylor, J. F. and Schenkel, F. S. (2009), 'Invited review: Reliability of genomic predictions for North American Holstein bulls', *J. Dairy Sci.* 92(1): 16–24.

VanRaden, P. M. and Wiggans, G. R. (1995), 'Productive life evaluations: calculation, accuracy and economic value', *J. Dairy Sci.* 78(3): 631–8.

VanRaden, P. M., Wiggans, G. R., and Ernst, C.A. (1991), 'Expansion of projected lactation yield to stabilize genetic variance', *J. Dairy Sci.* 74(12): 4344-49.

Van Tassell, C. P., Wiggans, G. R. and Misztal, I. (2003), 'Implementation of a sire-maternal grandsire model for evaluation of calving ease in the United States', *J. Dairy Sci.* 86(10): 3366–73.

Vollema, A. R. and Groen, A. F. (1996), 'Genetic parameters of longevity traits of an upgrading population of dairy cattle', *J. Dairy Sci.* 79(12): 2261–7.

Vollema, A. R. and Groen, A. F. (1997), 'Genetic correlations between longevity and conformation traits in an upgrading dairy cattle population', *J. Dairy Sci.* 80(11): 3006–14.

Vollema, A. R. and Groen, A. F. (1998), 'A comparison of breeding value predictors for longevity using a linear model and survival analysis', *J. Dairy Sci.* 81(12): 3315–20.

Weigel, K. A. and Barlass, K. A. (2003), 'Results of a producer survey regarding crossbreeding on US dairy farms', *J. Dairy Sci.* 86(12): 4148–54.

Weller, J. I. (1989), 'Genetic analysis of fertility traits in the Israeli dairy cattle population', *J. Dairy Sci.* 72(10): 2644–50.

Weller, J. I. (1994), *Economic Aspects of Animal Breeding*. Chapman & Hall, London. 244 pp.

Weller, J. I. (2009), *Quantitative Trait Loci Analysis in Animals*, second edition. CABI Publishing, London. 272 pp.

Weller, J. I. (2016), *Genomic Selection in Animals*. John Wiley & Sons, Inc., Hoboken, New Jersey. 175 pp.

Weller, J. I., Cole, J. B., VanRaden, P. M. and Wiggans, G. R. (2014), 'Application of the a posteriori granddaughter design to the Holstein genome', *Animal* 8(4): 511–19.

Weller, J. I. and Ezra, E. (1997), 'Genetic analysis of somatic cell concentration and female fertility of Israeli Holsteins by the individual animal model', *J. Dairy Sci.* 80(3): 586–93.

Weller, J. I. and Ezra, E. (2004), 'Genetic analysis of the Israeli Holstein dairy cattle population for production and non-production traits with a multitrait animal model', *J. Dairy Sci.* 87(5): 1519–27.

Weller, J. I. and Ezra, E. (2008), 'Genetic analysis of growth rate of Israeli Holstein calves', *Animal* 2(12): 1717–23.

Weller, J. I. and Ezra, E. (2015), 'Environmental and genetic factors affecting cow survival of Israeli Holsteins', *J. Dairy Sci.* 98(1): 676–84.

Weller, J. I. and Ezra, E. (2016), 'Genetic analysis of calving traits by the multi-trait individual animal model', *J. Dairy Sci.* 99(1): 427–42.

Weller, J. I. and Gianola, D. (1989), 'Models for genetic analysis of dystocia and calf mortality', *J. Dairy Sci.* 72(10): 2633–43.

Weller J. I., Kashi, Y. and Soller, M. (1990), 'Power of daughter and granddaughter designs for determining linkage between marker loci and quantitative trait loci in dairy cattle', *J. Dairy Sci.* 73(9): 2525–37.

Weller, J. I., Misztal, I. and Gianola, D. (1988), 'Genetic analysis of dystocia and calf mortality in Israeli-Holsteins by threshold and linear models', *J. Dairy Sci.* 71(9): 2491–501.

Weller, J. I. and Ron, M. (1992), 'Genetic analysis of fertility traits in Israeli Holsteins by linear and threshold models', *J. Dairy Sci.* 75(9): 2541–8.

Weller, J. I., Saran, A. and Zeliger, Y. (1992), 'Genetic and environmental relationships among somatic cell count, bacterial infection, and clinical mastitis', *J. Dairy Sci.* 75(9): 2532–40.

Weller, J. I., VanRaden, P. M. and Wiggans, G. R. (2013), 'Application of a posteriori granddaughter and modified granddaughter designs to determine Holstein haplotype effects', *J. Dairy Sci.* 96(8): 5376–87.

Westell, R. A., Quaas, R. L. and Van Vleck, L. D. (1988), 'Genetic groups in an animal model', *J. Dairy Sci.* 71(5): 1310–18.

Wiggans, G. R., Cole, J. B. and Thornton, L. L. M. (2008), 'Multiparity evaluation of calving ease and stillbirth with separate genetic effects by parity', *J. Dairy Sci.* 91(8): 3173–8.

Wiggans, G. R., Misztal, I. and Van Tassell, C. P. (2003), 'Calving ease (co)variance components for a sire-maternal grandsire threshold model', *J. Dairy Sci.* 86(5): 1845–8.

Wiggans, G. R. and Shook, G. E. (1987), 'A lactation measure of somatic cell count', *J. Dairy Sci.* 70(12): 2666–72.

Wiggans, G. R. and Weller, J. I. (2015), 'Revisiting the "A posteriori" granddaughter design', *Interbull Bull.* 49: 36–42. Orlando, Florida.

Yin, T., Pinent, T., Brügemann, K., Simianer, H. and König, S. (2015), Simulation, prediction, and genetic analyses of daily methane emissions in dairy cattle', *J. Dairy Sci.* 98(8): 5748–62.

Young, C. W. (1992), 'Breeding dairy cattle for disease resistance', 42 p. in *Large Dairy Herd Management*. Van Horn, H. H. and Wilcox, C. J., eds. Am. Dairy Sci. Assoc., Champaign, IL.

Zavadilová, L. and Štípková, M. (2012), 'Genetic correlations between longevity and conformation traits in the Czech Holstein population', *Czech J. Anim. Sci.* 57(3): 125–36.

Zwald, N. R., Weigel, K. A., Chang, Y. M., Welper, R. D. and Clay, J. S. (2004), 'Genetic selection for health traits using producer-recorded data. I. Incidence rates, heritability estimates, and sire breeding values', *J. Dairy Sci.* 87(12): 4287–94.

Chapter 8

Breeding and management strategies to improve reproductive efficiency in dairy cattle

D. J. Ambrose, Alberta Agriculture and Forestry[1], University of Alberta[2], Canada; and J. P. Kastelic, University of Calgary[3], Canada

1 Introduction

Approximately 150 million households around the world produce milk (FAO, 2013) with up to one billion people deriving their livelihood from the dairy sector (International Dairy Federation, 2016). Cattle, water buffaloes and other livestock contribute to global milk production, with >80% of milk from dairy cattle, ~15% from water buffaloes and <5% from goats, sheep, camels, yaks and so on. The global value of milk and milk products is estimated to exceed US$300 billion annually (International Dairy Federation, 2016). Clearly, the dairy industry is of huge importance to the global economy.

The global dairy industry is continually changing. China is making substantial investments to develop its dairy industry (Sharma and Rou, 2014); annual growth of domestic milk production was 12.8% from 2001 to 2010, and was projected to be 3.3% between 2011 and 2020, with a projected 38% increase in dairy consumption by 2022. Similarly, India, the

1. Livestock Research Branch, Alberta Agriculture and Forestry.
2. Department of Agricultural, Food and Nutritional Science, University of Alberta.
3. Department of Production Animal Health, Faculty of Veterinary Medicine, University of Calgary.

http://dx.doi.org/10.19103/AS.2016.0005.16

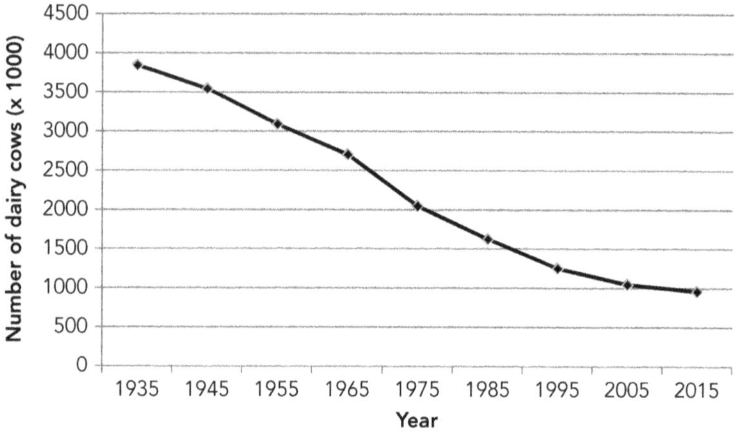

Figure 1 Changes in the number of dairy cows in Canada over the past 80 years. Source: Statistics Canada, accessed through Canadian Dairy Information Centre. http://dairyinfo.gc.ca/index_e.php?s1=dff-fcil&s2=farm-ferme&s3=nb.

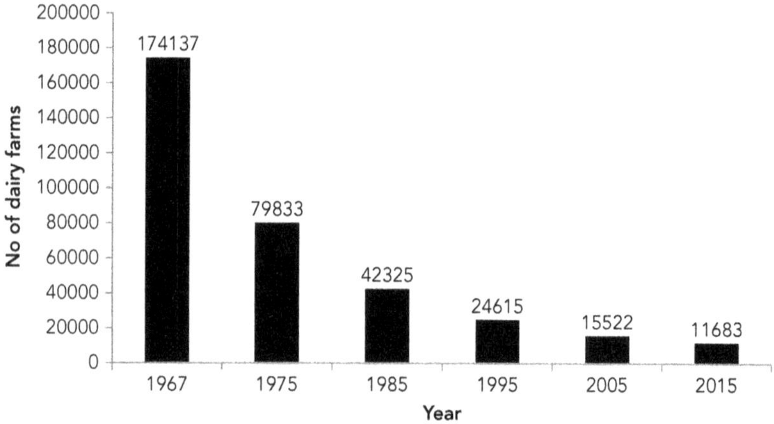

Figure 2 Change in the number of Canadian dairy farms over the past five decades. Source: Canadian Dairy Information Centre. http://dairyinfo.gc.ca/index_e.php?s1=dff-fcil.

largest milk-producing nation (mostly small farms) is modernizing its dairy industry (Swormink, 2014). Concurrently, global milk production has only a 1.9% annual growth rate, with 73% of the additional global milk production of 150 million tonnes this decade expected to come from developing countries, of which 38% is from India and China (Sharma and Rou, 2014).

Countries where most dairy cows are confined (e.g. Canada, the United States and Europe), greatly improved milk yield (per-cow basis), and increased herd size were observed. Changes in the number of dairy cows and dairy farms in Canada over the past several decades are shown in Fig. 1 and 2).

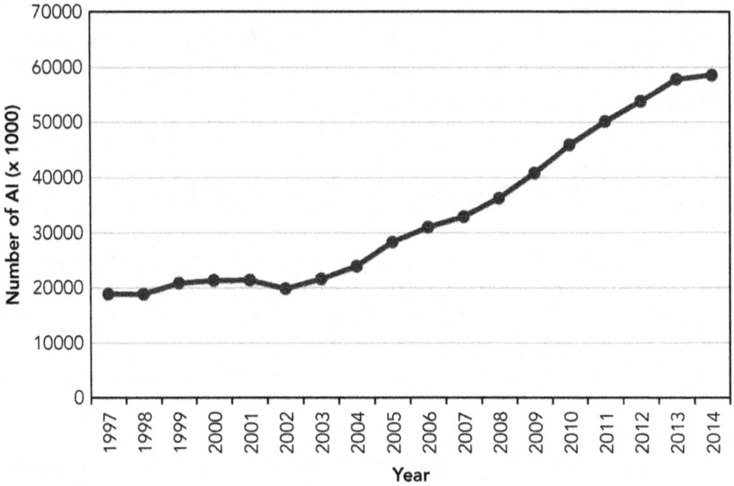

Figure 3 Growth in use of AI in India. Number (millions) of AI in cattle and water buffalo performed by government agencies in India between 1997 and 2014. Source: National Dairy Development Board.

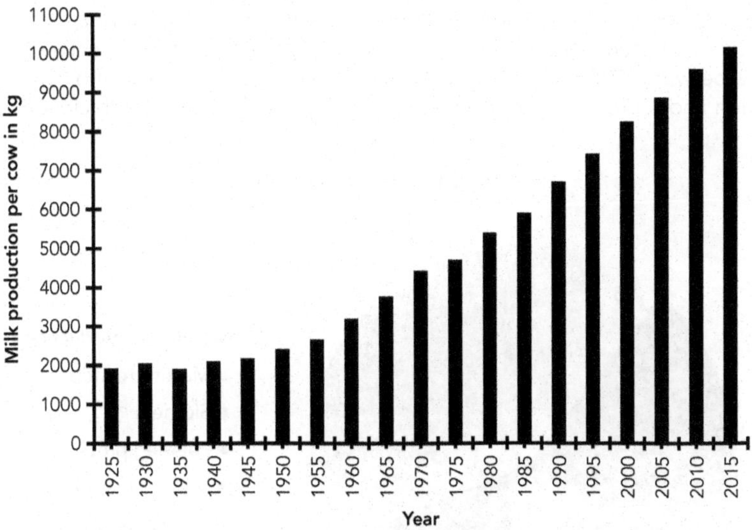

Figure 4 Milk production per cow (kg/y) in the United States from 1925 to 2015. Source: USDA, National Agricultural Statistics Service.

The use of artificial insemination (AI) to improve genetic merit of dairy cows started in the late 1930s (Foote, 2002). Approximately 90% of dairy cows in Europe, >80% in the United States (Gillespie et al., 2014) and >75% in Canada (Van Doormaal and Kistemaker, 2003) are bred by AI. In India, the use of AI tripled (~20 million to > 60 million) in the past decade (Fig. 3). Due to the resulting genetic progress (and improved management), annual milk production per cow has increased from <2000 kg in 1925 to >10000 kg as of 2016 in the United States (Fig. 4; USDA-NASS, 2016) and this currently exceeds 11400 kg in Israel (Flamenbaum and Galon, 2010).

2 Reproductive efficiency in dairy cattle

Although high reproductive efficiency is critical to sustainable dairy farming, reproductive failure is the primary reason for culling dairy cows in many countries (Seegers et al., 1998; Rozzi et al., 2007; Swedish Dairy Association, 2009–12; Ahlman et al., 2011; Ansari-Lari et al., 2012), and accounts for ~30 and 36.5% of all culling in North America (USDA, 2002; CanWest DHI, 2014; Fig. 5) and England (Esslemont and Kossaibati, 1997), respectively. Infertility is also the primary reason for culling dairy cattle managed on pasture (Crosse et al., 1999, cited by Maher et al., 2006).

Concurrent with increasing milk production, dairy cow fertility is generally declining in North America (Lucy, 2001; Westwood et al., 2002) and elsewhere (Macmillan et al., 1996; Royal et al., 2000; Lopez-Gatius, 2003; Kumaresan et al., 2009; Barbat et al., 2010; Dochi et al., 2010; Walsh et al., 2011). Annual decreases in conception rate (CR) in the United States (Beam and Butler, 1999) and Canada (Bosquet et al., 2004) were ~0.4% (mid-1970s to 1990s) and 0.5% (1990–2000) and even faster in Europe (Hoekstra et al., 1994; Jorritsma and Jorritsma, 2000). Royal et al. (2000) reported that the CR to first service (insemination) after calving in the United Kingdom declined from 56% to ~40% between 1975 and 1998. Although an antagonistic association between milk production and reproductive performance may be inferred, there is no clear evidence of

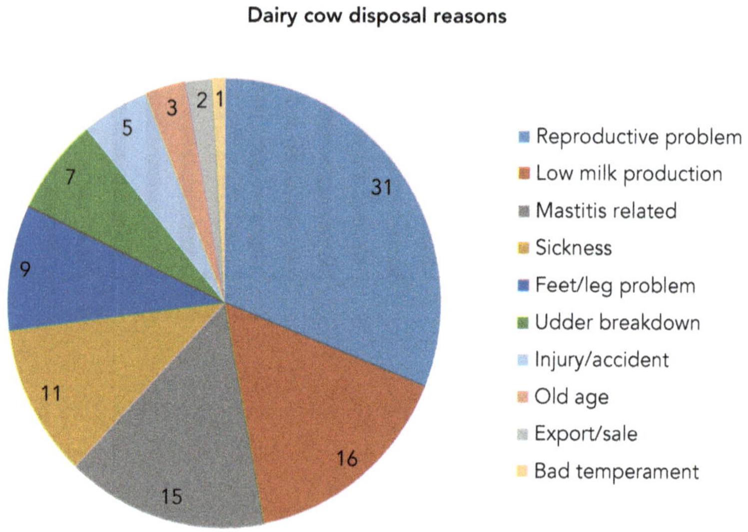

Figure 5 Reasons for culling dairy cows in Canada (expressed as per cent). At 31%, reproductive problems remain the primary reason for culling. Source: CanWest DHI Ontario Progress Report, 2014.

a cause-and-effect relationship (Raheja et al., 1989; Bello et al., 2012). In this regard, high milk production and high first-service CRs often co-exist in well-managed dairy herds (Peters and Pursley, 2002; Lopez-Gatius et al., 2006; Galon et al., 2010; Leblanc, 2010, 2013).

The cow-to-person ratio usually increases with herd size, which may reduce reproductive efficiency. Therefore, it is essential to assess reproductive performance, identify key problems and deficiencies, and plan and deliver corrective actions.

Due to the widespread use of AI in the dairy industry, accurate oestrus detection and timely and skilful performance of AI are of paramount importance to optimize reproductive efficiency if cattle are bred following detection of oestrus. Therefore, all persons involved in AI should have a very good understanding of the oestrous cycle and of both the primary and secondary signs of oestrus.

It is essential to sustain the motivation of those who are engaged in oestrus detection. An excellent attitude, interest and knowledge are extremely important for achieving reproductive success in dairy herds. Casual and new employees may not realize the importance of reproductive management and could have limited or no interest in oestrus detection. So, it becomes all the more important to send such staff to extension meetings and workshops dealing with dairy reproductive management so that they could gain first-hand knowledge.

3 The oestrous cycle and oestrus behaviour

The oestrous cycle of dairy cattle averages 21 days (range, 18–24) and has four phases (Fig. 6). Pro-oestrus, the interval from regression of the corpus luteum (CL) until manifestation of behavioural oestrus, lasts 3 to 4 days. Oestrus is the sexually receptive (and shortest) phase, and its primary sign is that cows 'stand' to be mounted by a bull or female herd mate (Fig. 7). For AI, breeding must be done during or shortly after the end of 'standing oestrus' to maximize CRs (Fig. 8). Duration of standing oestrus varies considerably and is longer in heifers than in cows (18–24 vs 8–12 h, respectively; O'Connor, 2007) with very short oestrus intervals and few mounts in high-producing cows (O'Connor, 2007). Metoestrus begins at the end of oestrus and lasts 3 to 5 days. Ovulation occurs during metoestrus, usually 24–32 h after the onset of standing oestrus (O'Connor, 2007). Vaginal bleeding during metoestrus is more common in heifers than in cows (Hansen and Asdell, 1952) and is attributed to extravasation of blood after oestrogen concentrations decrease (Hansen and Asdell, 1952; Peter et al., 2009a). Dioestrus, the longest phase of the cycle, is characterized by the CL actively secreting progesterone to prepare the uterus for implantation. If a viable embryo is present in the uterus, at ~15 d after oestrus, the elongating embryo produces interferon-tau, which suppresses the expression of oestrogen receptor alpha and oxytocin receptor, thereby suppressing the oxytocin-dependent pulsatile release of prostaglandin $F_{2\alpha}$ (PGF) and preventing luteolysis (Thatcher et al., 1989; Bazer, 2013). Concurrently, activation of numerous genes in the conceptus and maternal endometrium facilitate cross talk between the embryo and the uterus (Mamo et al., 2012), leading to maternal recognition of pregnancy (i.e. maintenance of CL). However, in the absence of a viable embryo, oestradiol from the dominant ovarian follicle binds to endometrial oestrogen receptors, resulting in the release of PGF, luteolysis and then oestrus.

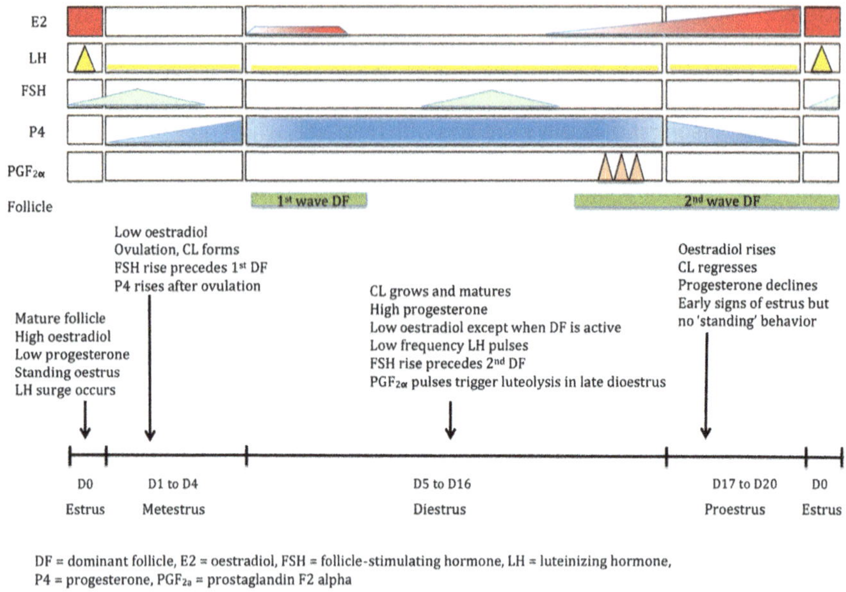

Figure 6 Phases of the oestrous cycle and events associated with each phase. Lighter and darker shades represent lower and higher hormone concentrations. The tapered ends represent increasing or decreasing concentrations. Upright triangles with narrow bases represent surge (LH) or pulsatile (PGF2a) release patterns. The thin horizontal bars represent basal LH concentrations.

Figure 7 'Standing oestrus', the primary sign of oestrus.

Key hormones in regulation of the oestrous cycle include oestradiol, oxytocin, PGF, GnRH, FSH, LH and progesterone. The relative presence and action of the various hormones at the four stages of the oestrous cycle are summarized (Fig. 6).

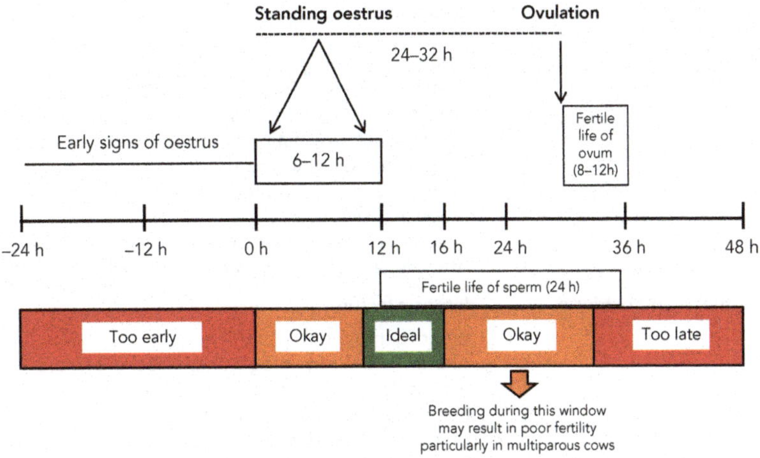

Figure 8 Optimum time for AI success. A schematic depiction of timelines associated with oestrus and ovulation. The duration of standing oestrus is only 6–12 h in lactating dairy cows. The interval from onset of standing oestrus to ovulation is 24–32 h, and the fertile life of an ovum after exiting the follicle at ovulation is only 8–12 h. Sperm may live in the female reproductive tract for up to 48 h, but viability is reduced beyond 24 h. The best time to breed a cow by AI is 12 h after the onset of standing oestrus, although acceptable CRs could be attained if AI occurred anytime from the onset of standing oestrus until up to 24 h. Breeding cows closer to ovulation time may result in poor fertility, particularly in multiparous cows (Stevenson et al., 2014).

4 Factors affecting reproductive efficiency

Many factors, either independently or through their interactions, can influence reproductive efficiency. The main factors affecting reproduction can be broadly grouped into four categories, namely human (managerial), animal (intrinsic and extrinsic) nutritional, and environmental. Specific examples under each category are summarized and briefly discussed in Table 1.

4.1 Human or managerial factors

Voluntary waiting period (VWP), also referred to as elective waiting period, is the minimum interval in days from calving to first insemination. In North America, the VWP is often 60 days (may range from 50 to >90), with variations among herds and even within herds (e.g. based on milk production). Clearly, VWP has great potential to affect reproductive efficiency.

Poor oestrus detection efficiency is a primary cause of reduced reproductive efficiency in dairy herds. In herds using AI, accurate detection of oestrus is extremely important for reproductive success. If protocols are used to synchronize ovulation for fixed-time AI without detection of oestrus, AI submission rate equals the oestrus detection rate (EDR). In Canadian studies, the EDR during the early 1990s was 48% (Kinsel and Etherington, 1998), but more recently, mean 21-d AI submission rates were 33% (Leblanc, 2005) and 38% (Ambrose and Colazo, 2007). In the latter study, only 42% of eligible cows had been inseminated by 80 days postpartum, and 23% had not been inseminated by 125 days postpartum, emphasizing the importance of oestrus detection.

Table 1 Factors affecting reproductive performance of dairy cows

Human (managerial factors)	Voluntary waiting period (VWP) Oestrus detection frequency and efficiency Use of oestrus detection aids Use of oestrus/ovulation synchronization protocols Semen storage, thawing and handling Insemination time and technique Pregnancy diagnosis Feeding, disease and environmental management Attitude, education, knowledge and skill
Animal (intrinsic and extrinsic factors)	*Intrinsic factors* Breed/genotype Anatomic and physiologic anomalies or barriers Age and parity (e.g. heifers vs cows) Energy status/body condition Level of milk production Expression of oestrus (behaviour) Low conception rate (CR) Embryonic loss Susceptibility to infectious disease and metabolic disorders Stress In natural service herds: Bull libido Semen quality Sperm survivability in female reproductive tract *Extrinsic factors* Infectious (e.g. metritis, mastitis, neosporosis, leptospirosis) Non-infectious (e.g. acidosis, ketosis, lameness, cystic ovary) Calving-related events (e.g. dystocia, retained placenta)
Nutritional	Fats and fatty acids Protein Starch Amino acids, minerals and vitamins
Environmental	Ambient temperature (heat and humidity, air quality) Extreme cold, wet and windy conditions Flooring (natural vs concrete) Light/photoperiod Contaminants in feed and water (e.g. mycotoxins) Stocking density Stray voltage

Whether oestrus detection aids such as tail chalk, tail paint, mount detectors, pedometers or other electronic activity monitors are used in addition to routine visual observation, or not used, will influence reproductive efficiency. Herds that rely only on visual observations are likely to have a lower EDR than herds using oestrus detection aids. Similarly, using oestrus or ovulation synchronization protocols increases AI submission rates and reproductive efficiency.

Other human factors that affect reproductive efficiency include storage conditions of frozen semen, semen thawing, pre- and post-thaw-handling of semen straws, inseminator

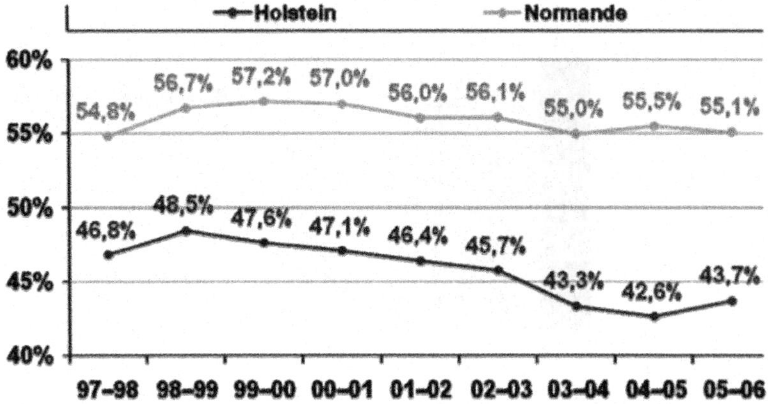

Figure 9 Conception rate changes in Holstein and Normande cows over 8 years in France (1997–8 to 2005–6). Reproduced with permission of the Society for Reproduction and Development from Barbat et al. (2010). Female fertility in French dairy breeds: Current situation and strategies for improvement. *J. Reprod. Dev.* 56 (Suppl): S15–S21.

skill, insemination technique and timing of insemination relative to oestrus and ovulation. Inadequacies in any of these factors can reduce CRs.

Yet another major determinant of reproductive efficiency is identification of non-pregnant cows as soon as possible after breeding. Although trans-rectal palpation of uterine contents for pregnancy diagnosis can be used reliably only beyond ~35 d, newer technologies make earlier pregnancy determination a reality (discussed in detail under Strategies to Improve Reproductive Efficiency).

Other managerial decisions relating to feeds and feeding, disease management (e.g. vaccination) and environmental management (e.g. heat abatement during hot summers) can have major impacts on reproductive efficiency. Finally, the attitude, skills and knowledge of personnel involved in reproductive management and their willingness to learn, can have a major influence on reproductive success. Unmotivated and unskilled workers should be considered a liability and if not willing to improve, ideally reassigned to chores not involving reproductive management.

4.2 Animal factors

Both intrinsic and extrinsic factors can influence reproductive outcomes in dairy cattle. Among intrinsic animal factors, breed and genotype have a tremendous impact. For example, Holsteins are reported to be less fertile than other dairy breeds like Jersey (Norman et al., 2009) and Normande (Barbat et al., 2010; Fig. 9).

4.3 Intrinsic factors

Anatomical defects in the reproductive tract (e.g. kinked or blind cervix, segmental aplasia of the uterus, freemartinism, hydrosalpinx, blocked oviduct) can either hinder AI or interfere with gamete transport, fertilization or pregnancy sustenance. Aberrant endocrine function could result in abnormal hormone concentrations disrupting normal reproductive

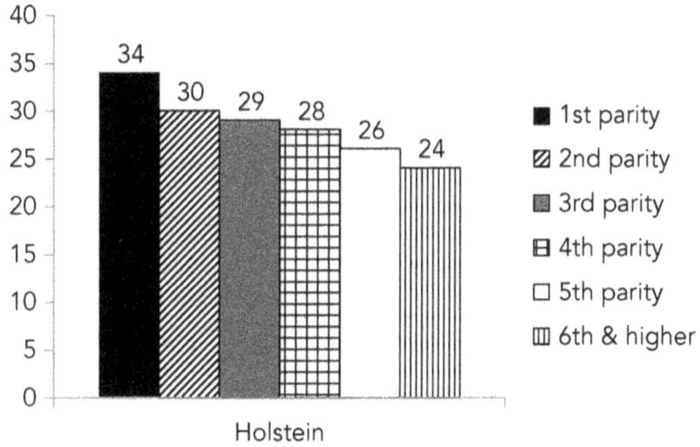

Figure 10 Influence of parity on first-service CRs in US Holstein cows (*n* = 1032506) in 2006. Source: Norman et al. (2009).

processes such as resumption of cyclicity postpartum (extended anoestrus), expression of oestrus behaviour and ovulation, or create a non-conducive environment for gamete transport, fertilization and pregnancy establishment (López-Gatius, 2012).

Age and parity exercise a major influence on fertility in dairy cattle. Whereas nulliparous heifers have high CRs of 65–75% (Pursley et al., 1997a; Ambrose et al., 2005; Balendran et al., 2008) following AI, CRs in cows are usually considerably lower (40% or lower; Pursley et al., 1997a; Dochi et al., 2010). Parity is also a significant factor affecting CRs in dairy cows, with primiparous cows having higher CRs than multiparous cows. Although this has been widely recognized (Tenhagen et al., 2004; Balendran et al., 2008), a report from the United States involving more than 1 million breeding records is one of the largest databanks to demonstrate the negative influence of parity on first-service CRs (Norman et al., 2009; Fig. 10). Furthermore, lactating cows have high embryonic losses (Santos et al., 2004) and greater susceptibility to infections, metabolic disorders and stress (Dobson and Smith, 2000; Dobson et al., 2008; Walker et al., 2008), all of which can reduce reproductive performance.

High milk production can negatively affect oestrus expression (Lucy, 2001; Van Eerdenburg et al., 2002; Lopez et al., 2004), as high-producing cows are more likely than first-lactation cows to have negative energy balance postpartum (Butler and Smith, 1989). Although this can impair fertility, well-managed herds with very high milk production can still maintain good fertility (Leblanc, 2010, 2013).

In herds exclusively using AI, semen should come from reliable sources and be of good quality and high fertility. Therefore, effects of male factor on reproductive efficiency in herds managed solely by AI will not be discussed. However, in herds using natural service, either exclusively or partially, the male factor is relevant. Breeding soundness evaluation should be performed by an experienced veterinarian before first using a bull and subsequently perhaps once annually (more frequently if there is reduced fertility).

4.4 Extrinsic factors

Calving disorders such as dystocia and retained foetal membranes predispose cattle to uterine infections (Kinsel and Etherington, 1998; Fourichon et al., 2000; Opsomer et al.,

2000). Cows that suffer from metritis, mastitis or other early postpartum diseases typically have lower fertility and require a much longer interval from calving to conception (Santos et al., 2009). Furthermore, non-infectious conditions and metabolic disorders such as acidosis, ketosis, lameness, cystic ovaries and so on are also known to delay the interval from calving to first service and CRs in dairy cows (López-Gatius, 2012).

If bulls are used for breeding, they should be tested and determined to be free of sexually transmitted diseases, particularly trichomoniasis and campylobacteriosis (Bondurant, 2005; Michi et al., 2016).

4.5 Nutritional factors

Inadequate energy intake during the early postpartum period is common in high-producing dairy cows, resulting in negative energy balance, with mobilization of fat and high concentrations of non-esterified fatty acids (NEFA). High NEFA concentrations have negative effects on oocyte function and embryo quality, which likely contribute to subfertility in dairy cows (Leroy et al., 2005; Van Hoeck et al., 2014). In addition, high-protein diets (which result in increased blood urea nitrogen concentrations), fats, long-chain polyunsaturated fatty acids, certain vitamins, amino acids and minerals also influence reproductive outcomes (Mattos et al., 2000; Ambrose et al., 2006; Bourne et al., 2007; Santos et al., 2010; Sinclair et al., 2014; Leroy et al., 2015).

4.6 Environmental factors

Extreme heat and humidity cause severe heat stress-induced fertility impairment in dairy cattle (Hansen, 1997; Jordan, 2003) and significant economic losses (St-Pierre et al., 2003). Although extreme cold and wet/windy conditions could have a detrimental effect on reproductive function, the impact of cold weather on reproductive outcomes is not yet extensively studied. However, it is known that severe cold stress can have adverse effects on reproductive function (Young, 1983; Gwasdauskas, 1985).

Other environmental factors that can have an indirect influence on reproduction in dairy cattle include:

1. Flooring (concrete vs dirt or bedded pack – better footing promotes expression of oestrus)
2. Type of housing (tie-stall or stanchion-barn vs free-stall or loose-housing – oestrus detection is much easier in barns where cows are free-roaming)
3. Stocking density (subordinate cows may not get access to good quality feed in overcrowded barns, putting them at risk of negative energy balance, affecting reproductive function)
4. Wet or dirty stalls (increased risk of mastitis, which can reduce reproductive performance)
5. Poor ventilation (reduces air quality, suppresses the immune system and increases susceptibility to disease, with indirect effects on reproduction)
6. Contaminants in feed and water – dairy cows may be exposed to various contaminants, environmental, bacterial or fungal, and toxins (e.g. mycotoxins, plant toxins) including gossypol and phytoestrogens (D'Mello, 2004), all of which can affect reproductive function.
7. Photoperiod (Dahl et al., 2000) and stray voltage (Appleman and Gustafson, 1985) may influence reproductive function in dairy cows, although strong evidence directly implicating these factors in poor reproductive efficiency is lacking.

5 Strategies to improve reproductive efficiency in cows

Various strategies can be used to improve reproductive efficiency in dairy herds. Managerial factors are the easiest to implement; therefore, they will be discussed first. Since poor oestrus detection efficiency is a major factor that reduces reproductive inefficiency, the first strategy should be to improve oestrus detection efficiency.

5.1 Improving oestrus detection

Among the many human factors that affect reproductive performance, inefficient oestrus detection is the most important; on average, up to 50% of oestrus events are undetected (Van Erdenberg et al., 2002). CR is defined as the per cent cows pregnant per insemination, whereas pregnancy rate (PR) is the per cent cows pregnant considering all eligible cows in the herd intended to be inseminated. For example, if 10 cows are eligible for breeding and 7 of the 10 cows are detected in oestrus and inseminated within a 21-day interval, then the 21-day EDR is 70% (7/10). Thereafter, if 3 of the 7 inseminated cows are confirmed pregnant, the CR is 43% (3/7) and the PR is 30% (3/10). Thus, PR is a product of EDR and CR. Increasing EDR can improve PR, even if CR remains constant. For example, if the CR is 50%, the PR will increase from 15 to 50% as EDR increases from 30 to 100%.

Based on blood or milk progesterone concentrations, 11–25% of cows are routinely inseminated when not in oestrus (Nebel et al., 1987; Ambrose and Colazo, 2007). They may be inseminated too early, too late or at a time when they are not even close to being in oestrus (e.g. dioestrus phase). This must be borne in mind when detecting oestrus in dairy cattle. Furthermore, 5–10% of all pregnant cows exhibit oestrus, usually from mid- to late gestation, although this can occur during any stage of pregnancy (Choudary et al., 1965; Thomas and Dobson, 1989; Dijkhuizen and Van Eerdenburg, 1997). Ovarian follicular growth occurs during pregnancy; large follicles increase blood oestradiol concentrations and occasionally trigger oestrus behaviour in pregnant cattle (O'Connor, 2007). Therefore, it is essential that farm personnel responsible for oestrus detection and AI are familiar with the breeding history of cows to minimize chances for erroneous inseminations. In this regard, colour-coded identification schemes (e.g. tail chalk or paint) can be used to identify cows that have recently calved, cows due for insemination, inseminated cows waiting for pregnancy diagnosis and cows confirmed pregnant. Such a system can help to quickly determine the status of cows in oestrus.

Allowing cows to interact with their herd mates in an open area on natural (dirt) footing rather than concrete floor improves the likelihood that cows will mount and oestrus detected (Table 2).

Table 2 Oestrus activity (mounting and standing to be mounted) of Holstein cows on dirt versus concrete flooring

Measure	Dirt floor	Concrete floor
Duration of oestrus (h)	13.8	9.4
Total number of mounts	7.0	3.2
Total number of stands	6.3	2.9

Source: Britt et al. (1986).

Efficiency of oestrus detection can be improved with aids (e.g. tail chalk, paint or Kamar heat mount detector). Applying chalk or paint to the tail head of cows and using a simple scoring system (e.g. 5 to 0; no change in colour = 5; complete loss of colour = 0) will enhance oestrus detection efficiency. Cows with a paint score of <2 may have been in standing oestrus after the last observation period; they should be watched carefully and bred upon confirming oestrus.

In larger herds where oestrus detection is poor, an electronic oestrus detection system may be cost-effective. The efficacy of HeatWatch, a radiotelemetric mount detecting system, has been reported (Bailey, 1997; Dransfield et al., 1998; Xu et al., 1998; Nebel et al., 2000), but the company is no longer in business. In studies that used the HeatWatch system, duration of oestrus in lactating Holstein cows was much shorter than previously documented. For instance, >70% of cows remained in oestrus for <12 h (30% for <6 h), with an average duration of oestrus of 8.5 h and ~10 mounts (Dransfield et al., 1998; Nebel et al., 2000). Some cows had less than three mounts, each lasting ~2 s. Most of the oestrus activity occurred when cows were on dirt surfaces, and travelling to and from the milking parlour, regardless of time of day.

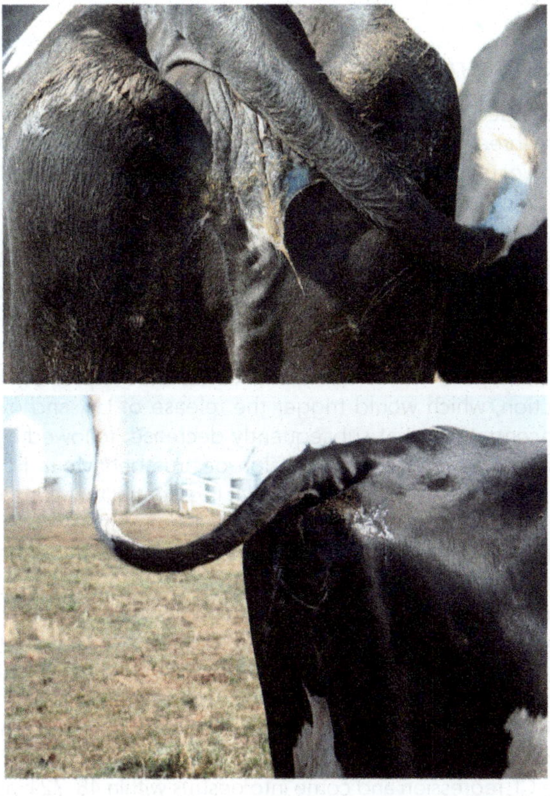

Figure 11 Presence of mucus on the vulva, tail and hindquarters are secondary signs of oestrus in cattle.

There are various activity-monitoring systems that are currently marketed for detection of oestrus in dairy cattle (Fricke et al., 2014; Madureira et al., 2015; Neves and LeBlanc, 2015). Although their efficacy varies, in one study, overall reproductive performance was similar or better than in herds that used timed-AI programmes (Neves and LeBlanc, 2015).

Oestrus detection matters! Improved oestrus detection efficiency can dramatically improve reproductive performance. Therefore, it is important to invest time into systematic detection of oestrus in dairy herds. Simple approaches such as increased frequency of observation, dedicated observation time and the use of oestrus detection aids can improve results. The recommended frequency of observation is 3 or 4 times a day, for 20 min at each occasion. While observing cows for oestrus, do not concurrently perform other tasks like cleaning, mixing or delivering feed. Observing cows for signs of oestrus from a short distance (preferably positioned above ground level to improve the view), or gently walking through the barn or animal pen without unduly disturbing the cows (encouraging cows to move slowly will generate new interactions that can stimulate mounting activity) provide the greatest opportunity to detect cows in oestrus. Walking through the animal pen facilitates visual inspection of the vulva, hindquarters, tail and tail head for secondary signs of oestrus such as vulvar oedema, mucus discharge, ruffled hair coat, fresh mucous on the tail and hindquarters (Fig. 11), excessively friendly cows and other tell-tale signs of imminent, current or recent oestrus.

5.2 Adopting oestrus synchronization

The next recommended strategy is to use oestrus synchronization. Inducing oestrus in groups of at least three cows will promote cows forming sexually active groups, which facilitates detection of oestrus. Although a single injection of PGF is the simplest procedure for induction and synchronization of oestrus, there are many approaches, including the use of progesterone and gonadotropin-releasing hormone (GnRH).

Giving PGF, either in its natural form (e.g. dinoprost) or as an analogue (e.g. cloprostenol), causes the CL to regress (i.e. luteolysis), leading to a rapid decline in progesterone followed by oestrus and ovulation. Progesterone (produced by the CL) keeps cows from expressing oestrus and sustains pregnancy. When blood progesterone concentrations decline, the viable dominant follicle that is present will grow, increase oestradiol production, which would trigger the release of LH, and ovulate. High blood progesterone concentrations that subsequently decrease, followed by increasing blood oestradiol concentrations, promote expression oestrus behaviour. For a cow to display oestrus from ~48 to 120 h after PGF treatment, it is essential that a mature CL and healthy follicle be present on her ovary at the time of PGF injection. Response to PGF is maximal between days 6 and 16 when the CL is active, whereas a newly formed CL (0–5 d post-ovulation) is generally refractory.

In cattle, ovarian follicles grow in a wave-like pattern (two or three waves) during each oestrus cycle (Adams et al., 2008). Each wave begins with a variable number of follicles (~10–20) 4–5 mm in diameter, with selection of a dominant follicle after approximately 3 d. The first wave emerges concurrent with ovulation, with the first dominant follicle reaching a maximum diameter at approximately 5 or 6 days after ovulation, and remaining viable for a few days. The second wave emerges at ~days 9 or 10. Cows receiving PGF on days 6, 7 or 8 are likely to have CL regression and come into oestrus within 48–72 h as the first dominant follicle is active, ready to grow and ovulate if progesterone declines. However, if PGF is given on days 9–11, the interval from treatment to oestrus is variable, depending on whether the

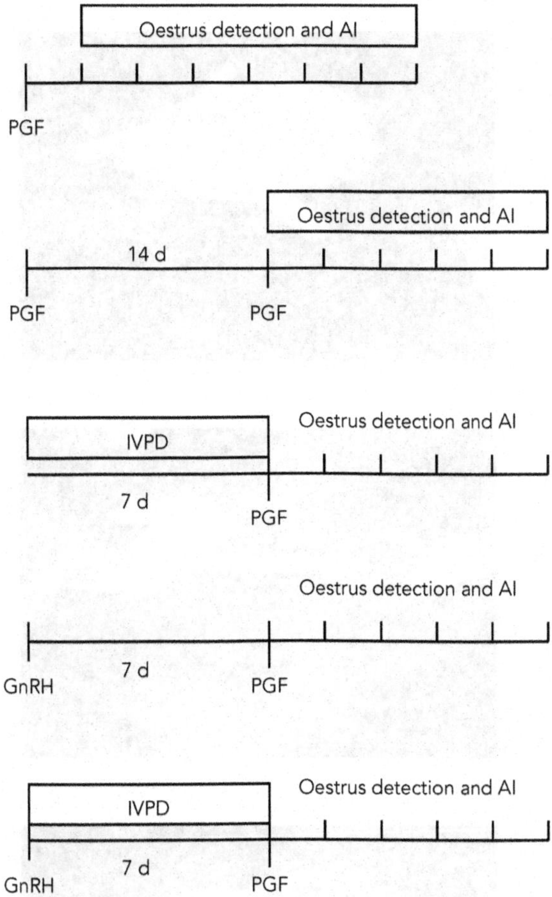

Figure 12 Various protocols for synchronization of oestrus for AI in cattle. PGF, prostaglandin F2 alpha; GnRH, gonadotropin releasing hormone; IVPD, intravaginal progesterone device.

first or second dominant follicle grows and ovulates (interval from treatment to ovulation ~2 and 5 d, respectively). In some cases an old (static) dominant follicle may ovulate, but the oocyte is aged, resulting in fertilization failure or early embryonic death (Roth et al., 2012). With PGF treatment on days 12–16, the CL is very responsive and the second dominant follicle will grow and ovulate. Beyond day 16, PGF is not consistently effective in inducing luteolysis, as spontaneous CL regression may have started, from endogenous PGF.

Several protocols are available for oestrus synchronization. A common protocol is a single injection of PGF given any day (with or without confirming the presence of a CL), followed by AI at detected oestrus. Another protocol involves giving two injections of PGF either 11 or 14 days apart to increase the proportion of cows coming into oestrus. Other protocols include the use of an intra-vaginal progesterone device (IVPD) in combination with PGF, the use of GnRH (to regulate follicle growth) with PGF and the use of GnRH with an IVPD and PGF (Fig. 12).

(a)

(b)

(c)

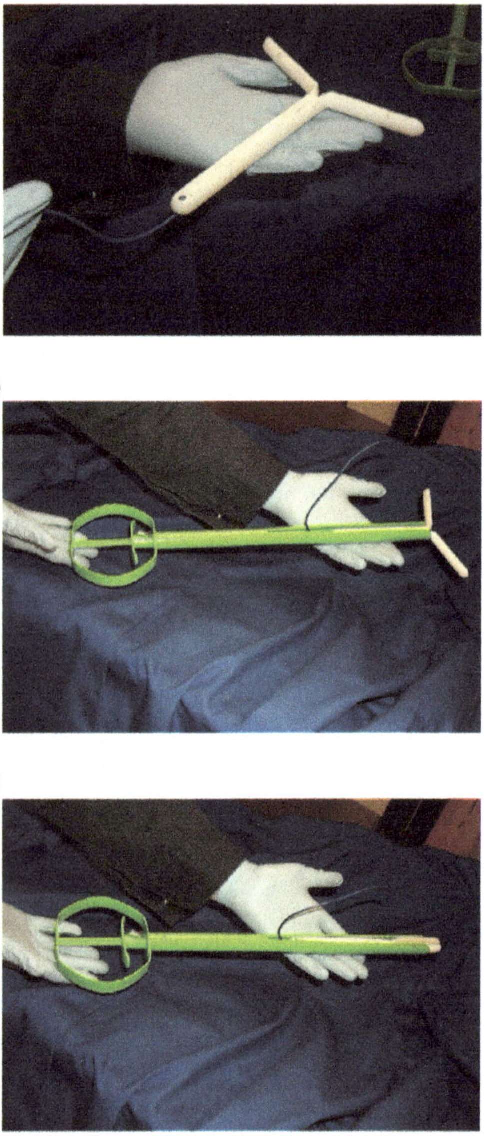

Figure 13 A controlled internal drug release (CIDR) device (one form of intravaginal progesterone device; IVPD) for estrus synchronization in cattle. This progesterone releasing device (a) is placed in an applicator (b), the T-shaped wings folded and positioned properly (c) for intravaginal insertion

The insertion of an IVPD (Fig. 13), usually for 7 days, will elevate blood progesterone concentrations and prevent cattle from coming into oestrus even if they undergo spontaneous CL regression when the IVPD is in place. The use of an IVPD alone for synchronization of oestrus is not recommended because oestrus will not be very synchronous if PGF is not used to induce CL regression at the end of the 7-day interval. The insertion of an IVPD for a longer interval (e.g. 14–21 d) may improve the synchrony of oestrus, but fertility will be very poor due to the release of aged oocytes (Revah and Butler, 1996) and potentially a higher incidence of anovulation. Insertion of an IVPD for 7 days and giving PGF at or 12–24 h before device removal will result in good synchronization of oestrus and high fertility, with cattle expected in oestrus 48–72 h after removal of the IVPD.

Another protocol for synchronization of oestrus involves giving GnRH to trigger growth of a new wave of follicles. In this protocol, GnRH is given at a random stage of the oestrous cycle, followed 7 days later by PGF. This reduces the variability in the interval from PGF to oestrus. In one of the early reports on the use of this protocol in beef cows, oestrus synchrony occurred in >70% of cyclic cows within a 4-day interval with a high fertility of 65–85% (Twagiramungu et al., 1995). Resumption of cyclicity and normal fertility in anoestrous cows was also reported. Furthermore, a protocol combining GnRH, an IVPD and PGF can also be used to increase the precision of oestrus, but at an added cost.

The above-described protocols to induce or synchronize oestrus in dairy cattle are presented in Fig. 12. Regardless of the protocol chosen, it must be easy to understand and implement, and yield consistent results. Farmers should choose a simple, yet effective, protocol in consultation with their veterinarian and use it aggressively to optimize reproductive efficiency.

5.3 Synchronizing ovulation (fixed-time AI)

In dairy herds where EDR is low, but CR is satisfactory to high, implementing synchronization of ovulation (Ovsynch) as a strategy for reproductive management is recommended, as it enables AI without oestrus detection. Although time and labour associated with oestrus detection can be redirected to implement the Ovsynch programme, the additional costs of pharmaceuticals, needles, etc. must be considered.

Early studies in Florida (Thatcher et al., 1989; Schmitt et al., 1996a), Wisconsin (Pursley et al., 1995) and Quebec (Twagiramungu et al., 1995) led to development of the Ovsynch protocol (Pursley et al., 1995) for synchronization of ovulation allowing fixed-timed AI without oestrus detection. This involves two injections of GnRH 9 d apart, with PGF 7 days after the first GnRH, and fixed-time AI 16–20 h after the second GnRH. Exogenous GnRH triggers release of pituitary gonadotropins luteinizing hormone (LH) and follicle-stimulating hormone (FSH). Therefore, the first GnRH-induced LH induces ovulation of an existing dominant follicle. If an LH-responsive dominant follicle is not present, the existing follicle will continue to regress. However, because of FSH, a new wave of follicles emerges 2 days later, from which one follicle will become the new dominant follicle. Giving PGF 7 days later causes CL regression, progesterone concentrations decline and a new dominant follicle emerges. The second GnRH given 2 days after PGF will once again release LH, inducing ovulation of the actively growing follicle. All events are programmed to occur at predetermined times and ovulation usually occurs within ~28 h after the second GnRH injection.

5.3.1 Variations of the Ovsynch protocol

There are several variations of the Ovsynch protocol. A common variation is to increase the interval between the PGF and second GnRH from 48 to 56 h, which may improve CRs.

In Cosynch, the second GnRH and AI occur concurrently. In Cosynch72, the second GnRH injection is given 72 h after PGF injection.

Incorporation of an IVPD into the Ovsynch protocol can prevent premature oestrus and increase the proportion of cows inseminated at optimal time. An IVPD would be inserted concurrent with the first GnRH injection and removed either at the time of PGF treatment or 12–24 h later.

5.3.2 Presynchronization

Presynchronization (Presynch) implies that ovarian status of cows were synchronized prior to starting an ovulation synchronization programme intended for breeding; the rationale is to make cattle more likely to respond to the protocol and increase CRs. There are various Presynch programmes (Colazo and Mapletoft, 2014), of which two are discussed here.

In a PGF-based Presynch protocol, cows are given two injections of PGF 14 days apart, followed by a standard Ovsynch protocol (Fig. 14a), with the first GnRH injection given 12 days after the second PGF of the Presynch protocol. The goal is to maximize the number of cows between days 5 and 12 of the oestrous cycle, the ideal stage for optimal response, when they receive their first GnRH injection of the Ovsynch protocol. In studies that have used this approach (Moreira et al., 2001; El-Zarkouny et al., 2004), dairy cows subjected to Presynch–Ovsynch had 9 (47 vs 38%) to 12 (49 vs 37%) percentage points of higher CRs than in cows subjected to Ovsynch.

In G6G, a GnRH-based Presynch protocol, cows receive a PGF injection, followed by GnRH 2 days later. A standard Ovsynch protocol is then started 6 days after the previous GnRH (Fig. 14b; Bello et al., 2006). The goal is to increase the proportion of cows that respond to the first GnRH injection of the Ovsynch programme, by increasing the probability of an ovulatory-sized follicle at the start of the Ovsynch programme (Colazo and Mapletoft, 2014).

When adopting controlled breeding protocols such as the ones described above, it is important to have strict compliance, that is, the correct dose/hormone/cow/time. Complex protocols, large herds and overlapping groups of cows increase the chance of errors. Therefore, depending on worker skill, level of knowledge and motivation, the most

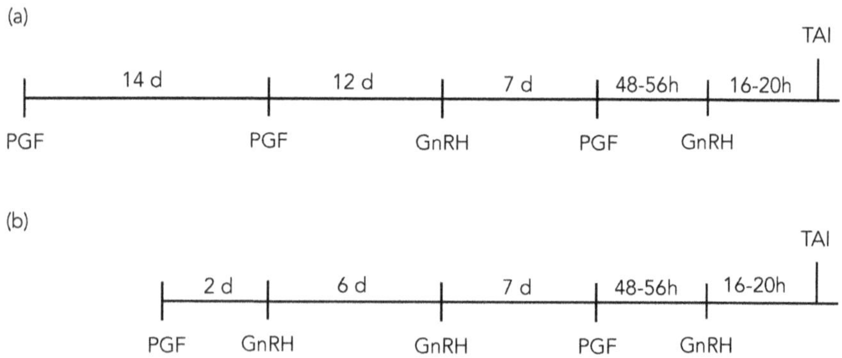

Figure 14 Standard Presynch protocol and G6G protocol (Panels a and b, respectively) for presynchronization of ovarian status in dairy cattle before implementation of an Ovsynch programme. PGF, prostaglandin F2 alpha; GnRH, gonadotropin releasing hormone; TAI, timed artificial insemination.

suitable protocol should be chosen and systematically implemented as sequential, rather than overlapping, breeding groups. An error of even of 5% at each event could have a compounding effect leading to disastrous results. For example, a Presynch–Ovsynch protocol involves five injections followed by AI. Thus, each cow needs to be handled six times for the successful completion of the programme. Assuming a 5% error in compliance at each of the six interventions, the cumulative error becomes 26.5%, which should be considered quite high and unacceptable. Therefore, it is very important for workers to understand the importance of each injection and ensure that treatments are administered correctly.

Systematic use of Ovsynch or other timed-AI programmes can significantly reduce the interval from calving to first conception, thus reducing days open and increasing overall reproductive efficiency. Ovsynch is most beneficial where EDR is poor, but is of limited value if EDR is >70%.

Even with the exclusive use of a programme for synchronization of ovulation, continuing to observe for oestrus, where possible, is recommended to improve the success of the programme. This is important, as a small proportion of cows subjected to ovulation synchronization programmes could be in oestrus during the course of the treatment, particularly when an IVPD is not used. In most cases, fertility in these cows will be improved if they are inseminated and subsequent injections are cancelled.

Worker safety and animal welfare should be given prime consideration. Although hormones used for synchronization treatments are considered safe, consumers may be uncomfortable with the use of hormones for reproductive management, either on the basis of food safety or on the basis of animal welfare. Further discussion of these issues is beyond the scope of this chapter.

5.4 Semen handling and AI technique

Seemingly simple tasks such as semen tank maintenance and the use of proper technique for thawing semen straws are vital to herd reproductive efficiency.

5.4.1 Semen tanks

The liquid nitrogen level must be examined periodically and filled as required. Accurate records of addition of liquid nitrogen and placement and removal of semen straws are very important.

The tank must be stored in a cool, dry, well-ventilated area, ideally on a wood platform (e.g. pallet) to keep it dry and prevent rust. The tank must be handled gently as dragging or sudden jarring movements can damage the insulated flask affecting temperature maintenance. If transporting in a vehicle, the tank must be secured and kept upright. Although narrow-mouthed liquid nitrogen tanks reduce evaporative losses, promptly replacing the lid is important. Persistent frost around the mouth and the neck region of the tank is an indication of loss of insulation, causing rapid evaporation of liquid nitrogen and impending tank failure.

Although liquid nitrogen is −196°C, there is a wide range of temperature zones, with temperatures as high as +7°C near the mouth of the tank when the lid is removed (Fig. 15). Removing a straw should not take more than 6 s; if it cannot be done in this interval, return the canister back into the tank, as even minor fluctuations in temperature can severely damage sperm. Exposing straws to the warm zone (closer to the mouth of the tank) for

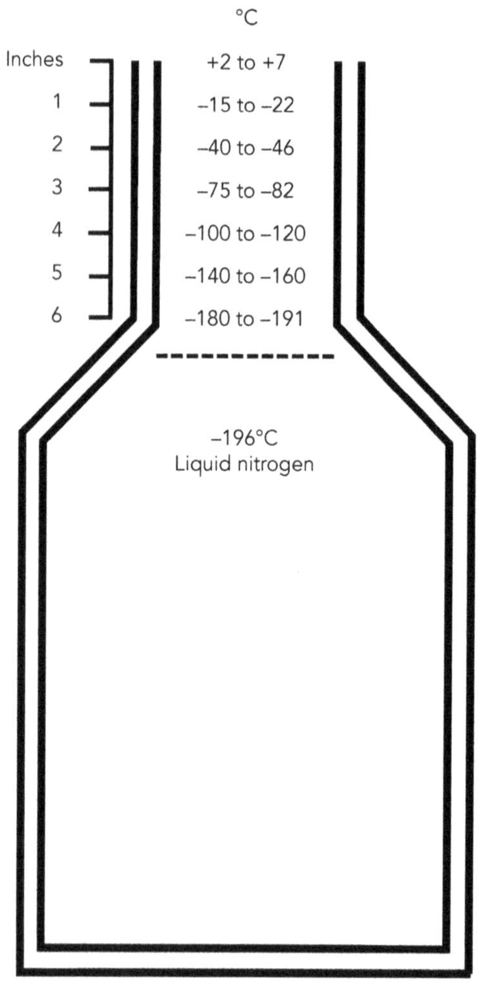

Figure 15 Temperature range within a 6-inch (15 cm) neck of semen storage tank. Source: Adapted from Saacke, 1974.

10–12 s could cause substantial damage to sperm. Furthermore, repeated exposure causes cumulative damage, which could become a hidden factor negatively affecting reproductive efficiency.

5.4.2 Semen thawing

The general recommendation is to thaw straws in a water bath for at least 40 s at 35–37°C (periodically verify if the temperature of the water bath is in the correct range). If a temperature-controlled thawing device is not available, use a thermometer to verify water temperature. Avoid thawing straws in batches. If batch thawing is unavoidable, thaw only

Table 3 Post-thaw temperature changes in semen packaged in 0.5-mL straws during loading a straw after removal from water bath*

	Semen temperature (°C)		
Ambient temperature (°C)	Initial	Final	Decrease
21	35	30	5
4	35	20	15
−16	35	13	22

*Average interval between initial and final temperature = 1 min

Source: Adapted from Saacke, 1977.

two or three straws at a time (ideally keep them separate from one another after initial placement in water) and leave thawed straws in water bath until ready for insemination. Upon removing from water, wipe the straw thoroughly, as even fine drops of water can cause drastic osmotic changes and damage sperm.

Never thaw straws in an open, windy area, especially in winter; on cold days, use a heated room protected from drafts. Pre-warm the AI catheter before loading the straw. Once loaded, the AI catheter must be wrapped in a clean paper towel, put under personal clothing and held close to the inseminator's body to minimize temperature decreases. Drying and loading the straw in the AI catheter must be performed quickly (ideally <1 min from removal from water bath to placement under clothing), as semen temperature is rapidly decreasing (more rapid decline with lower ambient temperature, as shown in Table 3).

The straw should be cut with clean scissors or straw-cutter (the latter is preferred as it consistently results in a straight cut without crimping the straw). If the cutting instrument is cold, warm it before cutting. It is recommended that the cutting instrument is washed periodically with mild detergent, rinsed thoroughly, disinfected and dried. Once thawed, insemination should be completed as soon as possible, preferably within 5 min. Discuss semen handling and insemination procedures with your semen supplier or veterinarian and ensure it is done correctly!

5.4.3 AI technique

Entrust AI to a trained, skilled inseminator. Always wipe the cow's external genitalia with a clean paper towel. If it is very soiled, wash with soap and water before AI. In any case, avoid soiling the AI catheter with faecal matter or other contaminants. Protective sheaths for use with AI catheters are available commercially and should be used to minimize contamination. With minimal time and trauma, traverse the cervix and deposit semen in the body of the uterus, no more than ~1.5 cm anterior to the cervix (as the uterine body is short).

5.4.4 Timing of insemination

The timing of insemination affects fertility. Before it is capable of fertilization, bull sperm must undergo a series of changes (capacitation), which takes 4–8 h after thawing (longer for fresh sperm). Capacitation occurs during sperm transport (sperm arrive at site of fertilization ~8–10 h after insemination). Although some sperm may live in the reproductive tract of a cow for over 48 h, after 24 h, they are likely to have reduced vigour and fertilizing ability. In contrast, the oocyte remains viable for only 8–12 h after ovulation.

Standing oestrus coincides with a surge in LH essential for oocyte maturation and ovulation. Ovulation occurs 24–30 h after the onset of standing oestrus (see Fig. 6). If a cow is inseminated much earlier than the onset of standing oestrus (i.e. more than 24 h before ovulation), only aged sperm will be available to fertilize the oocyte. Conversely, if AI is done >24 h after the onset of standing oestrus (i.e. close to, or after ovulation), sperm are unlikely to reach the site of fertilization while the oocyte is still viable. Thus, AI should be performed so that fertile sperm reach the site of fertilization before ovulation and are waiting for the arrival of the oocyte. In other words, the optimal time for AI is 12–16 h after the onset of standing oestrus (Fig. 8). Nonetheless, satisfactory CRs can be expected if AI is performed from the onset of standing oestrus up to 12 h after the end of standing oestrus (Stevenson et al., 2014). The long-standing AM–PM rule holds true. If a cow was detected in standing oestrus at 06:00, she may be bred at 18:00 and if seen standing at 18:00, she should be bred at 06:00 the next morning. If in doubt about the time of onset of standing oestrus, breed soon after detection of standing oestrus. In a recent study using automated activity monitoring in lactating dairy cows, Stevenson et al. (2014) found that the increased activity threshold is comparable with the onset of standing oestrus. In another experiment within the same report (19 herds, 4019 cows), satisfactory CRs were obtained when cows were subjected to AI up to 26 h after increased activity threshold. However, in multiparous cows, mean CR was significantly reduced when AI occurred between 17 and 26 h after the increase in activity threshold (i.e. after the onset of standing oestrus).

5.5 Reconsidering management factors

5.5.1 Voluntary waiting period

In most dairy cows the first ovulation occurs within 5 weeks after calving (Ambrose and Colazo, 2007), although it is often not preceded by oestrus (O'Connor, 2007). The more cycles a cow has after calving, the higher the chances for pregnancy to first insemination (Thatcher and Wilcox, 1973). Approximately 5 weeks are needed for the uterus to complete involution (return to normalcy) after calving. However, involution may not be complete in 5 weeks if calving was complicated and/or the uterus was infected. Cows that had uterine infections should not be bred before confirming that the infection is cleared. Furthermore, fertility can be reduced for at least 2 or 3 weeks after a uterine infection is apparently resolved. The ideal VWP should therefore be no sooner than 60 days to allow sufficient time for involution and resumption of cyclicity. Cows not detected in oestrus by 6 weeks after calving should be examined by a veterinarian to assess their reproductive status.

5.5.2 Considering body condition

Cows in a negative energy state often do not conceive. Poor body condition is associated with lower blood progesterone concentrations (Butler and Smith, 1989) and lower CRs (Butler and Smith, 1989; Moreira et al., 2000a). One study involving heat-stressed dairy cattle projected a 37% increase in CR when body condition score (BCS) increased from 2.25 to 3.25 (Ambrose et al., 1999). The rate at which body condition loss occurs has a greater detrimental effect on fertility rather than low BCS itself (Butler, 2003; Garnsworthy,

2006; Santos et al., 2009). For example, a loss of condition from 3.25 to 2.25 (1.0 loss) during the first 5 weeks after calving is likely to have a more damaging effect on fertility than a loss of condition from 2.50 to 2.25 (0.25 loss). Furthermore, to reduce the impact of negative energy balance on cow health and performance, BCS at calving should be no more than 0.5 units above target BCS (Garnsworthy, 2006). Therefore, monitoring relative changes in BCS can improve reproductive management. For example, cows in good body condition (BCS >3.0) can be selectively bred at the end of the VWP, whereas breeding might be delayed in cows in poor body condition (ideally, they should be placed on a higher-energy diet). It is recommended that BCS be assessed on a regular basis starting 3 weeks before expected calving.

5.5.3 Early determination of non-pregnant cows

Real-time B-mode trans-rectal ultrasonography has become an indispensable diagnostic tool. Pregnancy diagnosis can be conducted accurately using ultrasound as early as the fourth week of pregnancy in dairy cows (Kastelic et al., 1989), although diagnosis at 30 to 32 days of pregnancy is more common under field conditions. Other reliable methods of early pregnancy diagnosis include the detection of pregnancy-specific protein B (PSPB; Sasser et al., 1986, 1989) or pregnancy-associated glycoproteins (PAG; Sousa et al., 2006). The former is available as a blood test, whereas the latter is available either as a milk test or as a blood serum test. These tests are readily available to dairy farmers in Canada and the United States and in certain other countries. In North America, dairy farmers can collect blood or milk samples and have them sent to a lab that offers the service. Results are electronically relayed to the farmer the following day, making the service quite efficient. These tests are highly reliable with very high sensitivity and specificity (Ricci et al., 2015). The biggest value derived from these tests is early detection of non-pregnant cows, shortening the interbreeding intervals, as long interbreeding intervals are a constraint (Ambrose and Colazo, 2007) to achieving high reproductive efficiency in dairy cattle.

5.5.4 Resynchronization protocols

The active implementation of resynchronization programmes can also reduce interbreeding intervals in dairy cows. In a study by Ambrose and Colazo (2007) involving cows from 23 Canadian dairy farms, the mean interval between the first and second breeding was ~42 d, with only 28% of the cows rebred within 17 to 24 days after first AI. In the same study, the interval between the second and third breeding was 34 days. There were two apparent reasons: first, poor oestrus detection efficiency, and second, embryonic mortality occurring >21 d after the previous breeding.

There are several resynchronization protocols to reduce interbreeding intervals (Lucy, 2005; Dewey et al., 2010). These protocols include GnRH given 19 to 33 days after the first timed-AI (Moreira et al., 2000b; Fricke et al., 2003), followed by pregnancy diagnosis about 1 week later. Cows diagnosed as non-pregnant receive PGF at pregnancy diagnosis, followed by a second GnRH injection 2 days later and timed AI performed either at second GnRH (Fricke et al., 2003) or 16–20 h later (Moreira et al., 2000b). Another protocol is to insert a progesterone device 18 days after AI, remove it 7 days later (Day 25 post-AI) and concurrently give GnRH to all cows. Pregnancy is determined by ultrasonography on day 32 and non-pregnant cows are given PGF; 56 h later, the second GnRH is injected and cows reinseminated 16 h later (35 days after first AI; Thatcher et al., 2008). However, using GnRH to reprogramme the ovarian cycle in the third week after the previous AI is not recommended. It has been reported (Moreira et al., 2000b; Fricke et al., 2003) that initiating the resynchronization protocol by giving GnRH at 19 or 20 days after first timed-AI results in poor CRs possibly due to an increase in embryonic loss between 20 and 27 days after timed-AI (Moreira et al., 2000b), although no such detrimental effect was evident in another study (Chebel et al., 2003) when resynchronization was initiated at 21 days after previous AI. Nevertheless, starting a resynchronization programme by giving GnRH either 25 or 28 days after previous AI, followed by pregnancy diagnosis on days 32 or 35, is an acceptable and practical alternative. Other resynchronization protocols include the use of an IVPD inserted 21 days after the first timed-AI and removed 7 days later, followed by oestrus detection and breeding. Alternatively, a progesterone-releasing intra-vaginal device may be inserted from 14 to 21 days after the first timed-AI, and GnRH given 3 days later (Day 24), followed by pregnancy diagnosis (Day 31), with non-pregnant cows receiving PGF the same day and the second GnRH given 2 days later (day 33), with timed-AI at the second GnRH treatment or 16–20 h later. Variations of these protocols have been reported (Galvão et al., 2007) using oestradiol cypionate to synchronize ovulation, with acceptable CRs.

5.5.5 Shortened dry period

Shortening the dry period to minimize negative energy balance has been suggested with positive effects reported on reproductive function in dairy cows, with short to no dry period reducing the interval from calving to first ovulation, increasing first-service CR, reducing the incidence of anovulation and reducing days open. Shortening the dry period to either 28 or 0 days (continuous milking) resulted in fewer days to first ovulation, increased first-service CR and reduced days open (Grummer, 2007; Watters et al., 2009).

5.5.6 Extended lactation for better reproduction

There is one school of thought that extending the calving interval by reducing number of calving events during the lifetime of a dairy cow will improve reproductive performance by lowering the risk of diseases during the transition period including uterine infections and ovarian dysfunction commonly encountered during the early postpartum period (Dobson et al., 2007). Although this strategy might not be appealing to all dairy farmers, at least one study (Arbel et al., 2001) reported that high-yielding cows intentionally rebred approximately 60 days after the usual VWP improved profitability.

5.5.7 Minimizing embryo loss

Embryonic loss is an important cause of reproductive wastage (Thatcher et al., 1994; Santos et al., 2004; Inskeep and Dailey, 2005). Whereas only a 15% incidence of fertilization failure is documented, more than 35% of embryos die within 40 days after insemination (Peters and Ball, 1995). Embryonic losses of 22% (Pursley et al., 2008) to 27% (Ambrose et al., 2006) have been reported between about 4 weeks after AI and calving time. What could be the cause of such high rates of embryonic death? Progesterone insufficiency is one likely reason. This can be rectified by supplementing exogenous progesterone or by inducing the cow's ovaries to produce additional progesterone.

Supplemental exogenous progesterone

Intra-vaginal inserts of progesterone during various times have been attempted; in some cases, CRs have improved, but results have been mixed, often found beneficial only in cows lacking a CL, that is, those considered anoestrous (Colazo et al., 2013; Bisinotto et al., 2015). Therefore, indiscriminate use of this strategy in all cows is not recommended.

Boosting endogenous progesterone

Various options exist for improving peripheral progesterone concentrations in cattle without exogenous progesterone supplementation. One well-studied approach is to inseminate the cow (day 0) at either detected oestrus or by fixed-timed AI and then create an accessory CL by inducing ovulation of either the first-wave dominant follicle on days 5–7 of the oestrous cycle (Ambrose et al., 2000) or the second-wave dominant follicle on days 11–14 (Ambrose et al., 2000; Peters et al., 2000). Products such as human chorionic gonadotropin (hCG) and GnRH have been used to induce accessory CL which secretes additional progesterone (Sianangama and Rajamahendran, 1992; Schmitt et al., 1996b). Progesterone concentrations increase after such treatment and often result in improvements in CR (Sianangama and Rajamahendran, 1992; Peters et al., 2000; Santos et al., 2001). Increasing progesterone concentrations during early gestation is known to increase embryonic growth (Lonergan and Forde, 2015) and embryo survival.

Another approach is to replace the GnRH injection given to synchronize ovulation in Ovsynch/timed-AI protocols with porcine LH (Colazo et al., 2009) or hCG (Santos et al., 2001; Beltran and Vasconcelos, 2008). Such an approach has improved CRs and embryo survival in at least some studies.

Injection of GnRH or hCG on ~day 14 after AI may improve embryo survival through a reduction in estradiol from the dominant follicle, which is largely responsible for inducing PGF secretion from the uterus causing CL regression and onset of next oestrus. Elimination of the oestrogenic influence of the follicle leads to an extension in the lifespan of the CL. Since not all embryos grow at the same rate, this window of opportunity may allow slow-growing embryos some extra time to signal their presence, thereby facilitating maternal recognition of pregnancy.

A subcutaneous GnRH implant can substitute for injectable GnRH preparations to boost endogenous progesterone. This concept was tested by Ambrose et al. (1998) in lactating dairy cows subjected to an Ovsynch protocol. One group of cows was subjected to the standard Ovsynch protocol; another group received a GnRH (Deslorelin) implant instead of the second GnRH injection. A third group was allowed to ovulate spontaneously. In the presence of the implant, the growth of large follicles was restricted and establishment of a dominant follicle was delayed for several days. Concurrently, CL function was enhanced (significant increase in progesterone concentrations). As the absence of the influence of a large follicle may delay CL regression and help embryo survival, in a follow-up experiment, the influence of GnRH implant on CRs in lactating dairy cows was tested. Cows that were in poor body condition (2.25 BCS) were assigned to the study. All cows were synchronized subjected to the Ovsynch protocol. Of the 16 cows in the study, eight received the GnRH implant 16 h before insemination (instead of the second GnRH injection), whereas the eight control cows received an injection of GnRH as in a standard Ovsynch protocol. Only 1 (12.5%) of the 8 control cows became pregnant, whereas 5 (62.5%) of the 8 cows treated with GnRH implant were confirmed pregnant. The increased progesterone concentrations and the delayed emergence of a dominant follicle on ovaries of cows given GnRH implant likely contributed to increased embryo survival. That the use of a GnRH implant may increase CRs even in cows of poor body condition, although exciting, must be interpreted cautiously, due to the small number of animals used in this study. A later study (Bartolome et al., 2006), with more cows (~90 per group) tested the same concept during late embryonic period by giving a deslorelin implant on day 27 of pregnancy. On day 45, the number of accessory CL was greater in cows given deslorelin versus control, and pregnancy loss between days 45 and 90 was lower in cows with versus without an accessory CL (0 vs 16%).

Delaying manual pregnancy diagnosis may reduce embryo loss

Experienced dairy practitioners can accurately diagnose pregnancy by palpation of uterine contents per rectum before 6 weeks of pregnancy. However, there is a small risk of embryonic loss when palpation is performed very early in pregnancy (Franco et al., 1987). As a high percentage of early embryonic loss is being reported (Franco et al., 1987; Pursley et al., 1998) during early pregnancy stages, it is recommended to wait until approximately 6 weeks to confirm pregnancy by trans-rectal palpation of uterine contents. Even if pregnancy diagnosis is performed early, it is best to reconfirm around day 90 as embryonic losses beyond day 90 are relatively small except in cases of infectious abortions.

5.5.8 Managing 'problem cows'

By 'problem cows' we refer to cows that have either not been in oestrus (true anoestrus or unobserved oestrus), detected in oestrus too often at irregular intervals (possible follicular cyst), repeat breeders or those with other reproductive irregularities.

The term 'repeat breeder' describes cows that have no palpable abnormalities, no clinical evidence of an infection or ovarian dysfunction, detected in oestrus at regular intervals yet fail to conceive despite three or more inseminations. Animals identified as true repeat breeders have greater rates of defective eggs, fertilization failure and early embryonic loss. Ovulation failure is one possible reason for repeat breeding. GnRH given either 12–16 h before (or) at the time of insemination is one method by which ovulation can be induced. This method is likely to help cows that do not spontaneously ovulate. It must be remembered though that GnRH injections do not always ensure ovulation and ovulation failure is not the only cause for repeat breeding. Since GnRH is relatively inexpensive and likely to help at least some repeat-breeder cows, this is a recommended procedure. Repeat-breeder cows that ovulate normally but fail to conceive could have inadequate progesterone secretion. In such cases, administration of hCG (3000 IU) or GnRH (100 μg) 5 days after AI may be tried as a possible remedy. It was reported that hCG is more effective than GnRH in increasing progesterone concentration (Schmitt et al., 1996a). If the oviductal environment is not conducive to fertilization or early embryonic development, embryo transfer may be helpful (Dochi et al., 2008; Rodrigues et al., 2010), but using an expensive donor embryo is not recommended. For more details on this strategy, please refer to Ambrose et al. (2010).

Postpartum anoestrus is common in dairy cows (Peter et al., 2009b) but exogeneous GnRH can induce LH release as early as 10 days postpartum (Salehi et al., 2015). GnRH-based timed-AI protocols have been somewhat effective in achieving pregnancy to timed-AI in anoestrous cows (Moreira et al., 2000a; Kassa et al., 2002; Stevenson et al., 2008; Colazo et al., 2009). For anoestrous cows to conceive, ovulation to the first GnRH treatment seems crucial. Although anoestrous cows readily respond to an Ovsynch protocol and often conceive to timed-AI, they are at a much higher risk of losing their pregnancy than cyclic cows. A protocol for synchronization of ovulation and timed AI with the addition of an IVPD has been found useful in reducing such pregnancy losses in lactating dairy cows (Colazo et al., 2013).

5.5.9 Managing cystic ovarian disease

The incidence of cystic ovarian disease could be up to 15% in some herds (Brito and Palmer, 2004) and is an oft-identified factor in reproductive inefficiency. Although treatments involving the exogenous administration of GnRH, hCG and porcine LH are common, the outcomes are quite variable. Ovulation synchronization followed by timed AI has been reported to improve the success of pregnancy in cows diagnosed with cystic ovarian follicles by many authors (see Ambrose et al., 2010). In a study that followed treatment response to an Ovsynch protocol using ultrasonography in cystic cows, authors recorded excellent ovarian responses and CRs (Ambrose et al., 2004). In that study (Ambrose et al., 2004), all of the 18 cows followed by ultrasonography developed a new follicle and 15 of the 18 cows ovulated the newly developed follicle after the second GnRH treatment with 7 (41%) of the 17 inseminated cows confirmed pregnant. Although the cyst was detectable up to 40 days after AI in 65% of the cows, it had no detrimental effect on pregnancy establishment or sustenance. In a subsequent study (Ambrose et al. unpublished) conducted in research and commercial dairy herds, 109 cystic cows were assigned to Ovsynch-timed AI with (n = 53) or without (n = 56) the addition of an IVPD for the first 7 days. The addition of an IVPD did not improve ovarian response or CR. Overall, 62% of cows ovulated in response to the first GnRH treatment and 83% ovulated following the second GnRH and a CR of 42% was attained. These results clearly indicate that cows diagnosed with cystic ovarian follicles will respond well to an Ovsynch protocol and this is a recommended strategy to manage this condition.

5.5.10 Managing uterine inflammation

Cows with uterine inflammation such as endometritis, although seemingly normal, could have impaired fertility for an extended interval following bacteriological cure (Dourey et al., 2011; Gobikrushanth et al., 2016) and often become repeat breeders. Thus, endometritis should be considered a hidden factor that impacts reproductive efficiency and managed accordingly.

5.5.11 Nutritional management for improved fertility

Among the major dietary components that impact reproductive function are crude protein (Lean et al., 2012) and fat (Rodney et al., 2015). Although high-protein diets have a detrimental effect on fertility, supplemental fats (Staples et al., 1998) and specific polyunsaturated long-chain fatty acids (Mattos et al., 2000; Thatcher et al., 2011) have positive effects.

Feeding diets high in crude protein can lead to inefficient protein utilization and excess absorption of ammonia (Thatcher et al., 2011). In addition to the negative effects of high ammonia on reproductive function (Butler, 1998), detoxification of ammonia in the liver is an energy-expensive process, which could worsen negative energy balance in the postpartum dairy cow (Thatcher et al., 2011), which may already be in a negative energy state. High intake of rumen-degradable protein impairs embryo development (Rhoads et al., 2006). To reduce the likelihood of these negative effects, rumen-degradable proteins in excess of ~16% should not be included in dairy cow rations. However, feeding prepartum diets containing high protein levels and high energy density (1.6–1.65 Mcal NE_L/kg) have been recommended to maintain or increase prepartum dry matter intake (Butler, 2005) and to improve postpartum dry matter intake.

Dietary fat supplementation has several positive effects on reproductive function (Santos et al., 2008). Whereas reproductive efficiency may be improved through increased dietary energy and a resultant decrease in negative energy status, many positive effects on reproduction occur independent of improvements in energy status (Staples et al., 1998), including an increase in the number of ovarian follicles and size of the dominant follicle, increased progesterone concentrations, improvements in oocyte and embryo quality, and reductions in uterine PGF2-alpha (Mattos et al., 2000; Santos et al., 2008). Inclusion of the essential fatty acids, linoleic acid and alpha linolenic acid, and n-3 fatty acids contained in fish oils (eicosapentaenoic acid and docosahexaenoic acid) in diets of dairy cow rations and their effects on reproductive function have been studied extensively. Although dietary n-3 fatty acids have not consistently improved CRs, several studies reported reductions in pregnancy losses in cows given sources of supplemental n-3 fatty acids (Ambrose et al., 2006; Petit and Twagiramungu, 2006; Silvestre et al., 2011).

On the basis of a meta-analysis, Rodney et al. (2015) recently concluded that feeding fats is beneficial to reproductive function. Thus, supplementation of fat in the diet of lactating dairy cows is a promising strategy for improving reproductive performance, although the optimal reproductive window, duration and quantity of fats or type of fatty acids to be fed continue to be investigated.

Copper, magnesium, selenium, phosphorus and calcium, and vitamins A, D and E have important roles in regulating normal reproductive function. The role of vitamins and minerals in reproduction has been described in detail (Hurley and Doane, 1989).

5.5.12 Environmental management for improved fertility

Although environmental stressors, including extreme cold weather (Gwazdauskas, 1985; Young, 1983), poor air quality and stray voltage (Appleman and Gustafson, 1985) could affect reproductive performance indirectly, high environmental temperatures and resultant heat stress severely impair fertility in dairy cows (Gwazdauskas, 1985; Hansen, 1997; Al-Katanani et al., 1999). Cows must be protected from excessive solar radiation by providing adequate shade during summer and must be given shelter from extreme cold during winter months. Stray voltage is known to alter hormone profiles and intermittent electrical stimulation amplifies peak oxytocin response. Since oxytocin, a pituitary hormone responsible for milk ejection, also plays an important role in events leading to PG secretion and CL regression, chronic stray voltage may affect reproduction possibly through premature CL regression. Therefore, conscious efforts must be taken to minimize effects of these stressors as they could significantly affect overall production.

Heat-stress abatement strategies include provision of cool, clean drinking water, shade in outdoor lots and pastures, and timer-controlled sprinklers and fans for cows housed indoors. When sprinklers are used, cows should be sprayed with cool water, preferably wetting the back and sides without wetting the udder. As the oocyte and post-fertilization zygote are extremely sensitive to elevated temperatures (Edwards and Hansen, 1997), intrauterine transfer of 7-day-old embryos has been used as a strategy to circumvent negative effects of heat stress on early embryo development (Ambrose et al., 1999; Drost et al., 1999) with some success. Other strategies include increased oestrus detection through the use of visual and electronic oestrus detection aids. Working with a nutritionist to improve nutrition as a means to compensate for energy loss associated with inadequate dry matter intake, commonly encountered under heat stress conditions, and to combat acidosis, are also other recommended strategies. More details on these and other strategies are reported in Hansen and Arechiga (1999) and Flamenbaum and Galon (2010).

5.5.13 Cross-breeding

The North American Holstein breed has been aggressively selected for milk production traits over the years, with minimal attention given to functional traits including fertility-associated traits. As a result, although great strides have been made in milk production (Fig. 4), fertility, health and longevity have been compromised, presumably due, at least in part, to inbreeding. Heins et al. (2012) reported that when Holstein heifers were mated to Normande, Montbeliarde, Swedish Red or Norwegian Red sires, the resultant cross-breeds had superior performance compared to Holsteins for fertility across the first 5 lactations. Cross-bred cattle also had a shorter interval from calving to conception, greater longevity and greater overall profitability. Improvements in fertility have also been reported for Jersey × Holstein cross-breeds (Ferris et al., 2015). Thus, cross-breeding could be yet another strategy for improving reproductive efficiency, but it needs careful planning and implementation in consultation with a competent geneticist.

6 Future trends

A combination of AI with performance recording and genetic evaluations is critical for rapid genetic gains (Van Doormaal and Kistemaker, 2003). As indicated earlier, for decades, genetic selection of dairy cattle was largely performed with a focus on traits relating to milk production and conformation. The inclusion of fertility traits in genetic selection had been mostly ignored due to reports of low heritability of reproductive traits (Raheja et al., 1989). More recently, this trend has changed with the introduction of 'Daughter Pregnancy Rate' into Net Merit index in the United States in 2003 and 'Daughter Fertility Index' in Canada in 2004. Furthermore, the introduction of Sire Conception Rate (Kuhn and Hutchinson, 2008; Norman et al., 2011) as an additional reproductive index to replace the conventional 'Non-Return Rate' has also brought about improvements in dairy cow fertility. Recent reports indicate that the rate of decline in cow fertility has slowed down or halted and replaced with a slow upward trend (Garcia-Ruiz et al., 2016). Coincidentally, the adoption of fixed-timed AI programmes is perhaps at its peak, with many large North American herds heavily relying on these technologies. As fixed-timed AI programmes such as Ovsynch and Presynch/Ovsynch can increase herd PRs, the recent improvements seen in dairy cow fertility could well be due to a combination of both factors (genetics and adoption of timed AI). Regardless, halting further declines in fertility is important. With recent advances in genomic selection and the introduction of genomic selection of AI sires, fertility in dairy cattle, at least in the developed countries, is bound to improve in the coming years.

The PAG/PSPB tests for early pregnancy diagnosis will become available to dairy farmers in more nations, and possibly more affordable. Simply, the ability to detect non-pregnant cows early and prepare them for rebreeding will greatly reduce interbreeding intervals, improving reproductive efficiency. More advanced tests for early pregnancy diagnosis such as microRNA signatures of early pregnancy (Ioannidis and Donadeu, 2016) and interferon-stimulated gene products (e.g. ISG-15, Han et al., 2006) may become a reality, making even earlier detection of non-pregnancy possible.

In-line diagnostic tests for reproductive status assessment (progesterone test), ketosis (beta hydroxybutyric acid), mastitis (lactose dehydrogenase, somatic cell count) and dietary protein status (milk urea nitrogen) are already commercially available, revolutionizing farm decision-making processes.

Furthermore, the increased availability of more affordable portable ultrasound scanners and an increased adoption of the technology by veterinarians in both developed and developing countries will play a major role in improving reproductive efficiency.

7 Where to look for further information

Reproductive loss

Lucy, M. C. (2001). Reproductive loss in high-producing dairy cattle: where will it end? *J. Dairy Sci.* 84, 1277–93.

Terminology

Peter, A. T., Levine, H., Drost, M. and Bergfelt, D. R. (2009a). Compilation of classical and contemporary terminology used to describe morphological aspects of ovarian dynamics in cattle. *Theriogenology* 71, 1343–57.

Ovarian follicular development

Adams, G. P., Jaiswal, R., Singh, J. and Malhi, P. (2008). Progress in understanding ovarian follicular dynamics in cattle. *Theriogenology* 69, 72–80.

Heat stress

St-Pierre, N. R., Cobanov, B. and Schnitkey, G. (2003). Economic losses from heat stress by US livestock industries. *J. Dairy Sci.* 86(E. Suppl.), E52–E77.

Timing of insemination and conception risk

Stevenson, J. S., Hill, S. L., Nebel, R. L. and Dejarnette, J. M. (2014). Ovulation timing and conception risk after automated activity monitoring in lactating dairy cows. *J. Dairy Sci.* 97, 4296–308.

8 Acknowledgements

The authors thank Dr. Jason Lombard, United States Department of Agriculture, and graduate students, Drs. Tony C. Bruinjé, Mohanathas Gobikrushanth (Department of

Agricultural, Food and Nutritional Science, University of Alberta) and Guilherme Rizzoto (Department of Production Animal Health, University of Calgary) for their assistance during the preparation of this manuscript.

9 References

Adams, G. P., Jaiswal, R., Singh, J. and Malhi, P. (2008). Progress in understanding ovarian follicular dynamics in cattle. *Theriogenology* 69, 72–80.

Ahlman, T., Berglund, B., Rydhmer, L. and Strandberg, E (2011). Culling reasons in organic and conventional dairy herds and genotype by environment interaction for longevity. *J. Dairy Sci.* 94, 1568–75.

Al-Katanani, Y. M., Webb, D. W. and Hansen, P. J. (1999). Factors affecting seasonal variation in 90-day nonreturn rate to first service in lactating Holstein cows in a hot climate. *J. Dairy Sci.* 82, 2611–16.

Ambrose, D. J., Pires, M. F. A., Moreira, F., Diaz, T., Binelli, M. and Thatcher, W. W. (1998). Influence of Deslorelin (GnRH-agonist) implant on plasma progesterone, first wave dominant follicle and pregnancy in dairy cattle. *Theriogenology* 50, 1157–70.

Ambrose, D. J., Drost, M., Monson, R. L., Rutledge, J. J., Leibfried-Rutledge, M. L., Thatcher, M. J., Kassa, T., Binelli, M., Hansen, P. J., Chenoweth, P. J. and Thatcher, W. W. (1999). Efficacy of timed embryo transfer with fresh and frozen *in vitro* produced embryos to increase pregnancy rates in heat-stressed dairy cattle. *J. Dairy Sci.* 82, 2369–76.

Ambrose, D. J., Kastelic, J. P., Rajamahendran, R., Small, J. and Urton, G (2000). Pregnancy rates in dairy cows after GnRH treatment at 7, 14, or 7 and 14 days after timed insemination. *Can. J. Anim. Sci.* 80, 755.

Ambrose, D. J., Schmitt, E. J-P., Lopes, F. L., Mattos, R. C. and Thatcher, W. W. (2004). Ovarian and endocrine responses associated with the treatment of cystic ovarian follicles in dairy cows with gonadotropin releasing hormone and prostaglandin $F_{2\alpha}$, with or without exogenous progesterone. *Can. Vet. J.* 45, 931–7.

Ambrose, D. J., Day, P. A. and Small, J. A. (2005). Comparing timed AI to electronic estrus detection in Holstein heifers. *Proceedings of the Western Canadian Dairy Seminar in Advances in Dairy Technology* 17, 364.

Ambrose, D. J., Kastelic, J. P., Corbett, R., Pitney, P. A., Petit, H. V., Small, J. A. and Zalkovic, P. (2006). Lower pregnancy losses in lactating dairy cows fed a diet enriched in α-linolenic acid. *J. Dairy Sci.* 89, 3066–74.

Ambrose, D. J. and Colazo, M. G. (2007). Reproductive status of dairy herds in Alberta: a closer look. *Proceedings of the Western Canadian Dairy Seminar in Advances in Dairy Technology* 19, 227–44.

Ambrose, D. J., Colazo, M. G. and Kastelic, J. P. (2010). The applications of timed artificial insemination and timed embryo transfer in reproductive management of dairy cattle. *Revista Brasileira de Zootecnia* 39 (Special supplement), 383–92.

Ansari-Lari, M., Mohebbi-Fani, M. and Rowshan-Ghasrodashti, A. (2012). Causes of culling in dairy cows and its relation to age at culling and interval from calving in Shiraz, Southern Iran. *Vet. Res. Forum* 3, 233–7.

Appleman, R. D. and Gustafson, R. J. (1985). Source of stray voltage and effect on cow health and performance. *J. Dairy Sci.* 68, 1554–7.

Arbel, R., Bigun, Y., Ezra, E., Sturman, H. and Hojman, D. (2001). The effect of extended calving intervals in high lactating cows on milk production and profitability. *J. Dairy Sci.* 84, 600–8.

Bailey T (1997). Strategies for estrus detection to improve dairy reproductive performance. *Proceedings of the Society for Theriogenology, Annual Meeting* 264–73.

Balendran, A., Gordon, M., Pretheeban, T., Singh, R., Perera, R. and Rajamahendran, R (2008). Decreased fertility with increasing parity in lactating dairy cows. *Can. J. Anim. Sci.* 88, 425–8.

Barbat, A., Le Mezec, P., Ducrocq, V., Mattalia, S., Fritz, S., Boichard, D., Ponsart, C. and Humblot, P. (2010). Female fertility in French dairy breeds: current situation and strategies for improvement. *J. Reprod. Develop.* 56, 15–21.

Bartolome, J. A., Kamimura, S., Silvestre, F., Arteche, A. C., Trigg, T. and Thatcher, W. W. (2006). The use of a deslorelin implant (GnRH agonist) during the late embryonic period to reduce pregnancy loss. *Theriogenology* 65,1443–53.

Bazer, F. W. (2013). Pregnancy recognition signaling mechanisms in ruminants and pigs. *J. Anim. Sci. Biotech.* 4, 23.

Beam, S. W. and Butler, W. R. (1999). Effects of energy balance on follicular development and first ovulation in postpartum dairy cows. *J. Reprod. Fertil. Suppl.* 54, 411–24.

Bello, N. M., Steibel, J. P. and Pursley, J. R. (2006). Optimizing ovulation to first GnRH improved outcomes to each hormonal injection of Ovsynch in lactating dairy cows. *J. Dairy Sci.* 89, 3413–24.

Bello, N. M., Stevenson, J. S. and Tempelman, R. J. (2012). Milk production and reproductive performance: Modern interdisciplinary insights into an enduring axiom. *J. Dairy Sci.* 95, 5461–75.

Beltran, M. P. and Vasconcelos, J. L. M. (2008). Conception rate in Holstein cows treated with GnRH or hCG on the fifth day post artificial insemination during summer. *Arq Bras Med Vet Zootec* 60, 580–6.

Bisinotto, R. S., Lean, I. J., Thatcher, W. W. and Santos, J. E. P. (2015). Meta-analysis of progesterone supplementation during timed artificial insemination programs in dairy cows. *J. Dairy Sci.* 98, 2472–87.

Bondurant, R. H. (2005). Venereal diseases of cattle: natural history, diagnosis, and the role of vaccines in their control. *Vet. Clin. North Amer. Food Anim. Pract.* 21, 383–408.

Bourne, N., Laven, R., Wathes, D. C., Martinez, T. and McGowan, M. (2007). A meta-analysis of the effects of Vitamin E supplementation on the incidence of retained foetal membranes in dairy cows. *Theriogenology* 67, 494–501.

Bousquet, D., Bouchard, E. and Du Tremblay, B (2004). Decreasing fertility in dairy cows – myth or reality? *Proceeding 23rd World Buiatric Congress*, 11–16 July, Quebec, Canada.

Brito, L. F. C. and Palmer, C. W. (2004). Cystic ovarian disease in cattle. *Large Anim. Vet. Rounds* 4 (10), 6. https://www.researchgate.net/publication/213860186_Cystic_ovarian_disease_in_cattle. Accessed 16 August 2016.

Britt, J. H., Scott, R. G., Armstrong, J. D. and Whitacre, M. D. (1986). Determinants of estrous behavior in lactating Holstein cows. *J. Dairy Sci.* 69, 2195–202.

Butler, W. R. and Smith, R. D. (1989). Interrelationships between energy balance and postpartum reproductive function in dairy cattle. *J. Dairy Sci.* 72, 767–83.

Butler, W. R. (1998). Effect of protein nutrition on ovarian and uterine physiology in dairy cattle. *J. Dairy Sci.* 81, 2533–9.

Butler, W. R. (2003). Energy balance relationships with follicular development, ovulation and fertility in postpartum dairy cows. *Live Prod. Sci.* 83, 211–18.

Butler, W. R. (2005). Relationships of negative energy balance with fertility. *Proceedings, Western Canadian Dairy Seminar.* http://www.wcds.ca/proc/2005/Manuscripts/Butler.pdf. Accessed 9 November 2016.

CanWest DHI Report (2014). http://www.canwestdhi.com/west%20report/index.html#p=28. Accessed 11 April 2016.

Chebel, R. C., Santos, J. E. P., Cerri, R. L. A., Galvao, K. N., Juchem, S. O. and Thatcher, W. W. (2003). Effect of resynchronization with GnRH on day 21 after artificial insemination on pregnancy rate and pregnancy loss in lactating dairy cows. *Theriogenology* 60, 1389–99.

Choudhury, G., Luktukc, S. N. and Sharma, U. D. (1965). Studies on the pattern of oestrous cycle in Hariana heifers. *Ind. Vet. J.* 42, 581–9.

Colazo, M. G., Gordon, M. B., Rajamahendran, R., Mapletoft, R. J. and Ambrose, D. J. (2009). Pregnancy rates to timed-AI in dairy cows treated with gonadotropin releasing hormone or porcine luteinizing hormone. *Theriogenology* 72, 262–70.

Colazo, M. G., Dourey, A., Rajamahendran, R. and Ambrose, D. J. (2013). Progesterone supplementation before timed AI increased ovulation synchrony and pregnancy per AI, and supplementation after timed AI reduced pregnancy losses in lactating dairy cows. *Theriogenology* 79, 833–41.

Colazo, M. and Mapletoft, R. J. (2014). A review of current timed-AI (TAI) programs for beef and dairy cattle. *Can. Vet. J.* 55, 772–80.

Crosse, S., Haran, P. and Killen, L. (1999). Replacement strategies to maximize profitability in dairying. Teagasc, Irish Agriculture and Food Development Authority. Cited by Maher, P., Good, M. and More, S. J. (2008). Trends in cow numbers and culling rate in the Irish cattle population, 2003 to 2006. *Irish Vet. J.* 61, 455–63.

D'Mello, J. P. F. (2004). Contaminants and toxins in animal feeds. In *Assessing Quality and Safety of Animal Feeds.* FAO Animal production and health papers, Issue 160, 107–28.

Dahl, G. E., Buchanan, B. A. and Tucker, H. A. (2000). Photoperiodic effects on dairy cattle: A review. *J. Dairy Sci.* 83, 885–93.

Dewey, S. T., Mendonça, L. G. D., Lopes Jr., G, Rivera, F. A., Guagnini, F., Chebel, R. C. and Bilby, T. R. (2010). Resynchronization strategies to improve fertility in lactating dairy cows utilizing a presynchronization injection of GnRH or supplemental progesterone: I. Pregnancy rates and ovarian responses. *J. Dairy Sci.* 93, 4086–95.

Dijkhuizen, T. J. and Van Eerdenburg, F. J. C. M. (1997). Behavioural signs of oestrus during pregnancy in lactating dairy cows. *Vet. Quart.* 119, 194–6.

Dobson, H. and Smith, R. F. (2000). What is stress and how does it affect reproduction? *Anim. Reprod. Sci.* 60–1, 743–52.

Dobson, H., Walker, S. L., Morris, M. J., Routly, J. E. and Smith, R. F. (2008). Why is it getting more difficult to successfully artificially inseminate dairy cows? *Animal* 2, 1104–11.

Dobson, H., Smith, R. F., Royal, M. D., Knight, C. H., and Sheldon, I. M. (2007). The high producing dairy cow and its reproductive performance. *Reprod. Domest. Anim.* 42 (Suppl 2), 17–23.

Dochi, O., Takahashi, K., Hirai, T., Hayakawa, H., Tanisawa, M., Yamamoto, Y. and Koyama, H. (2008). The use of embryo transfer to produce pregnancies in repeat-breeding dairy cattle. *Theriogenology* 69, 124–8.

Dochi, O., Kabeya, S. and Koyama, H. (2010). Factors affecting reproductive performance in high milk producing Holstein cows. *J. Reprod. Develop.* 56 (Suppl), S61–S65.

Dourey, A., Colazo, M. G., Ponce Barajas, P. and Ambrose, D. J. (2011). Relationships between endometrial cytology and interval to first ovulation, and pregnancy in postpartum dairy cows in a single herd. *Res. Vet. Sci.* 91, 149–53.

Dransfield, M. B. G., Nebel, R. L., Pearson, R. E. and Warnick, L. D. (1998). Timing of insemination for dairy cows identified in estrus by a radiotelemetric estrus detection system. *J. Dairy Sci.* 81, 1874–82.

Drost, M., Ambrose, J. D., Thatcher, M. J., Cantrell, C. K., Wolfsdorf, K. E., Hasler, J. F. and Thatcher, W. W. (1999). Conception rates after artificial insemination or embryo transfer in lactating dairy cows during summer in Florida. *Theriogenology* 52, 1161–7.

El-Zarkouny, S. Z., Cartmill, J. A., Hensley, B. A. and Stevenson, J. S. (2004). Pregnancy in dairy cows after synchronized ovulation regimens with or without presynchronization and progesterone. *J. Dairy Sci.* 87, 1024–37.

Edwards, J. L. and Hansen, P. J. (1997). Differential responses of bovine oocytes and preimplantation embryos to heat shock. *Molec. Reprod. Develop.* 46, 138–45.

Esslemont, R. J. and Kossaibati, M. A. (1997). Culling in 50 dairy herds in England. *Vet. Rec.* 140, 36–9.

Ferris, C. P., Heins, B. J. and Buckley, F. (2015). Crossbreeding in dairy cattle: Pros and cons. *Western Canadian Dairy Seminar, Advances in Dairy Technology* 26, 223–43.

Flamenbaum, I. and Galon, N. (2010). Management of heat stress to improve fertility in dairy cows in Israel. *J. Reprod. Develop.* 56, 36–41.

Food and Agriculture Organization of the United Nations. www.fao.org/agriculture/dairy-gateway/milk-production. Accessed 9 November 2016.

Foote, R. H. (2002). The history of artificial insemination: Selected notes and notables. *J. Anim. Sci.* 80, 1–10.
Fourichon, C., Seegers, H. and Malher, X. (2000). Effect of disease on reproduction in the dairy cow: a meta-analysis. *Theriogenology* 53, 1729–59.
Franco, O. J., Drost, M., Thatcher, M. J., Shille, V. M. and Thatcher, W. W. (1987). Fetal survival in the cow after pregnancy diagnosis by palpation per rectum. *Theriogenology* 27, 631–44.
Fricke, P. M., Caraviello, D. Z., Weigel, K. A., Welle, M. L. (2003). Fertility of dairy cows after resynchronization of ovulation at three intervals following first timed insemination. *J. Dairy Sci.* 86, 3941–50.
Fricke, P. M., Carvalho, P. D., Giordano, J. O., Valenza, A., Lopes Jr., G. and Amundson, M. C. (2014). Expression and detection of estrus in dairy cows: the role of new technologies. *Animal* 8 (Suppl. 1), 134–43.
Galon, N., Zeron, Y. and Ezra, E. (2010). Factors affecting fertility of dairy cows in Israel. *J. Reprod. Dev.* 56, S8–S14.
Galvão, K. N., Santos, J. E. P., Cerri, R. L., Chebel, R. C., Rutigliano, H. M., Bruno, R. G. and Bicalho, R. C. (2007). Evaluation of methods of resynchronization for insemination in cows of unknown pregnancy status. *J. Dairy Sci.* 90, 4240–52.
Garcia-Ruiz, A., Cole, J. B., VanRaden, P. M., Wiggans, G. R., Ruiz-Lopez, F. J. and Van Tassel, C. P. (2016). Changes in genetic selection differentials and generation intervals in US Holstein dairy cattle as a result of genomic selection. PNAS, Published online 27 June 2016, E3995–E4004. www.pnas.org/cgi/doi/10.1073/pnas.1519061113
Garnsworthy, P. C. (2006). Body condition score in dairy cows: targets for production and fertility. In P. C. Garnsworthy and J. Wiseman (Eds), *Recent Advances in Animal Nutrition*, pp. 61–86. Nottingham University Press, Nottingham, UK.
Gillespie, J., Nehring, R. and Sitienei, I. (2014). The adoption of technologies, management practices, and production systems in U.S. milk production. *Agric. Food Econ.* 2(17), 24.
Gobikrushanth, M., Salehi, R., Ambrose, D. J. and Colazo, M. G. (2016). Categorization of endometritis and its association with ovarian follicular growth and ovulation, reproductive performance, dry matter intake, and milk yield in dairy cattle. *Theriogenology* 86, 1842–9.
Grummer, R. R. (2007). Strategies to improve fertility of high yielding dairy farms: Management of the dry period. *Theriogenology* 68, S281–S288.
Gwazdauskas, F. C. (1985). Effects of climate on reproduction in cattle. *J. Dairy Sci.* 68, 1568–78.
Han, H., Austin, K. J., Rempel, L. A. and Hansen, T. R. (2006). Low blood ISG-15 mRN. A. and progesterone levels are predictive of non-pregnant dairy cows. *J. Endocrinol.* 191, 505–12.
Hansel, W. and Asdell, S. A. (1952). The causes of bovine metestrous bleeding. *J. Anim. Sci.* 11, 346–54.
Hansen, P. J. (1997). Effects of environment on bovine reproduction. In R. S. Youngquist (Ed.), *Current Therapy in Large Animal Theriogenology*, pp. 403–15. W.B. Saunders Co., Philadelphia, PA.
Hansen, P. J. and Aréchiga, C. F. (1999). Strategies for managing reproduction in the heat-stressed dairy cow. *J. Dairy Sci.* 82 (Suppl. 2), 36–50.
Heins, B. J., Hansen, L. B. and De Vries, A. (2012). Survival, lifetime production, and profitability of Normande × Holstein, Montbé- liarde × Holstein, and Scandinavian Red × Holstein crossbreds versus pure Holsteins. *J. Dairy Sci.* 95, 1011–21.
Hurley, W. L. and Doane, R. M. (1989). Recent developments in the roles of vitamins and minerals in reproduction. *J. Dairy Sci.* 72, 784–804.
Hoekstra, J., Van der Lugt, A. W., Van der Werf, J. H. J. and Ouweltjes, W. (1994). Genetic and phenotypic parameters for milk production and fertility in upgraded dairy cattle. *Livestock Prod. Sci.* 40, 225–32.
Inskeep, E. K. and Dailey, R. A. (2005). Embryonic death in cattle. *Vet. Clin. Food Anim.* 21, 437–61.
International Dairy Federation. www.fil-idf.org. Accessed 9 November 2016.
Ioannidis, J. and Donadeu, F. X. (2016). Circulating miRNA signatures of early pregnancy in cattle.*BMC Genomics* 17, 184.
Jordan, E. R. (2003). Effects of heat stress on reproduction. *J. Dairy Sci.* 86, 104–14.

Jorritsma, H. and Jorritsma, R. (2000). An overview of fertility statistics and milk production data of 15 dairy operations in southeast Friesland. *Tijdschrift voor Diergeneeskunde* 15, 180–4.

Kassa, T., Ambrose, J. D., Adams, A. L., Risco, C., Staples, C. R., Thatcher, M. J., Van Horn, H. H., Garcia, A., Head, H. H. and Thatcher, W. W. (2002). Effects of whole cottonseed diet and recombinant bovine somatotropin on ovarian follicles in lactating dairy cows. *J. Dairy Sci.* 85, 2823–30.

Kastelic, J. P., Curran, S. and Ginther, O. J. (1989). Accuracy of ultrasonography for pregnancy diagnosis on Days 10 to 22 in heifers. *Theriogenology* 31, 813–20.

Kinsel, M. L. and Etherington, W. G. (1998). Factors affecting reproductive performance in Ontario dairy herds. *Theriogenology* 50, 1221–38.

Kuhn, M. T. and Hutchison, J. L. (2008). Prediction of dairy bull fertility from field data: Use of multiple services and identification and utilization of factors affecting bull fertility. *J. Dairy Sci.* 91, 2481–92.

Kumaresan, A., Prabhakaran, P. P., Bujarbaruah, K. M., Pathak, K. A., Chhetri, B. and Ahmed, S. K. (2009). Reproductive performance of crossbred dairy cows reared under traditional low input production system in the eastern Himalayas. *Trop. Anim. Health Prod.* 41, 71–8.

Lean, J. I., Celi, P., Raadsma, H., McNamara, J. and Rabiee, A. R. (2012). Effects of dietary crude protein on fertility: Meta-analysis and meta-regression. *Anim. Feed Sci. Tech.* 171, 31–42.

LeBlanc, S. (2005). Overall reproductive performance of Canadian dairy cows: challenges we are facing. *Adv. Dairy Technol.* 17, 137–47.

LeBlanc, S. (2010). Assessing the association of the level of milk production with reproductive performance in dairy cattle. *J. Reprod. Develop.* 56, 1–7.

LeBlanc, S. (2013). Is a high level of milk production compatible with good reproductive performance in dairy cows? *Anim. Frontiers* 3, 83–91.

Leroy, J. L., Vanholder, T., Mateusen, B., Christophe, A., Opsomer, G., de Kruif, A., Genicot, G. and Van Soom A (2005). Non-esterified fatty acids in follicular fluid of dairy cows and their effect on developmental capacity of bovine oocytes in vitro. *Reproduction* 130, 485–95.

Leroy, J. L., Valckx, S. D., Jordaens, L., De Bie, J., Desmet, K. L., Van Hoeck, V., Britt, J. H., Marei, W. F. and Bols, P. E. (2015). Nutrition and maternal metabolic health in relation to oocyte and embryo quality: critical views on what we learned from the dairy cow model. *Reprod. Fertil. Dev.* 27, 693–703.

Lonergan, P. (2011). Influence of progesterone on oocyte quality and embryo development in cows. *Theriogenology* 76, 1594–601

Lonergan, P. and Forde, N. (2015). The role of progesterone in maternal recognition of pregnancy in domestic ruminants. *Adv. Anat. Embryol. Cell Biol.* 216, 87–104.

Lopez, H., Satter, L. D. and Wiltbank, M. C. (2004). Relationship between level of milk production and estrous behavior of lactating dairy cows. *Anim. Reprod. Sci.* 81, 209–23.

Lopez-Gatius F (2003). Is fertility declining in dairy cattle? A retrospective study in northeastern Spain. *Theriogenology* 60, 89–99.

López-Gatius, F., Garcia-Ispierto, I., Santolaria, P., Yaniz, J., Nogareda, C. and Lopez-Bejar, M. (2006). Screening for high fertility in high-producing dairy cows. *Theriogenology* 65, 1678–89.

López-Gatius, F. (2012). Factors of a noninfectious nature affecting fertility after artificial insemination in lactating dairy cows. A review. *Theriogenology* 77, 1029–41.

Lucy, M. C. (2001). Reproductive loss in high-producing dairy cattle: where will it end? *J. Dairy Sci.* 84, 1277–93.

Lucy, M. C. (2005). Second insemination breeding strategies for dairy cows. *Adv. Dairy Technol.* 17, 149–57.

Lucy, M. C. (2007). Fertility in high-producing dairy cows: reasons for decline and corrective strategies for sustainable improvement. *Soc. Reprod. Fertil.* 64, 237–54.

Macmillan, K. L., Lean, I. J. and Westwood, C. T. (1996). The effects of lactation on the fertility of dairy cows. *Aust. Vet. J.* 73, 141–7.

Mamo, S., Mehta, J. P., Forde, N., McGettigan, P. and Lonergan, P. (2012). Conceptus-endometrium crosstalk during maternal recognition of pregnancy in cattle. *Biol. Reprod.* 87, 1–9.

Madureira, A. M., Silper, B. F., Burnett, T. A., Polsky, L., Cruppe, L. H., Veira, D. M., Vasconcelos, J. L. and Cerri, R. L. (2015). Factors affecting expression of estrus measured by activity monitors and conception risk of lactating dairy cows. *J. Dairy Sci.* 98, 7003–14.

Mattos, R., Staples, C. R. and Thatcher, W. W. (2000). Effects of dietary fatty acids on reproduction in ruminants. *Rev. Reprod.* 5, 38–45.

Michi, A. N., Favetto, P. H., Kastelic, J. and Cobo, E. R. (2016). A review of sexually transmitted bovine trichomoniasis and campylobacteriosis affecting cattle reproductive health. *Theriogenology* 85, 781–91.

Moreira, F., Risco, C., Pires, M. F. A., Ambrose, J. D., Drost, M., DeLorenzo, M. and Thatcher, W. W. (2000a). Effect of body condition on reproductive efficiency of lactating dairy cows receiving a timed insemination. *Theriogenology* 53, 1305–19.

Moreira, F., Risco, C., Pires, M. F. A., Ambrose, J. D., Drost, M. and Thatcher, W. W. (2000b). Use of bovine somatotropin in lactating dairy cows receiving timed artificial insemination. *J. Dairy Sci.* 83, 1237–47.

Moreira, F., Orlandi, C., Risco, C. A., Mattos, R., Lopes, F. and Thatcher, W. W. (2001). Effects of presynchronization and bovine somatotropin on pregnancy rates to a timed artificial insemination protocol in lactating dairy cows. *J. Dairy Sci.* 84, 1646–59.

Nebel, R. L., Whittier, W. E., Cassell, B. G. and Britt, J. H. (1987). Comparison of on-farm and laboratory milk progesterone assays for identifying errors in detection of estrus and diagnosis of pregnancy. *J. Dairy Sci.* 70, 1471–6.

Nebel, R. L., Dransfield, M. G., Jobst, S. M. and Bame, J. H. (2000). Automated electronic systems for the detection of oestrus and timing of AI in cattle. *Anim. Reprod. Sci.* 60(61), 713–23.

Neves, R. C. and LeBlanc, S. J. (2015). Reproductive management practices and performance of Canadian dairy herds using automated activity-monitoring systems. *J. Dairy Sci.* 98, 2801–11.

Norman, H. D., Hutchison, J. L. and VanRaden, P. M. (2011). Evaluations for service-sire conception rate for heifer and cow inseminations with conventional and sexed semen. *J. Dairy Sci.* 94, 6135–42.

Norman, H. D., Wright, J. R., Hubbard, S. M., Miller, R. H. and Hutchison, J. L. (2009). Reproductive status of Holstein and Jersey cows in the United States. *J. Dairy Sci.* 92, 3517–28.

O'Connor, M. L. (2007). Estrus detection. In R. S. Youngquist (Ed.), *Current Therapy in Large Animal Theriogenology*, pp. 270–8. W.B. Saunders Co., Philadelphia, PA.

Opsomer, G., Gröhn, Y. T., Hertl, J., Coryn, M., Deluyker, H. and De Kruif A (2000). Risk factors for postpartum ovarian dysfunction in high producing dairy cows in Belgium: A field study. *Theriogenology* 53, 841–57.

Peter, A. T., Levine, H., Drost, M. and Bergfelt, D. R. (2009a). Compilation of classical and contemporary terminology used to describe morphological aspects of ovarian dynamics in cattle. *Theriogenology* 71, 1343–57.

Peter, A. T., Vos, P. L. A. M. and Ambrose, D. J. (2009b). Postpartum anestrus in dairy cattle. *Theriogenology* 71, 1333–41.

Peters, A. R. and Ball, P. J. H. (1995). *Reproduction In Cattle*. 2nd Ed. Blackwell Science Ltd., London UK., p. 243.

Peters, A. R., Martinez, A. T. and Cook, A. J. C. (2000). A meta-analysis of studies of the effect of GnRH 11–14 days after insemination on pregnancy rates in cattle. *Theriogenology* 54, 1317–26.

Peters, M. W. and Pursley, J. R. (2002). Fertility of lactating dairy cows treated with Ovsynch after presynchronization injections of PGF2α and GnRH. *J. Dairy Sci.* 85, 2403–6.

Petit H. V. and Twagiramungu H (2006). Conception rate and reproductive function of dairy cows fed different fat sources. *Theriogenology* 66, 1316–24.

Pursley, J. R., Silcox, R. W. and Wiltbank, M. C. (1998). Effect of time of artificial insemination on pregnancy rates, calving rates, pregnancy loss, and gender ratio after synchronization of ovulation in lactating dairy cows. *J. Dairy Sci.* 81, 2139–44.

Pursley, J. R., Mee, M. O. and Wiltbank, M. C. (1995). Synchronization of ovulation in dairy cows using PGF2alpha and GnRH. *Theriogenology* 44, 915–23.

Pursley, J. R., Wiltbank, M. C., Stevenson, J. S., Ottobre, J. S., Garverick, H. A. and Anderson, L. L. (1997a). Pregnancy rates per artificial insemination for cows and heifers inseminated at a synchronized ovulation or synchronized estrus. *J. Dairy Sci.* 80, 295–300.

Raheja, K. L., Burnside, E. B. and Schaeffer, L. R. (1989). Relationships between fertility and production in Holstein dairy cattle in different lactations. *J. Dairy Sci.* 72, 2670–8.

Revah, I. and Butler, W. R. (1996). Prolonged dominance of follicles and reduced viability of bovine oocytes. *J. Reprod. Fertil.* 106, 39–47.

Rhoads, M. L., Rhoads, R. P., Gilberto, R. O., Toole, R. and Butler, W. R. (2006). Detrimental effects of high plasma urea nitrogen levels on viability of embryos from lactating dairy cows. *Anim. Reprod. Sci.* 91, 1–10.

Ricci, A., Carvalho, P. D., Amundson, M. C., Fourdraine, R. H., Vincenti, L. and Fricke, P. M. (2015). Factors associated with pregnancy-associated glycoprotein (PAG) levels in plasma and milk of Holstein cows during early pregnancy and their impact on the accuracy of pregnancy diagnosis. *J. Dairy Sci.* 98, 2502–14.

Rodney, R. M., Celi, P., Scott, W., Breinhild, K. and Lean, I. J. (2015). Effects of dietary fat on fertility of dairy cattle: A meta-analysis and meta-regression. *J. Dairy Sci.* 98, 5601–20.

Rodrigues, C. A., Teixeira, A. A., Ferreira, R. M., Ayres, H., Macilha, R. F., Souza, A. H. and Baruselli, P. S. (2010). Effect of fixed-time embryo transfer on reproductive efficiency in high-producing repeat-breeder Holstein cows. *Anim. Reprod. Sci.* 118, 110–17.

Roth, Z., Biran, D., Lavon, Y., Dafni, I., Yakobi, S. and Braw-Tal, R. (2012). Endocrine milieu and developmental dynamics of ovarian cysts and persistent follicles in postpartum dairy cows. *J. Dairy Sci.* 95, 1729–37.

Rozzi, P., Miglior, F. and Hand, K. J. (2007). A total merit selection index for Ontario organic dairy farmers. *J. Dairy Sci.* 90, 1584–93.

Royal, M. D., Darwash, A. O., Flint, A. P. F., Webb, R., Woolliams, J. A. and Lamming, G. E. (2000). Declining fertility in dairy cattle: changes in traditional and endocrine parameters on fertility. *Anim. Sci.* 70, 487–501.

Saacke, R. G. (1974). Concepts in semen package and use. *Proc. 8th Conf. AI Beef Cattle*, Nat Assoc Anim Breeders, p. 15.

Saacke, R. G. (1977). Procedures for handling frozen semen. *Proc. 12th Conf. Artific Insem. Beef Cattle*, Nat Assoc Anim Breeders, p. 55.

Salehi, R., Colazo, M. G., Oba, M. and Ambrose, D. J. (2015). A prepartum diet supplemented with oilseeds high in oleic or linoleic acid reduced GnRH-induced LH release in dairy cows during second week postpartum. *Reprod. Biol. Endocrinol.* 13, 69. https://www.ncbi.nlm.nih.gov/pmc/articles/PMC4490657/

Santos, J. E. P., Thatcher, W. W., Pool, L. and Overton, M. W. (2001). Effect of human chorionic gonadotropin on luteal function and reproductive performance of high-producing lactating Holstein dairy cows. *J. Anim. Sci.* 79, 2881–94.

Santos, J. E. P., Thatcher, W. W., Chebel, R. C., Cerri, R. L. A. and Galvao, K. N. (2004). The effect of embryonic death rates in cattle on the efficacy of estrus synchronization programs. *Anim. Reprod. Sci.* 82–3, 513–35.

Santos, J. E. P., Bilby, T. R., Thatcher, W. W., Staples, C. R. and Silvestre, F. T. (2008). Long chain fatty acids of diet as factors influencing reproduction in cattle. *Reprod. Dom. Anim.* 43, 23–30.

Santos, J. E. P., Rutigliano, H. M. and Sá Filho, M. F. (2009). Risk factors for resumption of postpartum cyclicity and embryonic survival in lactating dairy cows. *Anim. Reprod. Sci.* 110, 207–21.

Santos, J. E. P., Bisinotto, R. S., Ribeiro, E. S., Lima, F. S., Greco, L. F., Staples, C. R. and Thatcher, W. W. (2010). Applying nutrition and physiology to improve reproduction in dairy cattle. *Soc. Reprod. Fertil. Supp.* 67, 387–403.

Sartori, R., Haughian, J. M., Shaver, R. D., Rosa, G. J. and Wiltbank, M. C. (2004). Comparison of ovarian function and circulating steroids in estrous cycles of Holstein heifers and lactating cows. *J. Dairy Sci.* 87, 905–20.

Sasser, G. R., Ruder, C., Ivani, K. A., Butler, J. E. and Hamilton, W. C. (1986). Detection of pregnancy by radioimmunoassay of a novel pregnancy-specific protein in serum of cows and a profile of serum concentrations during gestation. *Biol. Repro.* 35, 936–42.
Sasser, R. G., Crock, J. and Ruder-Montgomery, C. A. (1989). Characteristics of pregnancy-specific protein B in cattle. *J. Reprod. Fertil.* 37, 109–13.
Schmitt, E. J., Diaz, T., Drost, M. and Thatcher, W. W. (1996a). Use of a gonadotropin-releasing hormone agonist or human chorionic gonadotropin for timed insemination in cattle. *J. Anim. Sci.* 74, 1084–91.
Schmitt, E. J., Diaz, T., Drost, M. and Thatcher, W. W. (1996b). Differential response of the luteal phase and fertility in cattle following ovulation of the first-wave follicle with human chorionic gonadotropin or an agonist of gonadotropin-releasing hormone. *J. Anim. Sci.* 74, 1074–83
Seegers, H., Beaudeau, F., Fourichon, C. and Bareille, N (1998). Reasons for culling in French Holstein cows. *Prev. Vet. Med.* 36, 257–71.
Sharma, S. and Rou, Z (2014). China's Dairy Dilemma: The Evolution and Future Trends of China's Dairy Industry. The Institute for Agriculture and Trade Policy, p. 27. http://www.iatp.org/files/2014_02_25_DairyReport_f_web.pdf. Accessed 31 March 2016.
Sianangama, P. C. and Rajamahendran, R (1996). Effect of hCG administration on day 7 of the estrous cycle on follicular dynamics and cycle length in cows. *Theriogenology* 45, 583–92.
Silvestre, F. T., Carvalho, T. S. M., Francisco, N., Santos, J. E. P., Staples, C. R., Jenkins, T. C. and Thatcher, W. W. (2011). Effects of differential supplementation of fatty acids during the peripartum and breeding periods of Holstein cows: I. Uterine and metabolic responses, reproduction and lactation. *J. Dairy Sci.* 94, 189–204.
Silvestre, F. T., Carvalho, T. S. M., Francisco, N., Santos, J. E. P., Staples, C. R., Jenkins, T. C. and Thatcher, W. W. (2011). Effects of differential supplementation of fatty acids during the peripartum and breeding periods of Holstein cows: I. Uterine and metabolic responses, reproduction, and lactation. *J. Dairy Sci.* 94, 189–204.
Sinclair, K. D., Garnsworthy, P. C., Mann, G. E. and Sinclair, L. A. (2014). Reducing dietary protein in dairy cow diets: implications for nitrogen utilization, milk production, welfare and fertility. *Animal* 8, 262–74.
Sousa, N. M., Ayad, A., Beckers, J. F. and Gajewski, Z (2006). Pregnancy-associated glycoproteins (PAG) as pregnancy markers in the ruminants. *J. Physiol. Pharmacol.* 57 (Suppl. 8), 153–71.
St-Pierre, N. R., Cobanov, B. and Schnitkey, G (2003). Economic losses from heat stress by US livestock industries. *J. Dairy Sci.* 86 (E-Suppl.), E52–E77
Staples, C. R., Burke, J. M. and Thatcher, W. W. (1998). Influence of supplemental fats on reproductive tissues and performance of lactating cows. *J. Dairy Sci.* 81, 856–71.
Stevenson, J. S., Hill, S. L., Nebel, R. L. and Dejarnette, J. M. (2014). Ovulation timing and conception risk after automated activity monitoring in lactating dairy cows. *J. Dairy Sci.* 97, 4296–308.
Stevenson, J. S., Tenhouse, D. E., Krisher, R. L., Lamb, G. C., Larson, J. E., Dahlen, C. R., Pursley, J. R., Bello, N. M., Fricke, P. M., Wiltbank, M. C., Brusveen, D. J., Burkhart, M., Youngquist, R. S. and Garverick, H. A. (2008). Detection of anovulation by heatmount detectors and transrectal ultrasonography before treatment with progesterone in a timed insemination protocol. *J. Dairy Sci.* 91, 2901–15.
Swedish Dairy Association. 2009–12. Cattle statistics. Swedish Dairy Association, Stockholm, Sweden (In Swedish). Cited by Lomander, H., Gustafsson, H., Svensson, C., Ingvartsen, K. L. and Frössling, J. (2012). Test accuracy of metabolic indicators in predicting decreased fertility in dairy cows. *J. Dairy Sci.* 95, 7086–96.
Swormink, B. K. (2014). Dairy farming sector in India experiences a rapid growth. Dairy Global. 7th November. http://www.dairyglobal.net/Articles/General/2014/11/Rapid-growth-in-Indias-dairy-farming-sector-1630621W/. Accessed 31 March 2016.
Tenhagen, B. A., Surholt, R., Wittke, M., Vogel, C., Drillich, M. and Heuwieser, W (2004). Use of Ovsynch in dairy herds – differences between primiparous and multiparous cows. *Anim. Reprod. Sci.* 81, 1–11.

Thatcher, W. W. and Wilcox, C. J. (1973). Post-estrus as indicator of reproductive status in dairy cows. *J. Dairy Sci.* 56, 608–10.

Thatcher, W. W., Macmillan, K. L., Hansen, P. J. and Drost, M (1989). Concepts for regulation of corpus luteum function by the conceptus and ovarian follicles to improve fertility. *Theriogenology* 31, 149–64.

Thatcher, W. W., Staples, C. R., Danet-Desnoyers, G., Oldick, B. and Schmitt, E. P. (1994). Embryo health and mortality in sheep and cattle. *J. Anim. Sci.* 72, 16–30.

Thatcher, W. W., Silvestre, F. T., Santos, J. E. P. and Staples, C. R. (2008). The impact of lactation on reproductive performance. *Adv. Dairy Technol.* 20, 17–31.

Thatcher, W. W., Santos, J. E. P. and Staples, C. R. (2011). Dietary manipulations to improve embryonic survival in cattle. *Theriogenology* 76, 1619–31.

Thomas, I. and Dobson, H (1989). Oestrus during pregnancy in the cow. *Vet. Rec.* 124, 387–90.

Twagiramungu, H., Guilbault, L. A. and Dufour, J. J. (1995). Synchronization of ovarian follicular waves with a gonadotropin-releasing hormone agonist to increase the precision of estrus in cattle: a review. *J. Anim. Sci.* 73, 3141–51.

USDA-NASS (2016). United States Department of Agriculture, National Agricultural Statistics Service. http://quickstats.nass.usda.gov/. Accessed 12 April 2016.

Van Doormaal, B. J. and Kistemaker, G. J. (2003). Dairy genetic improvement through artificial insemination, performance recording and genetic evaluation. *Can. J. Anim. Sci.* 83, 385–92.

Van Eerdenburg, F. J. C. M., Loeffler, S. H. and Vliet, J. H. van (1996). Detection of oestrus in dairy cows: A new approach to an old problem. *Vet. Quart.* 18, 52–4.

Van Eerdenburg, F. J. C. M., Karthaus, D., Taverne, M. A. M., Merics, I. and Scenzi, O (2002). The relationship between estrous behavioral score and time of ovulation in dairy cattle. *J. Dairy Sci.* 85, 1150–6.

Van Hoeck, V., Bols, P. E., Binelli, M. and Leroy, J. L. (2014). Reduced oocyte and embryo quality in response to elevated non-esterified fatty acid concentrations: a possible pathway to subfertility? *Anim. Reprod. Sci.* 149, 19–29.

Walker, S. L., Smith, R. F., Jones, D. N., Routly, J. E. and Dobson, H (2008). Chronic stress, hormone profiles and estrus intensity in dairy cattle. *Hormon. Behav.* 53, 493–501.

Walsh, S. W., Williams, E. J. and Evans, A. C. O. (2011). A review of the causes of poor fertility in high milk producing dairy cows. *Anim. Reprod. Sci.* 123, 127–38.

Watters, R. D., Wiltbank M. C. , Guenther, J. N., Brickner, A. E., Rastani, R. R., Fricke, P. M. and Grummer, R. R. (2009). Effect of dry period length on reproduction during the subsequent lactation. *J. Dairy Sci.* 92, 3081–90.

Westwood, C. T., Lean, J. J. and Garvin, J. K. (2002). Factors influencing fertility of Holstein cows: a multivariate description. *J. Dairy Sci.* 85, 3225–37.

Xu, Z. Z., McKnight, D. J., Vishwanath, R., Pitt, C. J. and Burton, L. J. (1998). Estrus detection using radiotelemetry or visual observation and tail painting for dairy cows on pasture. *J. Dairy Sci.* 81, 2890–6.

Young, B. A. (1983) Ruminant cold stress: effect of production. *J. Anim. Sci.* 57, 1601–7.

Chapter 9

Nutritional strategies to improve nitrogen efficiency and milk protein synthesis in dairy cows

James D. Ferguson, University of Pennsylvania, USA

1 Introduction: the importance of reducing nitrogen losses in dairy farming

1.1 The importance of dairy source proteins

The Food Agricultural Organization (FAO) estimates that 6 billion of the 7 billion people in the world consume milk and milk products, and 150 million households around the globe are engaged in milk production. In 2013 the global production of milk was 768 640 663 metric tonnes from 758 222 163 milk-producing animals (Table 1). The major dairy animals are cattle (82.69% of total production), buffalo (13.28% of total production), goats (2.34% of total production), sheep (1.32% of total production) and camels (0.38% of total production)

http://dx.doi.org/10.19103/AS.2016.0005.11

Table 1 Milk production in 2013[1]

Animal	Number count	Production tonnes	Total production %	Kilogram/ head	Protein %	Fat %
Buffalo[1]	61,239,507	102,041,460	13.28	1,666	4.50	8.00
Camel[1]	6,439,737	2,928,188	0.38	455	4.26	4.05
Cow[1]	270,848,210	635,575,895	82.69	2,347	3.20	3.90
Goat[1]	200,008,068	17,957,372	2.34	90	3.30	3.75
Sheep[1]	219,686,641	10,137,749	1.32	46	5.40	6.00
Lesser species						
Yak[2]	15,000,000	NA	NA	3.0	5.10	7.00
Reindeer[3]	5,000,000	NA	NA	0.3	10.00	15.00
Equine[1]	57,851,928	NA	NA		2.10	1.25
Donkey[1]	43,194,091	NA	NA	0.9	1.65	1.05

[1] Source: FAOSTATS. http://faostat3.fao.org/home/E. Total animal numbers.
[2] Source: http://www.fao.org/agriculture/dairy-gateway/milk-production/dairy-animals/other-animals/en/#.VsM-w8sm5dg.
[3] Source:http/reindeerherding.org. About 2000000 domesticated or partially domesticated.
NA = data not available.

(Table 1). In addition to the major dairy animals, yaks, horses, reindeer and donkeys also produce milk consumed by humans. Milk and milk products provide 20614 metric tonnes of protein globally (protein content across all milk produced, 3.316%) providing 8.2 g protein/capita/day of protein globally, which is 10.2% of daily available protein across the globe (FAOSTATS, Food Balance Sheets, 2011). It is expected that consumption of animal source proteins will increase over the next 20 years and dairy proteins will contribute significantly to this increase.

1.2 The environmental footprint of dairy farming

Dairy products are a source of high-quality protein and have been a significant component of human diets for millennia. However, dairy producers face various societal challenges with which they must comply to be sustainable. A major challenge is economic, which has been reflected in the significant decline in dairy farm numbers over the past 50 years in developed countries. Concomitant with the decline in farm numbers has been the increase in herd size and production per cow. The decline in farms and increase in size and production have been driven by market forces. From 1961 to 2013 dairy cow population has declined in the United States (17.2 million head to 9.3 million head) and in Europe (83.5 million head to 37.5 million head), and the production per cow has also increased (the United States: 1961, 3307 kg/hd; 2013, 9901 kg/hd; Europe: 1961, 2272 kg/hd; 2013, 5600 kg/hd). Herd size in the United States has increased from 10 cows in 1961 to 213 cows in 2015. The loss of number of farms and increase in farm size has been in response to market forces making margins per unit milk narrower, resulting in intensification of dairy farms.

The situation in the developing world is the opposite of the developed world. Dairy cattle numbers have increased, whereas production per cow has only slightly increased (dairy cows, 1961, 68.5 million; 2013, 216.5 million; production/animal, 1961, 688 kg/hd; 2013, 1372 kg/hd) (FAOSTAT). Herd size has remained small in many developing countries.

The increased herd size in developed countries and the increased animal numbers in developing countries have put pressure on the environmental footprint from dairy farming. Losses of nitrogen, phosphorus and greenhouse gases to the environment from livestock production units can create environmental degradation. Air emissions from animal facilities of ammonia released from urinary nitrogen (N) contribute to acid rain; atmospheric ozone and haze; air particulates of 2.5 μm and eutrophication of streams, lakes and estuaries. Nitrogen run-off and leachate from land applied animal manure increase nitrate concentration in groundwater, which contributes to algal blooms and eutrophication of lakes, ponds and estuaries. High nitrate concentrations in groundwater of agricultural regions present a risk for blue baby syndrome. Continuous manure application can increase soil phosphorus concentrations. Phosphorus (P) in soil sediment run-off also increases the risk of eutrophication of water systems. In addition, greenhouse gas emissions of enteric methane and nitrous oxide from manure storage and land disposal pose risks for climate alteration.

Environmental issues concerning air and water quality present regulatory challenges requiring that dairy farms become more nutrient efficient, transferring more feed N and P to milk and meat in order to reduce nutrient losses in animal manure. Current research is focused on reducing the environmental burden associated with dairy production. A major concern is the efficiency of conversion of feed nitrogen to milk protein nitrogen. Improving the proportion of feed N converted to milk protein N can significantly reduce environmental losses of N to air and water systems. Currently, only 10–30% of feed N is captured in milk protein N across all inputs and outputs on a dairy farm. For dairy production to be sustainable into the future to feed over 9 billion people, nutrient efficiencies must be improved and nutrient losses mitigated.

The primary means producers have in controlling nutrient losses of N is to improve capture of feed N into milk protein (Dou, 1996; Ferguson, 2001; Frank, 2002; Kohn, 1997). The efficiency of capture of feed N in milk increases when dietary N is reduced, but this often results in reduced milk production and farm income. The challenge for producers is to reduce dietary inputs yet maintain production and income. This review will explore the efficiency of conversion of dietary N to milk N with a view to reducing the risk of environmental pollution yet maintain production.

2 Protein in milk: protein content, determining factors and method of synthesis

2.1 Protein content and fractions in milk

Protein content in milk may be measured by three methods: (1) total N determination (using the Kjeldahl N or the Dumas (combustion) method), (2) Udy dye binding and (3) infrared reflectance (DePeters and Cant, 1992; Ng-Kwai-Hang, 2003). Milk crude protein (CP) content is estimated by multiplying total N by 6.38 the average N content of milk protein (Ng-Kwai-Hang, 2003; Stelwagon, 2000). Estimating milk protein from total N overestimates the true protein content as total N includes non-protein N (NPN) compounds. Nitrogen in milk is contained in true protein and NPN (Alston-Mills, 1995). NPN in milk varies primarily due to changes in urea content (De Peters and Ferguson, 1992). More precise methods measure true protein in milk, which is desired. Acid orange-12 dye (Udy dye method) binds only to protein (casein and whey proteins) so it measures true protein content of milk. The infrared near reflectance method detects peptide bonds and measures true protein content of milk when the reflectance spectrum is calibrated appropriately.

Milk protein consists of proteins produced in mammary epithelial cells (casein and whey proteins (lactalbumin, lactoglobulin)), proteins from blood (albumin, immunoglobulins) and enzymes and proteins arising from mammary cells. Milk fat globule membrane proteins comprise about 1–4% of total protein in milk and are a heterogeneous group of proteins (Bionaz, 2011). However, increasingly small peptides and minor bioactive proteins have been identified (935 in the study by Tacoma, 2016) in milk that have significant effects on a range of physiologic processes (Tacoma, 2016). Although breed has a major influence on the relative content of the major milk proteins (Table 3), it appears that when housed and fed similar diets, cows of different breeds have similar low-abundant proteins in milk and composition of amino acids (AAs) in major milk proteins (Ng-Kwai-Hang, 2003; Stelwagon, 2000; Tacoma, 2016).

Milk protein is heterogeneous with five main categories: caseins, whey proteins, milk fat globule proteins, enzymes and minor, miscellaneous proteins (Ng-Kwai-Hang, 2003; O'Mahoney, 2014). The nitrogenous fractions in bovine milk as a per cent of total N are casein (78%), lactalbumin (12%), lactoglobulin (5%), proteose peptones (2%) and NPN (3%) (Ng-Kwai-Hang, 2003; O'Mahony and Fox, 2014). The NPN component of milk N can be variable and is composed of urea, AAs, uric acid, creatine and creatinine (Alston-Mills, 1995; De Peters and Ferguson, 1992). The USDA Animal Improvement Laboratory estimates NPN as 0.19% of milk protein content (https://www.cdcb.us/reference/trueprot.htm). However, NPN is not fixed and can vary significantly depending on the N components of the diet that alter milk urea N (MUN) content (DePeters and Ferguson, 1992). MUN may be 50% or more of the NPN content in milk and can range from 8 mg/dl to 20 mg/dl (De Peters and Ferguson, 1992). The NPN content in milk may be quite variable depending on diet. The greater the proportion of urea in milk the greater the urinary N excretion and risk of ammonia emission (Jonker, 1999; Powell, 2011). In terms of human nutrition and cheese manufacturing, true protein content of milk is of primary importance, and feeding programmes on dairy farms should attempt to reduce the NPN component of milk and increase the yield of true protein. For most dairy nutritionists and producers, the production of total true protein content is the major focus of concern. For cheese manufacturers the proportion of casein in milk protein is of primary importance.

However, currently only milk reporting centres in the United States, Australia, France and Hungary report milk true protein content; other countries still use milk CP for reporting purposes, which includes true protein and NPN content, thus providing a challenge in interpreting changes in protein content of milk with changes in feeding practices.

Four genes control milk casein (α_{s1}, α_{s2}, β and κ alleles) each of which has several minor variants resulting in over 40 different AA variants of milk casein protein (Ng-Kwai-Hang, 2003). Typical casein fractions in bovine milk tend to have the following fractions: α_{s1} (38%, calcium sensitive), α_{s2} (10%, calcium sensitive), β (35%, calcium sensitive) and κ (12%, calcium insensitive). Casein alleles are genetically determined. The various casein alleles are associated with the manufacturing properties of milk protein and the protein percentage in milk. For example, the B-κ casein allele is associated with higher protein percentages in Holstein–Friesian breeds.

2.2 Factors influencing milk protein levels

Table 2 presents yield data for major dairy breeds in the United States from the Council on Dairy Cattle Breeding (https://www.cdcb.us), Holstein–Friesian, Jersey, Guernsey, Ayrshire, Brown Swiss, Milking Shorthorn and Red and White. Holstein–Friesian cows (Table 2) are the dominant milk-producing breeds. Protein yield per kilogram of milk ranges from 30.57 g/kg for Red and White cattle to 36.31 g/kg for Jersey cows (Table 2). Breed differences highlight the genetic influence on protein yield, but genetic factors also operate within breed, thus influencing protein yield (Oltenacu, 2005). Within breed, the protein content of milk varies by parentage, parity, season and days post-calving (DePeters and Ferguson, 1992; DePeters and Cant, 1994). Nutritional status and regional differences can further influence protein content of milk with summer months associated with lower protein content than winter months (DePeters and Ferguson, 1992). Milk volume and protein content of milk vary by days in milk (DIM), and the production curves for milk volume and protein content are the inverse of each other, whereas protein yield curves follow the same shape as milk yield curves (protein yield curves not shown, Fig. 1).

Table 2 Production data for cow records in the Council on Dairy Cattle Breeding for April 2015 (https:www.cdcb.us)

Breed	Number records	Production kg/lactation	Fat %	Protein content g/kg of milk
Ayrshire	5,559	8,381	3.90	31.47
Brown Swiss	17,997	10,038	4.06	33.42
Guernsey	5,532	7,905	4.48	32.91
Holstein–Friesian	2,175,009	12,083	3.73	30.66
Jersey	253,850	8,978	4.77	36.31
Milking Shorthorn	3,126	8,105	3.69	31.14
Red and White	2,269	10,797	3.74	30.57

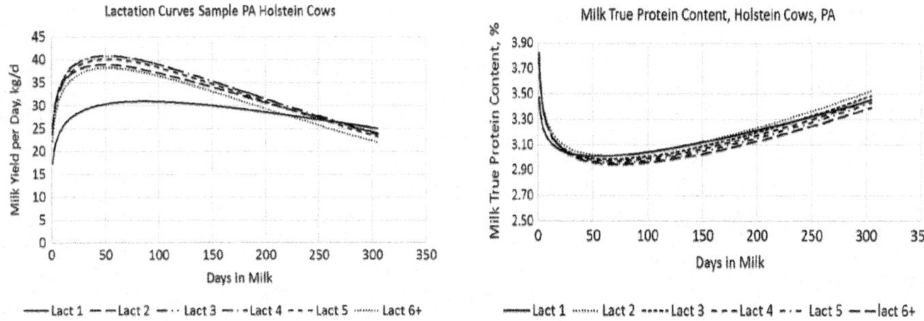

Figure 1 Milk production by days in milk for Pennsylvania Holstein cows for first lactation to 6 and greater lactation (left). Milk protein content by days in milk for Pennsylvania Holstein cows for first lactation to 6+ lactation (right).

Milk protein yield (MPY) and milk protein per cent are heritable traits (Hayes, 1984; Table 3). Protein yield has a heritability of 0.27 and milk protein content 0.47 (Table 3). Protein yield is phenotypically and genetically correlated with milk production (phenotypic correlation 0.76, genetic correlation 0.84) but protein content is negatively correlated with milk volume (phenotypic correlation: −0.35, genetic correlation: −0.28) (Table 3). Therefore, increasing milk volume will tend to decrease milk protein content while increasing MPY.

The sire influence on MPY across breeds is presented in Table 4. Red and White bulls are the intercept, and slope by breed is presented relative to the Red and White breed (Table 4). Sire predicted transmitting ability (PTA) for protein yield differs by breed and increases from 20.058 g/kg for Holstein bulls to 40.580 g/kg for Milking Shorthorn bulls, reflecting breed differences for protein yield (Table 4). Holstein bulls have a PTA for MPY per kilogram of milk of 20.058 g/kg (18.85–21.29 g/kg, 95% confidence limit, Table 4). As bulls of higher milk yield are selected by producers, protein yield from sire transmission of genes will increase by the mean amounts as shown in Table 4.

Table 3 Heritability, phenotypic and genetic correlations among milk components (Hayes et al., 1984)

Trait	Milk yield	Fat yield	Protein yield	Fat %	Protein %	Casein yield	Casein %
Milk, yield	0.26	0.83	0.76	−0.15	−0.35	0.87	−0.46
Fat, yield	0.81	0.23	0.70	0.30	−0.05	0.82	−0.25
Protein, yield	0.84	0.84	0.27	0.00	0.05	0.91	−0.30
Fat, %	−0.31	0.21	0.08	0.47	0.50	0.03	0.34
Protein, %	−0.28	0.14	0.23	0.58	0.47	0.00	0.52
Casein, yield	0.80	0.61	0.95	0.34	0.17	0.23	0.25
Casein, %	−0.76	−0.21	−0.28	0.31	0.96	0.23	0.45

Heritabilities are on the diagonal and underlined.
Phenotypic correlations are above the diagonal.
Genetic correlations are below the diagonal.

Table 4 Sire PTA for protein as a function of milk yield in the USDA Animal Improvement Program (*n* = 27280 bulls) (https://aipl.arsusda.gov/)

Breed	*N*	Intercept	SEM	Grams protein/kg of milk	SEM
Intercept		5960.805	180.880	24.359	0.517
Ayrshire	425	790.112	328.408	5.478	0.687
Brown Swiss	1190	−3381.945	270.973	2.956	0.903
Guernsey	209	−5569.992	844.241	−3.843	3.242
Holstein	22340	−3139.046	183.842	−4.301	0.527
Jersey	2551	−1844.067	235.943	−2.559	0.653
Milking Shorthorn	70	−2623.786	1243.807	16.221	4.209
Red and White	494	–	–	–	–

Data for protein yield per unit of milk for elite cows from the USDA Animal Production Laboratory and for Holstein cows from a set of Pennsylvania (PA)dairy farms (59) are shown in Table 5. The 305-day protein yield per kilogram of milk ranges from 26.849 g/kg to 31.625 g/kg for Red and White cows and Brown Swiss cows, respectively (Table 5, Red and White cows are the reference intercept). The data in Table 5 represent phenotypic yields of protein for each kilogram of milk produced. These values are greater than the PTA for bulls, reflecting phenotypic yield versus the genetic yield and heritability of the trait. For example, Holstein bull's PTA is 20.058 g/kg of milk, whereas the elite Holstein cows have an observed yield of 27.719 g/kg of milk (26.849 + 0.870, 95% confidence range, 26.23–29.22 g/kg, Table 5). Regressing MPY on test day milk records for PA Holsteins yields a slope of 26.613 g/kg, adjusted for effects of parity and DIM (Table 5). The 95% confidence range for the slope of the PA regression includes the 27.719 g/kg slope for complete lactation records for the elite Holstein cows. Therefore, for each kilogram of milk produced, we would expect an increase in protein yield consistent with the slopes in Table 5. About 70% of the yield appears associated with sire transmitting ability (Table 4). Increases in protein yield per unit of milk would need to have slopes greater than those in Table 5 to truly represent an increase in protein production in the mammary gland.

Table 5 Data for elite cows by breed in USDA Animal Production Laboratory for protein yield for lactation milk yields and for data for Holstein cows from 63 Pennsylvania dairy farms

		305 d protein yield/kg milk[1]		PTA[2] Protein Yield/PTA[2] kg milk[1]	
Breed	N	g/kg	sem	g/kg	sem
Ayrshire	248	3.075	0.867	9.500	0.946
Brown Swiss	361	4.776	0.809	2.251	0.998
Guernsey	131	1.961	0.964	4.984	1.001
Holstein	8690	0.870	0.650	2.282	0.773
Jersey	2090	2.851	0.692	−0.299	0.797
Milking Shorthorn	57	0.502	1.416	8.125	1.786
Red and White	230	26.849	0.638	16.270	0.758
Holstein cows in 63 Pennsylvania Herds[3]					
Holstein	101549	26.613	0.665		
Lactation 1		−0.493	0.092		
Lactation 2		−0.044	0.086		
Lactation 3		−0.153	0.092		
Lactation 4+		–	–		
DIM		0.355	0.004		

[1] Values are relative to the slope for Red and White and actual values are added to the slope for Red and White cows. Intercepts not shown.
[2] PTA = Predicted transmitting ability.
[3] Production data from monthly Dairy Herd Improvement Association from calving to 305 days of lactation. DIM is days in milk.

2.3 Amino acids in milk protein

There are 20 major AAs necessary for protein synthesis. Ten AAs are considered essential: lysine, methionine, threonine, isoleucine, leucine, histidine, phenylalanine, valine, arginine, tryptophan); and two may be considered semi-essential depending on physiologic conditions (cysteine and tyrosine) and eight are considered nonessential (NEAA, alanine, aspartic acid, asparagine, glutamic acid, glutamine, serine, glycine and proline) (Boison, 2000; Mepham, 1988; Swaisgood, 1995). Methionine and cysteine are sulphur-containing AAs and cysteine can meet 50% of the requirement for sulphur-containing AAs, reducing the requirement for methionine. Likewise, phenylalanine is a precursor for tyrosine and tyrosine can satisfy 50% of the need for tyrosine and phenylalanine (NRC, 2001).

Essential amino acids (EAAs) must be supplied in the diet; they are not synthesised in the body at sufficient rates to meet requirements. Semi-EAAs may be required in the diet under certain conditions (Haque, 2012, 2015). NEAAs may be synthesised at adequate rates as long as there are sufficient carbon skeletons and N donors available. Protein quality for human nutrition is defined based on the supply of EAA relative to NEAAs. An ideal protein has a perfect proportion of EAA for tissue synthesis and sufficient NEAA to meet tissue AA requirements.

Urinary losses are associated with efficiency of supply of EAA relative to requirement. The poorer the supply the greater the catabolism of AA and nitrogen losses in urine,

Table 6 Composition of bovine herd milk[1,2,3]

	Protein[1]	Protein[2]		NPN[3] (*N*)		
Item	g/kg	g/l	sd	Item	mg/l	sd
Total protein	35.1	32.71	1.80	Total NPN	296.4	37.7
Total casein	28.6	26.92	1.54	Urea N	142.1	32.6
Whey protein	6.1	5.79	0.32	Creatine N	25.5	6.4
α_{s1}–casein	11.5	10.25	0.57	Creatinine N	12.1	6.8
α_{s2}–casein	3.0	2.74	0.21	Uric acid N	7.8	3.3
β–casein	9.5	9.60	0.50	Orotic acid N	14.6	5.9
κ–casein	3.4	3.45	0.32	Hippuric acid N	4.4	1.2
γ–casein	1.2	0.88	0.15	Peptide N	32.0	14.9
α–lactalbumin	1.2	1.23	0.09	Ammonia N	8.8	6.1
β–lactoglobulin	3.1	3.14	0.19	α–amino N	44.5	8.2
Serum albumin	0.4	0.45	0.04			
Immunoglobulin	0.8					
Proteose peptones	1.0					
IPL		0.97	0.10			

[1] Swaisgood, H. E. (1992), assuming a density of 1.03 g/ml milk.
[2] Stelwagon, K. (2000), IPL = immunoglobulins, proteose peptones component 3, lactoferrin.
[3] Alston-Mills, B. (1992).

largely from urea excretion. Therefore, as AA absorbed from the diet complement tissue requirements more closely, urinary N excretion is diminished (Haque, 2012). In lactating dairy cows, milk protein synthesis is the major requirement for AA. Therefore, if diets can supply absorbed AA more closely to that required for milk protein synthesis, urinary N losses are reduced.

Milk protein is a high-quality protein, containing a high proportion of EAA (Table 6). The proportion of EAA in a protein may be described as a proportion of total AAs (Rulquin, 1993, 1998), as a proportion of EAA (Schwab, 1992a,b, 1996) or as a ratio to LYS, often considered the most limiting AA for growth (Boison, 2000). Protein synthesis is sensitive to the supply of EAA. When the AA pattern of a feed is similar to that required, it is considered an 'ideal' protein. Of utmost importance is that the absorbed AAs from the small intestine (the metabolisable protein, MP) contain the ideal proportion of EAA and sufficient nitrogenous compounds to meet the requirements for the NEAA. Often a particular EAA will be at a proportion of the EAA or total MP such that performance of the animal is limited by the content of that EAA. Providing more of that EAA will result in an increase in animal production. This EAA is referred to as being first limiting. For milk protein synthesis, methionine and lysine are often considered first limiting AAs (Doepel, 2004; Rulquin, 1993; Schingoethe, 1996; Schwab, 1992a,b; Tamminga,1980a,b). The challenge in dairy rations is to predict the supply of methionine and lysine and total MP for milk production as close to requirement as possible to minimise urinary nitrogen. Rumen degradation of feeds and microbial flows of protein make it challenging to predict methionine, lysine, and MP supply across diverse diets and feedstuffs (Clark, 1992; NRC 2001; Schwab, 2005; Smoler, 1998).

2.4 Mammary gland synthesis of protein

A detailed description of metabolic processes and regulation of protein synthesis in mammary cells is beyond this review, but an overview of processes will provide a context for evaluation of nutritional strategies to improve MPY and enhance dietary N efficiency. For a detailed overview the reader is referred to Bionaz (2012).

Briefly, regulation of milk protein synthesis is complex and includes endocrine, nutrient availability and mechanical influences (Finucane, 2008; Bionaz, 2012). Mechanical influences include milking and suckling stimuli, but also mechanical adhesion of epithelial cells in the gland which enhances cell differentiation. Supply of AAs and glucose influences protein synthesis. If energy is limiting, particularly glucose availability, milk protein synthesis is reduced, as is milk volume. There is a close association between energy availability to the gland and protein synthesis; thus, MPY is highly correlated with milk volume, a proxy for lactose yield (Tables 3, 4 and 5). Within a given volume of production, AA availability may modify protein yield, increasing or decreasing grams of protein per unit of volume produced, but the effect is constrained by energy protein relationships in the mammary cell.

Insulin, glucose and AAs seem to have inter-related roles influencing milk protein synthesis. Multiple hormones, prolactin, growth hormone, thyroid hormone, cortisol and insulin-like growth factor 1 (IGF1), influence milk protein synthesis, but insulin appears to be the most important controller. Milk protein synthesis is a coordinated process involving gene transcription, translation, elongation and packaging for secretion of milk protein micelles. Insulin appears to have two main points of control in protein synthesis, but it also stimulates a cascade of events to amplify milk protein synthesis (Bionaz, 2011,

2012). First, insulin, when it binds to the transmembrane insulin receptor, activates tyrosine kinase phosphorylation of signal transducer and activator of transcription 5 (STAT5). Activated STAT5 increases transcription of mRNA. Insulin potentiation of STAT5 also increases the expression of ELF5, a gene coding for a co-activator of STAT5, amplifying its activity. Secondly, insulin increases phosphorylation of rapamycin (mTOR), a key pathway for increasing translation of mRNA, necessary for protein synthesis. Furthermore, insulin's phosphorylation of mTOR decreases the formation of inhibitory factors that depress protein synthesis. Insulin binding ramps up protein synthesis in the mammary cell.

Energy availability modifies protein synthesis through a separate pathway, which intersects with insulin stimulation. Increased levels of adenosine monophosphate (AMP), reflecting diminished carbohydrate (CHO) energy, stimulate inhibitory factors of protein synthesis. Protein synthesis is an energy-demanding process, and limitation in energy substrates inhibits synthesis. Insulin depresses these inhibitors, which would dampen milk protein synthesis, to the extent energy precursors are available. Increasing cell concentrations of glucose and AAs are additive in reducing the expression of protein synthesis inhibitors. In addition to reducing inhibitory factors, insulin activation of mTOR upregulates factors which initiate translation of mRNA, and activation of mTOR inhibits the inhibitors of protein elongation, boosting translation. Polyribosomes attached to endoplasmic reticulum are upregulated to increase milk protein synthesis via the amplification of pathways associated with increased mRNA activation.

Activation of mTOR increases the glucose transporter, GLUT1 and AA transporters, increasing nutrient entry into mammary epithelial cells. Also, mTOR stimulates mitochondrial adenosine triphosphate (ATP) synthesis, increasing energy available for protein (and lactose and fat) synthesis. Nutrient supplies of glucose and branched chain AAs (BCAA), particularly LEU, also activate mTOR indirectly, enhancing protein synthesis. Combined actions of insulin and nutrient supplies of glucose and AAs integrate lactose and protein synthesis and increase production within the mammary cell.

Even though plasma insulin concentrations decline post-calving and remain low for two months following calving, insulin sensitivity in the mammary gland is increased by increasing key regulatory proteins, and the mammary cell may respond positively to small changes in insulin concentration. An increase in insulin stimulates protein synthesis primarily through the mTOR pathway. Providing more AAs for the gland stimulates phosphorylation of regulators involved in ribosomal synthesis, and BCAA can increase mTOR activation, but full activation of protein synthesis is blunted if energy supply is limited (Bionaz, 2011, 2012). In addition, provision of AAs which enhance formation of α-lactalbumin may enhance lactose synthesis and milk volume, as α-lactalbumin is an integral component of lactose synthase (DePeter and Cant, 1992). Protein synthesis is coordinated with AA and energy supply, reflected in the high correlation with protein yield and milk yield.

Post-calving, gene expression controlling the synthesis of casein is greatly increased, but overall protein synthesis in mammary cells is downregulated to direct more energy and AAs to milk protein synthesis (Bionaz, 2012). Prior to calving, genes related to ribosomal synthesis of other cellular proteins are upregulated, but do not increase further post-calving. This differential expression in cellular protein synthesis relative to milk protein synthesis allows for production of milk proteins at the expense of other cellular proteins. Even with this directed shift, typically milk protein synthesis represents about 50% of the protein metabolism occurring in the mammary cell (Bionaz, 2011, 2012). Limitations on protein translation can only be overcome by large availability of precursors, AAs and energy from CHO (Bionaz, 2012).

Mackle (1999) demonstrated the effects of increased insulin and increased AA supply in a study employing duodenal and intravenous infusion of AAs and insulin, respectively. Water or casein and BCAA were duodenally infused over a 4-day period with or without insulin given intravenously over the four days, employing a euglycaemic clamp technique to maintain plasma glucose concentrations. Casein infusion with BCAA increased milk volume +1 kg/d and protein yield +43 g/d compared to the water infusion. Insulin infusion alone almost doubled milk volume response (+1.8 kg/d) and increased milk protein content and yield (+102 g/d). When the insulin infusion was combined with casein and BCAA, milk volume increased +3.3 kg/d compared to the water infusion, and MPY increased +202 g/d above the control water infusion (Mackle, 1999). Insulin alone and insulin combined with increased AA supply increased milk protein synthesis by both increasing volume of milk and production of protein per unit volume. This study demonstrates the importance of insulin and AA availability for milk protein synthesis.

3 Abomasal and duodenal infusion studies

Although technically difficult, abomasal or duodenal infusion studies using casein or selected AAs combined with sampling of portal, mesenteric and hepatic veins and mesenteric artery along with estimates of blood flow and arterio-venous differences across specific organs enable clarification of AA delivery, uptake and metabolism of specific AAs by organs of the gastrointestinal tract (Reynolds, 1994). For a detailed description of technique and technical complications, the reader is referred to a review by Reynolds (1994). By also sampling the carotid artery, coccygeal artery and mammary vein delivery and uptake of AAs and glucogenic and lipogenic nutrients by the mammary gland may be estimated. Use of isotopes for N, glucose, propionate, bicarbonate and specific AAs enables calculation of fluxes and utilisation across organ systems. AA metabolism by portal-drained viscera (PDV), liver and mammary gland may be characterised under various dietary and metabolic situations. The reader is referred to papers by Armentano (1994), Reynolds (1994), Galindo (2002, 2015), Haque (2012, 2013, 2015), Kristensen (2010), Larsen (2014, 2015), Lemosquet (2009), Mackle (1999), Martineau (2016), Raggio (2006) and Ruis (2010a,b) for detailed descriptions of technique and responses to infusion of energy and AA nutrients. In brief the results of infusion studies reveal the following.

3.1 Liver and portal-drained viscera protein turnover

The PDV and liver have a high rate of protein turnover. Duodenal and abomasal supplies of AAs are not necessarily delivered to the mammary gland due to metabolism by these organs. It is estimated that 50% of total body oxygen consumption is due to PDV and liver metabolism, and these organs are not merely pass through interfaces (Reynolds, 1994). The liver is a major site of metabolism for HIS, MET, PHE and threonine (THR) absorbed from the PDV, and excess AA above that needed for anabolic processes are oxidised (Raggio, 2004). At low MP intakes, removal of HIS, MET, PHE and THR by the liver are decreased; thus, uptake is responsive to supply. The PDV oxidises LEU, ILE and VAL (the BCAA), whereas peripheral tissues catabolise LYS and BCAA (NRC, 2001). Liver uptake of LYS is influenced by supply, and only occurs at high intakes of LYS (Raggio, 2004). Catabolism of HIS, MET and PHE by the liver depends on mammary gland requirements, so there is co-ordination in utilisation between organs, but regulation is not clear at this time (Raggio, 2004).

3.2 Glucose

Glucose is the principal determinant of milk volume, associated with lactose production. Lactose is synthesised in the Golgi apparatus in mammary cells along with milk proteins (Stelwagen, 2000). The majority of glucose is supplied by hepatic gluconeogenesis from propionate, lactate, AAs and glycerol (Galindo, 2002, 2015) and a minor proportion (approximately 1%) from gluconeogenesis in the kidney (Galindo, 2002). The liver is a net exporter of glucose, and the mammary gland utilises about 70% of total body glucose, approximately 70 g of glucose is required for 1 kg of milk production (Reynolds, 1994).

3.3 Efficiency of AA use

Not all AAs taken up by the mammary gland are used with similar efficiencies (Mepham, 1994; Doepel, 2004; Table 7). Mepham categorised AAs into three groups based on uptake:output ratios (U:O). Group 1 AAs have an U:O ratio not different from one and include MET, HIS, PHE-TYR, tryptophan (TRP) and THR. Group 2 AAs have an U:O ratio greater than one (1.28–1.53) and include ILE, LEU, VAL, (the BCAA) and LYS. The remaining AAs are group 3 AAs and these can have an U:O ratio less than one. Group 2 EAAs are used by the mammary gland not only for milk protein synthesis but also for transamination interactions to form NEAA for milk protein synthesis. The ratio of Group 1, Group 2 and Group 3 AAs in plasma influence the overall efficiency of mammary uptake and output of AAs in milk protein.

3.4 Responses to increased amino acid supply

Cows in negative protein balance have greater responses to infused casein and AAs than cows in positive protein balance (Martineau, 2016). Cows in negative MP balance (<100 DIM) increased in dry matter intake (DMI), 0.28 kg/d and increased MPY 109 g/d with an average infusion of 333 g/d of casein (Martineau, 2016). For every additional 100 g/d of casein infused in cows in negative MP balance, DMI increased an additional 0.14 kg/d and MPY increased an additional 25 g/d. For cows in positive MP balance (>100 DIM), DMI decreased −0.20 kg/d with 333 g/day of casein infusion, and DMI decreased an additional −0.33 kg/d for every additional 100 g/d of casein infused. Milk protein increased, but only 65 g/d with 333 g/d of casein infusion and was not significant in cows in positive MP balance (Martineau, 2016). Plasma urea increased almost two fold in cows in positive energy balance with casein infusion compared with cows in negative MP balance, suggesting greater deamination of AAs in cows in positive MP balance than in cows in negative MP balance. Increased deamination may have reduced feed intake through satiety signals in cows in positive MP balance. Cows in negative MP balance responded with increased DMI possibly due to stimulatory effects of SER, THR and TYR on feed intake.

Table 7 Uptake of amino acids for 1 kg of milk containing 26.6 g of true protein

Amino acid	Abbrev.	g/100 g milk protein[1]	g/milk protein[2] 27.72 g/kg milk	Uptake: Output[3]	Estimate Supply g/kg
Essential					
Threonine	THR	4.3	1.11	1.2	1.34
Histidine	HIS	2.7	0.70	1.4	0.98
Arginine	ARG	3.4	0.88		
Valine	VAL	6.6	1.71	1.2	2.05
Methionine	MET	2.8	0.73	1.0	0.73
Isoleucine	ILE	5.9	1.53	1.81	2.77
Phenylalanine	PHE	4.9	1.27	1.05	1.33
Tryptophan	TRP	1.5	0.39	2.25	0.87
Leucine	LEU	9.8	2.54	1.3	3.30
Lysine	LYS	8.3	2.15	1.6	3.44
Nonessential amino acids					
Aspartate	ASP	3.4	0.88	0.1	0.09
Asparagine	ASN	4.2	1.09	0.4	0.44
Glutamate	GLU	12.1	3.13	2.5	7.84
Glutamine	GLN	9.4	2.44	0.85	2.07
Serine	SER	6.3	1.63	0.2	0.33
Glycine	GLY	1.8	0.47	0.2	0.09
Alanine	ALA	3.3	0.85	0.5	0.43
Tyrosine	TYR	5.6	1.45	0.93	1.35
Cysteine	CYS	0.7	0.18		
Proline	PRO	10.0	2.59	0.25	0.65
Total			27.72		30.09
Milk content		Mcal			
Lactose	4.8	0.1968			
Fat	3.7	0.3441			
TP	2.719	0.152			
Total/kg		0.693		g AA/Mcal	44.90
			Efficiency 0.67	g MP/Mcal	64.77

[1] Content of milk protein based on Swaisgood (1995).
[2] marginal yield of milk protein for Holstein Cows (Table 3).
[3] Cantalapiedra-Hijar (2015); Uptake:Output estimates across four diets.

Infusions of casein in the duodenum or abomasum increased supplies of AAs in blood, particularly EAA, increasing mammary gland uptake of EAA and decreasing uptake of NEAA (Doepel, 2004; Galindo, 2002; Larsen, 2014, 2015; Lemosquet, 2009; Raggio, 2006). Typically, the U:O ratio of Group 2 AAs increased to compensate the reduced uptake of NEAA with casein and AA infusions with similar profiles to casein. Milk volume and milk protein content and yield increased and blood urea concentrations decreased with improved supply of EAA, particularly in cows in negative MP balance (Haque, 2015).

Infusion studies demonstrate that the mammary gland can respond to increases in AA supply, especially in early lactation during negative MP balance. Casein or AA infusions in cows less than 29 DIM are associated with significant increases in milk volume (+6.4 kg/d (Galindo, 2015); +7.2 kg/d (Larsen, 2014); +7.8 kg/d (Larsen, 2015)) and MPY (+353 g/d (Galindo, 2015); +452 g/d (Larsen, 2014); +220 g/d (Larsen, 2015)). The increased milk yield continued to 29 DIM even as the amount of casein infusion declined as DMI increased. The casein infusion abolished the negative MP balance, but increased negative energy balance at 4 DIM. The cows responded with increased lipid mobilisation, which resulted in higher milk fat concentration but energy balance returned to control values by 29 DIM. The liver increased glucose output to meet the demand for increased lactose synthesis (Galindo, 2015; Larsen, 2014, 2015). Galindo (2002) reported that AA contributes from 5% to 30% of gluconeogenesis in the cow. The greater AA supply from casein and AA infusion that resulted in increased milk protein output was coordinated with liver glucose output and lipid mobilisation.

3.5 Responses to infusions of energy substrates

The mammary gland responds differently to infusions of carbon precursors (glucogenic compounds) for gluconeogenesis (propionate or glucose) compared with AA infusions, and these studies have had variable results on milk yield and milk protein output (Lemosquet, 2009; Raggio, 2006; Ruis, 2010). Effects of energy precursors, although increasing whole body glucose entry, as does casein, influence mammary gland response differently from AA infusions. Energy substrates increase mammary blood flow, whereas casein and AAs decrease mammary blood flow (Lemosquet, 2009; Raggio, 2006; Ruis, 2010). Energy substrates reduce arterial AA concentrations whereas casein and AA infusions increase arterial AA concentrations. Group 1 AA uptake increased to match output in milk for infusion studies with casein and glucogenic precursors. Group 2 AA uptake exceeded output to a greater extent for casein infusion than for propionate or starch infusion studies. Propionate and starch infusion increased uptake of Group 3 AA to a greater extent than casein infusion. Protein infusion increased Group 2 AA uptake and decreased Group 3 AA uptake. However, glucogenic precursor infusion studies has shown increased uptake of Group 3 AA and therefore has a smaller effect of Group 2 AA uptake. This suggests different mechanisms improve MPY with energy nutrients compared with AA nutrients (Raggio, 2006), but effects may be additive (Cantalapiedra-Hijar, 2015). Liver, peripheral tissues and mammary gland respond to increased glucogenic and aminogenic nutrients in a coordinated fashion depending on relative supply of carbon and nitrogen precursors (Cantalapiedra-Hijar, 2015).

3.6 Uptake efficiencies of individual amino acids

Uptake efficiencies of individual AA differ depending on arterial concentrations relative to each other and supply of glucogenic nutrients. Fixed efficiencies of utilisation by the mammary gland for all AA, and fixed efficiencies for individual AAs are not appropriate (Doepel, 2004). Models need to account for degree of negative protein balance, the proportion of EAA and NEAA, and glucose entry rates for adequate prediction of response to dietary inputs of AAs on milk protein synthesis.

Taken together, summaries indicate that AA supply and energy supply to the mammary gland need to be considered together. The French MP system, which utilises protein digested in small intestine (Jarrige, 1989; (PDI system, INRA); Haque, 2012) uses an estimate that 58.8 g of MP is required per Mcal of milk energy output if efficiency of MP utilisation for milk protein synthesis is 68% (ratio of MP to protein output in milk). The Dutch DVE/OEM system (DVE, MP protein, OEM, energy available for microbial synthesis) estimates that 66.9 g of DVE (MP) are needed for each Mcal of milk energy output if the efficiency of MP protein utilisation for milk protein output is 67% (Hof, 1994).

If glucose is a primary determinant of milk volume, and if 27.72 g of milk protein are synthesised for every 1 kg of milk (Holstein cows Table 5), then 41.37 g of MP would be required per kilogram of milk if the efficiency of utilisation is 67% (NRC, 2001). Using the proportion of AAs per gram of milk protein (Swaisgood, 1995) and estimating U:O based on Cantalapiedra-Hijar (2015) net uptake of AAs would be estimated at 30.081 g of milk protein (Table 7). If a kilogram of milk contains 48 g/kg of lactose, 37 g/kg of fat and 27.72 g/kg of true protein, the energy content would be 0.693 Mcal/kg (Table 7). The AA requirement, if used with a 67% efficiency (NRC, 2001) would need to be 64.77 g of AAs per Mcal of milk (Table 7). This is similar to INRA and Dutch estimates for optimal efficiency of MP to milk protein synthesis based on energy output in milk. Efficiencies of AA use at the mammary gland need to be considered in relation to energy supply, which is consistent with the interconnection of energy and AA supply considered in mammary cell metabolism discussed earlier.

Doepel (2004) found that increasing total energy supply and digestible duodenal supply from infused AAs were consistently associated with increases in milk protein. Energy supply was a dominant predictor of MPY. However, efficiency of conversion of AAs to milk protein decreased as the AA supply approached requirement. Doepel (2004) found the major determinant of MPY was net energy intake alone or net energy intake adjusted for AA supply of HIS, LYS and MET. Efficiency of capture of EAA in milk was influenced by supply of EAA, which improved as the ratio of EAA approached that for milk protein utilisation, but efficiencies decreased with the increase in EAA supply. Doepel (2004) developed a logistic model of AA efficiencies as supply approached requirement for each EAA. His analysis led to a suggested profile of an ideal EAA profile as a proportion of MP to achieve optimal efficiencies of uptake and utilisation (Table 8).

The ideal EAA pattern in Table 8 from several authors suggests LYS and LEU need to be a major component of supply, reflecting their higher proportion in milk protein. Most recommendations in Table 8 closely reflect mammary output of EAA as a proportion of milk protein and are in close agreement. Therefore, models are in close agreement as to the optimal pattern of EAA for milk production, but prediction of supply from diet and rumen microbial synthesis is the challenge in meeting this goal.

Table 8 Composition of milk protein and recommended ideal protein[1]

		g/100 g protein[2]			Ideal		% Lysine		
Amino acid	Abbrev.	Milk	Casein	Whey	PDIE	Boison	Doepel	Rulquin	Rohr
Essential									
Threonine	THR	4.3	4.43	4.09	4.5	5.2	71	55	75
Histidine	HIS	2.7	2.92	1.72	3.0	–	33	42	33
Arginine	ARG	3.4	3.73	1.99	4.1	–	67	43	63
Valine	VAL	6.6	6.96	4.18	5.2	5.3	85	73	81
Methionine	MET	2.8	2.98	1.89	2.4	2.0	35	34	31
Isoleucine	ILE	5.9	6.06	5.17	4.8	4.8	74	61	71
Phenylalanine	PHE	4.9	5.24	3.08	4.4	4.4	72	63	76
Tryptophan	TRP	1.5	1.21	2.25	–	–	–	–	–
Leucine	LEU	9.8	9.33	11.11	8.7	8.6	131	122	123
Lysine	LYS	8.3	7.98	9.15	7.3	6.7	100	100	100
Nonessential amino acids									
Aspartate	ASP	3.4	2.91	5.76					
Asparagine	ASN	4.2	4.07	3.50					
Glutamate	GLU	12.1	12.43	8.87					
Glutamine	GLN	9.4	10.46	4.78					
Serine	SER	6.3	6.88	3.24					
Glycine	GLY	1.8	1.82	1.53					
Alanine	ALA	3.3	3.16	4.39					
Tyrosine	TYR	5.6	5.98	3.27					
Cysteine	CYS	0.7	0.27	3.31					
Proline	PRO	11.23	10.0	3.13					

[1] PDIE, reported in Haque (2015), Doepel (2004), Rulquin (1998), Rohr reported in Doepel (2004).
[2] Based on Swaisgood (1995).

4 Ideal amino acid profile

Because rumen fermentation significantly alters feed inputs, the prediction of supply of AAs to the duodenum is not straightforward. Actual AA requirements have been difficult to define in the dairy cow (Lapierre, 2006; NRC, 2001). Dose response studies using infusion of casein and specific AA into the duodenum and abomasum have helped define milk production responses to increasing overall AA supply (NRC, 2001). Based on estimates of microbial protein AA composition and the AA composition of most feeds included in dairy rations, it was speculated that MET and LYS were co-first limiting AAs for milk production (Rulquin, 1993; Schwab, 1992a,b; Tamminga, 1980a,b). Dose response studies

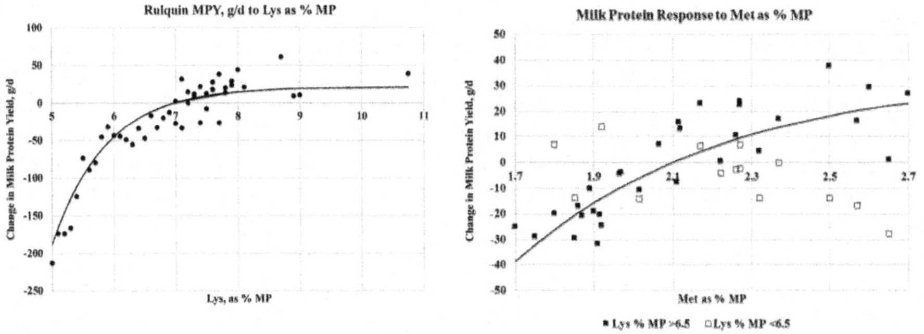

Figure 2 (a) MPY in response to increasing lysine as a proportion of metabolisable protein. Based on Rulquin (1993). (b) MPY in response to increasing methionine when lysine is >6.5% of metabolisable protein (•) and when lysine is <6.5% of metabolisable protein (□). Based on Rulquin (1993).

with infusions of MET and LYS have supported this observation. Both Schwab (1992a,b) and Rulquin (1993) identified that MET at about 2.5% of MP and LYS at about 7.3% of MP enhance yields of milk and milk protein (Fig. 2). The NRC (2001) calculated similar optimal duodenal concentrations for MET, 2.5%, and LYS, 7.2%, as a per cent of MP for maintenance and milk production. Socha (2008) observed that MET at 2.5% of MP seemed optimal for milk production. Target concentrations of MET and LYS have been suggested, but other than for total MP, content of other EAA to enhance milk protein efficiency has not been identified.

A problem with AA supplementation in dairy cows is rumen microbes, which readily degrade AAs (Chalupa, 1974). In order to deliver MET and LYS to the duodenum, they must be protected from rumen degradation (Chalupa, 1974). Various rumen-protected (RP) products are available commercially (RPMET, RPLYS). Responses to RPMET and RPLYS are greater when provided in diets with 14–19% CP than with lower CP concentrations (NRC, 2001). To be utilised effectively, prediction of duodenal flows of AA from microbial CP (MCP), rumen undegradable feed protein (RUP) and endogenous protein (EP) available in the duodenum are needed (Lapierre, 2006). Robinson (2010), in examining published papers which utilised RPMET and RPLYS, found that responses were positive, increasing milk energy output, MPY and improving the efficiency of capture of dietary N into milk protein, particularly when both RPMET and RPLYS are supplemented. However, Robinson (2010) concluded that responses were minor, and he questioned the economic benefit of using these products until improved models of AA flow to the duodenum were developed that would enable a more substantial response. Zanton (2014) evaluated different sources of commercial MET products and found MPY increased across all products but increases were not consistent across products. To use RP products efficiently, ration models need to account for small intestinal absorption of MET or LYS from the individual product, the cost of supplement and the value of anticipated production response. Value of RP EAA will be enhanced if they enable the CP content of diets to be reduced, enhancing efficiency of N capture in milk.

The study by Lee (2015) demonstrates an attempt to reduce CP content of diets and utilise supplemental RPLYS and RPMET to maintain yields relative to a higher CP diet. Lee (2015) fed a diet considered adequate in MP (15.5% CP) to a group of Holstein cows, and a diet deficit in MP (−281 g/d, 13.7% CP). One group of cows fed the deficient diet were supplemented with RPLYS (100 g/d), and another group of cows fed the deficient diet were supplemented with RPLYS (100 g/d) and RPMET (25 g/d). Reducing MP in cows on the deficient diet reduced milk yield (−1.3 kg/d), MPY (–49g/d) and MUN concentration to 6.9 mg/dl. The low MP diet increased N efficiency compared to the control diet. Supplementation with RPLYS alone and supplementation with RPLYS and RPMET together did not improve milk protein synthesis, or return production levels to that of the control diet. Although increasing the duodenal flow of LYS and MET, the efficiency of utilisation was not improved compared with the low CP diet. Increasing flow of duodenal EAA by supplementing with RP EAA may not result in positive production responses depending on the concentration of CP in the diet and DIM and the level of production of the cows.

Despite these challenges, Boisen (2000) and Doepel (2004) have proposed ideal AA profiles for EAA based on the analysis of milk protein AA content and the results of casein and AA infusion and feeding studies (Table 8). This has led to the concept of the ideal AA profile to meet milk production in dairy cows (Boisen, 2000; Doepel, 2004; Schingoethe, 1996; Tamminga, 1980a,b). By providing an ideal profile of EAA absorbed from the duodenum, CP content of diets should be reduced, the efficiency of N utilisation improved, and urinary N excretion reduced (NRC, 2001). However, the level of CP required in the diet would still be dependent on DIM, as responses to increases in EAA in early lactation are dramatic. Infusion studies of casein post-calving demonstrate that cows in negative protein balance respond significantly to an increase in EAA delivered to the small intestine compared with cows in positive MP balance (Martineau, 2016). In addition, specific AAs may have a positive influence on feed intake, improving all around nutritional balance. However, flows of AA must be predicted to the duodenum from MCP, RUP and EP given diverse dietary inputs and changing feed intake.

Despite these limitations, the evolution of ration formulation models has been to attempt to predict flows of microbial, feed and EP to the small intestine and the composition of AAs absorbed (Boston, 2000; O'Connor, 1993; NRC, 2001; Schwab, 2005; Thomas, 2004). The first component of the process is to predict rumen outflows of protein.

5 Central issues in estimating rumen microbial protein synthesis

5.1 Overview of rumen microbial protein synthesis

Early nutritional models were based solely on CP (NRC, 1978). Milk yield and MPY increased quadratically with increasing CP intake; maximal yields were achieved with CP content of 19% in lactating rations (NRC, 1978, 2001). However, urinary losses of N increased with increasing CP intake and dietary efficiency of N utilisation was low. Burroughs (1975) proposed that protein standards be defined on absorbed protein, or MP, and not CP. To describe the supply of MP, intestinal flow and digestion of rumen microbial protein and rumen undegraded feed protein need to be predicted. Through the 1980s, nutritional systems incorporated models of rumen fermentation of feedstuffs to account for microbial protein synthesis (MPS) and rumen-degradable and undegradable feed protein (RDP, RUP, respectively) to predict MP. Figure 3 presents a schematic of rumen metabolism of protein (Chalupa, 1984). MP is a summation of flows of MPS and RUP digested in the small intestine. Notice flow of EP is not considered in this model in Fig. 3.

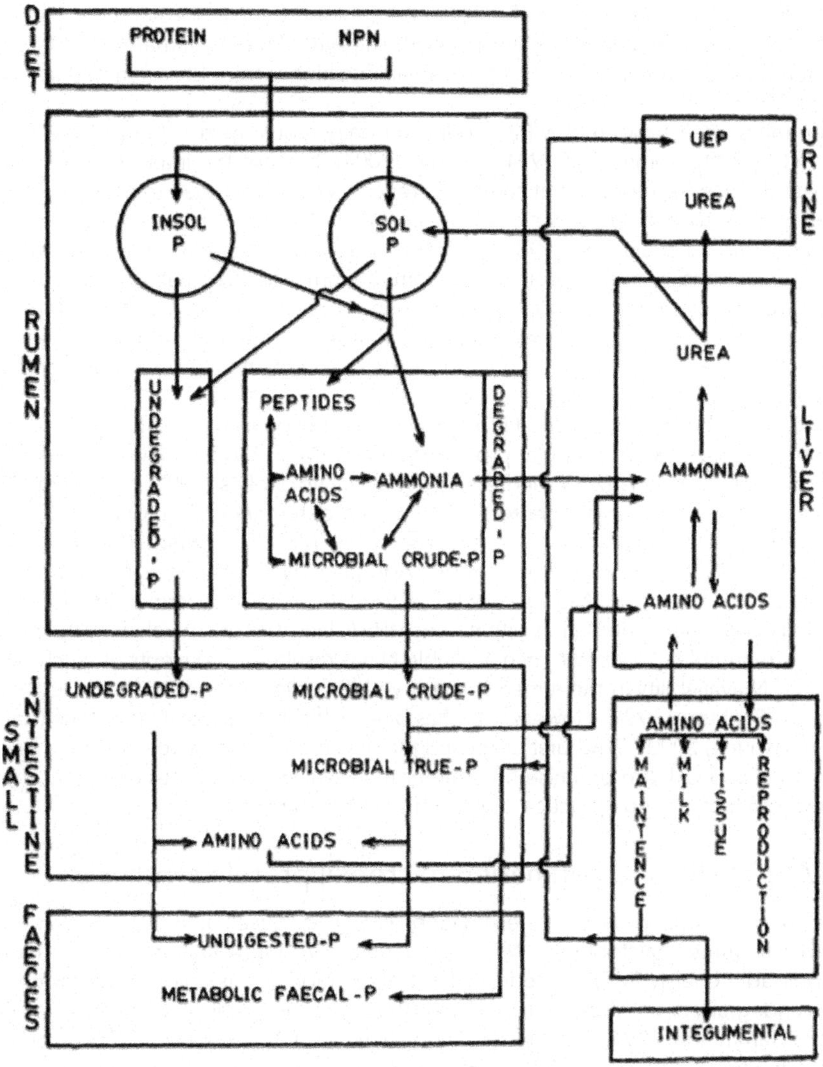

Figure 3 Schematic of protein digestion in lactating dairy cows (Chalupa, 1984). P = protein; Insol P = insoluble protein, Sol P = soluble protein; UEP = urinary endogenous protein.

Along with RUP, MCP and EP contribute to protein delivered to the small intestine (Lapierre, 2006; Marini, 2008). The amount of MCP delivered to the small intestine depends on the microbial growth, which depends on energy available in the rumen (Bach, 2005; Dewhurst, 2000; Hoover and Stokes, 1991; Stern, 1994; Thomas, 2004). Sources of EP to the small intestine include mucoproteins in saliva, epithelial cell debris from respiratory tract, mouth, oesophagus, reticulo-rumen, omasum and abomasum, and enzymes from the abomasum (Lapierre, 2006; NRC, 2001). EP from mouth, respiratory tract and reticulo-rumen may be degraded and utilised by microbes in the rumen; thus, some EP gets to the small intestine in MCP. The NRC (2001) estimates that 9–12% of non-ammonia nitrogen (NAN) flow to the small intestine is from EP, estimated as 1.9 g N EP/kg DMI.

Nutrition models now employ MP models for dairy requirements (AFRC, 1992; Jarrige, 1989; NRC, 2001; Tamminga, 1994; Thomas, 2004). It is not the intent of this review to compare the various models, but some differences will be discussed to provide a context of approaches to defining protein supply in the dairy cow. For more detailed reviews the reader should see Tuori (1998), Yu (2003) and Vermorel (1998). Although, there are three sources contributing to protein flow from the rumen to the small intestine: microbial protein, feed protein not degraded in the rumen (RUP) and EP, the majority of MP comes from rumen microbes. Clark (1992) estimated that MCP contributed on average to 59% of the MP in dairy cows, but with significant variation. Prediction of MCP flow is critical for assessment of protein flows to the small intestine.

Rumen microbes require C and N for growth, in addition to micronutrients. The reader is referred to reviews of microbial protein synthesis (MPS) by Bach (2005), Clark (1992), Dewhurst (2000), Firkins (2006 and 2007), Hoover and Stokes (1991) and Stern (1994). Organic matter (OM) truly fermented in the rumen determines MPS (Clark, 1992). CHO is the main organic substance fermented and drives MPS (Dewhurst, 2000; Hoover and Stokes, 1991; Russell, 1992; Stern, 1994). Adequate N is necessary to support CHO allowable microbial yield. Nitrogen deficiency relative to CHO fermentation results in energy spilling, energy utilisation with no net MPS (Clark, 1992). The relationship between rumen fermentable CHO and rumen available N has resulted in concepts of synchronising CHO and N availability in the rumen to improve MPS efficiency (Clark, 1992; Hoover and Stokes, 1991; Stern, 1994). However, this has been difficult to achieve in practice (Cabrita, 2011; Dewhurst, 2000). Dewhurst proposed that it is more practical to consider a balance of rumen N and CHO to optimise MPS rather than synchronising rates of supply and this has become the common practical approach to rumen MPS.

5.2 Approaches to estimating microbial protein synthesis

There are different approaches to estimating MPS. The critical factor is to predict energy available to rumen microbes for MPS. The Dutch DVE/OBM system uses fermentable metabolisable energy (ME) to predict MPS. Fermentable ME is adjusted for fat content, rumen undegraded protein and silage fermentation products, which do not provide energy for MPS (Tamminga, 1994). The French PDI system uses fermentable OM (FOM), adjusted for feed fractions not available for microbial energy, such as ether extract and silage acids, to predict MPS (Jarrige, 1989). The Feed into Milk (FinM) System uses rumen available ATP to predict MPS (Thomas, 2004). The NRC, 2001 uses adjusted total digestible nutrients (TDN) to predict MPS. The Cornell Net Carbohydrate Protein System (CNCPS) uses fermentable CHO to predict MPS (Sniffen, 1992; Russell, 1992; Tylutki, 2008). The Nordic system uses rumen-degradable CHO to predict rumen available energy (Madsen, 1985). Each system uses a factor times the rumen available energy to predict MPS.

Clark (1992) summarised studies which reported MPS and OM truly digested in the rumen (OMTDR). He produced the following equation:

$$\text{Microbial N, g/d} = -3.38 * (\text{OMTDR}^2) + 79.29 * (\text{OMTDR}) - 152.89\ (r^2\ 0.49)$$

Based on the above equation, maximal microbial N (MPS) occurs with an intake of 11.73 kg of OMTDR, which, based on other equations in Clark (1992), represents an intake of 25.27 kg of OM/day (Clark, 1992). The model from Clark highlighted that degradation of OM drives MPS. However, the model by Clark has a low fit to the data ($r^2 = 0.49$), indicating there were multiple factors influencing MPS in addition to total OMTDR. Total OMTDR is not constant across different feedstuffs; therefore, individual feed ingredients and amounts consumed should be considered to improve prediction of OMTDR for calculating MPS.

5.2.1 Rumen available energy

Two approaches have been taken to address the variation in MPS across feeds. The first approach is to calculate an energy value (ME, net energy (NE) or TDN) for each feed ingredient, sum the total energy in the diet based on the energy content of each feed times its proportional inclusion in the diet and multiply the total energy supply by a factor to predict MPS. The contribution for each feed is accounted for by its proportion of energy supplied. Coefficients of MPS times the energy value for the diet are based on the typical proportion of dietary energy fermented in the rumen developed from regression analysis. This approach is taken for the INRA, Dutch, NRC (1989), NRC (2001) and AFRC (1992) systems. The second approach predicts MPS based on each feed ingredient that is summed to estimate total supply of MCP. This approach is used by the CNCPS and FinM systems (Sniffen, 1992; Thomas, 2004).

The CNCPS partitions CHO into fractions soluble and insoluble in neutral detergent (Sniffen, 1992). Two broad classes of CHO are considered, neutral detergent fibre (NDF), residue remaining after digestion in neutral detergent, and non-fibre CHO (NFC), calculated as 100 – CP – Ether Extract (EE) – NDF – Ash. The NFC contains silage acids, sugars, starch and water-soluble fibre (WSF). Analysis of silage acids, sugar and starch is commonly done on feed samples to improve characterisation of NFC (Table 9). Each CHO fraction in a feed is assigned a degradation rate (%/h) for prediction of MPS. These two pools of CHO provide the major source of energy for fermentation of rumen microbes (Hoover and Stokes, 1991; Russell, 1992; Sniffen, 1992). The FinM programme partitions feed dry matter (DM) into soluble, small particle and large particle fractions. DM degradation of each feed fraction is used to predict MPS based on estimates of ATP available for each fraction (Table 9, Thomas, 2004).

The NRC 2001 committee felt prediction of MPS from OMTDR in the rumen was too imprecise and there was not sufficient data to utilise degradation of individual feed fractions to drive MPS as done with the CNCPS or FinM programmes. The NRC (2001) approach calculates a TDN value for each feed based on nutrient content for NDF, NFC, EE and CP (Table 9, NRC, 2001). This value is considered the TDN at 1x maintenance intake (TDN_{1x}). These feed values are multiplied by the intake for each feed as a proportion of the ration, and then summed to estimate TDN_{1x} for the total ration. The ration TDN_{1x} is then adjusted based on intake relative to maintenance. The adjusted TDN value is used to predict MPS.

Most systems consider a uniform efficiency of microbial growth to predict MPS. The CNCPS is the only system to consider subpopulations of microbes. The CNCPS divides microbial populations into two groups: those that ferment primarily structural CHO (NDF) and those that ferment NFC (Russell, 1992). Efficiencies of yield are treated similarly, but

Table 9 Feed analysis scheme for the Cornell Net Carbohydrate and Protein System (CNCPS) and the NRC (2001)

Scheme		Rumen Kd, %/h		
		Forage	Grains	Protein
CNCPS Scheme – laboratory chemical analysis				
Feed CP – N × 6.25				
Soluble in borate buffer	Soluble protein			
Non-protein nitrogen		10 000	10 000	10 000
Amino acids/peptides		10 000	10 000	10 000
Insoluble in borate buffer				
Acid detergent fibre bound CP (ADFCP)		0	0	0
Neutral detergent fibre bound CP (NDFCP)				
NDFCP – ADFCP		0.09–2.0	0.08–0.40	0–0.08
CP – SP – NDFCP		5.0–20.0	4.0–18.0	0–16.0
Carbohydrate				
Neutral detergent insoluble (NDF)				
Unavailable NDF (2.4 × Lignin)		0	0	0
Available NDF (NDF – 2.4 × lignin)		2.0–9.0	0–10.0	0–12.0
Neutral detergent insoluble				
Sugar		250–350	50–500	0–300
Starch		10–50	0–45	0–50
Silage acids		0	0	0
Water-soluble fibre (WSF)		10–50	0–45	0–50
NRC, 2001 and ARC, 2004 – in sacco dacron bag analysis				
CP	N × 6.25			
Immediate wash out of bag – soluble and small particles	–A	90	90	90
Residue after washing bag Residue end-point of degradation	–C	0	0	0
CP – A – C	–B	1.9–29.2	across feeds	
NRC, 2001 TDN_{1x}				
td_{NFC}	= 0.98 * (100–(NDF–NDFCP) – CP – EE – Ash) * processing factor			
td_{CP} forages	= CP * exp(–1.2 * (ADFCP/CP))			
td_{CP} concentrates	= [1–(0.4 * (ADFCP/CP))] * CP			
$td_{fattyacids}$	= FA (FA = EE–1; if EE < 1 FA = 0)			
td_{ndf}	= 0.75 * (NDF–Lignin) * (1–(Lignin/(NDF–NDFCP))^0.667)			
TDN_{1x}	= $td_{NFC} + td_{cp} + td_{NDF} + td_{NFC} + td_{FA} * 2.25 - 7$			
$TDN_{animalproteins}$	= CPdigest * CP + FA * 2.25 + 0.98 * (100 – CP – Ash – EE)–7			
$TDN_{fatglycerol}$	= (EE × 0.1) + [FAdigest × (EE × 0.9) * 2.25]			
$TDN_{fatnoglycerol}$	= (EE × FAdigest) * 2.25			
Discount	= [TDN – ((0.18 * TDN – 10.3) * ((0.18 * TDN – 10.3)–1))]/TDN			

maintenance energy requirements are different. The prediction of MPS is as follows for the systems used:

Aggregated by total diet:

NRC, 2001	130 g MCP/adjusted TDN, kg
Jarrige, 1989 (INRA)	145 g MCP/kg fermentable OM
Tamminga, 1994 (Dutch)	150 g MCP/kg fermentable OM
Madsen, 1985 (Nordic)	125 g MCP/kg rumen-degradable CHO

Predicted for each feed and summed:

Thomas, 2004 (AFRC) $[(ATP_{ssp} \times Y_{ATPssp}) + (ATP_{lp} \times Y_{ATPlp})] \times 0.0625$

Where: ssp = soluble and small particles
lp = large particles
ATP_{ssp} or ATP_{lp} = effective degradability of particles × DMI × ATPy
ATP_y = 27.34–0.0248 × CP (CP of the feed g/kg DM)

Russell, 1992 (CNCPS)
Fibre MPS = 1/((0.05/Kd) + (1/0.4)) × DMI x degraded available fibre
NFC MPS = 1/((0.15/kd) + (1/0.4)) × DMI x degraded NFC

Where: 0.05, 0.15 = maintenance energy for bacteria, g bact/g CHO
0.40 = maximum growth rate for bacteria, g bact/g CHO
Degraded fibre and NFC based on rate of CHO degradation, %/h
MPS is summed and as a function of rumen-degraded CHO

Once MCP is predicted, digestible true protein content from microbial supply is what the cow can utilise from the small intestine. The proportion of microbial true protein (MTP) content in MCP is 0.80 in INRA (Jarrige, 1989), 0.75 in the Dutch system (Tamminga, 1994), 0.85 in the Nordic system (Madsen, 1985), 0.75 in the ARC system (Thomas, 2004), 0.80 in the NRC system (NRC, 2001) and 0.80 in CNCPS (Russell, 1992; Tylutki, 2008). The digestibility of MTP in INRA is 0.80 (Jarrige, 1989), in the Dutch system 0.85 (Tamminga, 1994), in the Nordic system 0.82 (Madsen, 1985), in the FinM system 0.85 (Thomas, 2004), 0.80 in the NRC (2001) and 0.80 in the CNCPS (Russell, 1992; Tylutki, 2008). The proportion of MCP that is not MTP is N contained in nucleic acids. The indigestible component of MTP is protein largely contained in the microbial cell wall. The different coefficients of yield per unit rumen energy, different ratios of MTP to MCP and different digestibilities of MTP result in similar estimates for MP from microbial cells for each of the systems if feed inputs are similar.

5.2.2 Rumen available nitrogen

The CHO allowable growth of bacteria requires a source of N. Virtanen (1966) demonstrated years ago that cows could support up to 4000 kg of milk a year and produce about 10 kg/d of milk when fed purified diets that contain no protein sources but NPN from urea. Rumen microbes can utilise ammonia as an N source to produce microbial protein, which the cow then digests to meet her AA requirements. Rumen microbes require ammonia, but microbial growth is enhanced when AAs, peptides and isoacids are also available in the rumen (Russell, 1992; Firkins, 2007). Various authors have attempted to define a minimum rumen ammonia concentration for efficient microbial growth, and 5 mg/dl has been most cited (Clark, 1992), but ammonia concentrations in the rumen fluctuate throughout the day and an optimal concentration has been difficult to define (Firkins, 2007). Furthermore, rumen available N greater than the minimum amount required for MPS increases NDF digestion (Huhtanen, 2015). Therefore, optimal ruminal ammonia concentrations for MPS may be lower than concentrations that enhance overall rumen function.

Nitrogen available to the rumen is derived from degradation of feed proteins, salivary urea, blood urea absorbed across the rumen wall, EP sources of sloughed mucosal and epithelial cells, proteins in saliva and lysed microbial cells in the rumen (Lapierre, 2001, 2006). Recycled N to the rumen from endogenous supplies enables a reduction in dietary CP, which can be a strategy to reduce urinary N losses. Kristensen (2010) used urea IV infusions to examine N recycling from blood urea in cows fed a low CP (13.9%) and a moderate CP diet (15.8%). The low protein diet significantly reduced milk production by 6 kg/d and protein yield by 207 g/d. PDV increased extraction of the infused urea when cows were on the low CP diet, but the increased uptake of urea did not increase ventral rumen ammonia concentration. Ammonia in rumen venous drainage was increased suggesting the recycled N was not efficiently distributed in the rumen. Utilisation of recycled N on low CP diet was limited by entry into the rumen by rumen epithelial permeability (Kristensen, 2010). Therefore, a minimum amount of dietary protein needs to be included in the diet to meet the needs of microbial growth based on CHO fermentation, and the difficulty in using recycled N efficiently limit how low CP may be in the ration.

Hoover and Stokes (1991) in analysis of several experiments observed that MPS was maximal when diets contained 10–13% RDP (mean 12.3%) as a per cent of DM and suggested an optimal range of RDP may be 12–14%. Using Clark's (1992) data relating microbial nitrogen (MN) to OMTDR, the optimal RDP corresponding to maximal microbial efficiency would be 12% of DM. These values do not account for recycling of N to the rumen, which would reduce these estimates to approximately 9–11% of DM. Data from Baker (1995), Roseler (1993) and Hof (1994) suggest that the optimal RDP concentration would correspond to a blood urea concentration of 9–10 mg/dl, which would be associated with an RDP content of 9–11% of DM.

Reynal (2005) fed diets with a range of CP and RDP concentrations and concluded that an optimal RDP for milk production was approximately 12% of DM. However, an optimal RDP to minimise urinary excretion yet maintain a reasonable production level was 11% of DM. Cyriac (2008) found that reducing RDP below 8.8% of DM and CP below 15.2% of DMI significantly reduced DMI, milk yield and MPY. Kalscheur (2006) found linear decreases in DMI, milk yield and milk protein content with decreases in RDP from 11.0% DM to 9.6, 8.2 and 6.8% (with reductions in CP following suit: 17.2, 15.5, 13.9 and 12.3%, respectively). Olmos (2006a,b,c) observed that 16.5% CP was associated with maximal milk and protein yield across a range of CP diets offered (13.5, 15.0, 16.5, 17.9, 19.4% CP). Therefore, it seems RDP needs to be approximately 9–11% of DM and CP content above 15.0% to ensure reasonable MPS and milk production.

The NRC and FinM recommend RDP provide 1.15x the N in MPS for allowable microbial growth when based on energy available to rumen microbes. A positive rumen N balance is needed to safeguard predictions of MPS.

Nitrogen lost per day in the urine is influenced by the efficiency of nitrogen use in the rumen in addition to efficiency of tissue metabolism of AA. The amount of N lost may vary by 200 g/day depending on the level of CP in the ration and the balance of rumen-degradable CHO and protein. If CHO is provided in excess of nitrogen, microbial growth is limited because N is not sufficient to support MPS. The surplus energy from

CHO will produce VFA and heat rather than new microbial cells due to uncoupling of CHO and protein metabolism (Clark, 1992; Dewhurst, 2000; Stern, 1994). Conversely, if N is provided in excess of CHO or if CHO is limiting, microbial growth may be decreased and ammonia levels will rise in the rumen. Ammonia absorption will increase, more urea will be produced by the liver, and urea losses in the urine will increase. Balancing CHO and protein available in the rumen will optimise rumen microbial growth and minimise losses of urea nitrogen (De Boer, 2002; Dewhurst, 2000; Frank, 2002; Kabreab, 2002).

6 Additional factors in estimating microbial protein synthesis

6.1 Adjusting rumen available nitrogen: degradation and passage rates

Rumen degradation of feed proteins is not constant. Degradation of feed protein and CHO is dependent on competitive processes of rate of degradation and rate of passage out of the rumen. Passage rate equations are used to adjust potential RDP from feeds. To estimate rumen degradation rates for feed nutrients, most nutritional systems recommend suspending *in sacco* dacron bags in the rumen and measuring nutrient loss at sequential time points. Bags are approximately 10 × 20 cm and pore size is approximately 50 μm (Cyriac, 2008). Disappearance of feed DM, CP and CHO fractions over time are used to estimates degradation rates, %/h. Immediate loss of material from the bag when suspended in the rumen is from both soluble material and small particles, which may pass through pores, confounding the immediately degradable pool estimates. The FinM adjusts for a soluble fraction and small particle fraction by measuring soluble material in addition to immediate loss from the bag (Thomas, 2004). A rate of degradation may be calculated for total DM loss (Jarrige, 1989; Madsen, 1985; Tamminga, 1994; Thomas, 2004) and for each nutrient fraction in the feed (Tylutki, 2008).

The degradation rate and passage rate need to be considered to estimate total rumen degradation of a feed. The NRC 2001 calculates passage rates for wet forages, dry forages and concentrates as functions of DMI relative to body weight adjusted for the per cent concentrate in the ration. Thus, passage rate will vary with levels of feed intake and proportion of concentrate in the ration. The FinM predicts passage rates for liquid, forages and concentrates in the ration (Thomas, 2004). The CNCPS uses passage rates calculated for liquid, forage and concentrate components in the diet (Tylutki, 2008). The effective amount of energy and N available to rumen microbes is calculated as Kd/(Kd+Kp), the degradation rate of the feed or nutrient fraction (Kd) and the passage rate of the feed (Kp). This proportion is multiplied by the DMI and nutrient content to calculate the pool of available N in the rumen for each feed. The total is summed across feeds.

As feed intake increases, rates of passage change and the relative available rumen energy and nitrogen vary. Thus, systems become dynamic and non-linear and are not amendable to linear solutions. The FinM recommends fixing passage rates for liquid to 0.08/h, forages to 0.045/h and concentrates to 0.06/h to calculate a constant rumen availability for linear solutions (Thomas, 2004). An alternative is to use a non-linear solver, as in CPM-Dairy (Boston, 2000). Using Kd and Kp allows for a more dynamic estimate of MPS based on rumen available CHO and N as dietary ingredients and level of feed intake change.

6.2 Flows of amino acids to the small intestine

As discussed above, providing an optimal profile of EAA and total AA for absorption from the small intestine can improve the efficiency of N utilisation and reduce N excretion in urine. Faecal N varies over a much narrower range than urinary N and is in a more stable form. Diet influences urinary N to much greater extent than faecal N. The challenge is to predict the flow of AA absorbable from the small intestine from MCP, RUP and EP sources to minimise inefficiency of conversion of MP to milk protein.

There have been conflicting reports concerning the consistency of protein and AA content of rumen microbial protein. Some report a relatively consistent protein and AA content, whereas others report variable composition depending upon diet. Clark (1992) presented estimates of AA content in MTP and the standard deviation across studies (Table 10). Clark (1992) discussed the difficulty in partitioning variation in AA content across studies as that which is due to true variation in AA content versus that which is due to method of determination. Most have concluded that protein and EAA content of microbes is fairly constant, which makes it possible to predict AA content of MCP for use in ration models (Table 10, O'Connor, 1993). Protein and AA content of MCP are a constant in the CNCPS (Tylukti, 2008). MTP is considered a good source of EAA for milk protein synthesis, and microbial protein AA are considered readily digestible (Boisen, 2000). Factors influencing MCP composition are not well documented therefore constant composition is usually used in ration models.

The AA content and intestinal digestibility of RUP can be quite variable across feeds. The mobile dacron bag technique has been used to estimate CP and AA digestibility for RUP in feeds (NRC, 2001; Thomas, 2004). In general, digestibility of the CP in the RUP is assigned to the AAs present in the feed RUP (Boisen, 2000). Most systems utilise the AA content of the CP as an estimate of the AA content of RUP.

The NRC has utilised a regression approach to estimate AA flows from MCP and EP rather than predicting the AA content of microbial protein and EP. Measured duodenal flows of AA are adjusted based on the AA composition of the RUP of feeds in the study,

Table 10 Amino acid composition of rumen bacteria

	Clark, 1992		Boisen, 2000		O'Conner, 1993	
	Units					
	g/100 g AA		g AAN/100 g AAN		g/100 g CP	
Amino acid	Mean	CV	Mean	CV	Mean	CV
Lysine	7.9	11.9	10.6	7	7.90	NR
Methionine	2.6	25.6	1.6	9	2.60	NR
Threonine	5.8	8.9	4.6	6	5.80	NR
Isoleucine	5.7	7.4	4.6	5	5.70	NR
Leucine	8.1	10.3	6.1	5	8.10	NR
Valine	6.2	10.1	5.5	5	6.20	NR
Histidine	2.0	21.3	3.4	11	2.00	NR
Phenylalanine	5.1	6.4	3.2	9	5.10	NR
Arginine	5.1	13.2	11.7	8	5.10	NR

and regression analysis is used to predict the residual for each EAA, which is considered to arise from a combination of MCP and EP (NRC 2001). This approach allows a variable AA content of MCP depending on the residuals after adjusting for the AA flow from RUP of feeds. Error in prediction of the AA content of RUP will be adjusted in the regression of the residuals.

The CNCPS uses the AA composition of the insoluble feed fraction to estimate the composition of the RUP fraction of a feed (O'Connor, 1993). The RUP feed fraction is then multiplied by the EAA content to predict small intestinal supply (O'Connor, 1993).

Total profile of an ideal EAA in MP is difficult to achieve currently from nutritional models. Haque (2015) estimated an ideal EAA profile from the PDI system and used low protein and moderate protein diets with AA infusions to assess response to an ideal AA supply. The infusion of AA provided as much AA N supply as the moderate protein diet compared to the low protein diet. Providing an ideal EAA supply for each MP level improved production efficiency, and confirmed the importance of MET and LYS at 2.5% and 7.3% of MP (Haque, 2015). Haque (2105) concluded that increasing the EAA ratio in MP in dairy rations could improve MP efficiency before full knowledge of the exact requirement for each EAA is known. Pacheco (2006) identified the challenges in current models in predicting flows of AA to the duodenum. Despite difficulty in prediction, ration models should employ predictions of EAA supply and field responses evaluated in order to develop a more robust database of feed inputs, predicted supply and response (Swanepoel, 2010).

6.3 Nutrient fractions in feeds

Analytical methods are critical to accurately characterise feed CHO and protein fractions. Different ration programmes utilise various partitioning of nutrient fractions in feeds. NRC (2001) uses CP, soluble protein (SP), acid detergent insoluble CP (ADFCP), neutral detergent insoluble protein (NDFCP), ether extract, lignin, and ash to estimate TDN values. Recent versions of the CNCPS system (Boston, 2000; Tyluki, 2008) use similar analysis but further fractionate CHO into sugar, starch, silage acids and (WSF, by difference, WSF = NFC – silage acids – sugar – starch). Each feed standard system uses different approaches to describing feed inputs to account for rumen degradability of N, rumen available energy and rumen undegradable protein fractions. The advantage of more detailed description of feed inputs is that more mechanistic predictions of rumen metabolism and nitrogen utilisation may be made. The disadvantage is that the cost of feed analysis increases and more laboratory methods are required to characterise a feed.

Comparison of MP systems requires unique descriptions of feed inputs. They are different for different systems and make system comparisons difficult. Unifying chemical composition inputs across systems would improve the ability to compare systems.

7 The metabolisable protein requirements of dairy cows

MP provides AAs for body tissue maintenance, production, reproduction and growth. Schwab (2005) and Thomas (2004) describe differences in calculation of tissue requirement across the MP systems in use. Small intestinal digestion of MCP, RUP and EP comprise MP. Total MP requirement is calculated based on a factorial approach, summing MP requirements for each physiologic process. Net losses are evaluated and an efficiency factor is used to estimate MP requirement for net tissue utilisation. Systems differ in estimates of maintenance requirements and in efficiency of utilisation of MP for milk protein synthesis and other body processes. The FinM system has taken a regression approach to estimate maintenance energy requirements since maintenance and production effects are difficult to separate (Thomas, 2004).

Maintenance MP requirement includes endogenous urinary losses, scurf losses (hair, skin, keratin) and endogenous faecal losses from the gastrointestinal tract. The NRC (2001) and the FinM system (Thomas, 2004) estimate urinary and scurf losses from body weight ($4.1 * BWkg^{0.5}$, $0.3 * BWkg^{0.6}$, respectively). Endogenous secretions are trickier in estimation due to partial digestion and reabsorption during transit through the gastrointestinal tract and degradation and incorporation of N from endogenous secretions into bacteria in the rumen, the caecum and colon (Lapierre, 2006). Estimates of EP losses are handled differently in the various systems.

The NRC (2001), FinM (Thomas, 2004) and CNCPS (Tylutki, 2005) handle EP requirements similarly. Excreted EP is measured as metabolic faecal protein (MFP, g/d) and is estimated as MFP, g/d = 30 * DMI (kg/d). MFP is composed of bacteria from the large intestine and caecum, bacterial debris from the rumen, caecum and large intestine, sloughed intestinal cells and secretions of the gastrointestinal tract. The NRC (2001) assumes that 50% of indigestible MCP is in faeces and 50% is digested in the hindgut. MP needed to meet the MFP loss is adjusted to account for hindgut degradation and absorption of N as ammonia from degradation of indigestible rumen-synthesised microbial protein as follows:

$$\text{MP for MFP, g/d} = [30 * \text{DMI} - 0.50 * (\text{MPbacteria}/0.80 - \text{MPbacteria})]$$

Hindgut bacterial contribution to MFP is not constant and is a function of FOM reaching caecum and colon. Providing increasing amounts of starch to the hindgut can capture blood urea N in bacteria, increasing faecal N and reducing urinary N. An improvement in estimates of MP for MFP could be made with hindgut models of MPS. However, this would require predictions of fermentable NFC and NDF that reach the hindgut. Currently, models do not account for hindgut contribution to energy and N economy of the cow in a robust way.

Hindgut bacteria may degrade EP and rumen-synthesised microbial protein that comes from EP secreted in the rumen. EP entering the rumen from mucoproteins in saliva, epithelial cells from the respiratory tract, the mouth and oesophagus, and sloughed rumen-reticulum cells has N that can be utilised by rumen microbes and become incorporated into microbial protein. EP N can flow to the small intestine in the form of EP or as microbial protein.

The maintenance MP requirement used by the NRC, FinM and CNCPS is as follows:

$$MP_{maint} = 4.1 * BWkg^{0.50} + 0.3 * BWkg^{0.60} + [30 * DMI - 0.50 * (MPbacteria/0.80 - MPbacteria)] + 6.25 * 1.9 * DMI\ (kg)/0.67$$

The MP needed for milk production is the net true protein content produced in daily milk divided by an efficiency factor. The NRC (2001) uses an efficiency factor of 0.67. The efficiency of MP use for milk protein synthesis ranges from 0.64 in the PDI system to 0.73 in the Nordic system (AAT/PBV) whereas in the Dutch DVE system it is variable with milk production, decreasing with increasing yield (Subnel, 1994; Tamminga, 1994). The FinM system has adopted an efficiency factor of 0.68 based on studies by Metcalf (2008) and others across a range of MP supply at similar energy intakes. However, what most studies have shown is that the efficiency of utilisation of MP for milk protein synthesis is not static, but varies with quantity and quality of MP supply. As the ideal ratio of EAA in MP is optimal relative to that for milk protein synthesis, efficiency of MP utilisation increases. Secondly, as supply, quantity of AA, approaches and exceeds that needed for milk protein synthesis, efficiency decreases (Doepel, 2004; Haque, 2015; Metcalf, 2008). A dynamic approach assigning efficiency of utilisation of MP based on DIM, milk production and supply of EAA relative to an ideal profile is needed for more efficient capture of feed N in milk N.

Total MP requirement is the sum of maintenance, lactation, pregnancy and growth. For pregnancy and growth requirements the reader is referred to various publications of feeding standards. In brief, pregnancy requirement for MP is not significant until the last trimester, two-thirds of which is during the dry period. Growth requirements are greatest in the first year of life and lower during the second year of life and a small component of requirement during the first two years of lactation. The major MP requirement in lactation is for maintenance and milk protein production.

8 Milk urea nitrogen as a diagnostic tool

Presented in Fig. 4 is a graph of mean values of urinary N (UN, g/d) against mean values of MUN, mg/dl for Agle (2010), Broderick (2003), Hristov and Ropp (2003), Hristov (2004), Jonker (1998), Lee (2015), Olmos (2006b), Reynal (2005) and Sannes (2002). As seen in the figure, there is a linear relationship between UN and MUN. This has been reported

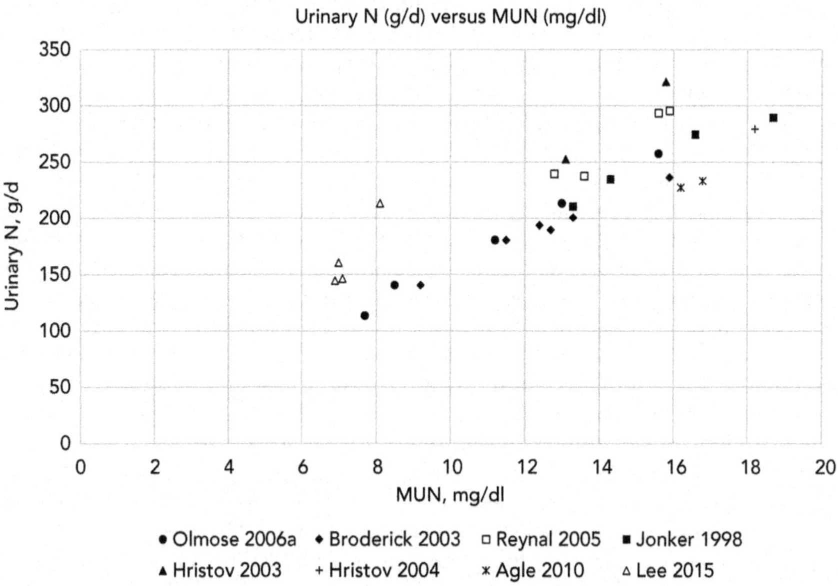

Figure 4 Urinary nitrogen (g/d) versus MUN concentration (MUN, mg/dl) from mean values for 7 studies: Agle (2010), Olmos Calermo (2004a), Broderick (2003), Hirstov (2004), Hristov and Ropp (2003), (Reynal (2005), Jonker (1998), Lee (2015), Sannes (2002).

by Kauffman and St-Pierre (2001), Jonker (1998), Nousiainen (2004), Sannes (2002), Spek (2013) and Wauttiaux and Karg (2004). Jonker (1998) reported that there were 12.54 g/d of UN per unit of MUN (mg/dl). Kauffman and St-Pierre (2001) revised this equation to account for differences in body weight across breed (UN (g/d) = 0.0283 * MUN (mg/dl) * BW (kg), similar to an equation reported in Wauttiaux and Karg (2004) (UN = 0.026 * MUN (mg/dl) * BW (kg)). Kohn (2002) in a modelling approach, also used BW to adjust UN (g/d) losses based on MUN (mg/dl) concentration (0.026 * BW (kg) * MUN (mg/dl)). Sannes (2002) in an evaluation of current prediction equations based on a study employing 4 CP diets with varying CHO sources (CP content 17.0, 17.4, 18.5 and 19.6%) that MUN was a useful tool to estimate N losses in urine, which is apparent in Fig. 4.

Spek (2013) in a meta-analysis found UN prediction increased 11.92 g/d for each unit increase in MUN (mg/dl) for data from Northern European studies compared with 14.08 g/d for each increase in 1 unit of MUN (mg/dl) for North American publications. Prediction equations improved when CP (%) was included in the models. Huhtanen (2015) in a meta-analysis for mean production data from studies using individual cow data, rather than population means, found UN (g/d) was increased 5.57 g/d per unit increase in MUN (mg/dl), substantially lower for individual cows than for population means. Adding DMI and BW improved prediction, and lowered the estimate per unit of MUN to 4.3 g/d UN per MUN, mg/dl. Based on Huhtanen's analysis, 1 kg of milk would require 8 g of dietary CP for gluconeogenesis at a fixed DMI of 20 kg/d. This would increase MUN 0.069 mg/dl, assuming the increase in MUN was due to deamination for gluconeogenesis. This would correspond to about 8.2 g of glucose produced from deamination of AAs, about 12% of the 70g of glucose needed for a kilogram of milk production (Reynolds, 1994).

Mean MUN (mg/dl) concentrations recommended for bulk tank milk are 10–14 mg/dl (De Peters and Ferguson, 1992) and 8–12 mg/dl (Kohn, 2002). Mean MUNs for a group of cows consuming the same diet should also be between 8 mg/dl and 14 mg/dl and variation in cows in the group is about 2.8 mg/dl for a coefficient of variation of about 28% (personal observation). Therefore, within the group 99% of cows will have MUN values between 1.6 mg/dl and 18.4 mg/dl. Evaluation of MUN concentrations should be made on a group basis and not on an individual cow basis. However, if the coefficient of variation is greater than 35% for a group, then uniform access to the feed bunk must be questioned or other factors examined contributing to the increase in variation.

Monitoring herd MUN provides a diagnostic management tool for assessing N efficiency and predicting UN excretion. Urinary nitrogen is the most volatile N component on dairy farms. However, bulk milk MUN is an average across cattle groups and may mask individual group inefficiencies in MUN. Mean MUN for feeding groups on the farm should also be monitored. Variation of MUN in a group can only be determined if MUN concentrations are determined for each cow in a group. Individual MUN concentrations in test day milk costs about $0.10/cow, which on a monthly basis can be costly if the information does not contribute to informed management decisions. Most bulk milk MUN analysis is done by processors as part of component analysis and is not an extra cost for the producer. Bulk tank MUN is currently the most frequently available information. Dietary changes will result in changes in mean MUN within days. However, we observe mean MUN (mg/dl) for a herd varies 0.553 (sd 0.088) mg/dl for every 1% deviation in CP content of the total mixed rations (TMR). If CP of a TMR remains constant at 16%, the bulk tank MUN could vary from 9.3 mg/dl to 14.9 mg/dl (95% confidence range, sd 1.44, mean MUN 12.07 mg/dl). Deviation of bulk tank MUN from expected ranges by 2.5 mg/dl should result in a diagnostic plan to account for factors that may influence MUN. Factors include changes in milk production, altering DMI, changes in NDF content of the diet, altering CHO fermentation, changes in CP content of the TMR, altering N intake and rumen starch fermentability, altering capture of rumen N in MPS. There is not one factor that influences bulk tank MUN. Significant changes in bulk tank MUN require a diagnostic evaluation of potential causes. However, population models based on bulk tank MUN are useful predictors of UN excretion independent of causes for the change in MUN.

9 Designing rations to improve N efficiency in dairy cows

The goal of precision protein feeding programmes is to capture as much dietary N into milk N as possible and reduce urinary N losses. Urea in urine comprises the greatest proportion of urinary N and increases as efficiency of dietary N capture in milk decreases (Baker, 1995; Broderick, 2003; Hristov, 2004; Hof, 1997; Jonker, 1998; Roseler, 1993). Urinary urea rapidly decomposes to ammonia when mixed with faeces due to bacterial urease activity (Muck, 1982). Urease activity is temperature and pH sensitive (Muck, 1981, 1982, 1983) so season (summer vs. winter higher rate of conversion) and bedding (limestone slows activity) influences the rate of conversion of urea to ammonia. Cattle faeces typically have a pH around 7 at which urease has a high activity. Urine is alkaline, so barn floor manure has a pH around 7.0 to 7.4 (personal observation), so urea conversion to ammonia occurs rapidly in cattle houses. Using limestone bedding or spreading limestone on alleys can increase pH to 10 and decrease enzyme activity, but the effect is short lived as the limestone decomposes (Muck, 1983). Once ammonia is formed, it rapidly volatilises. Urine can be collected separately from faeces in some barn systems to reduce the conversion of urea to ammonia, but such systems require special designs. Most manure storage systems collect urine, faeces, bedding and waste parlour water. Urea will be converted to ammonia, and to prevent emission from storage structures, some type of surface seal is needed. This may be a crust of OM or an impermeable sheet. Stored material is agitated when ready to field apply, which disrupts the sealed surface, causing some ammonia emission. When applied on fields, manure needs to be injected into the soil to reduce further losses of ammonia. The best approach to reducing ammonia emissions from dairy farms is to ensure feeding programmes reduce N inefficiencies in the cow and reduce urinary N excretion.

An early strategy to improve N efficiency in the cow was to increase RUP and decrease RDP and therefore increase MP and milk production. Less N would be lost from rumen degradation of feed protein. However, in many cases, milk production was not enhanced, nor was N efficiency improved (Santos, 1998; Ipharraguerre, 2014). In a meta-analysis, Santos (1998) observed that 73% of studies had a reduction in the flow of MCP to the small intestine because N supply to the rumen was inadequate due to the reduction in RDP. MPS was compromised by N deficiency in rumen. Rumen undegraded protein was increased, but RDP was decreased, reducing MPS. Ipharraguerre (2014) came to the same conclusion and further observed that MPS was reduced to the greatest extent when protein supplements had the lowest rates of rumen degradation. Flow of non-ammonia non-microbial N (NANMN, RUP) was increased by RUP protein supplements (Santos, 1998; Ipharraguerre, 2014) but often flow of non-ammonia microbial nitrogen (NAMN,

MCP) was reduced. Total NAN (MP) flow was a function of nitrogen intake (NI) across the studies. Thus, CP intake was still the best predictor of NAN flow and production responses. Ipharraguerre (2014) found that NAMN flow to the small intestine was best predicted when apparent rumen N balance was included in a regression model with OMTDR. Rumen N balance must be accounted for based on energy available in the rumen to ensure MPS is not compromised when using MP protein systems. This requires matching dietary supply to meet requirements for rumen N and tissue AA as closely as possible. Merely reducing RDP and replacing it with RUP does not result in more efficient N utilisation. Rumen microbial synthesis must be adequately accounted in order for RUP to compliment MPS and improve N efficiency.

Recommendations for RDP are for 9.2–10.5% of diet DM (NRC, 2001). Diets with RDP below 8.8% and 15.2% CP significantly reduced milk yield, DMI and tended to reduce MCP yield. RDP content in rations appears to be adequate at 9.0 to 11.0% of DM. Concentrations of RDP above 11% of DM will provide rumen ammonia concentrations above that needed, increasing UN. The total CP content of diets will depend on the amount of RUP needed above RDP to meet the needs of milk production. Responses to casein and AA infusions significantly increase milk yield and MPY in early lactation cows. Thus, RUP will need to be greater in early lactation cows than in later lactation cows, which will increase the total CP content of the diet. Peak milk production is sensitive to energy and MP supply. Total lactation yield in 305 days is about 250 x peak yield; therefore, increasing peak yield has been an important strategy to increase total milk yield. Wu and Satter (2000) in a complete lactation trial compared feeding strategies with different CP diets for cows less than 150 DIM and those greater than 150 DIM. Diets were 15.4–16.0% CP; 17.4%–16.0% CP; 17.4–17.9% CP and 19.3% and 17.9% CP, for cows <150 DIM and cows >150 DIM, respectively. Wu and Satter (2000) found that 19.3% CP in early lactation maximised peak milk yield, but overall, considering production and N excretion, 17.4% diets in early lactation and 16.0% CP diets in late lactation provided maximal milk yield at reasonable N excretion in urine and faeces. Cows will respond with greater peak milk production to increasing CP in early lactation, but diets, which optimise peak yield, may not be the optimal diets for controlling N losses in urine and faeces. Early lactation cows need a greater concentration of CP to achieve peak milk production, and late lactation cows can be fed lower CP diets when milk production is declining and DMI is greater relative to production.

Predictions of RDP from protein fractions in feeds combined with predictions of rates of passage provide a robust system to predict rumen available N pools along with rumen available energy. However, Huhtanen (2008) found that a constant rumen degradability factor for grass silages was a better predictor of production responses than accounting for protein fractions and adjusting for NPN. In a subsequent analysis of grass silage diets, Huhtanen (2008b, 2009) observed that the best means to increase efficiency of N capture in milk and reduce urinary losses was to minimise CP and RDP content of diets. This concept is consistent with first minimising losses of N from excessive rumen degradation of protein. Thus the first step in ration balancing is to ensure adequate N for rumen MPS without oversupplying N to the rumen.

Approaches to ration formulation should maximise rumen MPS, which is a function of rumen available energy. CHO fermented in the rumen drives MPS (Hoover and Stokes, 1991). Increasing fermentation of rumen CHO increases ammonia capture in MPS and can increase the flow of MCP to the small intestine (Broderick, 2003; Hristov, 2003). The starch and sugar content of the diet provides the most fermentable supply of rumen fermentable CHO (Russell, 1992). Increasing starch content of rations increases milk yield and MPY (Caccamo, 2012; Cantalapiedra-Hijar, 2015). Caccamo (2012) used a test day model to account for genetic, herd, year of calving, season of calving, age at calving and stage of pregnancy effects on test day records to examine the effects of dietary CP, NDF and starch on milk production. Samples of TMR from Sicilian herds were collected quarterly for compositional analysis, which was merged with cow test day production records. Caccamo (2012) observed more pronounced effects of dietary NDF (increasing NDF reduced milk and protein yields) and starch (increasing starch increased milk and protein yields) relative to CP on production traits throughout lactation. This supports the observation that rumen fermentable CHO is critical in driving MPS. However, excessive rumen fermentable CHO can negatively affect MPS through decreased rumen pH and rumen acidosis therefore a minimum amount of NDF (26–28% NRC, 2001) and a maximum of starch (20% to 30%) need to be considered in formulating diets.

Once MPS is optimal, then requirements for MP for milk production must be met with RUP. Although CP intake is positively correlated with milk yield and protein yield, MP supply has improved model predictions and should be the basis of ration formulation (Hristov, 2004, 2005). Content of MET and LYS in the expected amount of absorbed MP should be estimated to ensure these EAA are not limiting milk production and protein synthesis. Various approaches to calculating EAA content of MP have been taken and appear to be sufficient to consider MET and LYS content of MP. Rulquin (1993), Schwab (1996, 2005) and NRC (2001) all seem to agree that 2.5% MET and 7.2% LYS are optimal. However, based on Rulquin's curves (1993), 2.05% MET and 6.57% LYS may be sufficient. RPAA can be supplemented to complement RUP protein sources.

10 From research trials to real farm applications

10.1 Complications in developing rations for real dairy farm situations

Balancing a ration for a dairy farm presents complications not attendant in research trials. Research trials typically utilise a few animals with a narrow range in DIM and milk production. Dairy farms feed a ration to a group of cows with a coefficient of variation of 25–30% for milk production (Fig. 5). Figure 5 presents milk production for a group of second and older lactation cows from a herd in PA. This group would receive a ration balanced to meet a target level of production. Production ranges from less than 40–70 kg/d making it difficult to select an appropriate level of production to formulate a ration. Cows in the group would be under- and overfed CP if the ration was based on average yield for the group. In addition, if cows are less than 70 days post-calving production will be increasing and underfeeding this group of cows will reduce peak milk yield, as observed by Wu and Satter (2000). DMI will vary with production, reducing some of the inefficiencies in nutrient supply, but the ratio of energy requirement and MP requirement do not change at similar rates as production changes; thus, an imbalance in nutrient requirement will exist as production deviates from ration target. Underfeeding protein will reduce UN, but it will also reduce milk production. Overfeeding protein will increase production but also increase UN. Finding an appropriate balance is critical for managing production and N excretion on dairy farms.

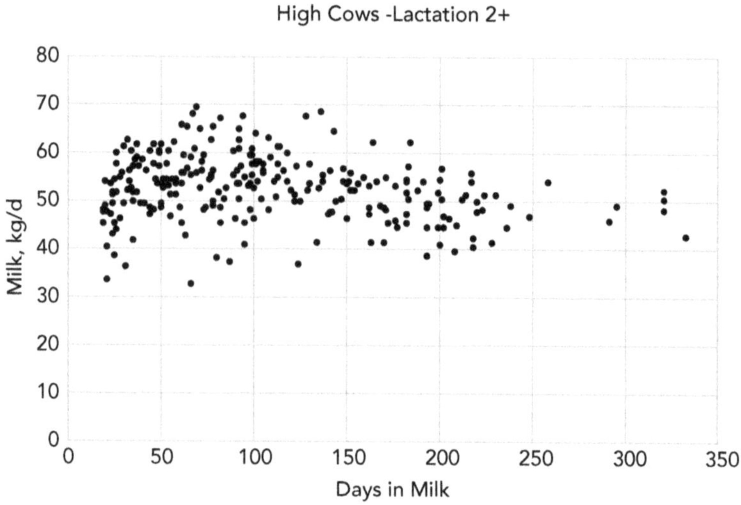

Figure 5 Range of milk production in a group of cows in a PA dairy herd. One diet is fed to this group balanced to 55 kg/d of milk. The diet formulation and the range of production in the group will determine N efficiencies for the group of cows.

10.2 Field application

Selecting a ration optimal for the group is difficult. Most research trials desire efficacy, accurately describing N flows and utilisation in the cows studied. Nutritionists desire efficacious models, models that are easy to use and enable efficient feeding of a population of cows across a wide range of production. The CNCPS was one of the first systems to attempt to predict MP supply, AA content, and rumen CHO degradability and MPS. The CNCPS (version 5.0, Tylutki, 2008) was the basis for CPM-Dairy (CPM, Boston, 2000) a field usable programme employing these concepts. Although several papers have been critical of the MP predictions in CPM compared the NRC model, it should be noted that over 50% of the papers used for the assessment were also used to develop the NRC model; therefore, it would be expected that the evaluations would favour the NRC (2001). Furthermore, description of feed inputs is critical in CPM and estimates of CHO and protein fractions may bias the predictions. Our nutrition section at the University of Pennsylvania has used the CPM model or earlier versions for almost 20 years and find it a very useful tool for evaluating and formulating diets for dairy herds. It is an efficacious model.

Production data and nutritional records across multiple dairy herds should be used to assess ration models in addition to university trials. The study by Swanepoel (2010) examined rations from a group of California dairy farms and used three different ration models to evaluate predictions of MP and milk performance. More studies need to employ this approach for a more robust evaluation of protein–AA systems. A field nutritional record needs to incorporate region, ration ingredients and composition, model prediction, and production for the cows receiving the ration in a test day model. Aggregating information across herds would facilitate model development for more precise ration formulation.

A project with the Chesapeake Bay foundation was undertaken to evaluate precision of feeding of N and P on 62 Holstein dairy farms in the Susquehanna River basin in Pennsylvania. These farms were located in Northcentral to Southcentral PA where 75% of dairy farms are located in PA. The Susquehanna River is the major tributary to the Chesapeake Bay, which has been under intense study to control eutrophication since the early 1980s. The Susquehanna is the major tributary contributing to N and P loadings into the bay, particularly from agricultural activities in the region. All farms on the project fed TMR. Samples of the TMR were collected on a quarterly basis over a one to three-year period for a total of 507 TMR samples. The TMR feed samples were analysed for nutrient fractions used in the CPM model by Cumberland Valley Laboratory, Maugansville, Maryland. Milk production records for volume, fat and protein content, and somatic cell and MUN concentration were collected from Dairy Records Management Systems (DRMS), Raleigh, North Carolina, and were merged with monthly TMR composition data. Not all farms tested MUN on monthly dairy herd improvement association (DHIA) milk samples. Herds ranged in size from 42 to 1003 adult dairy cows with a mean herd size of 308 cows (sd 266) and a 305-day production mean of 10569 kg (sd 1373 kg).

Corn silage was the predominant forage fed on the farms. The second main forage was alfalfa silage on approximately 75% of farms, followed by grass or small grain silage on approximately 29% of the farms, and a mixed silage fed on approximately 19% of farms. Dry hay was offered in small amounts, usually as Alfalfa hay (17% of farms), a grass hay (12% of farms) or a mixed hay (8% of farms). Three herds utilised pasture during summer months and were Jersey or mixed breed farms and are not included in this summary, resulting in 59 Holstein herds. Ground shell corn was the primary concentrate source, which may be offered as either dry corn or high moisture ground ear corn. Soybean meal was the main CP supplement with canola meal occasionally replacing soybean meal, but not commonly. Protein supplements were included to increase the RUP, and these supplements typically included blood meal, Distiller's Dried Grains, soybean meal products processed to reduce rumen degradation of protein, roasted soy beans, or a protein blend from a commercial mill. Urea was a common NPN supplement included in rations on the farms. Each farm worked with a professional nutritionist on a regular basis who prepared rations for the farms. Feeds were typical for PA dairy farms. Data on use of RPMET or RPLYS was not collected across the farms surveyed, but these supplements were often in the lactating diets.

CP content of the quarterly TMR samples were used to group herds into CP classes (CPCLASS): 14 (<15.0% CP), 15 (15.0–15.9% CP), 16 (16.0–16.9% CP), 17 (17.0–17.9% CP), and 18 (CP> = 18.0% CP). Herds moved across classes based on quarterly samples, so a herd could have multiple classifications based on CP content of the sampled TMR, and for most herds this was true. No herd remained in the same CP class for each quarterly sampling. This change in herd classification to a different CPCLASS could have been due to changes in forages available on the farm, changes in ration formulation in response to supplement costs or perceived production issues, sampling error in collecting the TMR, or lab error in analysis. TMR samples were from the main high production group on the farm. Many farms were feeding one TMR to all lactating cows. The goal of the project was to characterise feeding practices from a sample of farms in the Chesapeake Bay drainage basin with respect to N and P content of rations.

Mean diet composition and standard deviation is presented in Table 11. Nutrient content of NFC, EE, NDF, starch and sugar were similar across CPCLASS, except for differences in CP and SP content of the TMRs. CPCLASS was nested in herd to examine

Table 11 Mean total mixed ration composition for 507 samples from 59 Holstein dairy herds in PA

	Protein class									
	14		15		16		17		18	
Item,%DM	Mean	SD	Mean	SD	Mean	SD	Mean	SD	Mean	SD
Num. samples	66		109		143		128		61	
DM	48.33	5.83	47.92	4.35	47.38	4.21	47.56	4.92	47.80	3.97
CP	14.10	0.76	15.58	0.28	16.54	0.27	17.49	0.28	18.92	0.71
SP	5.93	0.95	6.44	1.02	6.61	0.97	6.97	1.12	7.30	1.21
RDP	8.86	0.76	9.55	0.77	10.01	0.63	10.71	0.76	11.53	0.84
NDFCP	2.13	0.86	2.13	0.77	2.23	0.62	2.36	0.78	2.37	0.71
ADFCP	1.04	0.32	1.07	0.44	1.08	0.27	1.08	0.28	1.09	0.27
Sugar	4.50	1.94	4.61	1.51	4.55	1.55	4.62	1.48	5.11	1.79
Starch	24.90	5.16	24.42	4.10	23.94	3.48	23.62	3.45	22.88	4.43
NFC	40.00	4.37	39.90	3.37	39.70	3.11	39.28	3.27	38.50	4.15
WSF	9.98	3.43	10.78	2.62	11.10	2.37	11.03	2.18	10.52	2.47
NDF	34.42	4.14	32.64	3.21	31.72	2.93	31.00	2.91	30.01	3.42
ADF	22.81	2.46	22.34	2.17	21.69	2.21	21.29	2.26	20.26	2.59
Lignin	3.60	0.80	3.68	0.71	3.58	0.67	3.59	0.68	3.41	0.72
EE	4.44	0.72	4.58	0.93	4.64	0.79	4.74	0.89	4.89	0.94
Ash	7.15	1.28	7.30	0.85	7.40	0.93	7.54	0.82	7.68	0.90
TDN	71.29	2.50	71.81	2.39	72.25	2.27	72.50	2.20	73.18	2.15
NeL	0.75	0.03	0.75	0.03	0.76	0.03	0.76	0.03	0.77	0.02

Num = number of samples in each protein class from the herds.
RDP = estimated based on SP, NDFCP and ADFCP and calculated moderately degradable fraction after Sniffen, 1992.
NDFCP = protein content of NDF residue.
ADFCP = protein content of ADF residue.
NFC = non-fibre carbohydrate calculated as 100-NDFCP-EE-Ash.
WSF = water-soluble fibre, calculated as NFC – sugar – starch (includes silage acids, which ranged from 3 to 6% of TMRs).
EE = ether extract as estimate of crude fat.
NeL = Mcal/lb DM, value calculated by laboratory.

milk production (kg/d), milk fat and protein content (%), MPY, MUN (mg/dl), and MPY compared to predicted yield (PRODIF = MPY – predicted yield = 29.4 + 27.179*milk/kg), predictions of urinary nitrogen (Wattiaux and Karg, 2004, UN (g/d) = 0.0283 * MUN (mg/dl) * BW (kg); using BW, kg for lactation 1 = 545 kg, lactation 2 = 612 kg, lactation 3+ = 650 kg), predictions of faecal N (Faecal N, g/day = intake N (g/d) – urinary N (g/d) – milk N (g/d)) and N efficiency (N milk (g/d)/N intake (g/d)).

Milk production, milk protein content and yield, and milk fat content are presented in Fig. 6. Production increased linearly with increasing CPCLASS, but production of CPCLASS 16 and 17 was very similar. Yield differences were most apparent between test day 2

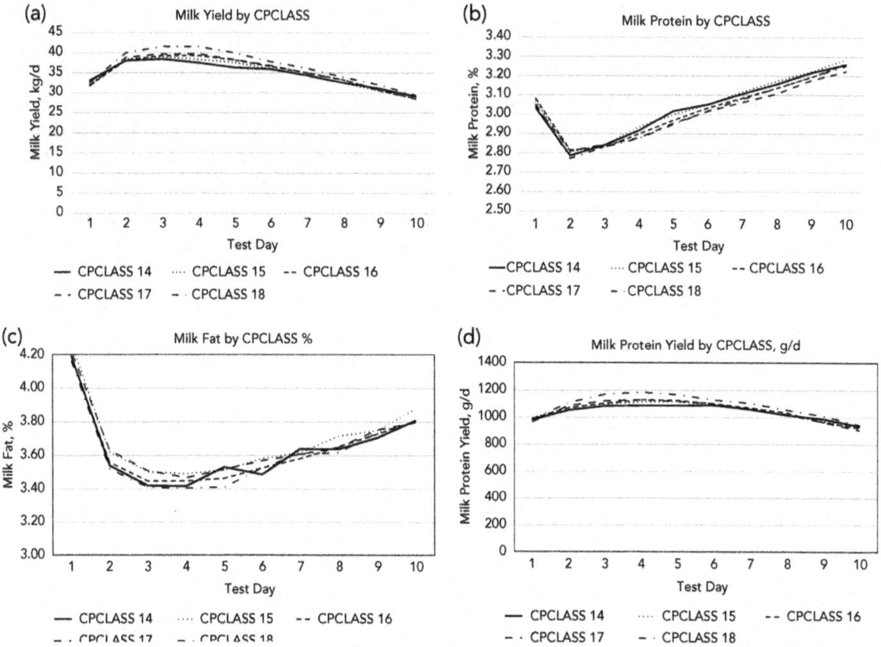

Figure 6a–d Mean values from 59 PA Holstein farms over a three-year period for milk (Figure 6a, kg/d), milk protein content (Figure 6b, %), milk fat content (Figure 6c, %) and MPY (Figure 6d, g/d) for test day calculated as (Integer(DIM/30.3)+1) for monthly test day production records matched with quarterly sample of the lactation TMR and classified based on the CP content as follows: CPCLASS 14 (▬ CP < 15.0% DM); CPCLASS 15 (… 14.9% < CP < 16.0%); CPCLASS 16 (- - - 15.9% < CP < 17.0%); CPCLASS 17 (-- -- -- 16.9% < CP < 18.0%); CPCLASS 18 (_. _. 17.9% < CP).

(about 47 DIM) and test day 6 (about 180 DIM). CPCLASS most influenced peak milk production and MUN (Table 12). Mean milk production was as follows for CPCLASS, kg/d (sem): 14, 34.5 (0.77); 15, 34.6 (0.57); 16, 34.9 (0.52); 17, 35.1 (0.54); 18, 36.3 (0.72). MUN was numerically lowest for CPCLASS 15 and highest for CPCLASS 18. However, averages for CPCLASS were within a range of 10–14 mg/dl, which is generally a recommended range. Kohn (2002) suggested a range of 8–12 mg/dl may be a more efficient range based on modelling of N excretion, and mean MUN values for some herds across the trial were less than 10 mg/dl (Fig. 7 and 8).

Milk protein content and yield varied slightly across CPCLASS (Table 12). Yield of protein was greater in the CPCLASS 18 herds, but only by approximately 50 g/d compared to the CPCLASS 14 herds, with other CPCLASS's intermediate. Regression of MPY on milk volume resulted in similar slopes of protein yield per unit of production, 26.6 g/kg. Therefore, CPCLASS did not alter the relationship between volume and protein yield. In fact, using predicted protein yield in Table 5 (27.719 g/kg of milk) to compare expected yields with actual yields found little differences in expected yields versus observed yields across the CPCLASS groups. Overall, N efficiency was influenced by CPCLASS and declined with increasing CPCLASS, which would be expected based on all the literature published till date.

Figure 7 presents MUN, urinary N, faecal N and N efficiency by test day and mean MUN and standard deviation of MUN for each herd, and mean MUN by CPCLASS for each herd. MUN was fairly flat across test days and reflected CPCLASS, increasing with CP content of rations. Mean values ranged from 10 mg/dl to 14 mg/dl across the CPCLASS groups. Correspondingly, urinary N, predicted from MUN concentrations in test day milk, was increased across CPCLASS, consistent with MUN. Faecal N increased with CPCLASS and was curvilinear with test day, which was due to estimated DMI, influenced by parity, production and DIM. Nitrogen efficiency was greatest in early lactation, partly due to tissue mobilisation of protein, which was not considered in the protein balance calculations. If included, MP balance tends to be negative the first four weeks post-calving, which would have primarily influenced the efficiency for test day 1, reducing it approximately by 25%.

Mean herd MUN and standard deviation ac the trial are presented in Fig. 8a. Mean MUN for herds ranged from 8 mg/dl to just under 16.0 mg/dl. Recommendations have been to

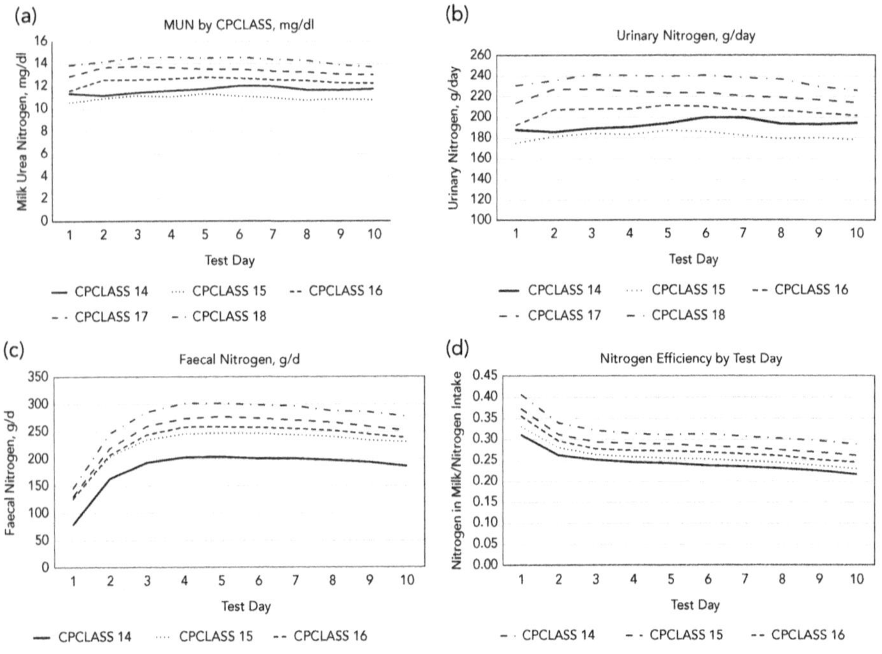

Figure 7a–d Mean values from 59 PA Holstein farms over a three-year period for MUN (Figure 7a, mg/dl), urinary nitrogen excretion (Figure 7b, g/d), faecal nitrogen excretion (Figure 7c, g/d) and nitrogen efficiency (milk true protein N (MPY/6.38/NI) (CP intake/6.25, Figure 7d, g/d)) for test day calculated as (Integer(DIM/30.3)+1) for monthly test day production records matched with quarterly sample of the lactation TMR and classified based on the CP content as follows: CPCLASS 14 (▬ CP < 15.0% DM); CPCLASS 15 (... 14.9% < CP < 16.0%); CPCLASS 16 (- - - 15.9% < CP < 17.0%); CPCLASS 17 (-- -- -- 16.9% < CP < 18.0%); CPCLASS 18 (_. _. 17.9% < CP). Urinary nitrogen calculated based on prediction of urinary N from MUN concentration from Wattiaux and Karg (2004); UN (g/d) = 0.0283*MUN (mg/dl) * BW (kg); using BW, kg for lactation 1 = 545 kg, lactation 2 = 612 kg, lactation 3+ = 650 kg). Predictions of faecal N (Faecal N, g/day = intake N (g/d) – urinary N (g/d) – milk N (g/d)). N efficiency (N milk (g/d)/N intake (g/d)).

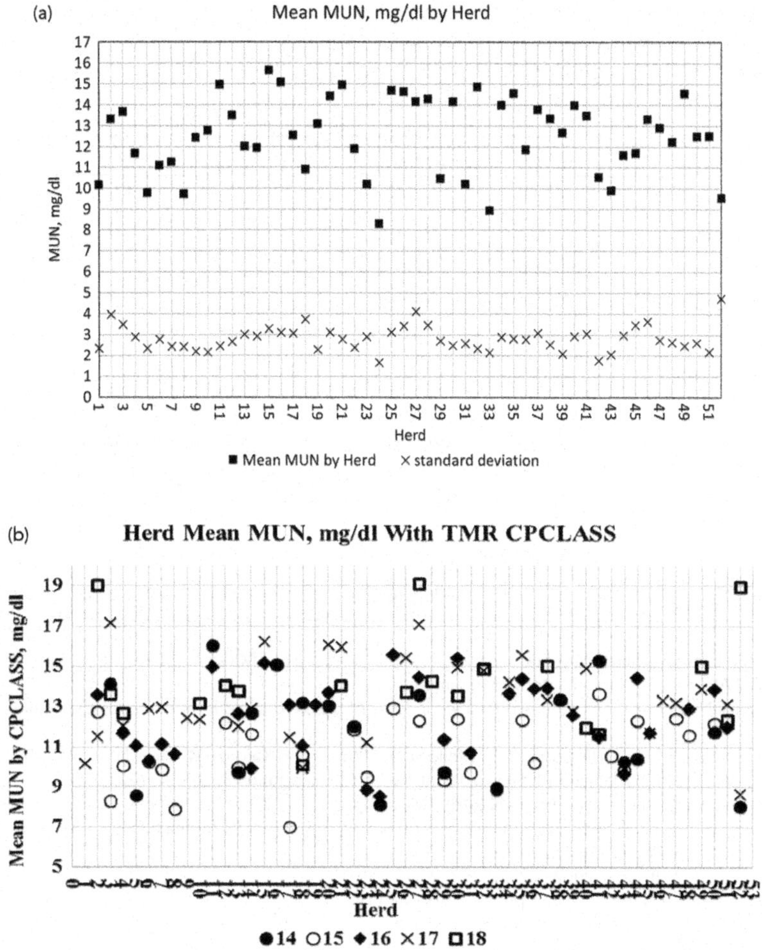

Figure 8 (a) Herd mean MUN and standard deviation, mg/dl, for 52 PA Holstein Herds which were part of a group of 59 herds which participated in a three project involving quarterly samples of lactation TMR and milk production records. (b) Herd mean MUN for monthly DHIA test records classified based on the CP content of quarterly TMR samples as follows: CPCLASS 14 (▬ CP < 15.0% DM); CPCLASS 15 (… 14.9% < CP < 16.0%); CPCLASS 16 (- - - 15.9% < CP < 17.0%); CPCLASS 17 (-- -- -- 16.9% < CP < 18.0%); CPCLASS 18 (_. _. 17.9% < CP). Not all farms tested MUN on each monthly DHIA test.

target MUN concentrations between 10 mg/dl and 14 mg/dl. Thirteen herds had mean MUN concentrations greater the 14.0 mg/dl, but were less than 15 mg/dl in all but two cases. Six herds were less than 10 mg/dl and above 8 mg/dl, which is acceptable based on Kohn (2002). Standard deviation for MUN for test day was usually between 2 mg/dl and 3 mg/dl, which is typical for cows consuming the same TMR.

Mean herd MUN was also examined based on quarterly TMR CP classification for each herd (Fig. 8b). Within herd mean MUN fluctuated quite a bit as CP changed in the

Table 12 Mean milk production, fat and protein content, MUN, faecal N, and urinary N, and N efficiency for 59 PA herds classified bases on TMR CP content at month of tested milk production over a one to three-year period with samples collected quarterly from each farm

	CPCLASS				
Item (sem)	14	15	16	17	18
Herd count[1], *n*	24	42	50	47	26
TMR samples[2], *n*	66	109	143	128	61
Milk, kg/d	34.47 (0.77)	34.59 (0.57)	34.92 (0.52)	35.08 (0.54)	36.30 (0.54)
True protein, %	3.04 (0.02)	3.05 (0.01)	3.03 (0.01)	3.01 (0.01)	3.01 (0.02)
Milk fat, %	3.63 (0.05)	3.69 (0.04)	3.64 (0.03)	3.67 (0.04)	3.62 (0.05)
Protein yield, kg/d	1.03 (0.03)	1.04 (0.02)	1.05 (0.02)	1.04 (0.02)	1.08 (0.02)
MUN, mg/dl	11.59 (0.48)	10.90 (0.39)	12.39 (0.36)	13.31 (0.37)	14.20 (0.47)
Urinary N[3], g/d	192.51 (7.97)	181.42 (6.43)	205.36 (5.94)	220.75 (6.19)	235.65 (7.83)
Faecal N[4], g/d	181.39 (5.42)	224.69 (4.06)	233.66 (3.69)	247.73 (3.84)	272.07 (5.11)
Proportion cows>=50 kg/d					
DIM 60–120 DIM (sep)[5]	19.43[a] (0.03)	25.95[ab] (0.02)	27.78[b] (0.01)	30.00[b] (0.01)	36.77[c] (0.02)
Nitrogen efficiency[6]	0.32 (0.003)	0.29 (0.003)	0.28 (0.002)	0.26 (0.002)	0.25 (0.003)

[1] Herd count: Number of herds classified in each CPCLASS based on the TMR sample CP content; a herd could move across CPCLASS based on the quarterly TMR sample.

[2] TMR samples: Number of TMR samples classified in CPCLASS; a herd could have been classified multiple times in a CPCLASS.

[3] Urinary N estimated from MUN for each cow following Wattiaux and Karg (2004) Urine N output = 0.0283 x MUN (mg/dl) x BW (kg); BW, kg: lactation 1, 545 kg; lactation 2, BW, kg: lactation 3+, 650 kg 620 kg.

[4] Faecal N estimated as NI (g/d) – Urinary N (g/d) – Milk N (g/d) for each cow on Test day. NI was estimate for each cows based on milk production, days in milk and body weight using NRC formula. BW, kg: lactation 1, 545 kg; lactation 2, 620 kg; lactation 3+, 650 kg.

[5] numbers with different superscripts differ by $P < 0.05$.

[6] Nitrogen efficiency = Nitrogen in milk protein (g/d)/NI (g/d).

TMR samples. Herds were not rigid in MUN or TMR CP content across the trial. Mean MUN tended to be lowest for test days when CPCLASS was 14 or 15, but even some observations for mean MUN were 15–16 mg/dl when CPCLASS was 14, which suggests sampling of the TMR was in error. MUN is a screening tool, influenced by efficiency of N utilisation, but values may be elevated due to rumen excess N or excess MP, resulting in AA oxidation. Cows over 100 DIM across all CPCLASS levels were in positive MP balance, more so for the higher CP diets than for the lower CP diets. Therefore, MUN may be elevated due to post-ruminal supply as well as inefficiencies in rumen capture of feed N in MPS. Martineau (2016) found MUN increased with casein infusions in cows over 100 DIM.

A sensitive period to evaluate production in dairy herds is between 60 and 120 DIM. During this period, cows are transitioning from negative energy balance to positive energy balance and energy intake usually matches closely milk energy output and maintenance requirements. Milk production will be maximal for a group of cows in the herd, and maximal level of production in this period reflects feeding management and nutritional nutrient supply. Figure 9 presents the proportion of cows >=50 kg/d milk production during this

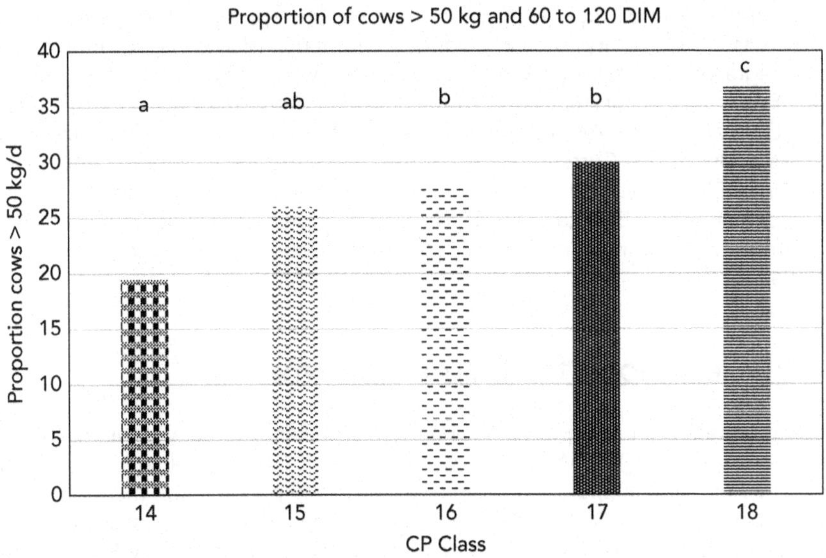

Figure 9 The proportion of cows greater than 50 kg/d in milk production by CPCLASS (CPCLASS 14 (CP < 15.0% DM); CPCLASS 15 (14.9% < CP < 16.0%); CPCLASS 16 (15.9% < CP < 17.0%); CPCLASS 17 (16.9% < CP < 18.0%); CPCLASS 18 (17.9% < CP)) for cows between 60 and 120 DIM. Labels a, b, c differ by $p < 0.05$.

period for the CPCLASS's in Table 12. Significantly, more cows achieved greater production as cows increased in CPCLASS from 14 to 15 and higher. CPCLASS 16 and 17 had significantly more cows >=50 kg then CPCLASS 14, and CPCLASS 15 was intermediate in proportion of cows >50 kg/d production. CPCLASS 18 had the greatest proportion of cows >=50 kg but also had greater concentrations of MUN. Early lactation cows will reach peak production with rations >16% CP, but keeping rations <18% CP will minimise UN. It appears that after 150 DIM cows can be fed 14–15% CP rations with no loss of production.

What the data from these PA farms demonstrate is that dietary CP concentrations can be significantly reduced, probably to 15–17% CP without significantly reducing production. However, across DIM, diets need to be tailored based on changes in N efficiency to improve overall herd efficiency of N utilisation. Milk protein content across the herds was not different in terms of grams of protein per kilogram of milk; therefore, yield of protein was not influenced by CP content. Milk protein content tended to decline with increasing yield (Fig. 3 and 6), but this is expected. In order to feed low protein diets, the forage base of the ration needs to have a low CP content, especially since CP in forages tends to be more rumen degradable. Diets which are composed of significant amounts of legume haylage (most of the TMR samples in CPCLASS 18) are challenged to balance a low CP ration.

11 Conclusion

It is possible to significantly reduce dietary CP and maintain reasonable milk production levels in dairy herds. Reasonable CP concentrations for cows less than 100 DIM currently appear to be 15.5–17.5% CP. Forage quality, appropriate protein and energy supplements are necessary to ensure adequate rumen available energy and N for MPS. Rumen-degradable protein appears to be adequate when concentrations are 9.0–11.0% of DM. Predictions of protein flow to the duodenum needs to account for RUP, MTP and EP. Estimates of the AA content of each protein group can provide a starting point to assess EAA supply. Current ration models are sufficient to supplement diets with RPMET and RPLYS to achieve 2.5% and 7.3% concentrations in predicted MP to enhance production efficiency. Bulk tank MUN for a herd appears adequate between 8 mg/dl and 12 mg/dl. Improving the efficiency of N utilisation in cows reduces urinary N and ammonia emissions.

12 Where to look for further information

Further information on milk protein and nitrogen efficiency in dairy cows may be found in Alston-Mills, B. (1995), O'Mahony, J. A. and Fox, P. F. (2014) and Frank, B., Persson, M. and Gustafsson, G. (2002) among the references in the paper. Trade magazines such as *Hoard's Dairyman, Dairy Today,* and *Progressive Dairyman* will have articles that provide general overview of protein feeding in dairy cows. The *Journal of Dairy Science, Animal, Animal Feed and Technology,* and *Livestock Production,* have scientific articles which deal with the topics of milk protein, protein utilisation in dairy rations, and nitrogen excretion.

13 Glossary of abbreviations

AA	Amino acid
AMP	Adenosine monophosphate
ATP	Adenosine triphosphate
BCAA	Branched chain amino acids
CHO	Carbohydrate
CNCPS	Cornell net carbohydrate protein system
CP	Crude protein
CPCLASS	Crude protein classification
DIM	Days in milk
DMI	Dry matter intake
EAA	Essential amino acids
EE	Ether extract
ELF5	E74-like factor 5 (ETS domain transcription factor)
EP	Endogenous protein
FAO	Food and Agriculture Organization
FOM	Fermentable organic matter
GLUT1	Glucose transporter 1
HIS	Histidine
IGF1	Insulin-like growth factor 1
ISE	Isoleucine
LEU	Leucine
LYS	Lysine

MCP	Microbial crude protein
ME	Metabolisable energy
MET	Methionine
MFP	Metabolic faecal protein
MN	Microbial nitrogen
MP	Metabolisable protein
MPS	Microbial protein synthesis
MPY	Milk protein yield, g/day
mTOR	Rapamycin
MTP	Microbial true protein
MUN	Milk urea nitrogen
N	Nitrogen
NAN	Non-ammonia nitrogen
NE	Net energy
NEAA	Nonessential amino acids
NDF	Neutral detergent fibre
NFC	Non-fibre carbohydrate
NPN	Non-protein nitrogen
OM	Organic matter
OMTD	Organic matter truly digested
OMTDR	Organic matter truly digested in the rumen
P	Phosphorus
PDI	Protein digested in small intestine
PDV	Portal-drained viscera
PHE	Phenylalanine
PTA	Predicted transmitting ability
RDP	Rumen-degraded (degradable) protein
RPMET	Rumen-protected methionine
RPLYS	Rumen-protected lysine
RUP	Rumen undegraded (undegradable) protein
SP	soluble protein
SEM	Standard error of the mean
SER	Serine
STAT5	Signal transducer and activator of transcription 5
TDN	Total digestible nutrients
THR	Threonine
U:O	Uptake to output ratios
UN	Urinary nitrogen
VAL	Valine
WSF	Water-soluble fibre

14 References

Agle, M., Hristov, A. N., Zaman, S., Schneider, C., Ndegwa, P. M. and Vaddella, V. K. (2010). Effect of dietary concentrate on rumen fermentation, digestibility and nitrogen losses in dairy cows. *J. Dairy Sci.* 93: 4211–22.

Agricultural and Food Research Council. (1992). Nutrient Requirements of Ruminant Animals: Protein. Technical Committee on Responses to Nutrients. Report No. 9. *Nutr. Abstr. Rev.* 62 (Ser B): 787–835.

Allison, R. D. and Garnsworthy, P. C. (2002). Increasing the digestible undegraded protein intake of lactating dairy cows by feeding fishmeal or a rumen protected vegetable blend. *Anim. Feed. Sci. Tech.* 96: 69–81.

Alston-Mills, B. (1995). G. Nonprotein nitrogen compounds in bovine milk. In R. G. Jenssen (ed.), *Handbood of Milk Composition*, 468–72. New York, NY: Academic Press.

Apelo Arriola, S. I., Bell, A. L., Estes, K., Ropelewski, J., de Veth, M. J. and Hanigan, M. D. (2014). Effects of reduced dietary protein and supplemental rumen-protected essential amino acids on the nitrogen efficiency of dairy cows. *J. Dairy Sci.* 97: 5688–99.

Armentano, L. E. (1994). Impact of metabolism by extragastrointestinal tissues on secretory rate of milk proteins. *J. Dairy Sci.* 77: 2809–20.

Bach, A., Calsamiglia, S. and Stern, M. D. (2005). Nitrogen metabolism in the rumen. *J. Dairy Sci.* 88 (E. Suppl.): E9–E21.

Baker, L. D., Ferguson, J. D. and Chalupa, W. (1995). Responses in urea and true protein of milk to different protein feeding schemes for dairy cows. *J. Dairy Sci.* 78: 2424–34.

Bionaz, M. and Loor, J. J. (2011). Gene networks driving the bovine mammary protein synthesis during the lactation cycle. *Bioinform. Biol. Insights.* 5: 83–98.

Bionaz, M. Hurley and Loor, J. (2012). Milk protein synthesis in the lactating mammary gland: insights from transcriptomics analyses. INTECH. http://dx.doi.org/10.5772/46054. http://creativecommons.org/licenses/by/3.0.

Boisen, S., Hvelplund, T. and Weisbjerg, M. R. (2000). Ideal amino acid profiles as a basis for feed protein evaluation. *Live. Prod. Sci.* 64: 239–51.

Boston, R. C., Fox, D. G., Sniffen, C. J., Janczewski, R., Munsen, R. and Chalupa, W. (2000). The conversion of a scientific model describing dairy cow nutrition and production to an industry tool: the CPM Dairy project. In J. P. McNamara, J. France and D. Beever (eds), *Modeling Nutrient Utilization in Farm Animals*, 361–77. Oxford: CABI Publishing.

Broderick, G. (2003). Effects of varying dietary protein and energy levels on the production of lactating dairy cows. *J. Dairy Sci.* 86: 1370–81.

Burroughs, W., Nelson, D. K. and Mertens, D. R. (1975). Protein physiology and its application in the lactating cow: the metabolic protein feeding standard. *J. Anim. Sci.* 41: 933–44.

Cabrita, A. R. J., Dewhurst, R. J., Abreu, J. M. F. and Fonseca, A. J. M. (2006). Evaluation of the effects of synchronizing the availability of N and energy on rumen function and production responses of dairy cows – a review. *Anim. Res.* 55: 1–24.

Cabrita, A. R. J., Bessa, R. J. B., Alves, S. P., Dewhurst, R. J. and Fonseca, A. J. M. (2007). Effects of dietary protein and starch on intake, milk production, and milk fatty acid profiles of dairy cows fed corn silage-based diets. *J. Dairy Sci.* 90: 1429–39.

Cabrita, A. R. J., Dewhurst, R. J., Melo, D. S. P., Moorby, J. M. and Fonseca, A. J. M. (2011). Effects of dietary protein concentration and balance of absorbable amino acids on productive responses of dairy cows fed corn silage-based diets. *J. Dairy Sci.* 94: 4647–56.

Cabrita, A. R. J., Fonseca, A. J. M. and Dewhurst, R. J. (2014). Short communication. Relationship between the efficiency of utilization of feed nitrogen and ^{15}N enrichment in casein from lactating dairy cows. *J. Dairy Sci.* 97: 7225–9.

Caccamo, M., Veerkamp, R. F., Licitra, G., Petriglieri, R., La Terra, F., Pozzebon, A. and Ferguson, J. D. (2012). Association of total-mixed-ration chemical composition with milk, fat, and protein yield lactation curves at the individual level. *J. Dairy Sci.* 95: 6171–83.

Cantalapiedra-Hijar, G., Ortigues-Marty, I. and Lemosquet, S. (2015). Diets rich in starch improve the efficiency of amino acids use by the mammary gland in lactating Jersey cows. *J. Dairy Sci.* 98: 6939–53.

Chalupa, W. (1974). Rumen bypass and protection of amino acids. *J. Dairy Sci.* 58: 1198–218.

Chalupa, W. (1984). Discussion of the protein symposium. *J. Dairy Sci.* 67: 1134–46.

Clark, J. H., Klusmeyer, T. H. and Cameron, M. R. (1992). Microbial protein synthesis and flows of nitrogen fractions to the duodenum of dairy cows. *J. Dairy Sci.* 75: 2304–23.

Cyriac, J., Rius, A. G., McGilliard, M. L., Pearson, R. E., Bequette, B. J. and Hanigan, M. D. (2008). Lactation performance of mid-lactation dairy cows fed ruminally degradable protein at concentrations lower then National Research Council recommendations. *J. Dairy Sci.* 91: 4704–13.

De Boer, I. J. M., Smits, M. C. J., Mollenhorst, H., van Duinkerken, G. and Monteny, G. J. (2002). Prediction of ammonia emission from dairy barns using feed characteristics part 1: relation between feed characteristics and urinary urea concentration. *J. Dairy Sci.* 85: 3382–8.

DePeters, E. J. and Cant, J. P. (1992). Nutritional factors influencing the nitrogen composition of bovine milk: a review. *J. Dairy Sci.* 75: 2043–70.

DePeters, E. J. and Ferguson, J. D. (1992). Nonprotein nitrogen and protein distribution in the milk of cows. *J. Dairy Sci.* 75: 3192–209.

Dewhurst, R. J., Davies, D. R. and Merry, R. J. (2000). Microbial protein supply from the rumen. *Anim. Feed. Sci. Tech.* 85: 1–21.

Doepel, L., Pacheco, D., Kennelly, J. J., Hanigan, M. D., López, I. F. and Lapierre, H. (2004). Milk protein synthesis as a function of amino acid supply. *J. Dairy Sci.* 87: 1279–97.

Dou, Z., Kohn, R. A., Ferguson, J. D., Boston, R. C. and Newbold, J. D. (1996). Managing nitrogen on dairy farms: an integrated approach 1. Model description. *J. Dairy Sci.* 79: 2071–80.

FAOStats. Accessed January, 2016. http://faostat3.fao.org/home/E.

Ferguson, J. D., Dou, Z. and Ramberg, C. F. (2001). An assessment of ammonia emissions from dairy facilities in Pennsylvania. Proceedings of the 2nd International Nitrogen Conference on Science and Policy. *TheScientificWorld.* DOI 10.1100/tsw.w001.277. www.thescientificworld.com.

Finucane, K. A., McFadden, T. B., Bond, J. P., Kennelly, J. J. and Zhao, F-Q. (2008). Onset of lactation in the bovine mammary gland: gene expression profiling indicates a strong inhibition of gene expression in cell proliferation. *Funct. Integr. Genomics.* 8: 251–64.

Firkins, J. L., Hristov, A. N., Hall, M. B., Varga, G. A. and St-Pierre, N. R. (2006). Integration of ruminal metabolism in dairy cattle. *J. Dairy Sci.* 89 (E. Suppl.): E31–E51.

Firkins, J. L. and Morrison, M. (2007). Ruminal nitrogen metabolism: perspectives for integration of microbiology and nutrition for dairy. *J. Dairy Sci.* 90 (E. Suppl.): E1–E16.

Frank, B., Persson, M. and Gustafsson, G. (2002). Feeding dairy cows for decreased ammonia emissions. *Live. Prod. Sci.* 171–9.

Galindo, C. E., Ouellet, D. R., Pellerin, D., Lemosquet, S., Ortigues-Marty, I. and Lapierre, H. (2011). Effect of amino acid or casein supply on whole-body, splanchnic, and mammary glucose kinetics in lactating dairy cows. *J. Dairy Sci.* 94: 5558–68.

Galindo, C. E., Larsen, M., Ouellet, D. R., Maxin, G., Pellerin, D. and Lapierre, H. (2015). Abomasal amino acid infusion in postpartum dairy cows: effect on whole-body, splanchnic, and mammary glucose metabolism. *J. Dairy Sci.* 98: 7962–74.

Haque, M. N., Rulquin, H., Andrade, A., Faverdin, P., Peyraud, J. L. and Lemosquet, S. (2012). Milk protein synthesis in response to the provision of an 'ideal' amino acid profile at 2 levels of metabolizable protein supply in dairy cows. *J. Dairy Sci.* 95: 5876–87.

Haque, M. N., Rulquin, H. and Lemosquet, S. (2013). Milk protein responses in dairy cows to changes in postruminal supplies of arginine, isoleucine, and valine. *J. Dairy Sci.* 96: 420–30.

Haque, M. N., Guinard-Flament, J., Lamberton, P., Mustière, C. and Lemosquet, S. (2015). Changes in mammary metabolism in response to the provision of an ideal amino acid profile at 2 levels of metabolizable protein supply in dairy cows: consequences on efficiency. *J. Dairy Sci.* 98: 3951–68.

Hayes, J. F., Ng-Kwai-Hang, K. F. and Moxley, J. E. (1984). Heritability of milk casein and genetic and phenotypic correlations with production traits. *J. Dairy Sci.* 67: 841–6.

Higgs, R. J., Chase, L. E., Ross, D. A. and Van Amburgh, M. E. (2015). Updating the cornell net carbohydrate and protein system feed library and analyzing model sensitivity to feed inputs. *J. Dairy Sci.* 98: 6340–60.

Hoover, W. H. and Stokes, S. R. (1991). Balancing carbohydrate and proteins for optimum rumen microbial yield. *J. Dairy Sci.* 74: 3630–44.

Hristov, A. N. and Ropp, J. K. (2003). Effect of dietary carbohydrate composition and availability on utilization of ruminal ammonia nitrogen for milk protein synthesis in dairy cows. *J. Dairy Sci.* 86: 2416–27.

Hristov, A. N., Etter, R. P., Ropp, J. K. and Grandeen, K. L. (2004). Effect of dietary crude protein level and degradability on ruminal fermentation and nitrogen utilization in lactating dairy cows. *J. Anim. Sci.* 82: 3219–29.

Hristov, A. N., Price, W. J. and Shafli, B. (2005). A meta-analysis on the relationship between intake of nutrients and body weight with milk volume and milk protein yield in dairy cows. *J. Dairy Sci.* 88: 2860–9.

Huhtanen, P. and Hristov, A. N. (2009). A meta-analysis of the effects of dietary protein concentration and degradability on milk protein yield and milk N efficiency in dairy cows. *J. Dairy Sci.* 92: 3222–32.

Huhtanen, P., Cabezas-Garcia, E. H., Krizsan, S. J. and Shingfield, K. J. (2015). Evaluation of between-cow variation in milk urea and rumen ammonia nitrogen concentrations and the association with nitrogen utilization and diet digestibility in lactating cows. *J. Dairy Sci.* 98: 3182–96.

Hof, G., Tamminga, S. and Lenaers, P. J. (1994). Efficiency of protein utilization in dairy cows. *Live. Prod. Sci.* 38: 169–78.

Hof, G., Vervoorn, M. D., Lenaers, P. J. and Tamminga, S. (1997). Milk urea nitrogen as a tool to monitor the protein nutrition of dairy cows. *J. Dairy Sci.* 80: 3333–40.

Jarrige, R. (1989). *Ruminant Nutrition. Recommended Allowances and Feed Tables*. Paris, France: INRA. John Libbey Eurotext.

Jonker, J. S., Kohn, R. A. and Erdman, R. A. (1998). Using milk urea nitrogen to predict nitrogen excretion and utilization efficiency in lactating dairy cows. *J. Dairy Sci.* 81: 2681–92.

Kalscheur, K. F., Vandersall, J. H., Erdman, R. A., Kohn, R. A. and Russek-Cohen, E. (1999). Effects of dietary crude protein concentration and degradability on milk production responses of early, mid, late lactation dairy cows. *J. Dairy Sci.* 82: 545–54.

Kauffman, A. J. and St-Pierre, N. R. (2001). The relationship of milk urea nitrogen and urine nitrogen excretion in Holstein and Jersey cows. *J. Dairy Sci.* 84: 2284–94.

Ipharraguerrre, I. and Clark, J. H. (2005). Varying protein and starch in the diet of dairy cows. II. Effects on performance and nitrogen utilization for milk production. *J. Dairy Sci.* 88: 2556–70.

Ipharraguerrre, I. and Clark, J. H. (2014). A Meta-analysis of ruminal outflow of nitrogen fractions in dairy cows. *Adv. Dairy Res.* 2: 2. http://dx.doi.org/10.417/2329-888x.1000122.

Kebreab, E., France, J., Mills, J. A. N., Allison, R. and Dijkstra, J. (2002). A dynamic model of N metabolism in the lactating dairy cow and an assessment of impact of N excretion on the environment. *J. Anim. Sci.* 80: 248–59.

Kohn, R. A., Dou, Z., Ferguson, J. D. and Boston, R. C. (1997). A sensitivity analysis of nitrogen losses from dairy farms. *J. Envir. Man.* 50: 417–28.

Kohn, R. A., Kalscheur, K. F. and Russek-Cohen, E. (2002). Evaluation of models to estimate urinary nitrogen and expected milk urea nitrogen. *J. Dairy Sci.* 85: 227–33.

Kristensen, N. B., Storm, A. C. and Larsen, M. (2010). Effect of dietary nitrogen content and intravenous urea infusion on ruminal and portal-drained visceral extraction of arterial urea in lactating Holstein cows. *J. Dairy Sci.* 93: 2670–83.

Lapierre, H. and Lobley, G. E. (2001). Nitrogen recycling in the ruminant: a review. *J. Dairy Sci.* 84 (E. Suppl.): E223–36.

Lapierre, H., Pacheco, D., Berthiaume, R., Ourllet, D. R., Schwab, C. G., Dubreuil, P., Holtrop, G. and Lobley, G. E. (2006). What is the true supply of amino acids for a dairy cow? *J. Dairy Sci.* 89 (E. Suppl.): E1–E14.

Larsen, M., Lapierre, H. and Kristensen, N. B. (2014). Abomasal protein infusion in postpartum transition dairy cows: effect on performance and mammary metabolism. *J. Dairy Sci.* 97: 5608–22.

Larsen, M., Galindo, C., Ouellet, D. R., Maxin, G., Kristensen, N. B. and Lapierre, H. (2015). Abomasal amino acid infusion in postpartum dairy cows: effect on whole-body, splanchnic, and mammary amino acid metabolism. *J. Dairy Sci.* 98: 7944–61.

Lee, C., Giallongo, F., Hristov, A. N., Lapierre, H., Cassidy, T. W., Heyler, K. S., Varga, G. A. and Parys, C. (2015). Effect of dietary protein level and rumen-protected amino acid supplementation on amino acid utilization for milk protein in lactating dairy cows. *J. Dairy Sci.* 98: 1885–1902.

Lemosquet, S., Raggio, G., Lobley, G. E., Rulquin, H., Guinard-Flament, J. and Lapierre, H. (2009). Whole-body glucose metabolism and mammary energetic nutrient metabolism in lactating dairy cows receiving digestive infusions of casein and propionic acid. *J. Dairy Sci.* 92: 6068–82.

Mackle, T. R., Dwyer, D. A., Ingvartsen, K. L., Chouinard, P. Y., Lynch, J. M., Barbano, D. M. and Bauman, D. E. (1999). Effects of insulin and amino acids on milk protein concentration and yield from dairy cows. *J. Dairy Sci.* 82: 1512–24.

Madsen, J. (1985). The basis for the Nordic protein evaluation system of ruminants. The AAT/PBV system. *Acta. Agric. Scand. Suppl.* 25: 9–20.

Marini, J. C., Fox, D. G. and Murphy, M. R. (2008). Nitrogen transactions along the gastrointestinal tract of cattle: a meta-analytical approach. *J. Anim. Sci.* 86: 660–79.

Martineau, R., Ouellet, D. R., Kebreab, E. and Lapierre, H. (2016). Casein infusion rate influences feed intake differently depending on metabolizable protein balance in dairy cows: a multilevel meta-analysis. *J. Dairy Sci.* 99: 1–14.

Mepham, T. B. (1988). Amino acid utilization by lactating mammary gland. *J. Dairy Sci.* 65: 287–98.

Metcalf, J. A., Mansbridge, R. J., Blake, J. S., Oldham, J. D. and Newbold, J. R. (2008). The efficiency of conversion of metabolizable protein into milk true protein over a range of metabolizable protein intakes. *Animal* 2: 1193–202.

Muck, R. E. and Steenhuis, T. S. (1981). Nitrogen losses is dairy barns. In *Livestock Waste: A Renewable Resource*, 406–09. St. Lous, MO: American Societ of Agricultural Engineers.

Muck, R. E. (1982). Urease activity in bovine feces. *J. Dairy Sci.* 65: 2157–63.

Muck, R. E. and Richards, B. K. (1983). Losses of manorial nitrogen in free-stall barns. *Agric. Wastes.* 7: 1–15.

Ng-Kwai-Hang, K. F. (2003). Milk proteins: heterogeneity, fractionation and isolation. In H. Roginski, J. W. Fuquay and P. F. Fox (eds), *Encyclopedia of Dairy Sciences*, 1881–94. New York, NY: Academic Press.

Nousiainen, J. K., Shingfield, J. and Huhtanen, P. (2004). Evaluation of milk urea nitrogen as a diagnostic of protein feeding. *J. Dairy Sci.* 87: 386–98.

Nutritional Research Council. (1978). *Nutrient Requirements of Dairy Cattle.* 5th rev. ed. Washington, DC: National Academy Sciences.

Nutritional Research Council. (1989). *Nutrient Requirements of Dairy Cattle.* 6th rev. ed. Washington, DC: National Academy Sciences.

Nutritional Research Council. (2001). *Nutrient Requirements of Dairy Cattle.* 7th rev. ed. Washington, DC: National Academy Sciences.

O'Connor, J. D., Sniffen, C. J., Fox, D. G. and Chalupa, W. (1993). A net carbohydrate and protein system for evaluating cattle diets. IV. Predicting amino acid adequacy. *J. Anim. Sci.* 71: 1298–311.

Olmos Colmenero, J. J. and Broderick, G. A. (2006a). Effect of amount and ruminal degradability of soybean meal protein on performance of lactating dairy cows. *J. Dairy Sci.* 89: 1635–43.

Olmos Colmenero, J. J. and Broderick, G. A. (2006b). Effect of dietary crude protein concentration on ruminal nitrogen metabolism in lactating dairy cows. *J. Dairy Sci.* 89: 1694–703.

Olmos Colmenero, J. J. and Broderick, G. A. (2006c). Effect of dietary crude protein concentration on milk production and nitrogen utilization in lactating dairy cows. *J. Dairy Sci.* 89: 1704–12.

O'Mahony, J. A. and Fox, P. F. (2014). Milk: an overview. In *Milk Proteins.* DOI: 10.1016/B978-0-12-405171-3.00001-5. Copyright © 2014. Elsevier Inc.

Pacheco, D., Schwab, C. G., Berthiaume, R., Raggio, G. and Lapierre, H. (2006). Comparison of net portal absorption with predicted flow of digestible amino acids: scope for improving current models. *J. Dairy Sci.* 89: 4747–57.

Powell, J. M., Wattiaux, M. A. and Broderick, G. A. (2011). *Short Communication:* evaluation of milk urea nitrogen as a management tool to reduce ammonia emissions from dairy farms. *J. Dairy Sci.* 94: 4690–4.

Raggio, G., Pacheco, D., Berthiaume, R., Lobley, G. E., Pellerin, D., Allard, G., Dubreuil, P. and Lapierre, H. (2004). Effect of level of metabolizable protein on splanchnic flux of amino acids in lactating dairy cows. *J. Dairy Sci.* 87: 3461–72.

Raggio, G., Lemosquet, S., Lobley, G. E., Rulquin, H. and Lapierre, H. (2006). Effect of casein and propionate supply on mammary protein metabolism in lactating dairy cows. *J. Dairy Sci.* 89: 4340–51.

Reynal, S. M. and Broderick, G. A. (2005). Effect of dietary level of rumen-degraded protein on production and nitrogen metabolism in lactating dairy cows. *J. Dairy Sci.* 88: 4045–64.

Renolds, C. K., Harmon, D. L. and Cecava, M. J. (1994). Absorption and delivery of nutrients for milk protein synthesis by portal-drained viscera. *J. Dairy Sci.* 77: 2787–808.

Robinson, P. H. (2010). Impacts of manipulating ration metabolizable lysine and methionine levels on the performance of lactating cows: a systematic review of the literature. *Live. Sci.* 127: 115–26.

Roseler, D. K., Ferguson, J. D., Sniffen, C. J. and Herrema, J. (1993). Dietary protein degradability effects on plasma and milk urea nitrogen and milk nonprotein nitrogen in Holstein cows. *J. Dairy Sci.* 76: 525–34.

Rulquin, H., Pisulewski, P. M., Vérité, R. and Guinard, J. (1993). Milk production and composition as a function of postruminal lysine and methionine supply: a nutrient-response approach. *Livest. Prod. Sci.* 37: 69–90.

Rulquin, H., Guinard, J. and Vérité. (1998). Variation of amino acid content in the small intestine digesta of cattle: development of a prediction model. *Live. Prod. Sci.* 53: 1–13.

Ruis, A. G., McGilliard, M. L., Umberger, C. A. and Hanigan, M. D. (2010a). Interactions of energy and predicted metabolizable protein in determining nitrogen efficiency in the lactating dairy cow. *J. Dairy Sci.* 93: 2034–43.

Ruis, A. G., Appuhamy, J. A. D. R. N., Cyriac, J., Kirovski, D., Becvar, O., Escobar, J., McGilliard, Bequette, B. J., Akers, R. M. and Hanigan, M. D. (2010b) Regulation of protein synthesis in mammary glands of lactating dairy cows by starch and amino acids. *J. Dairy Sci.* 93: 3114–27.

Russell, J. B., O'Connoer, J. D., Fox, D. G., Van Soest, P. J. and Sniffen, C. J. (1992). A net carbohydrate and protein system for evaluating cattle diets: I. Ruminal Fermentation. *J. Anim. Sci.* 70: 3551–61.

Sannes, R. A., Messman, M. A. and Vagnoni, D. B. (2002). Form of rumen-degradable carbohydrate and nitrogen on microbial protein synthesis and protein efficiency of dairy cows. *J. Dairy Sci.* 85: 900–8.

Santos, F. A. P., Santos, J. E. P., Theurer, C. B. and Huber, J. T. (1998). Effects of rumen-undegradable protein on dairy cow performance: a 12-year literature review. *J. Dairy Sci.* 81: 3182–213.

Schingoethe, D. J. (1996). Balancing the amino acid needs of the dairy cow. *Anim. Feed. Tech.* 60: 153–60.

Schwab, C. G., Bozak, C. K., Whithouse, N. L. and Mesbah, M. M. A. (1992a). Amino acid limitation and flow to the duodenum at four stages of lactation. 1. Sequence of lysine and methionine limitation. *J. Dairy Sci.* 75: 3486–502.

Schwab, C. G., Bozak, C. K., Whithouse, N. L. and Olson, V. M. (1992b). Amino acid limitation and flow to the duodenum at four stages of lactation. 2. Extent of lysine limitation. *J. Dairy Sci.* 75: 3503–18.

Schwab, C. G. (1996). Rumen-protected amino acids for dairy cattle: progress towards determining lysine and methionine requirements. *Anim. Feed Tech.* 59: 87–101.

Schwab, C. G., Huhtanen, P., Hunt, C. W. and Hvelplund, T. (2005). Nitrogen requirements of cattle. In E. Pfeffer and A. Hristov (eds), *Nitrogen and Phosphorus Nutition of Cattle*, 13–70. Cambridge, MA: CABI Publishing.

Smoler, E., Rook, A. J., Sutton, J. D. and Beever, D. E. (1998). Prediction of milk protein concentration from elements of the metabolizable protein system. *J. Dairy Sci.* 81: 1619–23.

Sniffen, C. J., O'Connor, J. D., van Soest, P. J., Fox, D. G. and Russell, J. B. (1992). A net carbohydrate and protein system for evaluating cattle diets. II. Carbohydrate and protein availability. *J. Anim. Sci.* 70: 3562–77.

Socha, M. T., Schwab, C. G., Putnam, D. E., Whitehouse, N. L., Garthwaite, B. D. and Ducharme, G. A. (2008). Extent of methionine limitation in peak, early, and mid-lactation dairy cows. *J. Dairy Sci.* 91: 1996–2010.

Spek, J. W., Dijkstra, J., van Duinker, K. G., Hendriks, W. N. and Bannink, A. (2013). Prediction of urinary nitrogen and urea nitrogen excretion by lactating dairy cattle in Northern Europe and North America: a meta-analysis. *J. Dairy Sci.* 96: 4310–22.

Stelwagen, K. (2000). Milk protein. Milk biochemistry and secretion. In H. Roginski, J. W. Fuquay and P. F. Fox (eds), *Encyclopedia of Dairy Sciences. Volume Three*, 1835–47. New York, NY: Academic Press.

Stern, M. D., Varga, G. A., Clark, J. H., Firkins, J. L., Huber, J. T. and Pamquist, D. L. (1994). Evaluation of chemical and physical properties of feeds that affect protein metabolism in the rumen. *J. Dairy Sci.* 77: 2762–86.

Subnel, A. P. J., Meijer, R. G. M., van Straalen, W. M. and Taminga, S. (1994). Efficiency of milk protein production in the DVE protein evaluation system. *Live. Prod. Sci.* 40: 215–24.

Swanepoel, N., Robinson P. H. and Erasmus, L. J. (2010). Amino acid needs of lactating dairy cows: predicting limiting amino acids in contemporary rations fed to high producing dairy cattle in California using metabolic models. *Anim. Feed Sci. Tech.* 161: 103–20.

Swaisgood, H. E. (1995). F. protein and amino acid composition of bovine milk. In R. G. Jenssen (ed.), *Handbood of Milk Composition*, 464–8. New York, NY: Academic Press.

Tacoma, R., Fields, J., Ebenstein, D. B., Lam, Y-W. and Greenwood, S. L. (2016). Characterization of bovine milk proteome in early-lactation Holstein and Jersey breeds of dairy cows. *J. Proteomics.* 130: 200–10.

Tamminga, S. and Oldham, J. D. (1980a). Amino acid utilization by dairy cows. I. Methods of varying amino acid supply. *Live. Prod. Sci.* 7: 437–52.

Tamminga, S. and Oldham, J. D. (1980b). Amino acid utilization by dairy cows. II. Concept of amino acid requirements. *Live. Prod. Sci.* 7: 453–63.

Tamminga, S., Van Straalen, W. M., Subnel, A. P. J., Meijer, R. G. M., Steg, A., Wever, C. J. G. and Blok, M. C. (1994). The Dutch protein evaluation system: the DVE/OEB-system. *Live. Prod. Sci.* 40: 139–55.

Tamminga, S. (2003). Pollution due to nutrient losses and its contrl in European animal production. *Live. Prod. Sci.* 84: 101–11.

Thomas, C. ed. (2004). *Feed Into Milk.* Nottingham, UK: Nottingham University Press.

Tuori, M., Kaustell, K. V. and Huhtanen, P. (1998). Comparison of the protein evaluation systems of feeds for dairy cows. *Live. Prod. Sci.* 55: 33–46.

Tylutki, T. P., Fox, D. G., Durbal, V. M., Tedeschi, L. O., Russell, J. B., Van Amburgh, M. E., Overton, T. R., Chase, L. E. and Pell, A. N. (2008). Cornell net carbohydrate and protein system: a model of precision feeding of dairy cattle. *Anim. Feed Sci. Tech.* 143: 174–202.

Vermorel, M. and Coulon, J. B. (1996). Comparison of the national research council energy system for lactating cows with four European systems. *J. Dairy Sci.* 81: 846–55.

Virtanen, A. I. (1966). Milk production of cows on protein-free feeds. *Science* 153: 1603–8.

Wattiaux, M. A. and Karg, K. L. (2004). Protein level of alfalfa and corn silage based diets. II. Nitrogen balance and manure characteristics. *J. Dairy Sci.* 87: 3492–502.

Wu, Z. and Satter, L. D. (2000). Milk production during the complete lactation of dairy cows fed diets containing different amounts of protein. *J. Dairy Sci.* 83: 1042–51.

Yu, P., Christensen, D. A. and McKinnon. (2003). Comparison of the national research Council-2001 model and the Dutch system (DVE/OEB) in the prediction of nutrient supply to dairy cows from forages. *J. Dairy Sci.* 86: 2178–92.

Zanton, G. I., Bowman, G. R., Vázquez-Añón, M. and Rode, L. M. (2014). Meta-analysis of lactation performance in dairy cows receiving supplemental dietary methionine sources or postruminal infusion of methionine. *J. Dairy Sci.* 97: 7085–101.

Index

Lightning Source UK Ltd.
Milton Keynes UK
UKOW06n0910140917
309119UK00001B/6/P